Basic Physics for Graduate Students

Hidenori Sonoda • Kayhan Ülker

Basic Physics for Graduate Students

Springer

Hidenori Sonoda
Kobe, Hyogo, Japan

Kayhan Ülker
Department of Physics
Mimar Sinan Fine Arts University
Istanbul, Türkiye

ISBN 978-981-95-6606-8 ISBN 978-981-95-6607-5 (eBook)
https://doi.org/10.1007/978-981-95-6607-5

Translation from the Japanese language edition: "Daigakuinsei No Tame No Kiso Butsurigaku" by Hidenori Sonoda, © Hidenori Sonoda 2011. Published by Kodansha Publishing Co. All Rights Reserved.

This Springer imprint is published by the registered company Springer Nature Singapore Pte Ltd.
The registered company address is: 152 Beach Road, #21-01/04 Gateway East, Singapore 189721, Singapore

Preface

This book is an English translation of the book originally written by one of the authors (H.S. to be referred to as Hidenori below) and published in Japan from Kodansha in 2011. The intention behind the original publication is explained in the preface to the original.

The idea of translation came first to the other author (K.Ü. to be referred to Kayhan below) during one of the pandemic lockdowns, who had been teaching both graduate and undergraduate physics courses of various kinds at his home institution in Istanbul. We have been close collaborators and friends, and Hidenori had told Kayhan about his Japanese book. Kayhan could get a pretty good idea of what the book is about by glancing over the equations scattered throughout without any knowledge of Japanese.

We thought that the handbook nature of the book would appeal to the physics students all over the world if it were published in English. The book was originally meant to be used as the textbook for a review course taught by an instructor, but it can also serve as a handy book not only for those beginning graduate students preparing for qualifying exams but also for the seasoned physics researchers and instructors who want to remind themselves quickly of what they have learned but forgotten.

We sat on the idea of translation for a year or two while Kayhan used occasionally Google Translate™and the like for some parts of the book. To discuss the translation project Hidenori flew from Japan to Istanbul in September 2023; unfortunately he was infected by COVID-19 on the way. Kayhan, left alone, did not waste his time and went ahead to try his hands on ChatGPT™for translating the Japanese LaTeXsource files into English LaTeXsource files while both were in quarantine. It was a learning process for both Kayhan and ChatGPT™, since ChatGPT™often tries to "improve" on the original text and does not hesitate to make changes. It took Kayhan some efforts to rein on the process. By February 2024, Kayhan finished the first draft of translation while he was again infected by COVID-19, and we were ready to submit a book proposal to Springer.

The first draft was revised by Hidenori. The first section of Electromagnetism was rewritten due to the revision of the SI units in 2019. Kayhan revised the draft further to improve the readability. For the English edition, we have also introduced a couple of additional supplementary notes, such as the summary of mathematics essential for quantum mechanics, for the readers' convenience. Overall, however, the first draft was very helpful, and without it we would not have been able to complete the translation. We thank OpenAI™for making ChatGPT™available for public use.

Some comments for the readers. When you read the book, it is a very good idea to go through all the examples given at the end of each section. The examples come with solutions, not necessarily given in details. The exercises that follow the examples have no solutions, however; it is because the instructor, if any, can pick some of them as homework. Supplementary notes at the end of each chapter give depths but can be skipped.

May the readers find the book useful.

Kobe, Japan
Istanbul, Türkiye
September 2025

Hidenori Sonoda
Kayhan Ülker

Competing Interests The authors have no competing interests to declare that are relevant to the content of this manuscript.

Preface to the Original Japanese Edition

This book is intended for the first year graduate students in physics, and summarizes all the basic **undergraduate level** physics necessary before taking graduate-level physics courses. It provides a physics minimum for all physics graduate students regardless of their intended specialization.

The book covers all the standard **undergraduate physics**, from mechanics, electromagnetism, thermodynamics, statistical mechanics, and quantum mechanics. A half of the book is dedicated to quantum mechanics due to its larger importance.

Although the content is at the undergraduate level, it does not mean that the student should simply repeat what they learned in undergraduate school. There is a style of understanding more appropriate for graduate students, even for undergraduate physics. This book gives a review from a higher point of view. Therefore, we assume the knowledge of undergraduate-level mathematical methods such as linear algebra, complex analysis, vector analysis, ordinary differential equations, and Fourier transforms, and these are used freely.

In American graduate schools, students typically spend at least their first year studying basic physics. In particular, it is common to study graduate-level quantum mechanics for a full year. In Japanese graduate schools, there is still a strong belief that "students should learn on their own." This may have been acceptable when graduate schools were small and elite, but it is no longer feasible. It is not appropriate for graduate education to be called "education" if a master's degree can be obtained by simply writing a master's thesis. I think it ideal for the students to receive a solid foundation of the basics for a year as in the United States, but it is impossible given the realities of Japan where the importance of doing research is overemphasized. This book is meant to be a compromise.

This book was prepared as a textbook for the first-year graduate course "Theoretical Physics I" at Kobe University (and its predecessor, "The Fundamentals of Theoretical Physics"). Each section of a chapter corresponds to a 90-minute lecture; it would be impossible to teach everything in it unless you rush. However, if you choose and select appropriately what you cover, you can go over the entire book in 14 weeks with 90 minutes of a lecture and 90 minutes of an exercises session per week.

I would like to thank the students who have taken this course at Kobe University for providing feedback. I would also like to thank Tomoaki Nagasawa, Hiroshi Tsujimoto, Yu Koyama, Daisuke Kitagawa, Kenji Nishiwaki, Takaaki Sugiura, Mutsumi Nishimura, and Tetsuya Sugimura for creating solutions to the homework assignments as TAs. I would like to thank Atsushi Keiyama of Kodansha Scientific for his generous assistance in the planning and printing of the book. Finally, I would like to thank Prof. Hiroyuki Shiba for his encouragement over the past few years. Without his encouragement, I would not have been able to publish this book.

Kobe, Japan
September 2011

Hidenori Sonoda

References

In writing this book, I (H.S.) have consulted with the following references. These are also my favorite textbooks. The readers may prefer more recent reference books.

- **Mechanics**
  - **L. Landau and E. Lifshitz**, *Mechanics* (Butterworth-Heinemann)
  - **H. Goldstein**, *Classical Mechanics* (Addison-Wesley)
- **Electromagnetism**
  - **M. Schwartz**, *Principles of Electrodynamics* (Dover)
  - **E. Purcell**, *Electromagnetism* (McGraw-Hill)
- **Thermodynamics and Statistical Mechanics**
  - **A. B. Pippard**, *The Elements of Classical Thermodynamics* (Cambridge University Press)
  - **E. Fermi**, *Thermodynamics* (Dover)
  - **Ryogo Kubo**, *Statistical Mechanics: an advanced course with problems and solutions*, in cooperation with H. Ichimura, T. Usui, N. Hashitsume (North-Holland)
  - **L. Landau and E. Lifshitz**, *Statistical Physics* (Butterworth-Heinemann)
- **Quantum Mechanics**
  - **P. A. M. Dirac**, *Quantum Mechanics* (Cambridge University Press)
  - **L. Schiff**, *Quantum Mechanics* (McGraw-Hill)
  - **D. Griffiths**, *Introduction to Quantum Mechanics* (Prentice Hall)
  - **E. Abers**, *Quantum Mechanics* (Prentice Hall)

The author also recommends that everyone studying physics read through

- **R. P. Feynman, R. B. Leighton, M. Sands**, *The Feynman Lectures on Physics* Volumes I, II, III (Addison-Wesley)

These have been edited, and the edited version is freely available at

https://www.feynmanlectures.caltech.edu

The site makes all relevant materials available, such as recording of the classroom lectures.

Contents

About the Authors

Hidenori Sonoda received his Ph.D. in physics from Caltech (1985) under the supervision of Prof. H. David Politzer and Prof. Mark B. Wise. After his postdoc years at Lawrence Berkeley National Laboratory and Harvard University, he was assistant professor of physics at UCLA till 1996. From 1997 to 2024, he was associate professor at Kobe University, Japan, where he wrote the original Japanese edition of this book. Currently, he is visiting research associate at the Department of Physics and Astronomy, the University of Iowa, USA. His main research interest is in quantum field theory, especially the exact renormalization group (ERG). He has also written a Japanese textbook on Wilsonian renormalization groups *Renormalization Theory Demystified* (Kodansha, 2014).

Kayhan Ülker received his Ph.D. in Physics from Istanbul Technical University (ITU) in 2002 under the supervision of Mahmut Hortaçsu (ITU) and Rainald Flume (University of Bonn). Afterward, he joined the Feza Gürsey Institute, where he worked as a researcher until its closure by a controversial decision of TÜBİTAK in 2011. He then moved to the Department of Physics at Mimar Sinan Fine Arts University, where he is currently a Full Professor. Between 2013 and 2019, he also served as Vice Rector (Provost) of the university. His main research area is quantum field theory, with particular emphasis on its supersymmetric and/or noncommutative extensions. He received the Young Scientist Award of the Turkish Academy of Sciences (TÜBA) in 2009.

About the Authors

Chapter 1
Mechanics

Mechanics is the foundation of physics. It provides essential templates not only for classical physics but also for various branches of quantum physics. It is important to study it well.

1.1 Mechanics 1

Three Points to Remember

1. For one-dimensional motions, you can solve the equations of motion using the conservation of energy.
2. A motion under a central force can be reduced to that under a one-dimensional potential.
3. Deriving the equations of motion from a Lagrangian is both convenient and less prone to errors.

1.1.1 For One-Dimensional Motions, You Can Solve the Equations of Motion Using the Conservation of Energy

In a one-dimensional motion, where the force depends only on the coordinate, the total energy is conserved. The motion is entirely determined by the total energy and the initial coordinate.

The equation of motion for a point particle with mass m on the x-axis is given by

$$m\ddot{x} = F(x) \tag{1.1}$$

H. Sonoda, K. Ülker, *Basic Physics for Graduate Students*,
https://doi.org/10.1007/978-981-95-6607-5_1

where the force $F(x)$ depends solely on the coordinate x and is assumed to be independent of the velocity $\dot{x}$.[1] Let us first show that under this assumption the total energy is conserved.

In one dimension, the potential energy $U(x)$ can be defined as

$$U(x) \equiv -\int^{x} dx' F(x'), \tag{1.2}$$

and can be expressed as

$$F(x) = -\frac{d}{dx}U(x) \tag{1.3}$$

This means that any force is a conservative force. The equation of motion (1.1) becomes

$$m\ddot{x} = -\frac{d}{dx}U(x) \tag{1.4}$$

Multiplying both sides by $\dot{x}$, we obtain

$$m\dot{x}\ddot{x} = -\dot{x}\frac{dU}{dx} \tag{1.5}$$

Using

$$\frac{d}{dt}\frac{m\dot{x}^2}{2} = m\dot{x}\ddot{x}, \quad \dot{x}\frac{dU}{dx} = \frac{d}{dt}U(x) \tag{1.6}$$

we derive the **law of conservation of energy** as

$$\boxed{\frac{d}{dt}\left(\frac{1}{2}m\dot{x}^2 + U(x)\right) = 0} \tag{1.7}$$

Setting the total energy as a constant E,

$$E = \frac{1}{2}m\dot{x}^2 + U(x), \tag{1.8}$$

we can solve this for velocity in terms of E, which determines the time dependence of x as

$$\dot{x} = \pm\sqrt{\frac{2}{m}(E - U(x))} \Longrightarrow dt = \pm\frac{dx}{\sqrt{\frac{2}{m}(E - U(x))}} \tag{1.9}$$

[1] We use $\dot{x} \equiv \frac{d}{dt}x$, $\ddot{x} \equiv \frac{d^2}{dt^2}x$.

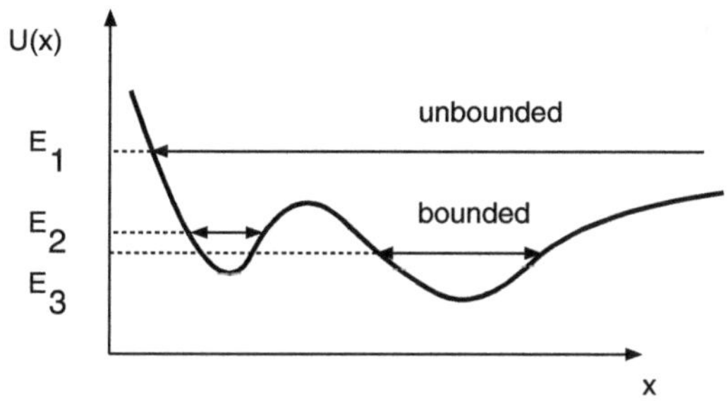

Fig. 1.1 For energy values E_2 and E_3, the motion is bound

Therefore, integrating this equation, we get

$$\boxed{t - t_0 = \pm \int_{x(t_0)}^{x(t)} \frac{dx}{\sqrt{\frac{2}{m}(E - U(x))}}} \tag{1.10}$$

This equation gives the time dependence of $x(t)$ implicitly. In principle we can invert the equation to determine the time dependence of x. Depending on the value of energy, the range of motion can be either finite or infinite (see Fig. 1.1).

Example: Harmonic Oscillator $U(x) = \frac{m\omega^2}{2}x^2$

We consider the simplest example of a harmonic oscillator (Fig. 1.2). If we take the amplitude of oscillation as x_0, the total energy is given by

$$E = \frac{m\omega^2}{2}x_0^2\,, \tag{1.11}$$

and we obtain

$$\frac{2}{m}(E - U(x)) = \omega^2(x_0^2 - x^2) \tag{1.12}$$

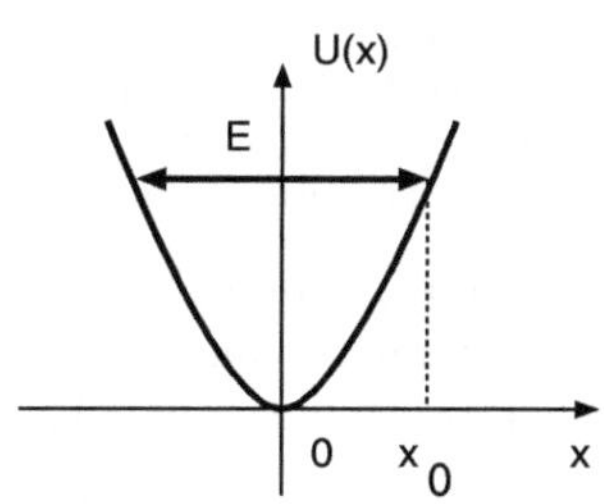

Fig. 1.2 A harmonic oscillator

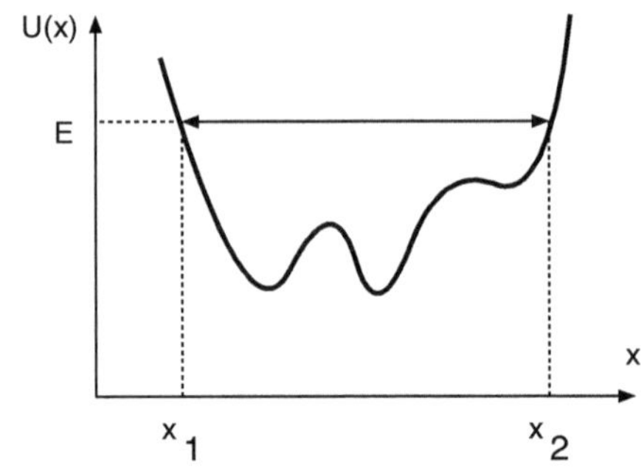

Fig. 1.3 Periodic motion between x_1 and x_2

which leads to

$$t = \int_0^x \frac{dx}{\omega\sqrt{x_0^2 - x^2}} = \frac{1}{\omega}\arcsin\frac{x}{x_0} \iff x(t) = x_0 \sin\omega t \tag{1.13}$$

Consider a bound motion in general (Fig. 1.3). Let x_1 and x_2 be the solutions to $E = U(x)$. Then the particle moves periodically between x_1 and x_2. The period is given by

$$T = 2\int_{x_1(E)}^{x_2(E)} \frac{dx}{\sqrt{\frac{2}{m}(E - U(x))}} \tag{1.14}$$

We need the factor of 2 because the particle goes from x_1 to x_2 and then from x_2 back to x_1.

1.1.2 A Motion Under a Central Force Can Be Reduced to that Under a One-Dimensional Potential

Under central forces, the angular momentum and total energy are both conserved. The conservation of angular momentum implies that the motion is confined to a plane, and the areal velocity remains constant. Therefore, the distance from the center determines the angular velocity. Since the radial equation of motion is one dimensional, we can solve it using the conservation of energy.

A force originating from a potential that depends only on the distance from a certain point (which we take as the origin of coordinates) is called a **central force**. Let $U(r)$ be the potential at an arbitrary point $\vec{r}$ at a distance $r \equiv |\vec{r}|$ from the center. The force vector at that point is given by

$$\vec{F}(\vec{r}) = -\nabla U(r) = -\hat{r}\, U'(r) \tag{1.15}$$

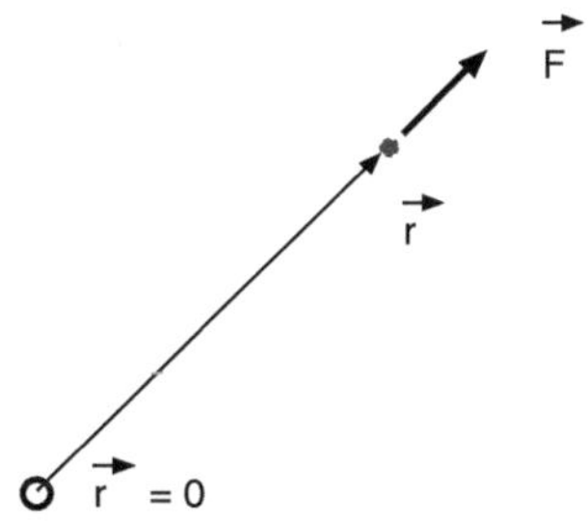

Fig. 1.4 Central force acts in the radial direction

(where $\hat{r} \equiv \vec{r}/r$ is the unit vector in the direction of $\vec{r}$). Hence, the force always points in the radial direction, with its magnitude depending solely on the distance (See Fig. 1.4). The equation of motion is then given by

$$m\ddot{\vec{r}} = -\hat{r}\, U'(r) \tag{1.16}$$

The angular momentum, defined by

$$\vec{M} \equiv \vec{r} \times m\dot{\vec{r}}\,, \tag{1.17}$$

is conserved as follows:

$$\begin{aligned} \frac{d}{dt}\vec{M} &= \underbrace{\dot{\vec{r}} \times m\dot{\vec{r}}}_{=0} + \vec{r} \times m\ddot{\vec{r}} \\ &= -\vec{r} \times \hat{r}\, U'(r) = 0 \end{aligned} \tag{1.18}$$

Note that $\vec{M}$ is orthogonal to both $\vec{r}$ and $\dot{\vec{r}}$. Geometrically, this means that the plane formed by $\vec{r}$ and $\dot{\vec{r}}$ is orthogonal to $\vec{M}$. As shown in Fig. 1.5, this plane passes through the origin. The plane passing through the origin and orthogonal to $\vec{M}$ does not depend on time. Thus, under central forces, we have **motions restricted to a fixed plane passing through the origin**.

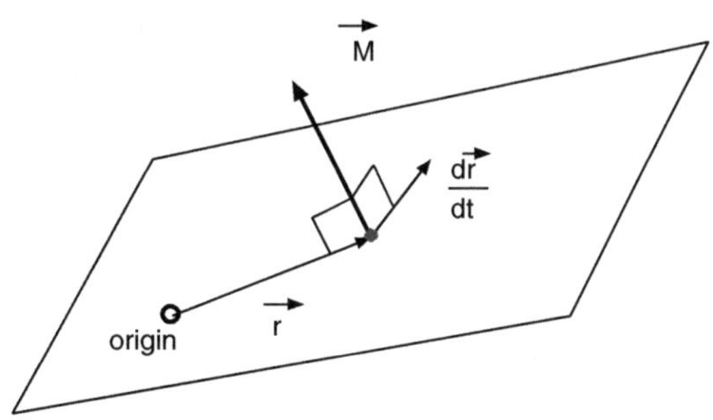

Fig. 1.5 Motion is confined to a plane through the origin and orthogonal to $\vec{M}$

The conservation of total energy can be demonstrated similarly to the one-dimensional case. Taking the dot product of the equation of motion and the velocity vector $\dot{\vec{r}}$, we get

$$\underbrace{\dot{\vec{r}}\cdot m\ddot{\vec{r}}}_{=\frac{d}{dt}\left(\frac{m}{2}\dot{\vec{r}}^2\right)} = -\underbrace{\dot{\vec{r}}\cdot\hat{r}\,U'(r)}_{=\dot{r}U'(r)=\frac{d}{dt}U(r)} \tag{1.19}$$

This leads to the conservation of energy:

$$E \equiv \frac{m}{2}\dot{\vec{r}}^{\,2} + U(r) \tag{1.20}$$

To solve the equations of motion, it is convenient to take the plane of motion as the xy-plane and express the particle's position in polar coordinates (r, θ) (Fig. 1.6). Then, the angular momentum is expressed as

$$M = mr^2\dot{\theta}\,, \tag{1.21}$$

and the total energy as

$$E = \frac{m}{2}\left(\dot{r}^2 + r^2\dot{\theta}^2\right) + U(r) \tag{1.22}$$

For $M > 0$, the motion is counterclockwise. Using Eq. (1.21), we can express the total energy as

$$E = \frac{m}{2}\dot{r}^2 + U_{\text{eff}}(r) \tag{1.23}$$

using the **effective potential**

$$\boxed{U_{\text{eff}}(r) \equiv U(r) + \frac{M^2}{2mr^2}} \tag{1.24}$$

The term $\frac{M^2}{2mr^2}$ in the effective potential is the centrifugal potential, implying an outward force due to angular momentum (Fig. 1.7).

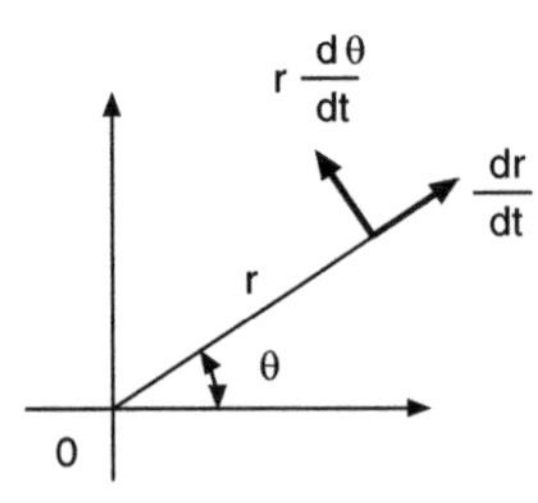

Fig. 1.6 Polar coordinates r, θ

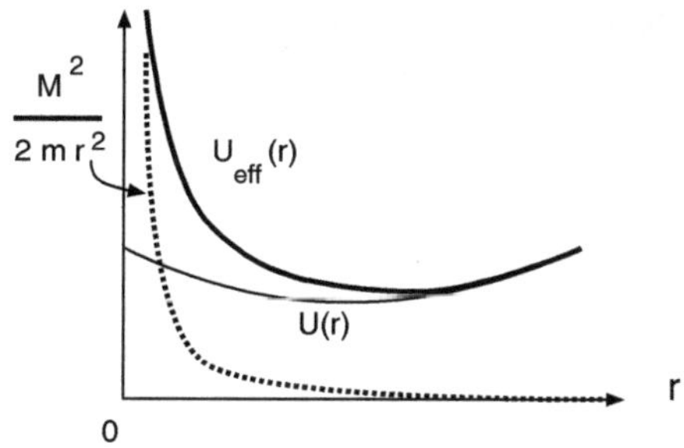

Fig. 1.7 The effective potential is the sum of the central potential and the centrifugal potential

The conservation of energy (1.23) implies further that the motion at a distance r is equivalent to a one-dimensional motion under the potential energy $U_{\rm eff}(r)$. The time dependence of the distance is given by

$$\dot{r} = \pm\sqrt{\frac{2}{m}\left(E - U_{\rm eff}(r)\right)}\,, \tag{1.25}$$

and the time dependence of the angle is given by the conservation of angular momentum (1.21):

$$\dot{\theta} = \frac{M}{mr^2} \tag{1.26}$$

Combining these two equations, we obtain the equation that determines the orbit:

$$\frac{dr}{d\theta} = \frac{\dot{r}}{\dot{\theta}} = \pm\frac{mr^2}{M}\sqrt{\frac{2}{m}\left(E - U_{\rm eff}(r)\right)} \tag{1.27}$$

Integrating this, we get

$$\theta - \theta_0 = \int_{\theta_0}^{\theta} d\theta' = \int_{\theta_0}^{\theta} \frac{dr'}{\frac{dr'}{d\theta'}} = \pm\int_{r_0}^{r} dr'\,\frac{M}{mr'^2}\,\frac{1}{\sqrt{\frac{2}{m}\left(E - U_{\rm eff}(r')\right)}} \tag{1.28}$$

where $r_0 \equiv r(\theta_0)$ is the angle at $t = 0$.

Now let us try to understand the above results a little more geometrically. In the case where the effective potential is as shown in Fig. 1.8, the distance r oscillates periodically between the maximum distance $r_{\rm max}(E)$ and the minimum distance $r_{\rm min}(E)$. Here, the maximum and minimum distances are defined as solutions to

$$U_{\rm eff}(r) = E\,, \tag{1.29}$$

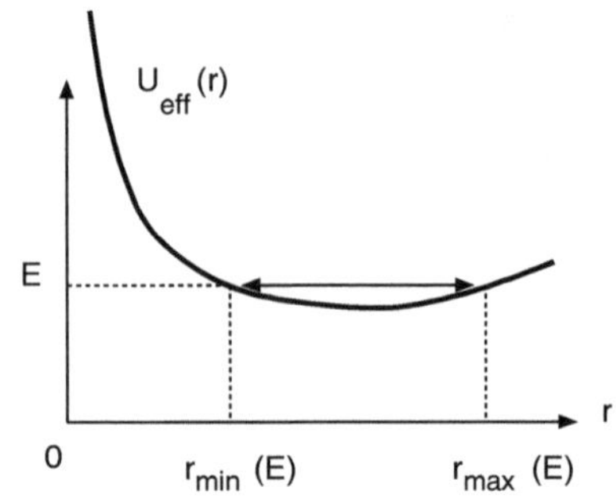

Fig. 1.8 Periodic motion in distance

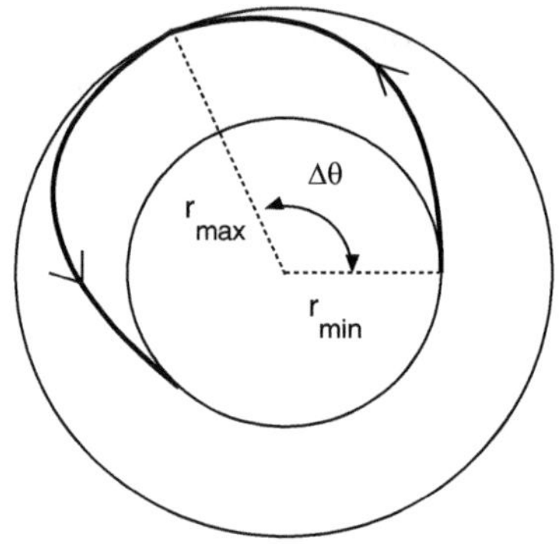

Fig. 1.9 Orbit of bound motion

and are functions of energy E. Let $\Delta\theta$ be the angular change of the particle as it moves from $r_{\min}$ to $r_{\max}$. From Eq. (1.28), we find

$$\Delta\theta = \int_{r_{\min}}^{r_{\max}} dr' \, \frac{M}{mr'^2} \frac{1}{\sqrt{\frac{2}{m}\left(E - U_{\text{eff}}(r')\right)}} \tag{1.30}$$

Therefore, the orbit looks like Fig. 1.9.

The closure of the orbit requires that when the distance returns to $r_{\min}$, the angle θ must be a rational multiple of π. This is because if we assume that on the n-th return the radius becomes $r_{\min}$, and if the angle at this point is $2\pi n'$, we obtain

$$n \cdot 2\Delta\theta = 2\pi n' \Longrightarrow \Delta\theta = \pi \frac{n'}{n} \tag{1.31}$$

In general, $\Delta\theta$ is a function of energy E, and this condition is only satisfied for special values of energy. Therefore, in general, the orbit is open, and the annular region bounded by $r_{\max}$ and $r_{\min}$ is filled by the orbit. The exceptions are the two cases:

$$U(r) = \frac{k}{2} r^2 \text{ (harmonic oscillator)}, \quad \frac{\alpha}{r} \text{ (Coulomb-type potential)} \tag{1.32}$$

In these cases, regardless of the value of energy, we find

$$\Delta\theta = \frac{\pi}{2} \text{ (harmonic oscillator)}, \quad \pi \text{ (Coulomb-type potential)} \tag{1.33}$$

This result is obvious for the harmonic oscillators, but not so for the Coulomb-type potential, and calculations are necessary (Example 1-7).

Law of Equal Areas

The areal velocity dA/dt refers to the area swept out by the radius vector per unit time:

$$\frac{dA}{dt} \equiv \frac{1}{2}r^2\dot{\theta} = \frac{M}{2m} \tag{1.34}$$

This is proportional to the angular momentum (Fig. 1.10).

Therefore, the conservation of angular momentum implies Kepler's Second Law, the "Law of Equal Areas." Note that **the larger (smaller) the distance, the smaller (larger) the angular velocity.**

1.1.3 Deriving the Equations of Motion from a Lagrangian Is Both Convenient and Less Prone to Errors

Deriving the equations of motion from a Lagrangian is the most straightforward method. Though you can choose any set of coordinates, it is preferable to select those reflecting the symmetry of the system. To find the kinetic and potential energies, it is the easiest to express them first in Cartesian coordinates and then express them using preferred coordinates.

Let $\vec{x}(t)$ denote the coordinates of a particle with mass m. The kinetic energy is given by

$$T = \frac{m}{2}\dot{\vec{x}}^{\,2}, \tag{1.35}$$

and the potential energy by

$$U = U(\vec{x}) \tag{1.36}$$

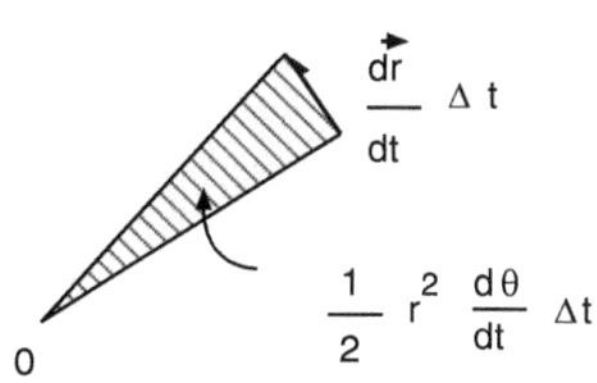

Fig. 1.10 Areal velocity

The **Lagrangian** (or Lagrangian function) is defined as the difference between the kinetic and potential energies:

$$\boxed{L(\vec{x}, \dot{\vec{x}}) \equiv T - U = \frac{m}{2}\dot{\vec{x}}^{\,2} - U(\vec{x})} \tag{1.37}$$

We then define the **action** as the time integral of the Lagrangian:

$$S[\vec{x}] \equiv \int_{t_0}^{t_1} dt\, L(\vec{x}, \dot{\vec{x}}) \tag{1.38}$$

This is a **functional** that depends on the entire time dependence of $\vec{x}(t)$ for $t_0 < t < t_1$. Let $\vec{x}_0 = \vec{x}(t_0)$ be the initial coordinates and $\vec{x}_1 = \vec{x}(t_1)$ be the final coordinates.

We now show that $\vec{x}(t)$ that minimizes the action $S[\vec{x}]$ satisfies the equation of motion. If $\vec{x}(t)$ minimizes the action, then the action for a slightly perturbed path $\vec{x}\,'(t) = \vec{x}(t) + \delta\vec{x}(t)$ is only slightly larger, with a small difference second-order in $\delta\vec{x}$. Therefore, expanding $S[\vec{x} + \delta\vec{x}]$ in powers of $\delta\vec{x}$, the terms linear in $\delta\vec{x}$ must vanish for the action to be minimized (Fig. 1.11).

To first order in $\delta\vec{x}$, we find

$$\begin{aligned} S[\vec{x} + \delta\vec{x}] - S[\vec{x}] &= \int_{t_0}^{t_1} dt \left(\frac{m}{2} \left(\dot{\vec{x}} + \delta\dot{\vec{x}} \right)^2 - U\left(\vec{x} + \delta\vec{x} \right) \right) \\ &\quad - \int_{t_0}^{t_1} dt \left(\frac{m}{2}\dot{\vec{x}}^{\,2} - U(\vec{x}) \right) \\ &= \int_{t_0}^{t_1} dt \left(m\delta\dot{\vec{x}} \cdot \dot{\vec{x}} - \delta\vec{x} \cdot \nabla U(\vec{x}) \right) \\ &= \int_{t_0}^{t_1} dt\, \delta\vec{x} \cdot \left(-m\ddot{\vec{x}} - \nabla U(\vec{x}) \right) + m \left[\delta\vec{x} \cdot \dot{\vec{x}} \right]_{t_0}^{t_1} \\ &= \int_{t_0}^{t_1} dt\, \delta\vec{x} \cdot \left(-m\ddot{\vec{x}} - \nabla U(\vec{x}) \right) \end{aligned} \tag{1.39}$$

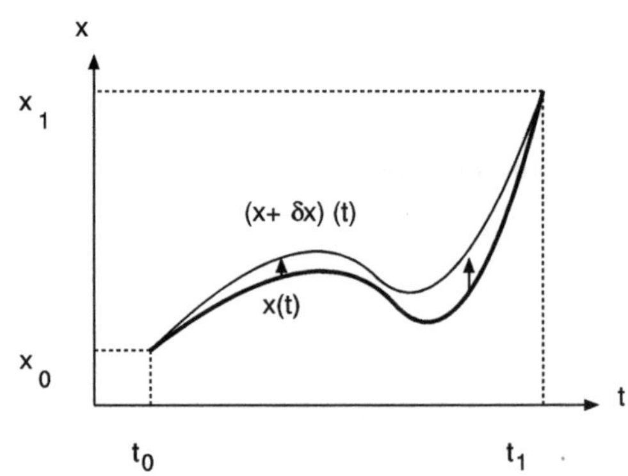

Fig. 1.11 Vary the path slightly and compare the values of the action

In the last step we have used $\delta\vec{x}(t_0) = \delta\vec{x}(t_1) = 0$; the values of $\vec{x}$ at $t = t_0, t_1$ are fixed.

In order for the above expression to be zero for any $\delta\vec{x}(t)$, the integrand must vanish for all t. This leads to the equation of motion:

$$m\ddot{\vec{x}} + \nabla U(\vec{x}) = 0 \tag{1.40}$$

Hence, the equation of motion is equivalent to **Hamilton's principle** of least action.

The Lagrangian method is particularly powerful for systems with multiple degrees of freedom. The strength of the principle of least action is that it does not depend on the choice of coordinates to calculate the action. As long as the coordinates specify uniquely the state of the system, any choice is permissible. For a system with N degrees of freedom, represented by generalized coordinates $q_1, \ldots, q_N$, with kinetic energy $T(q, \dot{q})$ and potential energy $U(q)$, the Lagrangian is given by

$$L(q, \dot{q}) = T(q, \dot{q}) - U(q) \tag{1.41}$$

We can use the same technique to derive the **Euler-Lagrange equations** for generalized coordinates $q_1, \ldots, q_N$. Let us vary $q_i(t)$ infinitesimally, $q_i'(t) = q_i(t) + \delta q_i(t)$. Expanding $S[q + \delta q]$ in powers of $\delta q_i(t)$, the terms linear in $\delta q_i(t)$ must vanish for the action to be minimized:

$$\begin{aligned}
\delta S &\equiv S[q_1, q_2, \cdots, q_i + \delta q_i, \cdots, q_N] - S[q_1, q_2, \cdots, q_i, \cdots, q_N] \\
&= \int_{t_0}^{t_1} L(q_1, \dot{q}_1, \cdots, q_i + \delta q_i, \dot{q}_i + \delta\dot{q}_i \cdots, q_N, \dot{q}_N) dt \\
&\qquad - \int_{t_0}^{t_1} L(q_1, \dot{q}_1, \cdots, q_i, \dot{q}_i, \cdots, q_N, \dot{q}_N) dt \\
&= \int_{t_0}^{t_1} \left(\frac{\partial L}{\partial q_i}\delta q_i + \frac{\partial L}{\partial \dot{q}_i}\delta\dot{q}_i \right) = \int_{t_0}^{t_1} \left(\frac{\partial L}{\partial q_i}\delta q_i + \frac{\partial L}{\partial \dot{q}_i}\frac{d}{dt}\delta q_i \right)
\end{aligned} \tag{1.42}$$

Integrating the second term by parts, we get

$$\delta S = \frac{\partial L}{\partial \dot{q}_i}\delta q_i \bigg|_{t_0}^{t_1} + \int_{t_0}^{t_1} \left(\frac{\partial L}{\partial q_i} - \frac{d}{dt}\frac{\partial L}{\partial \dot{q}_i} \right) \delta q_i dt = 0 \tag{1.43}$$

Since the values of $q_i(t)$ at $t = t_0, t_1$ are fixed and must vanish for any $\delta q_i(t)$, we obtain the equations of motion

$$\boxed{\frac{d}{dt}\frac{\partial L}{\partial \dot{q}_i} = \frac{\partial L}{\partial q_i} \quad (i = 1, \ldots, N)} \tag{1.44}$$

Fig. 1.12 Coupled springs

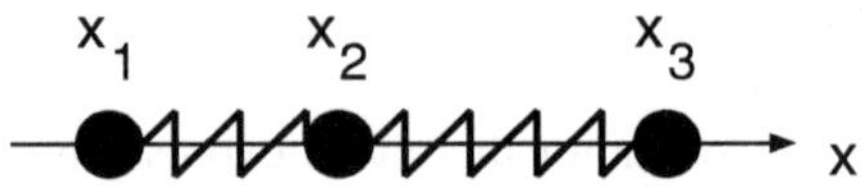

where the quantities

$$p_i \equiv \frac{\partial L}{\partial \dot{q}_i} \tag{1.45}$$

are called the momenta **conjugate** to q_i, and their time derivatives are equal to the forces $\frac{\partial L}{\partial q_i}$.

Example: Coupled Springs

Assuming the natural length of the springs is l, we have the following:

$$T = \frac{m_1}{2}\dot{x_1}^2 + \frac{m_2}{2}\dot{x_2}^2 + \frac{m_3}{2}\dot{x_3}^2 \tag{1.46a}$$

$$U = \frac{k}{2}\left\{(x_2 - x_1 - l)^2 + (x_3 - x_2 - l)^2\right\} \tag{1.46b}$$

$$L \equiv T - U \tag{1.46c}$$

(See Fig. 1.12)
These give us the equations of motion:

$$m\ddot{x_1} = -k(x_1 - x_2 + l) \tag{1.47a}$$

$$m\ddot{x_2} = -k(x_2 - x_1 - l) - k(x_2 - x_3 + l) \tag{1.47b}$$

$$m\ddot{x_3} = -k(x_3 - x_2 - l) \tag{1.47c}$$

Noether's Theorem

Let us assume that the Lagrangian L depends on the time derivatives $\dot{q}$ but not on the coordinates q themselves. Coordinates of this kind are called **cyclic coordinates**. Therefore, we find

$$\frac{\partial L}{\partial q} = 0 \tag{1.48}$$

which means the equations of motion for q are

$$\frac{d}{dt}\frac{\partial L}{\partial \dot{q}} = 0 \tag{1.49}$$

This implies that the momentum conjugate to a cyclic coordinate is a conserved quantity. This is Noether's theorem. The existence of cyclic coordinates reflects the symmetry of the system. If a certain motion is allowed, then a motion obtained by shifting q by a constant is also allowed. One of the greatest advantages of formulating mechanics with a Lagrangian is the clear relationship it reveals between conservation laws and symmetries of the system.

Example: Central Force

The Lagrangian for the motion under a central force

$$L = \frac{m}{2}\left(\dot{r}^2 + r^2\dot{\theta}^2\right) - U(r) \tag{1.50}$$

depends on the angular velocity $\dot{\theta}$, but not on the angular coordinate θ. Geometrically, this means L is invariant under rotations. In this case, the momentum conjugate to the angle, which is the angular momentum M, is conserved:

$$M \equiv \frac{\partial L}{\partial \dot{\theta}} = mr^2\dot{\theta} \tag{1.51}$$

Energy

The total energy E can be interpreted as the momentum conjugate to the time variable. In general, the total energy is given by

$$E \equiv \sum_{i=1}^{N} \dot{q}_i \frac{\partial L}{\partial \dot{q}_i} - L \tag{1.52}$$

If L does not depend on time t explicitly, meaning it only depends on q and $\dot{q}$, then E is conserved (Exercise 1-9).

Examples 1

1. **Qualitative understanding of motion**—Describe qualitatively all possible motions under the potential energy shown in the figure below:

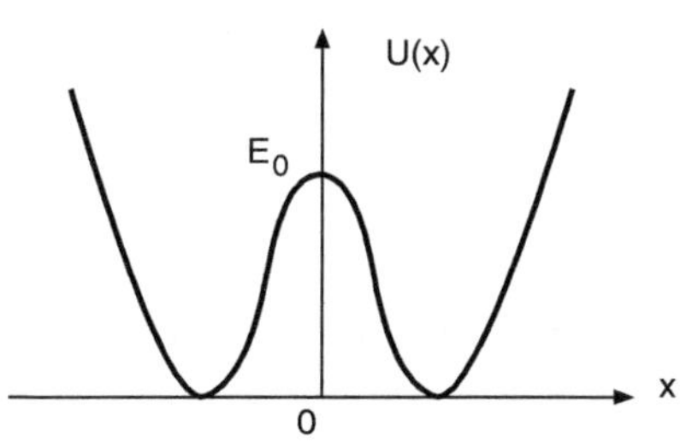

2. **Periodic motion**—Consider a particle of mass m defined by the Lagrangian

$$L = \frac{m}{2}\dot{x}^2 - A|x|^n$$

where A and n are positive constants.

(a) Find the period $T(E)$ of a bound motion as a function of energy $E > 0$. (You do not need to calculate the definite integral giving a multiplicative constant.)
(b) How much can be determined about the E dependence of $T(E)$ using dimensional analysis alone?

3. **Action variables and period**—Consider a particle with energy E, moving along the x-axis under the potential energy $U(x)$. Define the action variable as

$$S(E) \equiv 2\int_{x_1}^{x_2} m\dot{x}dx = 2\int_{x_1}^{x_2} dx\,\sqrt{2m\,(E - U(x))}$$

where the motion is bound within the region $x_1 < x < x_2$.

(a) Show that the period $T(E)$ of the motion is given by

$$T(E) = \frac{dS(E)}{dE}$$

Hint: Note that x_1 and x_2 depend on E, but this dependence can be ignored.
(b) Bohr's quantization condition is given by $S(E) = nh$ (where $n = 1, 2, \ldots$) with h as Planck's constant. Assuming

$$T(E) \propto E^a$$

where a is a constant, derive the n-dependence of quantized energy levels.

4. **Conservative force**—Consider a force field $\vec{f}(x, y)$ in a two-dimensional space.

(a) What condition must $\vec{f}$ satisfy so that it is the gradient of a potential energy $V(x, y)$, i.e., $\vec{f} = -\nabla V$?
(b) Show that curves of constant potential $V(x, y) =$ constant are orthogonal to $\vec{f}$ everywhere along the curve.

5. **Spiral trajectory**[2]—Consider a particle whose orbit on a plane is given by

$$r = k\mathrm{e}^{\alpha\theta}$$

where α is a constant.

(a) Find the corresponding central force potential $U(r)$.
(b) Determine the time dependence of r and θ.
(c) Find the kinetic energy of the particle as $r \to \infty$.

[2] See Sect. 1.3.2.

6. **Conservation of energy**—Given a Lagrangian $L(q_1, \ldots, q_N, \dot{q}_1, \ldots, \dot{q}_N)$, the energy of the system is defined as

$$E \equiv \sum_{i=1}^{N} \dot{q}_i \frac{\partial L}{\partial \dot{q}_i} - L$$

   (a) For $L = \frac{m}{2}\dot{x}^2 - U(x)$, find E.
   (b) For $L = \frac{1}{2}\sum_{i=1}^{N}\sum_{j=1}^{N} \dot{q}_i M_{ij}(q)\dot{q}_j - U(q_1, \ldots, q_N)$, find E.
   (c) Using the Lagrange equations for q_i, show that E defined above is conserved.

7. **Kepler motion**—Consider a mass m moving under the central force

$$U(r) = -\frac{\alpha}{r} \quad (\alpha > 0)$$

   (a) Using the angular momentum M, find the equivalent one-dimensional mechanical system. What is the effective potential $U_{\text{eff}}(r)$?
   (b) For bound motions with negative total energy $E < 0$, show that the trajectory is given by

$$\frac{p}{r} = 1 + e\cos\theta$$

   where

$$p = \frac{M^2}{m\alpha}, \quad e = \sqrt{1 + \frac{2EM^2}{m\alpha^2}}$$

   (c) Show that the trajectory described above is an ellipse with major axis $a = \frac{p}{1-e^2}$ and minor axis $b = \frac{p}{\sqrt{1-e^2}}$.

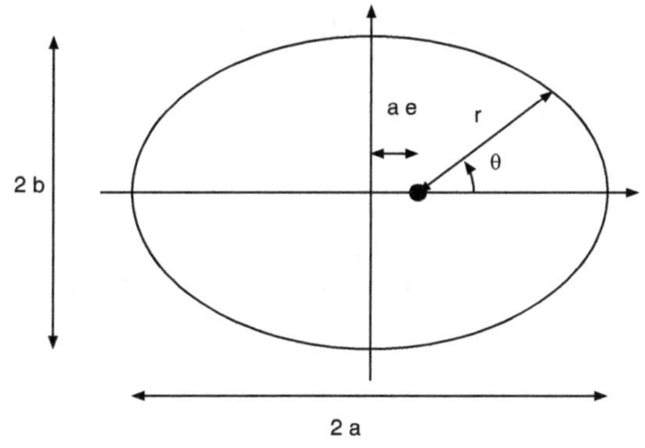

   (d) Express a and b as functions of M and E. In particular, demonstrate that a is independent of M.

Solutions to Examples 1

1. When $E \neq E_0$, the motion is always oscillatory. When $E = E_0$, x approaches $x = 0$ asymptotically (taking an infinite amount of time). Viewing the period $T(E)$ as a function of time, it diverges as $E \to E_0$.

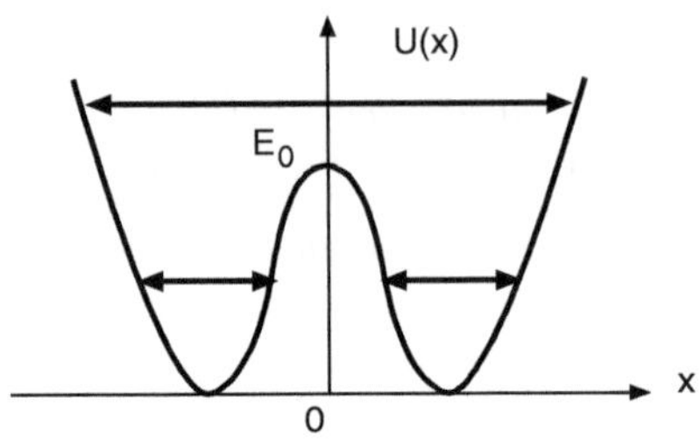

2. (a) Keeping the integral (expressible in terms of the beta function) as it is, we obtain

$$T(E) = \sqrt{2} \int_0^1 \frac{ds}{\sqrt{1 - s^n}} \cdot \sqrt{\frac{m}{E}} \left(\frac{E}{A}\right)^{\frac{1}{n}}$$

   (b) Note the dimensionality

$$[m] = \mathrm{kg}, \quad [E] = \mathrm{kg\,m\,s^{-2}}, \quad [A] = \mathrm{kg\,m^{2-n}\,s^{-2}}$$

   Since $[T] = \mathrm{s}$, dimensional analysis yields $T \propto m^{\frac{1}{2}} E^{\frac{1}{n} - \frac{1}{2}} A^{-\frac{1}{n}}$. The only undetermined factor is the constant coefficient.

3. (a) The integral $\int_a^b f(x)dx$ changes by $f(b)\epsilon$ under a small change ϵ in b. Since $f(a) = f(b) = 0$, the integral does not change under small changes of a, b.
   (b) From $S(E) \propto E^{a+1}$, we obtain $E \propto n^{\frac{1}{a+1}}$. In the case of a harmonic oscillator ($a = 0$, where the period is independent of energy), we get $E \propto n$.

4. (a) $\partial_x f_y = \partial_y f_x$ (This is Maxwell's relation, which plays an important role in thermodynamics. See Sect. 3.1.1.)
   (b) A small change in coordinates $\Delta\vec{x}$ along equipotential lines satisfies

$$\Delta\vec{x} \cdot \nabla V(x, y) = 0$$

5. (a) $U(r) = -\frac{A}{r^2}$, where A is a positive constant. This represents the trajectory with total energy $E = 0$. Given mass m and angular momentum M, α is given by

$$\frac{1 + \alpha^2}{2m} M^2 = A$$

   It also implies the constraint $M^2 < 2mA$.

(b) If we choose $r = 0$ at $t = 0$, we obtain

$$\begin{cases} r = \sqrt{\frac{2M\alpha}{m}}\, t \\ \theta = \frac{1}{2\alpha} \ln \frac{t}{t_0} \end{cases}$$

whcrc $t_0 \equiv \frac{k^2 m}{2M\alpha}$.

(c) Both the kinetic and potential energies approach zero as $r \to \infty$, behaving like $\frac{1}{r^2}$.

6. (a) $E = \frac{m}{2}\dot{x}^2 + U(x)$

(b) $E = \frac{1}{2}\sum_{i,j} \dot{q}_i M_{ij} \dot{q}_j + U$

(c)

$$\frac{d}{dt}E = \sum_i \left(\ddot{q}_i \frac{\partial L}{\partial \dot{q}_i} + \dot{q}_i \frac{d}{dt}\frac{\partial L}{\partial \dot{q}_i} \right) - \sum_i \left(\dot{q}_i \frac{\partial L}{\partial q_i} + \ddot{q}_i \frac{\partial L}{\partial \dot{q}_i} \right)$$
$$= \sum_i \dot{q}_i \left(\frac{d}{dt}\frac{\partial L}{\partial \dot{q}_i} - \frac{\partial L}{\partial q_i} \right) = 0$$

7. (a) $U_{\text{eff}}(r) = \frac{M^2}{2mr^2} - \frac{\alpha}{r}$. Thus, for $E < 0$, the motion is bound.

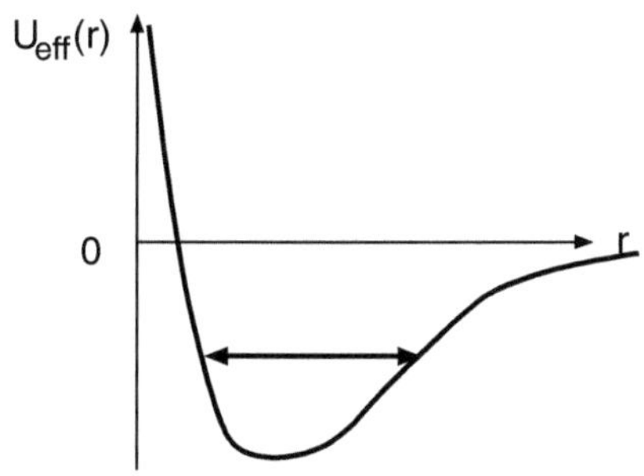

(b) The equation determining the trajectory can be rewritten in terms of p and e as

$$\frac{d}{d\theta}\left(\frac{p}{r} - 1\right) = -\sqrt{e^2 - \left(\frac{p}{r} - 1\right)^2}$$

This gives $\frac{p}{r} - 1 = e\cos\theta$.

(c) With $x = ae + r\cos\theta$ and $y = r\sin\theta$, we obtain $x^2/a^2 + y^2/b^2 = 1$.

(d)

$$\begin{cases} a = \frac{\alpha}{-2E} \\ b = \frac{M}{\sqrt{-2mE}} \end{cases}$$

a depends on E, but not on M. When $M = 0$, we get $b = 0$, i.e., a line through the origin.

Exercises 1

1. **Period**—Consider a particle of mass m moving along the x-axis under the potential

$$U(x) = g|x| \qquad (g > 0)$$

   Find the period of motion as a function of energy.
2. **Simple harmonic motion**—Assume a gravitational acceleration g acting in the negative y-direction. Determine the function $f(x)$ for which the period of motion along the curve $y = f(x) > 0$ is independent of the amplitude. Show that this curve is a cycloid. **Hint**: The kinetic energy $\frac{m}{2}\left(1 + f'(x)^2\right)\dot{x}^2$ should be proportional to $\left(\frac{d}{dt}\sqrt{f(x)}\right)^2$.

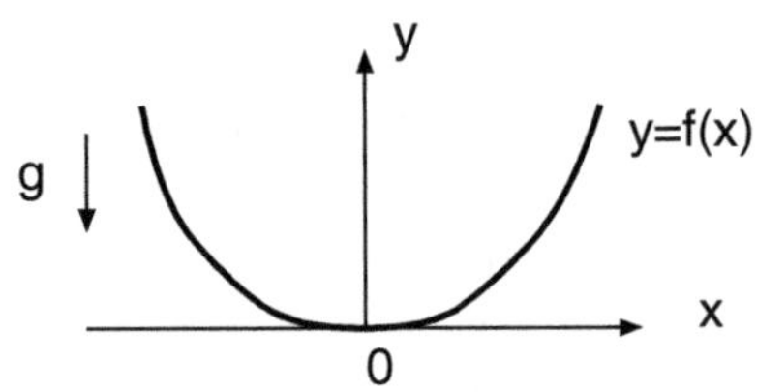

3. **Central force $\frac{1}{r^2}$**—Consider a particle of mass m moving under the central force potential

$$U(r) = -\frac{\alpha}{r^2} \qquad (\alpha > 0)$$

   Assuming that angular momentum M satisfies

$$\frac{M^2}{2m} > \alpha\,,$$

   determine the trajectory and its time dependence.
4. **Central force r^4**—Describe semi-quantitatively the trajectories of a particle moving under the potential

$$U(r) = kr^4 \qquad (k > 0)$$

   (Leave a definite integral as a constant. It is interesting to find the trajectories numerically.)
5. **Lagrangian**—For the mechanical system shown below, derive the Lagrangians and the equations of motion.

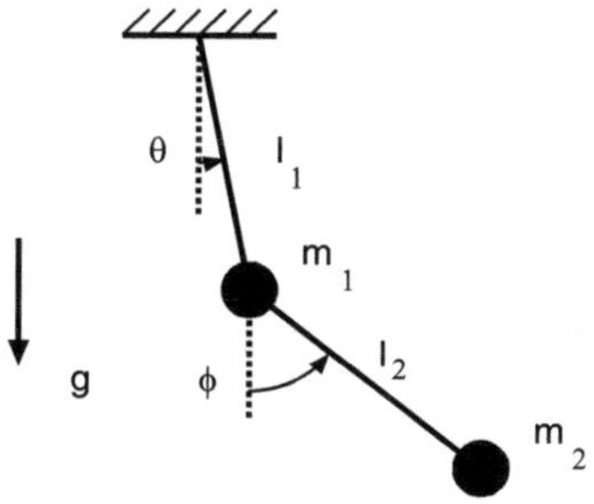

(i) |θ|, |ϕ| << 1

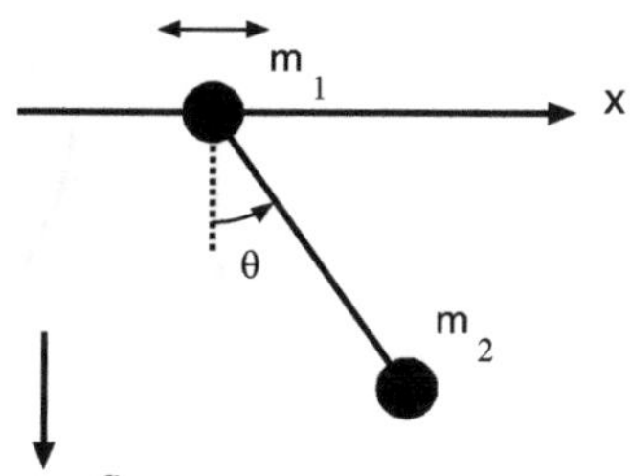

(ii) |θ| << 1

6. **Particle on a curved surface**—Consider a particle of mass m moving on a surface defined by the equation $z = f(x, y)$ in three-dimensional space. There is no gravity.

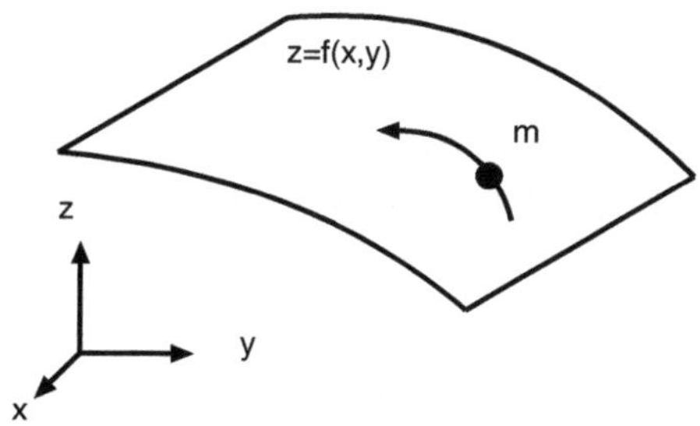

(a) Express the Lagrangian L as functions of the coordinates x and y, and their time derivatives.
(b) Show that energy and speed $v \equiv \sqrt{\dot{x}^2 + \dot{y}^2}$ are conserved.
(c) The length $K[\vec{x}(t)]$ of an arbitrary curve $\vec{x}(t)(t_1 < t < t_2)$ on the surface is given by

$$K[\vec{x}(t)] \equiv \int_{t_1}^{t_2} dt\, v(t)$$

where $v(t)$ is the speed at time t. Show that the curve that minimizes the action $S[\vec{x}] \equiv \int_{t_1}^{t_2} dt\, L(x, y, \dot{x}, \dot{y})$ also minimizes the length K.
(d) Use the above result to describe the motion of a particle on the sphere of radius R.

7. **Brachistochrone**—Consider two points $(x_1, y_1) = (0, 0)$ and (x_2, y_2) $(y_2 > 0)$ on the xy-plane. A particle of unit mass is allowed to slide without friction along a smooth curve connecting the two points. The potential energy is given by

$$U(y) = -gy \quad (g > 0)$$

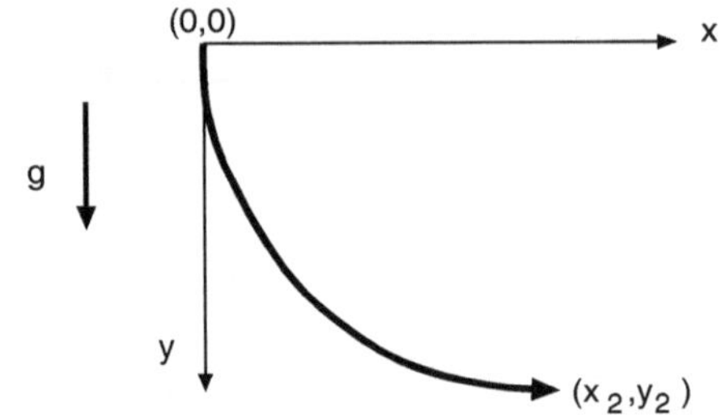

(a) Assuming the particle starts at rest at point 1, express the speed $v(y)$ as a function of y using the conservation of energy.
(b) Show that the time for the particle to move from point 1 to point 2 is given by

$$t_{12} = \int_1^2 \frac{\sqrt{(dx)^2 + (dy)^2}}{v(y)}$$

(c) Find the equation for the curve $y = f(x)$ that minimizes the time t_{12}.
(d) Show that the solution to the above equation is the cycloid generated by a circle of radius r, given parametrically as

$$x = r(\theta - \sin\theta), \quad y = r(1 - \cos\theta)$$

(e) How does the curve change if the initial speed is non-zero?

8. **Conserved quantities**—Determine all the conserved quantities for the following Lagrangians:

(a) $L_1 = \frac{m_1}{2}\dot{x}_1^2 + \frac{m_2}{2}\dot{x}_2^2 - U(x_1 - x_2)$
(b) $L_2 = \frac{m}{2}\left(\dot{\rho}^2 + \rho^2\dot{\theta}^2 + \dot{z}^2\right) - U(\kappa z - \theta)$

9. **Cyclic coordinates** —Consider a system of N particles in three-dimensional space.

(a) When the total momentum is conserved, what are the corresponding cyclic coordinates? (The answer is not unique.)
(b) When the total angular momentum around a certain point P is conserved, what are the corresponding cyclic coordinates? (This is also not unique.)

The conservation of energy can also be understood by introducing a corresponding cyclic coordinate. For simplicity, consider a one-dimensional mechanical system with coordinate x. Both x and t are considered as functions of a parameter τ. (Thus, x is a function of τ and not explicitly a function of t.) In this case, the Lagrangian of the system is defined as

$$L \equiv \frac{m}{2}\left(\frac{dx}{d\tau}\right)^2 \frac{1}{\frac{dt}{d\tau}} - U(x)\frac{dt}{d\tau},$$

and the action is defined as

$$S[x,t] \equiv \int_{\tau_1}^{\tau_2} d\tau L\left(x(\tau), t(\tau), \frac{dx}{d\tau}, \frac{dt}{d\tau}\right)$$

(c) Use the principle of least action to derive the equations of motion, and show that they are equivalent to the standard equation of motion

$$m\ddot{x} = -U'(x)$$

The Lagrangian depends on the τ-derivative of t but not explicitly on t. Therefore, t is a cyclic coordinate, and the corresponding conjugate momentum

$$p_t \equiv \frac{\partial L}{\partial\left(\frac{dt}{d\tau}\right)}$$

should be conserved.

(d) Confirm that p_t is the total energy.

1.2 Mechanics 2

Two Points to Remember

1. Small oscillations can be expressed as linear combinations of eigenmodes.
2. Small oscillations in continuous media propagate as waves.

1.2.1 Small Oscillations Can Be Expressed as Linear Combinations of Eigenmodes

The equation of motion for small oscillations is a linear differential equation, and we first need to find the eigenmodes. Any oscillation can be expressed as a sum of these eigenmodes. The coefficients for each eigenmode are determined by the initial conditions.

Before introducing a general theory, let us consider a specific example given by the following equations of motion:

$$\ddot{x} + \omega^2 x - \Omega^2 y = 0 \tag{1.53a}$$

$$\ddot{y} + \omega^2 y - \Omega^2 x = 0 \tag{1.53b}$$

Fig. 1.13 Coupled oscillators

Here, we assume $\omega^2 > \Omega^2$. These equations of motion represent a coupling of two harmonic oscillators with the same natural frequency ω.

For example, in the case of the coupled springs on the left-hand side of Fig. 1.13, the equations of motion are

$$m\ddot{x} = -Kx + k(y - x) = -(K + k)x + ky \tag{1.54a}$$

$$m\ddot{y} = -Ky + k(x - y) = -(K + k)y + kx \tag{1.54b}$$

Thus, we have

$$\omega^2 = \frac{K + k}{m}, \quad \Omega^2 = \frac{k}{m} \tag{1.55}$$

The right-hand side of the same figure gives coupled pendulums, and the equations of motion are given by

$$ml\ddot{\theta} = -mg\theta + kl(\phi - \theta) \tag{1.56a}$$

$$ml\ddot{\phi} = -mg\phi - kl(\phi - \theta) \tag{1.56b}$$

These result in

$$\omega^2 = \frac{g}{l} + \frac{k}{m}, \quad \Omega^2 = \frac{k}{m} \tag{1.57}$$

In both cases, the condition $\omega^2 > \Omega^2$ is satisfied.

Since Eqs. (1.53a) and (1.53b) have two degrees of freedom, they have two eigenmodes (Fig. 1.14).

- **Mode 1** with angular frequency ω_1

$$x_1(t) = y_1(t) = \mathrm{e}^{-i\omega_1 t} \tag{1.58a}$$

$$\text{where} \quad -\omega_1^2 + \omega^2 - \Omega^2 = 0 \Longrightarrow \omega_1^2 = \omega^2 - \Omega^2 \tag{1.58b}$$

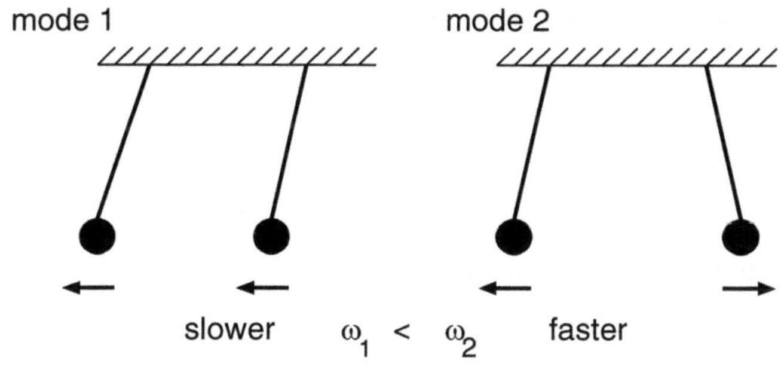

Fig. 1.14 There are two eigenmodes

- **Mode 2** with angular frequency ω_2

$$x_2(t) = -y_2(t) = \mathrm{e}^{-i\omega_2 t} \tag{1.59a}$$

$$\text{where} \quad -\omega_2^2 + \omega^2 + \Omega^2 = 0 \Longrightarrow \omega_2^2 = \omega^2 + \Omega^2 \tag{1.59b}$$

Note that $\omega_2 > \omega_1$. (Think why $\omega_2 > \omega_1$ for the coupled pendulums.)

The general solution to the equations of motion can be expressed as a **linear** combination of these two eigenmodes:

$$x(t) = \Re[C_1 x_1(t) + C_2 x_2(t)] \tag{1.60a}$$

$$y(t) = \Re[C_1 y_1(t) + C_2 y_2(t)] = \Re[C_1 x_1(t) - C_2 x_2(t)] \tag{1.60b}$$

where the coefficients $C_{1,2}$ are complex numbers with 2 complex degrees of freedom and $\Re$ represents taking the real part. The values of $C_{1,2}$ can be determined from the initial conditions $x(0)$, $\dot{x}(0)$, $y(0)$, $\dot{y}(0)$ with 4 real degrees of freedom.

Moving on to the general theory, let us consider the Lagrangian of a system with N degrees of freedom:

$$L = \frac{1}{2} \sum_{i,j=1}^{N} \dot{q}_i M_{ij} \dot{q}_j - \frac{1}{2} \sum_{i,j=1}^{N} q_i \Omega_{ij} q_j \tag{1.61}$$

Here, M and Ω are constant symmetric matrices. They are positive definite so that both kinetic and potential energies are positive. A system described by coordinates q and velocities $\dot{q}$ in this quadratic form is called a **linear system**. Any mechanical system sufficiently close to a stationary equilibrium point can be approximated as a linear system; this makes linear systems quite general.

The Euler–Lagrange equations for the linear system are a set of linear differential equations:

$$\sum_{j=1}^{N}\left(M_{ij}\ddot{q}_j + \Omega_{ij}q_j\right) = 0 \qquad (i = 1, \cdots, N) \tag{1.62}$$

In the vector and matrix notation, this becomes

$$M\ddot{\vec{q}} + \Omega\vec{q} = 0 \tag{1.63}$$

where

$$M = (M_{ij}), \quad \Omega = (\Omega_{ij}), \quad \vec{q} = \begin{pmatrix} q_1 \\ \vdots \\ q_N \end{pmatrix}. \tag{1.64}$$

To find the eigenmodes, we first assume that the entire system oscillates at a single frequency:

$$\vec{q}(t) = \vec{u}\,\mathrm{e}^{-i\omega t} \tag{1.65}$$

Substituting this into the equation of motion (1.63), we get

$$\left(-\omega^2 M + \Omega\right)\vec{u} = 0 \tag{1.66}$$

For this to have a non-trivial solution, i.e., $\vec{u} \neq 0$, we must find

$$\det(-\omega^2 M + \Omega) = 0 \tag{1.67}$$

This gives an N-th degree algebraic equation in terms of ω^2, which, taking multiplicities into account, yields N solutions (eigenvalues). Denoting the i-th eigenvalue as ω_i^2 and its corresponding solution to (1.66) as $\vec{u}_i$, the general solution to Eq. (1.63) is expressed as

$$\boxed{\vec{q}(t) = \Re\left(\sum_{i=1}^{N} C_i\,\mathrm{e}^{-i\omega_i t}\,\vec{u}_i\right)} \tag{1.68}$$

where C_i are complex constants independent of time.

Example: Coupled Oscillators

Let us apply the general results obtained above to the coupled oscillators considered at the beginning of this section. First, let us rewrite the equations of motion using matrices. (To save the symbol ω for eigen frequencies, we replace the ω appearing in the equations of motion by ω_0.)

$$\begin{cases} \ddot{x} + \omega_0^2 x - \Omega^2 y = 0 \\ \ddot{y} + \omega_0^2 y - \Omega^2 x = 0 \end{cases} \tag{1.69}$$

$$\Longleftrightarrow \frac{d^2}{dt^2}\begin{pmatrix} x \\ y \end{pmatrix} + \begin{pmatrix} \omega_0^2 & -\Omega^2 \\ -\Omega^2 & \omega_0^2 \end{pmatrix}\begin{pmatrix} x \\ y \end{pmatrix} = 0 \tag{1.70}$$

Let

$$\begin{pmatrix} x(t) \\ y(t) \end{pmatrix} = \mathrm{e}^{-i\omega t}\begin{pmatrix} u \\ v \end{pmatrix} \quad (u, v \text{ constants}) \tag{1.71}$$

so that

$$-\omega^2\begin{pmatrix} u \\ v \end{pmatrix} + \begin{pmatrix} \omega_0^2 & -\Omega^2 \\ -\Omega^2 & \omega_0^2 \end{pmatrix}\begin{pmatrix} u \\ v \end{pmatrix} = 0 \tag{1.72}$$

Consequently, we obtain

$$\begin{pmatrix} -\omega^2 + \omega_0^2 & -\Omega^2 \\ -\Omega^2 & -\omega^2 + \omega_0^2 \end{pmatrix}\begin{pmatrix} u \\ v \end{pmatrix} = 0 \tag{1.73}$$

To obtain non-trivial solutions, we must find

$$\det\begin{pmatrix} -\omega^2 + \omega_0^2 & -\Omega^2 \\ -\Omega^2 & -\omega^2 + \omega_0^2 \end{pmatrix} = 0, \tag{1.74}$$

which leads to a quadratic equation for ω^2:

$$\left(\omega^2 - \omega_0^2\right)^2 = (\Omega^2)^2 \Longrightarrow \omega^2 = \omega_0^2 \pm \Omega^2 \tag{1.75}$$

The eigenmodes are obtained as follows:

- Slow mode: $\omega_1^2 \equiv \omega_0^2 - \Omega^2 > 0$

$$\begin{pmatrix} \Omega^2 & -\Omega^2 \\ -\Omega^2 & \Omega^2 \end{pmatrix}\begin{pmatrix} u_1 \\ v_1 \end{pmatrix} = 0 \Longrightarrow u_1 = v_1 \tag{1.76}$$

- Fast mode: $\omega_2^2 \equiv \omega_0^2 + \Omega^2 > \omega_1^2$

$$\begin{pmatrix} \Omega^2 & \Omega^2 \\ \Omega^2 & \Omega^2 \end{pmatrix}\begin{pmatrix} u_2 \\ v_2 \end{pmatrix} = 0 \Longrightarrow u_2 = -v_2 \tag{1.77}$$

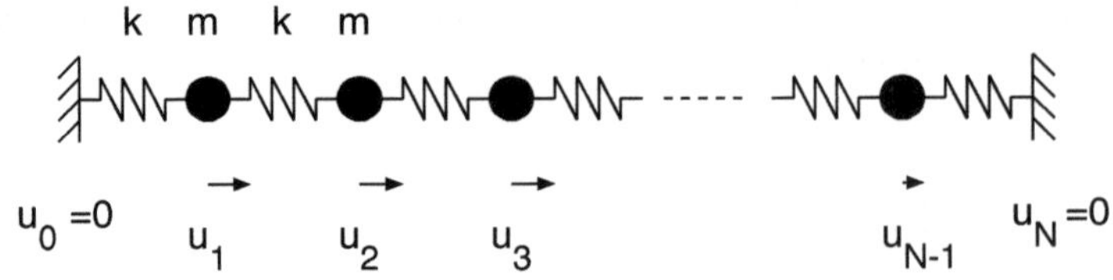

Fig. 1.15 Vibrations of a system with a large number of degrees of freedom $N - 1 \gg 1$

1.2.2 Small Oscillations in Continuous Media Propagate as Waves

A uniform medium can be treated as an extension of the linear system discussed in the previous section to a system with an infinite number of degrees of freedom. Small oscillations in the system appear as waves, and the various eigenmodes, equal in number to the degrees of freedom, are distinguished by their wave numbers. Boundary conditions have a significant impact on the behavior of these eigenmodes. Under fixed (or free) boundary conditions, the eigenmodes are standing waves; under periodic boundary conditions, they are traveling waves.[3]

Let us consider a system composed of many springs and particles as shown in Fig. 1.15.

Here, u_i $(i = 1, \cdots, N-1)$ represents the displacement of the i-th particle from its equilibrium position.[4] The Lagrangian for this system, with $k = m\omega^2$, is given by

$$L = \frac{m}{2}\sum_{i=1}^{N-1}\dot{u}_i^2 - \frac{m\omega^2}{2}\left(u_1^2 + \sum_{i=2}^{N-1}(u_i - u_{i-1})^2 + u_{N-1}^2\right) \tag{1.78}$$

The resulting equations of motion are

$$\ddot{u}_i = -\omega^2\left(u_i - u_{i-1} + u_i - u_{i+1}\right) \qquad (i = 1, \cdots, N-1) \tag{1.79}$$

with the boundary conditions

$$u_0 = u_N \equiv 0 \tag{1.80}$$

The number of degrees of freedom for this equation is $N - 1$, so there should be $N - 1$ eigenmodes, and the solution to the equation of motion is obtained as the sum of these eigenmodes. In the matrix notation, the vector u for the eigenmode becomes

[3] Away from the vicinity of the boundaries, however, the boundary conditions hardly affect the physics.

[4] We use u_i instead of q_i and reserve x for the space coordinate.

$$\Omega^2 u = \omega^2 \begin{pmatrix} 2 & -1 & 0 & \cdots & & & \\ -1 & 2 & -1 & 0 & \cdots & & \\ 0 & -1 & 2 & -1 & 0 & \cdots & \\ \vdots & & \ddots & \ddots & \ddots & & \\ & & & & -1 & 2 & -1 \\ 0 & \cdots & \cdots & & 0 & -1 & 2 \end{pmatrix} u \tag{1.81}$$

By diagonalizing the $N-1$ dimensional square matrix on the right-hand side, we obtain Ω^2 as eigenvalues. Instead, let us first consider the **continuum limit** while keeping the total length of the system as $L = Na$; the eigenvalues are easier to find this way. In the limit of $a \to 0$ and $N \to \infty$, with i denoting the particle number, $x = ia$ represents the coordinate of the i-th particle. We express the displacement $u_i(t)$ at that point as a function $u(x,t)$. The Lagrangian becomes, in the limit of $a \to 0$,[5]

$$L = \int_0^L dx \, \frac{1}{2} \left[\rho (\partial_t u(x,t))^2 - \kappa (\partial_x u(x,t))^2 \right] \tag{1.82}$$

where ρ and κ represent the mass per unit length and the elastic constant (Young's modulus), respectively, given by

$$\rho \equiv \frac{m}{a}, \qquad \kappa \equiv m\omega^2 a \tag{1.83}$$

In the limit $a \to 0$, m and ω^2 must depend on a so that ρ and κ remain finite. This means that m is proportional to a and ω is inversely proportional to a:

$$m \propto a, \qquad \omega \propto \frac{1}{a} \tag{1.84}$$

Table 1.1 summarizes the correspondence between a discrete system and its continuum limit.

Similarly, in the continuum limit, the equation of motion (1.79) yields the **wave equation**:[6]

$$\boxed{\rho \partial_t^2 u(x,t) - \kappa \partial_x^2 u(x,t) = 0} \tag{1.85}$$

---

[5] Remember

$$\partial_x u(x,t) \equiv \frac{\partial u(x,t)}{\partial x} = \lim_{a \to 0} \frac{u(x+a,t) - u(x,t)}{a}$$

[6] See Sect. 1.3.5 for a direct derivation from the principle of least action.

Table 1.1 Procedure for taking the continuum limit

Coordinate	ia	$\to$	x
Displacement	$u_i(t)$	$\to$	$u(x,t)$
Change in displacement	$(u_{i+1}-u_i)/a$	$\to$	$\partial_x u(x,t)$
Change in change in displacement	$(u_{i+1}-2u_i+u_{i-1})/a^2$	$\to$	$\partial_x^2 u(x,t)$
Sum	$a\sum_{i=1}^{N-1}$	$\to$	$\int_0^L dx$
Mass density	m/a	$\to$	ρ
Elastic constant	$ka = m\omega^2 a$	$\to$	κ

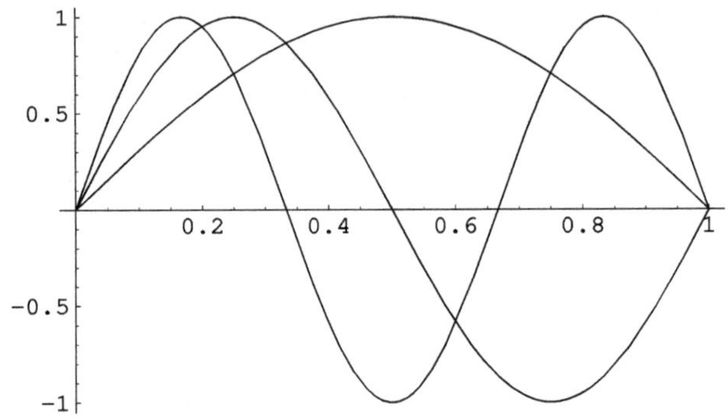

Fig. 1.16 Eigenmodes for $n = 1, 2, 3$. The horizontal axis is $\frac{x}{L}$. The maximum value of u_n is normalized to 1

Here, $u(x,t)$ is subject to the boundary conditions of fixed ends:

$$u(0,t) = u(L,t) = 0 \tag{1.86}$$

The equation for the eigenmode with eigenfrequency ω becomes

$$\left(\omega^2\rho + \kappa\frac{d^2}{dx^2}\right)u(x) = 0 \tag{1.87}$$

Thus, the solution satisfying the boundary condition (1.86) is given by

$$u_n(x) = \sqrt{\frac{2}{L}}\sin\frac{n\pi x}{L} \quad (n = 1, 2, \cdots) \tag{1.88}$$

This is depicted for $n = 1, 2, 3$ in Fig. 1.16. By substituting this into the equation of motion, ω is given as

$$\omega_n = \sqrt{\frac{\kappa}{\rho}}\frac{\pi}{L}\cdot n \tag{1.89}$$

The relation between the wave number $\frac{\pi n}{L}$ and the angular frequency ω_n is generally called the **dispersion relation**. (In this case, the angular frequency is proportional to the wave number.) Note that the angular frequency becomes larger as the wave number increases. In the formula (1.88), the eigenmode is normalized to satisfy the

orthonormality (orthogonality and normalization)

$$\int_0^L dx\, u_n(x)u_{n'}(x) = \delta_{nn'} \tag{1.90}$$

The general solution to the equation of motion is

$$u(x,t) = \Re\left[\sum_{n=1}^{\infty} C_n \mathrm{e}^{-i\omega_n t} \cdot u_n(x)\right] \quad (C_n \in \mathbf{C}) \tag{1.91}$$

As we will see in Exercise 2-5, the general solution can be expressed using any function $f(x)$ with a period of $2L$ as

$$u(x,t) = f(-x+vt) - f(x+vt) \qquad \left(v \equiv \sqrt{\frac{\kappa}{\rho}}\right) \tag{1.92}$$

This implies two things:

1. Waves propagate with a constant phase velocity v regardless of the wave number.
2. The wave gets reflected continuously at the fixed end while changing its phase by π.

We now leave the continuous case here and return to the original discrete equation of motion (1.79). By analogy with the continuum limit, the n-th eigenmode is given as

$$u_j^{(n)}(t) = \sin\frac{\pi n}{N}j \cdot \mathrm{e}^{-i\Omega_n t} \qquad (j = 0, \cdots, N) \tag{1.93}$$

We sketch the eigenmodes for $n = 1, 2$ in Fig. 1.17.
From the equation of motion, we obtain

$$\begin{aligned} -\Omega_n^2 \sin\frac{\pi n}{N}j &= -\omega^2\left(2\sin\frac{\pi n}{N}j - \sin\frac{\pi n}{N}(j+1) - \sin\frac{\pi n}{N}(j-1)\right) \\ &= -\omega^2 2\left(1-\cos\frac{\pi n}{N}\right)\sin\frac{\pi n}{N}j \\ &= -4\omega^2\sin^2\frac{\pi n}{2N}\cdot\sin\frac{\pi n}{N}j \end{aligned} \tag{1.94}$$

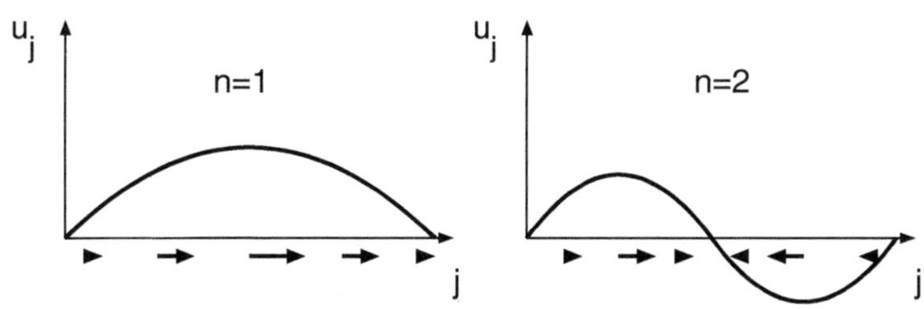

Fig. 1.17 Eigenmodes for $n = 1, 2$

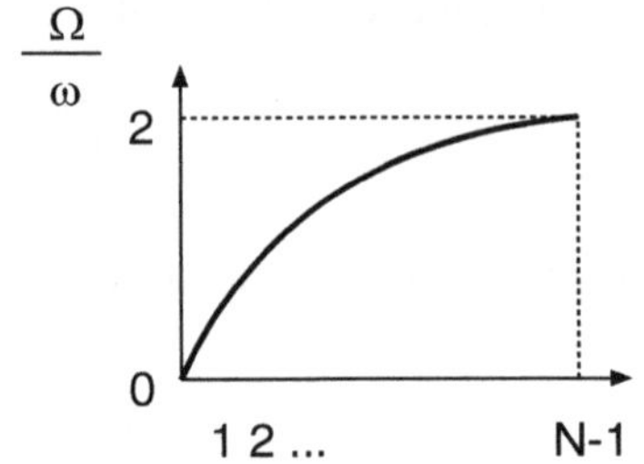

Fig. 1.18 Dispersion relation

From this, we obtain the dispersion relation as

$$\Omega_n = 2\sin\frac{\pi n}{2N}\cdot\omega \quad (n = 1, 2, \cdots, N-1) \tag{1.95}$$

(Fig. 1.18) The general solution to the equation of motion, similar to the continuous case, is given as the sum over the eigenmodes:

$$u_j(t) = \Re\left[\sum_{n=1}^{N-1} C_n\, u_j^{(n)}(t)\right] \tag{1.96}$$

Finally, we want to diagonalize the symmetric matrix of order $N-1$ on the right side of Eq. (1.81). Calling this matrix M, the normalized eigenvector is

$$u^{(n)} = \sqrt{\frac{2}{N}}\begin{pmatrix} \sin\frac{\pi n}{N} \\ \sin\frac{\pi n}{N}2 \\ \vdots \\ \sin\frac{\pi n}{N}(N-1) \end{pmatrix} \tag{1.97}$$

We therefore obtain

$$\begin{aligned} &M(u^{(1)}, u^{(2)}, \cdots, u^{(N-1)}) \\ &= \mathrm{diag}(\Omega_1^2, \Omega_2^2, \cdots, \Omega_{N-1}^2)\,(u^{(1)}, u^{(2)}, \cdots, u^{(N-1)}) \end{aligned} \tag{1.98}$$

where $\mathrm{diag}(a_1, \cdots, a_{N-1})$ is the diagonal matrix with diagonal elements $a_1, \cdots, a_{N-1}$. From the orthogonality of the eigenvectors $u^{(n)}$

$$u^{(n),T}u^{(n')} = \delta_{nn'}\,, \tag{1.99}$$

we obtain

$$\begin{pmatrix} u^{(1),T} \\ u^{(2),T} \\ \vdots \\ u^{(N-1),T} \end{pmatrix} M(u^{(1)}, u^{(2)}, \cdots, u^{(N-1)}) = \mathrm{diag}(\Omega_1^2, \Omega_2^2, \cdots, \Omega_{N-1}^2) \tag{1.100}$$

This means that M can be diagonalized by the orthogonal matrix

$$O \equiv (u^{(1)}, u^{(2)}, \cdots, u^{(N-1)}) \Longrightarrow O^T M O = \text{diagonal}$$

Up to this point, we have only dealt with the case of fixed boundaries. In physics, however, there are other commonly used boundary conditions such as the free boundary and periodic boundary conditions. We will cover these cases in the Examples and the Exercises below.

Examples 2

1. **Eigenmodes**—Determine the eigenmodes of a double pendulum. (You can take the angles θ, ϕ as small.) In addition to determining the eigenfrequencies, illustrate the mode of each eigenfrequency. Explain physically which eigenfrequency is larger.

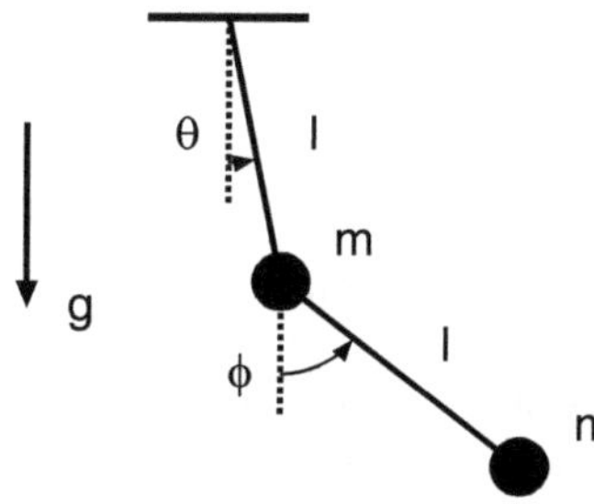

2. **Eigenmodes**—Determine the eigenmodes of the small oscillations of the mechanical system shown below.

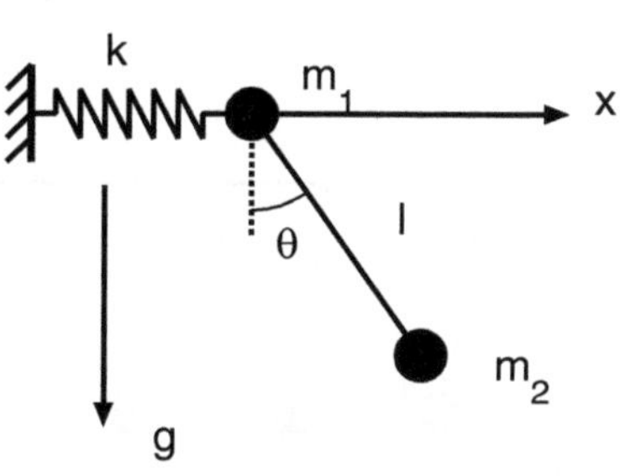

3. **Coupled oscillators**—Consider the mechanical system shown below. Here, x_1, x_2 give deviations from the equilibrium positions. Assume that the spring constant K is much smaller than the spring constant k, i.e., $K \ll k$.

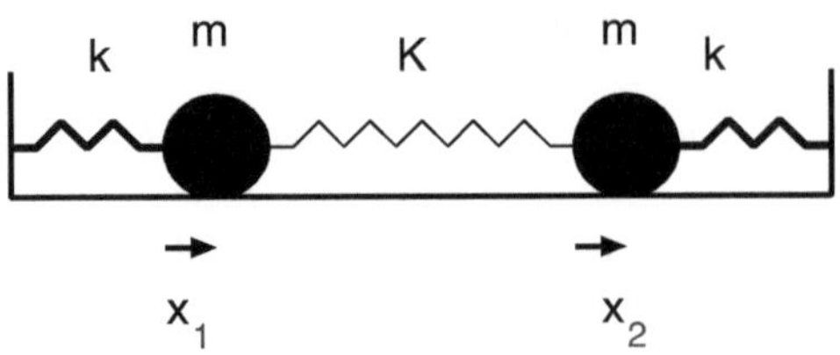

(a) Derive the Lagrange equations of motion.
(b) Determine the eigenmodes.
(c) Find the solution that satisfies the initial conditions

$$x_1(0) = A, \quad x_2(0) = 0, \quad \dot{x}_1(0) = \dot{x}_2(0) = 0.$$

(d) Illustrate $x_1(t), x_2(t)$ for $t \in \left[0, \text{an integral multiple of } \frac{\sqrt{mk}}{K}\right]$.

4. **Periodic boundary conditions**—Consider the equations of motion

$$m\ddot{x}_n = k(x_{n+1} - x_n) - k(x_n - x_{n-1}) \quad (n = 0, \cdots, N-1)$$

under the periodic conditions

$$x_N = x_0, \; x_{-1} = x_{N-1}.$$

We assume $N \gg 1$ is even. Determine all the eigenfrequencies, and understand each eigenmode physically. **Hints**: Find the solution in the form of $x_n = e^{i\frac{2\pi n}{\lambda} - i\omega t}$. The boundary conditions will constrain the wavelength λ. Consider the range $|\lambda| \geq 2$; why?

5. **Coupled pendulums**—Consider the mechanical system below with an infinite number of degrees of freedom. g is the gravitational acceleration.

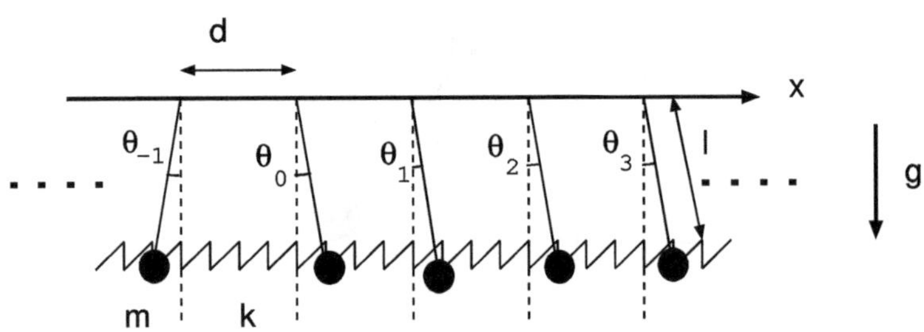

(a) Derive the Lagrangian. (Assume $|\theta_i| \ll 1$.)
(b) Derive the equations of motion.

We next consider the continuum limit. Let

$$\kappa \equiv kd, \quad \Omega^2 \equiv \frac{g}{l}.$$

Taking the limit $d \to 0$ such that the above quantities remain finite, define a continuous function $u(x,t)$ at $x = nd (n \in \mathbf{Z})$ as

$$u(x,t) \equiv l\theta_n(t).$$

(c) Determine the equation of motion that $u(x,t)$ satisfies in the continuum limit.
(d) Assuming the x, t dependence of the eigenmode as $e^{ikx-i\omega t}$, determine the dispersion relation between the wave number k and angular frequency ω.
(e) Determine the relation between the wave number k and the group velocity $\frac{d\omega}{dk}$.

6. **Transverse waves of a rope**—Consider the lateral displacement of a rope with a total length X located on the x-axis. Let ρ be the mass per unit length of the rope and T be the tension. The small displacement of the rope in the y-direction at x is denoted by $y(x,t)$.

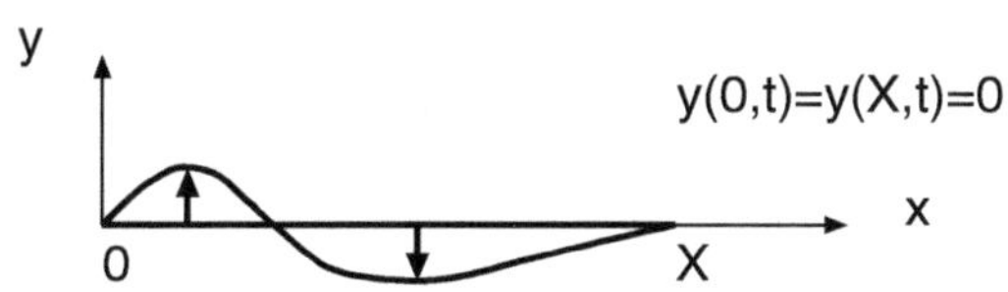

(a) Given $y(x,t)$, express the change in total length of the rope $\Delta X[y]$ as a functional of y. (Use the approximation formula $\sqrt{(dx)^2 + (dy)^2} \simeq dx(1 + \frac{1}{2}\left(\frac{dy}{dx}\right)^2)$.)
(b) Explain that the potential energy of the rope is given by

$$U[y] \equiv T \cdot \Delta X[y]$$

(c) Determine the kinetic energy of the rope $K[\partial_t y]$. **Hint**: Since you only need to determine K up to the second order in $\partial_t y$, you can ignore $(\partial_x y)^2$ and take the mass of the rope in $[x, x+dx]$ as ρdx.
(d) From the Lagrangian $L[y, \partial_t y] \equiv K[\partial_t y] - U[y]$, derive Lagrange's equation of motion. **Hint**: You may refer to Sect. 1.3.5 for the infinitesimal variation of functionals.
(e) Express the wave velocity v as a function of ρ and T. **Hint**: The general solution to the wave equation is given as $f(x \pm vt)$, where $f(x)$ is an arbitrary function. When determining v, you do not need to consider the boundary conditions $y(0,t) = y(L,t) = 0$.

Solutions to Examples 2

1. The x, y coordinates of the two masses are

$$\begin{cases} x_1 &= l\sin\theta \simeq l\theta \\ y_1 &= -l\cos\theta \simeq l\left(-1+\frac{\theta^2}{2}\right) \end{cases} \qquad \begin{cases} x_2 &= x_1 + l\sin\phi \simeq l(\theta+\phi) \\ y_2 &= y_1 - l\cos\phi \simeq l\left(-2+\frac{\theta^2+\phi^2}{2}\right) \end{cases}$$

Thus, the Lagrangian for small angles is

$$\begin{aligned} L &= \frac{m}{2}\left(\dot{x}_1^2+\dot{y}_1^2+\dot{x}_2^2+\dot{y}_2^2\right) - mg\,(y_1+y_2) \\ &\simeq \frac{ml^2}{2}\left(\dot{\theta}^2+(\dot{\theta}+\dot{\phi})^2\right) - \frac{mgl}{2}\left(2\theta^2+\phi^2\right) + \text{const} \end{aligned}$$

Hence, the equations of motion are

$$\begin{cases} 2\ddot{\theta}+\ddot{\phi} &= -2\frac{g}{l}\theta \\ \ddot{\theta}+\ddot{\phi} &= -\frac{g}{l}\phi \end{cases}$$

Therefore, the eigenfrequencies ω are determined by

$$\det\begin{pmatrix} 2\left(\omega^2-\frac{g}{l}\right) & \omega^2 \\ \omega^2 & \omega^2-\frac{g}{l} \end{pmatrix} = 0$$

The solution is

$$\omega_\pm^2 = (2\pm\sqrt{2})\frac{g}{l}$$

For the ω_+ mode, θ and ϕ have opposite signs, and the ratio $|\theta/\phi|$ of the amplitude is $1+\sqrt{2}$. For the ω_- mode, θ and ϕ have the same sign, and the ratio is $\sqrt{2}-1$.

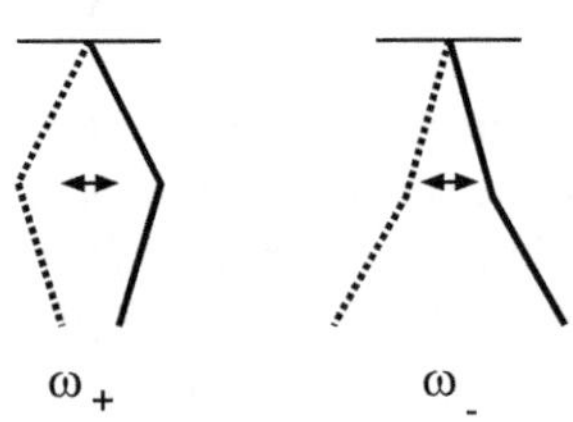

2. Let x_1 be the x coordinate of m_1. The x, y coordinates of m_2 are

$$x_2 = x_1 + l\sin\theta \simeq x_1 + l\theta, \quad y_2 = -l\cos\theta \simeq l\left(-1+\frac{\theta^2}{2}\right)$$

for a small angle θ. Thus, the Lagrangian is

$$L = \frac{m_1}{2}\dot{x}_1^2 + \frac{m_2}{2}\left(\dot{x}_2^2 + \dot{y}_2^2\right) - \frac{k}{2}x_1^2 - m_2 g y_2$$
$$\simeq \frac{m_1}{2}\dot{x}_1^2 + \frac{m_2}{2}\left(\dot{x}_1 + l\dot{\theta}\right)^2 - \frac{k}{2}x_1^2 - \frac{m_2 g l}{2}\theta^2 + \text{const}$$

The equations of motion are

$$(m_1 + m_2)\ddot{x}_1 + m_2 l\ddot{\theta} = -kx_1$$
$$m_2 l\ddot{x}_1 + m_2 l^2\ddot{\theta} = -m_2 g l\theta$$

Hence, the eigenfrequencies are obtained from

$$\det\begin{pmatrix} (m_1 + m_2)\omega^2 - k & m_2 l\omega^2 \\ m_2 l\omega^2 & m_2 l^2\omega^2 - m_2 g l \end{pmatrix} = 0$$

as

$$\omega_\pm^2 = \frac{1}{2m_1 l}\Big((m_1 + m_2)g + kl \pm \sqrt{g^2(m_1 + m_2)^2 + 2(m_2 - m_1)gkl + k^2 l^2}\Big)$$

The eigenmodes can be obtained from

$$\frac{\theta}{x_1} = \frac{\omega^2}{g - l\omega^2}$$

3. (a) From the Lagrangian

$$L = \frac{m}{2}\left(\dot{x}_1^2 + \dot{x}_2^2\right) - \frac{k}{2}\left(x_1^2 + x_2^2\right) - \frac{K}{2}\left(x_1 - x_2\right)^2$$

we obtain the equations of motion

$$\begin{cases} \ddot{x}_1 = -\frac{k}{m}x_1 + \frac{K}{m}(x_2 - x_1) \\ \ddot{x}_2 = -\frac{k}{m}x_2 - \frac{K}{m}(x_2 - x_1) \end{cases}$$

(b) Given the angular frequency ω, the equations of motion become

$$\underbrace{\begin{pmatrix} \frac{k+K}{m} & -\frac{K}{m} \\ -\frac{K}{m} & \frac{k+K}{m} \end{pmatrix}}_{\equiv A}\begin{pmatrix} x_1 \\ x_2 \end{pmatrix} = \omega^2\begin{pmatrix} x_1 \\ x_2 \end{pmatrix}$$

The eigenvalues of the matrix A are $\omega_-^2 \equiv \frac{k}{m}$ for the eigenmode $(x_1, x_2) = (1, 1)$ and $\omega_+^2 \equiv \frac{k+2K}{m}$ for the eigenmode $(x_1, x_2) = (1, -1)$.

(c) The general solution is then given by

$$\begin{pmatrix} x_1(t) \\ x_2(t) \end{pmatrix} = C_- \cos(\omega_- t + \phi_-) \begin{pmatrix} 1 \\ 1 \end{pmatrix} + C_+ \cos(\omega_+ t + \phi_+) \begin{pmatrix} 1 \\ -1 \end{pmatrix}$$

Hence, the solution that satisfies the given initial conditions is

$$\begin{pmatrix} x_1(t) \\ x_2(t) \end{pmatrix} = \frac{1}{2} A \begin{pmatrix} \cos \omega_- t + \cos \omega_+ t \\ \cos \omega_- t - \cos \omega_+ t \end{pmatrix}$$

$$= \begin{pmatrix} \cos \frac{1}{2}(\omega_+ + \omega_-)t \cdot \cos \frac{1}{2}(\omega_+ - \omega_-)t \\ \sin \frac{1}{2}(\omega_+ + \omega_-)t \cdot \sin \frac{1}{2}(\omega_+ - \omega_-)t \end{pmatrix}$$

(d) Since $K \ll k$, we find

$$\omega_+ \simeq \omega_- \left(1 + \frac{K}{k}\right) = \omega_- + \frac{K}{\sqrt{mk}}$$

Therefore, the solution obtained in part (c) can be approximated as

$$\begin{pmatrix} x_1(t) \\ x_2(t) \end{pmatrix} \simeq A \begin{pmatrix} \cos \left(\omega_- + \frac{K}{2\sqrt{mk}}\right) t \cdot \cos \frac{K}{2\sqrt{mk}} t \\ \sin \left(\omega_- + \frac{K}{2\sqrt{mk}}\right) t \cdot \sin \frac{K}{2\sqrt{mk}} t \end{pmatrix}$$

This represents a "beat" with a period of $2\pi \frac{\sqrt{mk}}{K}$. Note the two points:

(i) The amplitudes of x_1 and x_2 become zero alternatively. In other words, the two oscillators exchange their energies. You can compare this with Exercise 2-3 where the exchange of energy is incomplete.

(ii) The angular frequencies of each oscillator increase by $\frac{K}{2\sqrt{mk}}$ from ω_-.

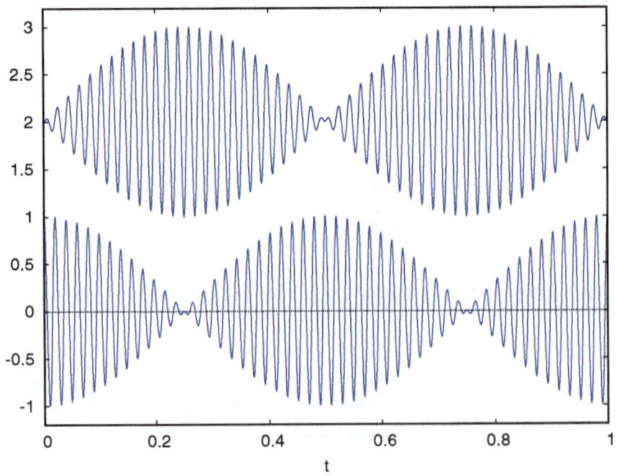

$x_1(t)$ and $x_2(t)+2$(above) are plotted for $A=1$, $\frac{1}{2\pi}\omega_- = 50$, and $2\pi\frac{\sqrt{mk}}{K} = \frac{1}{2}$

4. For $x_n = \mathrm{e}^{2\pi i\frac{n}{\lambda}-i\omega t}$ to satisfy $x_N = x_0$, we must find

$$\frac{N}{\lambda} = 0, 1, 2, \cdots, N-1$$

($\lambda = 1$ is the same as $\lambda = 0$). Alternatively, it can be

$$\frac{N}{\lambda} = 0, \pm 1, \cdots, \pm\left(\frac{N}{2}-1\right), \frac{N}{2}$$

where a positive $\frac{N}{\lambda}$ represents a wave traveling in the positive direction and a negative one that in the negative direction. From the equations of motion, we get

$$-m\omega^2 = k\left(\mathrm{e}^{\frac{2\pi i}{\lambda}} + \mathrm{e}^{-\frac{2\pi i}{\lambda}} - 2\right) = -2k\left(1-\cos\frac{2\pi}{\lambda}\right) = -4k\sin^2\frac{\pi}{\lambda}$$

Therefore, the dispersion relation is

$$\omega = 2\sqrt{\frac{k}{m}}\left|\sin\frac{\pi}{\lambda}\right|$$

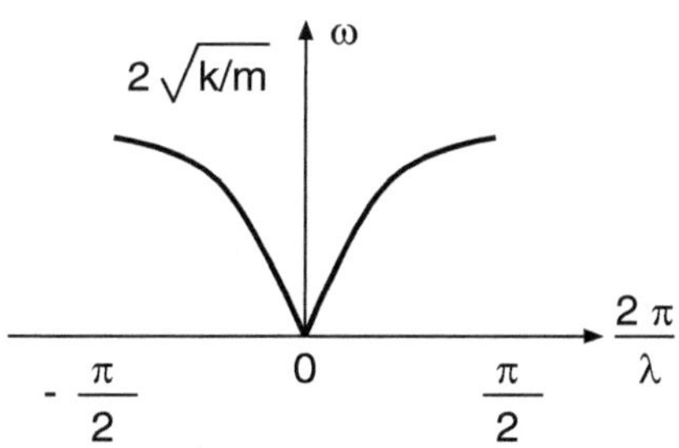

5. (a) The Lagrangian is given by

$$L = \sum_i\left(\frac{m}{2}l^2\dot{\theta}_i^2 - \frac{mgl}{2}\theta_i^2 - \frac{k}{2}l^2\left(\theta_i - \theta_{i-1}\right)^2\right)$$

(b) The equation of motion for θ_i is

$$l\ddot{\theta}_i = -g\theta_i - \frac{kl}{m}\left(2\theta_i - \theta_{i-1} - \theta_{i+1}\right)$$

(c) Using $\rho \equiv \frac{m}{d}$ as the linear mass density, we find

$$\partial_t^2 u(x,t) = -\Omega^2 u(x,t) + \frac{\kappa}{\rho}\partial_x^2 u(x,t)$$

(d) Let $v^2 \equiv \frac{\kappa}{\rho}$. We then obtain

$$\omega^2 = \Omega^2 + k^2 v^2 \Longrightarrow \omega = \sqrt{\Omega^2 + k^2 v^2}$$

(e) The dispersion relation is given by

$$\frac{d\omega}{dk} = \frac{\kappa}{\rho}\frac{k}{\omega} = \frac{\kappa}{\rho}\frac{k}{\sqrt{\Omega^2 + k^2 v^2}}$$

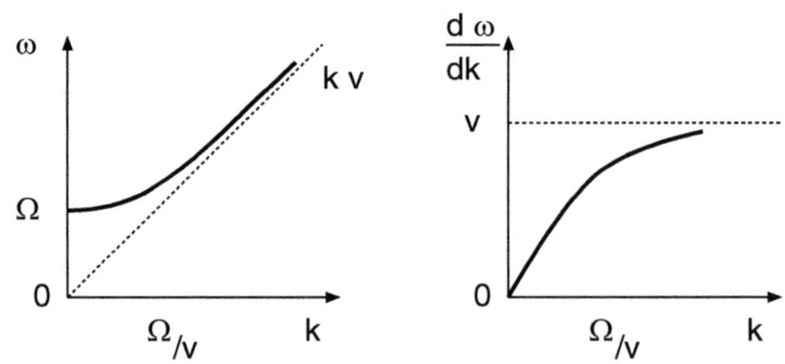

6. (a)

$$\Delta X[y] = \frac{1}{2}\int_0^X dx \left(\frac{dy}{dx}\right)^2$$

(b) This is the work required to stretch the string by ΔX.
(c)

$$K[\partial_t y] = \frac{\rho}{2}\int_0^X dx \, (\partial_t y)^2$$

(d) $\rho \partial_t^2 y = T \partial_x^2 y$
(e) $v = \sqrt{\frac{T}{\rho}}$

Exercises 2

1. **Three-atom molecule**—Find the eigenmodes of small oscillations in the x-direction for the following mechanical system. Let $x_{1,2,3}$ be the deviations from the equilibrium positions.

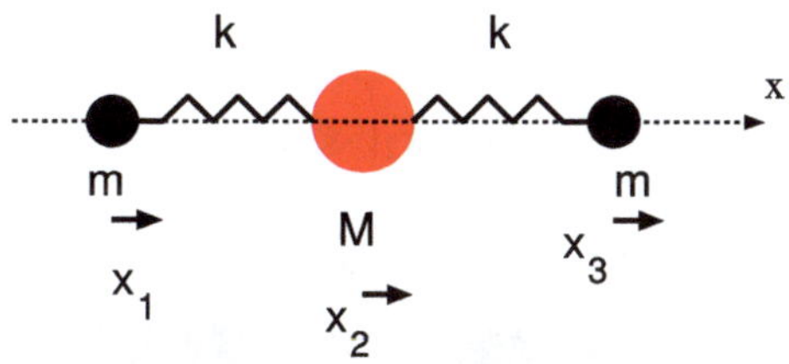

2. **Standard form of the Lagrangian**—Consider a Lagrangian that describes small oscillations with N degrees of freedom:

$$L(q, \dot{q}) \equiv \frac{1}{2} \sum_{i,j=1}^{N} M_{ij} \dot{q}_i \dot{q}_j - \frac{1}{2} \sum_{i,j=1}^{N} U_{ij} q_i q_j$$

where M and U are positive definite symmetric matrices.

(a) Using a coordinate transformation $\vec{Q} = W\vec{q}$, where W is an N-by-N matrix, demonstrate that the Lagrangian can be expressed as

$$L = \frac{1}{2} \frac{d}{dt} \vec{Q} \cdot \frac{d}{dt} \vec{Q} - \frac{1}{2} \sum_{i,j=1}^{N} V_{ij} Q_i Q_j$$

where V is a positive definite symmetric matrix.

(b) Furthermore, by an orthogonal transformation of $\vec{\tilde{Q}} = O\vec{Q}$ using an orthogonal matrix O, show that the Lagrangian can be expressed as

$$L = \frac{1}{2} \frac{d}{dt} \tilde{Q} \cdot \frac{d}{dt} \tilde{Q} - \frac{1}{2} \sum_{i=1}^{N} \tilde{Q}_i \omega_i^2 \tilde{Q}_i$$

(c) Derive the equations of motion for $\tilde{Q}_i$ and solve them.

(d) Express the original coordinates q_i as a linear combination of $\tilde{Q}_j$ ($j = 1, \cdots, N$) as

$$q_i = \sum_{j=1}^{N} A_{ij} \tilde{Q}_j$$

Show that A satisfies

$$A^T M A = \mathbf{1}_N \quad (N\text{-dimensional identity matrix})$$

3. **Coupled oscillators**—Let us consider a mechanical system similar to the one discussed in Example 2-3, but with a slight difference in the spring constants k_i

at both ends:

$$|k_1 - k_2| \ll k_1 \text{ or } k_2$$

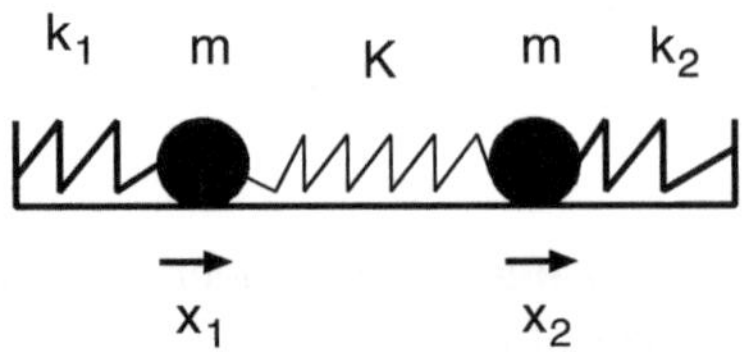

(a) Find the equations of motion for small oscillations.
(b) Solve the equations of motion under the initial conditions:

$$x_1(0) = A, \quad x_2(0) = 0, \quad \dot{x}_1(0) = \dot{x}_2(0) = 0$$

(c) Use a graph ($x_{1,2}$ vs t) to explain the characteristics of the solution. **Hint**: The amplitudes of x_1, x_2 never vanish if $k_1 \neq k_2$.

4. **Solution to the 1-D wave equation** —Let us consider the one-dimensional wave equation:

$$\left(\partial_t^2 - v^2\partial_x^2\right) f(x, t) = 0$$

(a) Rewrite the wave equation using coordinates $\xi \equiv x - vt$ and $\eta \equiv x + vt$, and demonstrate that the general solution to the equation can be expressed as

$$f(x, t) = f_+(\xi) + f_-(\eta)$$

where $f_\pm$ are arbitrary functions of a single variable.
(b) Explain the physical significance of the solution.
(c) Solve the wave equation under the initial conditions:

$$f(x, 0) = F(x), \quad \partial_t f(x, 0) = G(x)$$

Hint: You can express $f_\pm$ in terms of F, G.

5. **Standing waves**—Consider the wave equation

$$\left(\partial_t^2 - v^2\partial_x^2\right) u(x, t) = 0$$

subject to the boundary conditions

$$u(0, t) = u(L, t) = 0 \quad (L > 0) \tag{1.101}$$

In Sects. 1.1 and 1.2, we learned that the general solution is given by the Fourier series

$$u(x,t) = \sum_{n=1}^{\infty} C_n \cos\left(\frac{\pi n}{L} vt + \phi_n\right) \sin\frac{\pi n}{L} x \tag{1.102}$$

Now, let us explore an alternative derivation using the results from Exercise 2-4.

(a) Demonstrate that under the boundary condition $u(0,t) = 0$ at $x = 0$, the general solution is given by

$$u(x,t) = f(-x+vt) - f(x+vt)$$

(b) Explain the general solution physically.
(c) Consider the additional boundary condition $u(L,t) = 0$ at $x = L$. Show that the function f must be periodic with a period of $2L$.
(d) According to the theorem of Fourier's series, any periodic function $f(x)$ with a period of $2L$ is given by

$$f(x) = \sum_{n=-\infty}^{\infty} f_n e^{i\pi n \frac{x}{L}}$$

Using this representation, solve the wave equation (1) and explain the relation to (2).

6. **Reflection**—Consider a string lying along the x-axis in its equilibrium state. For $x < 0$, it has mass ρ_1 per unit length, and for $x > 0$, it has ρ_2. The tension of the string is denoted as T. We express the small displacement in the y-direction as $y(x,t)$.

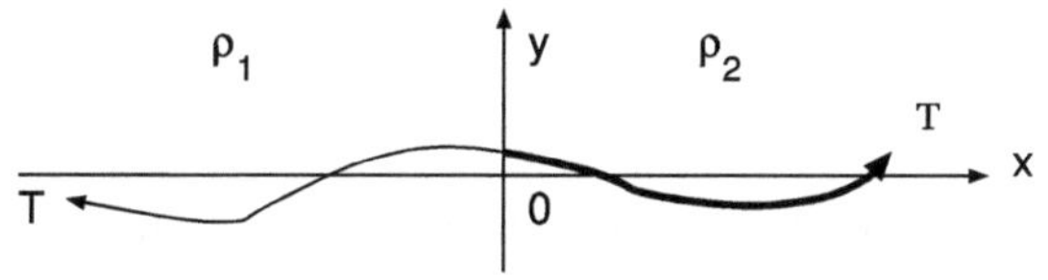

(a) Explain why both $y(x,t)$ and $\partial_x y(x,t)$ must be continuous at $x = 0$.
(b) Show that the following expression for $y(x,t)$ is a solution to the wave equation:

$$y(x,t) = \begin{cases} f(x - v_1 t) - \frac{v_1 - v_2}{v_1 + v_2} f(-x - v_1 t) & \text{for } x < 0 \\ \frac{2v_2}{v_1 + v_2} f\left(\frac{v_1}{v_2} x - v_1 t\right) & \text{for } x > 0 \end{cases}$$

Here, $v_{1,2}^2 \equiv \frac{T}{\rho_{1,2}}$, and $f(x)$ is an arbitrary function.

(c) Interpret the above solution physically.

(d) Solve the wave equation for $y(x,0)$ given in the figure, and $\partial_t y(x,0) = 0$.

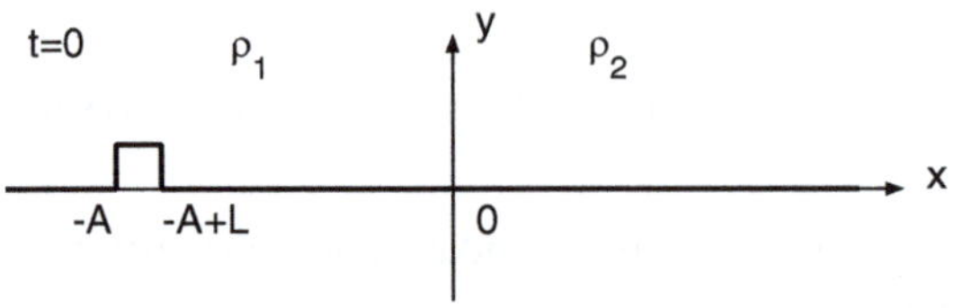

$y(x,0)$

7. **Model of friction**—Consider a simple harmonic oscillator with mass m and spring constant k, oscillating along the y-axis without friction.
We attach a string with mass density ρ and tension T to this oscillator. As has been shown in Example 2-6, the small vertical displacement $y(x,t)$ of the string in the y-direction satisfies the wave equation:

$$(\rho\partial_t^2 - T\partial_x^2)y(x,t) = 0$$

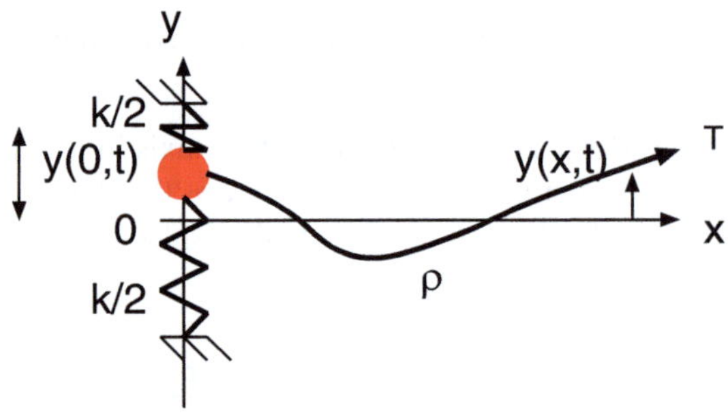

(a) Given $\partial_x y(0,t)$, find the equation of motion for the simple harmonic oscillator $y(0,t)$. **Hint**: Assume $|\partial_x y| \ll 1$.

From Exercise 2-4, we know that the general solution for $y(x,t)$ can be expressed using arbitrary one-variable functions f and g as

$$y(x,t) = f(-x+vt) + g(x+vt) \qquad \left(v \equiv \sqrt{\frac{T}{\rho}}\right)$$

(b) If $y(x, 0) = \partial_t y(x, 0) = 0$ for $x > L$, show that $g'(x) = 0$ for $x > L$.
(c) If $y(x, 0) = \partial_t y(x, 0) = 0$ for $x > L$, demonstrate that for $t > \frac{L}{v}$, $y(0, t)$ satisfies the equation of motion:

$$m\partial_t^2 y(0, t) + m\gamma\partial_t y(0, t) + ky(0, t) = 0$$

What is the friction coefficient γ? **Hint**: There is only the right-moving mode. The oscillator loses its energy by sending a right-moving wave. You find $m\gamma = \sqrt{T\rho}$.

1.3 Supplementary Notes

1.3.1 Resonance

The amplitude of a forced oscillation is large when the frequency of the external force is close to the natural frequency of the oscillator.

In the main part of Chap. 1, we did not get a chance to solve the important equation of motion

$$\boxed{m\ddot{x} + m\gamma\dot{x} + m\omega_0^2 x = f(t)} \tag{1.101}$$

which is the equation of motion of an oscillator with friction under an external force $f(t)$. The friction constant $\gamma > 0$ has the dimension of a frequency, and the friction force is proportional to the velocity. This equation appears often in varied circumstances (electromagnetism and quantum mechanics), and we summarize the main features here.

We first solve the homogeneous equation $f(t) = 0$. Substituting

$$x(t) = e^{i\alpha t} \tag{1.102}$$

into the equation, we obtain a quadratic equation that determines α:

$$-\alpha^2 + i\alpha\gamma + \omega_0^2 = 0 \tag{1.103}$$

To get damping oscillations, we assume[7]

$$\omega_0 > \frac{\gamma}{2} \tag{1.104}$$

[7] Note that the system is overdamped if $\omega_0 < \frac{\gamma}{2}$ and that the system is critically damped if $\omega_0 = \frac{\gamma}{2}$.

and we obtain

$$\alpha = \alpha_\pm \equiv i\frac{\gamma}{2} \pm \sqrt{\omega_0^2 - \frac{\gamma^2}{4}} \tag{1.105}$$

Hence, a homogeneous solution is an arbitrary linear combination of two solutions $e^{i\alpha_\pm t}$. The amplitude decays exponentially as $e^{-\frac{\gamma}{2}t}$. The natural angular frequency decreases from ω_0 to $\sqrt{\omega_0^2 - \gamma^2/4}$.

We now solve the equation of motion for a particular external force

$$f(t) = f_0 e^{i\omega t} \tag{1.106}$$

where f_0 is an arbitrary complex constant. It is the real part of $x(t)$ that is physical. Substituting $x(t) = Ae^{i\omega t}$ into the equation, we obtain

$$\left(-\omega^2 + i\omega\gamma + \omega_0^2\right) A = \frac{f_0}{m} \tag{1.107}$$

This gives

$$A = \frac{1}{-\omega^2 + i\omega\gamma + \omega_0^2} \frac{f_0}{m} \tag{1.108}$$

An arbitrary solution is the linear combination of $Ae^{i\omega t}$ together with a homogeneous solution. We find

$$|A|^2 = \frac{1}{\left(\omega^2 - \omega_0^2\right)^2 + \omega^2\gamma^2} \left|\frac{f_0}{m}\right|^2 \tag{1.109}$$

This is large only for $\omega \simeq \pm\omega_0$. For $\omega \simeq \omega_0$, we can approximate

$$|A|^2 \simeq \frac{1}{(\omega - \omega_0)^2 + \frac{\gamma^2}{4}} \frac{1}{4\omega_0^2} \left|\frac{f_0}{m}\right|^2 \tag{1.110}$$

The ω-dependence of $|A|^2$ and the phase θ of $A = |A|e^{i\theta}$ are shown in the figure below.

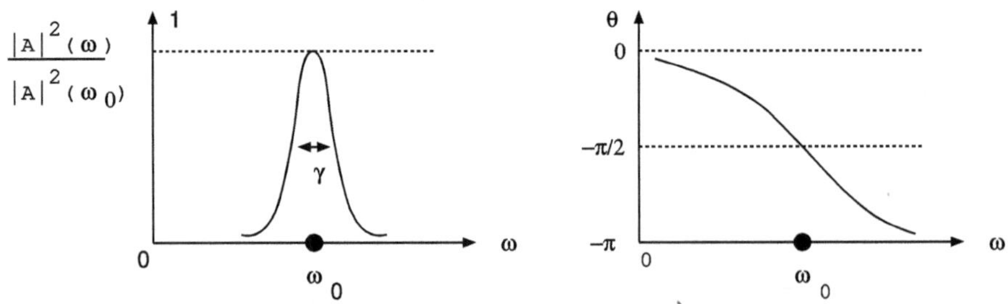

The full width at half maximum (FWHM) is given by γ. We define the Q-value of the resonance by

$$Q \equiv \frac{\omega_0}{\gamma} \tag{1.111}$$

We have assumed $Q \gg 1$ in the above.

Finally, let us solve the equation of motion for an arbitrary $f(t)$. We define the **retarded Green function** by

$$K_R(t-t') \equiv \theta(t-t') \frac{1}{\sqrt{\omega_0^2 - \frac{\gamma^2}{4}}} e^{-\frac{\gamma}{2}(t-t')} \sin\left(\sqrt{\omega_0^2 - \frac{\gamma^2}{4}}\,(t-t')\right) \tag{1.112}$$

where $\theta(t-t')$ is a step function, equal to 1 for $t > t'$ and zero otherwise. This is a solution to the homogeneous equation, multiplied by the step function. Using

$$\partial_t \theta(t-t') = \delta(t-t') \tag{1.113}$$

we can show

$$\left(\partial_t^2 + \gamma \partial_t + \omega_0^2\right) K_R(t-t') = \delta(t-t') \tag{1.114}$$

Therefore,

$$X(t) \equiv \int_0^t dt'\, K_R(t-t') \frac{1}{m} f(t') \tag{1.115}$$

satisfies the equation of motion and the boundary condition

$$X(0) = \dot{X}(0) = 0 \tag{1.116}$$

Note that only the force $f(t')$ with $t' < t$ can affect $X(t)$, i.e., the influence of the force is retarded. Given the initial condition

$$x(0) = x_0, \quad \dot{x}(0) = v_0 \tag{1.117}$$

the solution to the equation of motion is given by

$$x(t) = e^{-\frac{\gamma}{2}t} \left\{ x_0 \cos\left(\sqrt{\omega_0^2 - \frac{\gamma^2}{4}}\, t\right) + \frac{v_0 + \frac{\gamma}{2} x_0}{\sqrt{\omega_0^2 - \frac{\gamma^2}{4}}} \sin\left(\sqrt{\omega_0^2 - \frac{\gamma^2}{4}}\, t\right) \right\} + X(t) \tag{1.118}$$

For $\gamma t \gg 1$, the memory of the initial condition is lost, and we are left with

$$x(t) \simeq X(t) \tag{1.119}$$

1.3.2 Determining Central Force Potential from the Orbit

When the force acting on a point mass is known to be a central force, the potential energy can be determined from the trajectory.

The trajectory of a particle of mass m with energy E and angular momentum M under the central force potential $U(r)$ is given by

$$\frac{dr}{d\theta} = \frac{mr^2}{M}\sqrt{\frac{2}{m}(E - U_{\text{eff}}(r))} \tag{1.120}$$

as we learned in Sect. 1.1. Interpreting this result in reverse, we can determine the potential from the knowledge of the trajectory.

Let us assume that the trajectory is given by

$$\frac{dr}{d\theta} = f(r) \tag{1.121}$$

Then, from Eq. (1.120) we obtain

$$f(r)^2 = \frac{m^2r^4}{M^2}\frac{2}{m}(E - U_{\text{eff}}(r)) = \frac{2mr^4}{M^2}(E - U_{\text{eff}}(r)) \tag{1.122}$$

Since the effective potential $U_{\text{eff}}(r)$ is related to the central force potential $U(r)$ by

$$U_{\text{eff}}(r) \equiv U(r) + \frac{M^2}{2mr^2} \tag{1.123}$$

we find

$$f(r)^2 = \frac{2mr^4(E - U(r))}{M^2} - r^2 \tag{1.124}$$

Thus, the central force potential is determined as

$$\boxed{U(r) = E - \frac{M^2}{2mr^4}\left(f(r)^2 + r^2\right)} \tag{1.125}$$

Let us consider two examples:

1. Example 1-5—From $r = ke^{\alpha\theta}$, we have $f(r) = \alpha r$. Hence, the potential is proportional to $1/r^2$. We cannot determine α since we do not know M or m.
2. Kepler Motion—The trajectory is an ellipse given by

$$\frac{p}{r} = 1 + e\cos\theta \tag{1.126}$$

where e is a positive number less than 1.
Differentiating this with respect to r, we get

$$-\frac{p}{r^2} = -e\sin\theta\frac{1}{\frac{dr}{d\theta}} \tag{1.127}$$

Using

$$e^2\sin^2\theta = e^2 - \left(\frac{p}{r} - 1\right)^2 \tag{1.128}$$

we have

$$f(r)^2 \equiv \left(\frac{dr}{d\theta}\right)^2 = \frac{r^4}{p^2}\left(e^2 - \left(\frac{p}{r} - 1\right)^2\right) = \frac{r^4}{p^2}\left(e^2 - 1 + \frac{2p}{r}\right) - r^2 \tag{1.129}$$

From (1.125), we obtain

$$U(r) = E - \frac{M^2}{2mr^4}\frac{r^4}{p^2}\left(e^2 - 1 + \frac{2p}{r}\right) = E + \frac{M^2(1-e^2)}{2mp^2} - \frac{M^2}{mp}\frac{1}{r} \tag{1.130}$$

Using the convention that the potential goes to zero at infinity, the energy is given by a positive constant α as

$$U(r) = -\frac{\alpha}{r} \tag{1.131}$$

and the angular momentum M and energy E are given by the parameters $p,\ e$ of the orbit as

$$M^2 = \alpha p m, \quad E = -\frac{M^2(1-e^2)}{2mp^2} = -\frac{\alpha(1-e^2)}{2p} \tag{1.132}$$

1.3.3 Huygens's Pendulum

Huygens invented a pendulum whose period is independent of the amplitude.

Exercise 1-2 explains the principle behind Huygens's improvement of pendulums. The Lagrangian for a simple pendulum of length l is given as

$$L = \frac{ml^2}{2}\dot{\theta}^2 + mgl(\cos\theta - 1) \tag{1.133}$$

For a small angle θ, we find

$$\cos\theta \simeq 1 - \frac{\theta^2}{2} \tag{1.134}$$

The Lagrangian is reduced to the case of a harmonic oscillator.

$$L = \frac{ml^2}{2}\dot{\theta}^2 - \frac{mgl}{2}\theta^2 \tag{1.135}$$

However, for slightly larger θ, higher-order terms change the potential to

$$mgl\left(\frac{\theta^2}{2} - \frac{\theta^4}{24} + \cdots\right) \tag{1.136}$$

In fact the exact period, as a function of the maximum angle θ_0, is

$$T(\theta_0) = 2\sqrt{\frac{2l}{g}}\int_0^{\theta_0} \frac{d\theta}{\sqrt{\cos\theta - \cos\theta_0}} \tag{1.137}$$

which is an increasing function of θ_0. (The integral is called an elliptic integral, and cannot be expressed in terms of elementary functions.)

Huygens designed a pendulum with a period independent of the maximum angle θ_0 by using a cycloid of radius a and a pendulum of length $4a$. As is shown in Exercise 1-2, if the end of the pendulum's string traces a cycloid, the period becomes independent of the maximum angle.

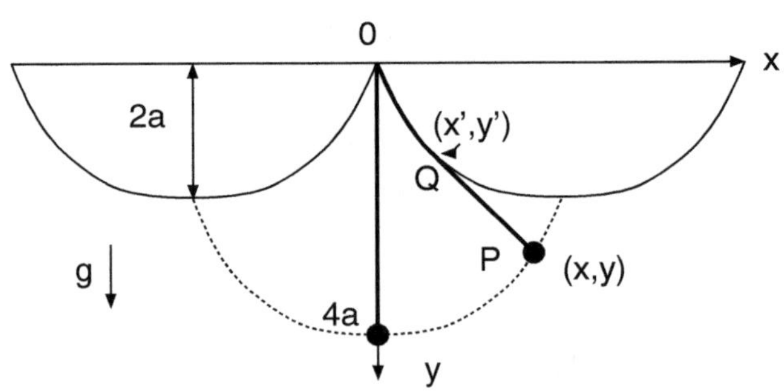

We show below that the point P in the above figure traces a cycloid. It is a matter of geometry.

Let the coordinates of the point Q, where the pendulum's string deviates from the cycloid, be (x', y'). This lies on the cycloid, and we can write

$$x' = a(\theta - \sin\theta), \quad y' = a(1 - \cos\theta) \tag{1.138}$$

The length of the cycloid from the origin to Q is $4a(1 - \cos\frac{\theta}{2})$ so that the length from Q to P is $4a\cos\frac{\theta}{2}$. Hence, the coordinates of P are

$$x = a(\theta + \sin\theta), \quad y = 2a + a(1 + \cos\theta) \tag{1.139}$$

Using $\theta + \pi$ as an angle, we can rewrite

$$x = -\pi a + a\left(\theta + \pi - \sin(\theta + \pi)\right), \quad y = 2a + a\left(1 - \cos(\theta + \pi)\right) \tag{1.140}$$

Hence, P lies on a cycloid.

1.3.4 Alternative Variational Principles

Hamilton's variational principle states that the time integral (action) of the Lagrange function be minimized. There are several other variational principles that are physically equivalent. Here, we introduce two.

Variational Principle Using the Hamilton Function

Instead of using the Lagrange function, we can formulate a variational principle using the Hamilton function or Hamiltonian. This provides the most general variational principle.

We consider the simplest one-dimensional case as an example. Let the Hamilton function be given as a function $H(q, p)$ of the coordinate q and the momentum p. We define the action as a functional of the two functions of time $q(t)$, $p(t)$ as

$$S[q, p] \equiv \int_{t_1}^{t_2} dt \left(p(t)\frac{dq(t)}{dt} - H(q(t), p(t)) \right) \tag{1.141}$$

We require that the action $S[q, p]$ be minimized, given the initial value of the coordinate $q(t_1) = Q_1$ and the final value $q(t_2) = Q_2$. Note we do not specify the initial or final values of the momentum. If $S[q, p]$ is minimized, then the action is invariant under arbitrary infinitesimal changes in $q(t)$ and $p(t)$:

$$\delta S[q, p] = 0$$

$$
\begin{aligned}
\Longleftrightarrow \int_{t_1}^{t_2} dt \Bigg(\delta p(t) \frac{dq(t)}{dt} + p(t) \frac{d\delta q(t)}{dt} \\
- \delta q(t) \cdot \frac{\partial H(q(t), p(t))}{\partial q(t)} - \delta p(t) \cdot \frac{\partial H(q(t), p(t))}{\partial p(t)} \Bigg) \\
= \int_{t_1}^{t_2} dt \Bigg[\delta p(t) \left(\frac{dq(t)}{dt} - \frac{\partial H(q(t), p(t))}{\partial p(t)} \right) \\
+ \delta q(t) \left(-\frac{dp(t)}{dt} - \frac{\partial H(q(t), p(t))}{\partial q(t)} \right) \Bigg] = 0
\end{aligned}
\tag{1.142}
$$

For this to vanish for arbitrary $\delta q(t), \delta p(t)$, the Hamilton equations of motion

$$
\frac{dq(t)}{dt} = \frac{\partial H(q, p)}{\partial p(t)}, \quad \frac{dp(t)}{dt} = -\frac{\partial H(q, p)}{\partial q(t)} \tag{1.143}
$$

must be satisfied.

It is interesting that only the initial and final values of the coordinate, $q(t_1) = Q_1, q(t_2) = Q_2$, need to be specified. The momentum at either t_1 or t_2 does not need to be specified. Actually, given Q_1, a particular value of the momentum $p(t_1) = P_1$ is determined by Q_2; for the equations of motion to give $q(t_2) = Q_2$ under the initial condition (Q_1, P_1), a particular value of P_1 is required. Hence, specifying (Q_1, Q_2) amounts to specifying (Q_1, P_1).

This principle is slightly more general than Hamilton's principle of minimizing the time integral of the Lagrange function. This is because Hamilton's principle is obtained as a special case of this principle, where the time dependence of the momentum is constrained. In the usual Hamilton's principle, it is the Lagrange function $L(q, \dot{q})$ that is given initially. The momentum is then defined by

$$
p(t) \equiv \frac{\partial L(q(t), \dot{q}(t))}{\partial \dot{q}(t)} \tag{1.144}
$$

Finally the Hamilton function is defined as a function of q and p by the Legendre transformation

$$
H(q, p) \equiv \dot{q} p - L(q, \dot{q}) \tag{1.145}
$$

Under the variational principle using the action (1.141), we can consider any time dependence of the coordinate $q(t)$ and the momentum $p(t)$. If we restrict the time dependence of the momentum using (1.144), the action in Eq. (1.141) becomes

$$
S[q, p] = \int_{t_1}^{t_2} dt \, (\dot{q} p - (\dot{q} p - L(q, \dot{q})) = \int_{t_1}^{t_2} dt \, L(q(t), \dot{q}(t)) \tag{1.146}
$$

This reduces the action $S[q, p]$ to the one used in Hamilton's principle. Thus, the usual Hamilton's principle corresponds to minimizing the action $S[q, p]$ under the restriction that the time dependence of the momentum is given by Eq. (1.144). It is in this sense we have said that the variational principle using the Hamilton function is more general than Hamilton's principle.

Principle of Maupertuis

In the above we have derived the equations of motion by minimizing the action (1.141) without any constraints on the coordinates and momenta except for the initial and final values of the coordinates. When the Hamilton function has a special form given by

$$H(q, p) = \frac{1}{2} \sum_{i,j=1}^{N} M_{ij}^{-1}(q) p_i p_j + U(q), \tag{1.147}$$

we can introduce an alternative variational principle that gives only the trajectory without time dependence. We do this by considering a restricted class of variations for the time dependence of $q_i(t)$, $p_i(t)$.

Instead of considering all possible time dependence of momenta and coordinates, we consider only the trajectories satisfying the condition of constant energy, which is given by

$$E = \frac{1}{2} \sum_{i,j=1}^{N} M_{ij}^{-1}(q) p_i p_j + U(q), \tag{1.148}$$

where E is an arbitrary constant. Solving the Hamilton equations of motion

$$\frac{dq_i}{dt} = \frac{\partial H(q, p)}{\partial p_i} = \sum_{j=1}^{N} M_{ij}^{-1}(q) p_j \tag{1.149}$$

we obtain

$$p_i = \sum_{j=1}^{N} M_{ij}(q) \dot{q}_j \tag{1.150}$$

The constant energy equation can then be rewritten as

$$E = \frac{1}{2} \sum_{i,j=1}^{N} M_{ij}(q) \dot{q}_i \dot{q}_j + U(q) \tag{1.151}$$

Denoting the small change in the coordinates q_i during a small time dt as dq_i, we can rewrite Eq. (1.151) as

$$\sqrt{E - U(q)}\, dt = \sqrt{\frac{1}{2} \sum_{i,j=1}^{N} M_{ij}(q) dq_i dq_j} \tag{1.152}$$

Hence, the action (1.141) is obtained as a functional of $q(t)$ and a function of energy E as follows:

$$\begin{aligned} S[q; E] &\equiv \int_{t_1}^{t_2} dt \left(\sum_{i=1}^{N} p_i \dot{q}_i - H(q, p) \right) \\ &= \int_{q(t_1)}^{q(t_2)} \sum_{i=1}^{N} p_i dq_i - E(t_2 - t_1) \\ &= \int_{q(t_1)}^{q(t_2)} \sum_{i,j=1}^{N} M_{ij}(q) \frac{dq_i dq_j}{dt} - E(t_2 - t_1) \\ &= \int_{q(t_1)}^{q(t_2)} \sqrt{2(E - U(q)) \sum_{i,j=1}^{N} M_{ij}(q) dq_i dq_j} \\ &\quad - E(t_2 - t_1) \end{aligned} \tag{1.153}$$

The last transformation used Eq. (1.152). To summarize so far, we have assumed Eq. (1.150) for momenta and that the energy takes a constant value E as in Eq. (1.151).

We can vary the action $S[q; E]$ with respect to E and also the functions $q_i(t)$. First, we vary with respect to E to get

$$\begin{aligned} t_2 - t_1 &= \frac{\partial}{\partial E} \int_{q(t_1)}^{q(t_2)} \sqrt{2(E - U(q)) \sum_{i,j=1}^{N} M_{ij}(q) dq_i dq_j} \\ &= \int_{q(t_1)}^{q(t_2)} \sqrt{\frac{\sum_{i,j=1}^{N} M_{ij}(q) dq_i dq_j}{2(E - U(q))}} \end{aligned} \tag{1.154}$$

Note that the integral on the right side is determined only by the trajectory and does not depend on the time dependence of the coordinates. In fact it is the other way round: this equation itself becomes the equation that gives the time dependence of the coordinates.

To determine the trajectory itself, we can take the action as

$$S_M[q] \equiv \int_{q(t_1)}^{q(t_2)} \sqrt{2(E-U(q)) \sum_{i,j=1}^{N} M_{ij}(q) dq_i dq_j} \tag{1.155}$$

which depends only on the trajectory from $q(t_1)$ to $q(t_2)$ in the N-dimensional coordinate space. $S_M[q]$ should be minimized with respect to the trajectory.

This is called the principle of Maupertuis.[8]

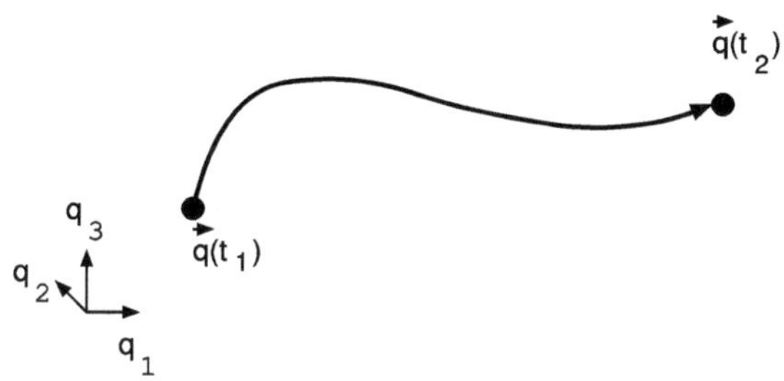

Trajectory from $\vec{q}(t_1)$ to $\vec{q}(t_2)$ in $\vec{q}$-space

Example: A Particle on a Curved Surface

Consider a particle moving on a curved surface defined by $z = f(x, y)$ in three-dimensional space. Assuming a uniform gravitational acceleration g acting in the negative z direction, Maupertuis's action is given by

$$S_M[x, y] = \int_{(x_1,y_1)}^{(x_2,y_2)} \sqrt{2(E - mgf(x, y))m}\, ds\,, \tag{1.156}$$

where ds is the line element on the surface, given by

$$\begin{aligned}(ds)^2 &\equiv (dx)^2 + (dy)^2 + (dz)^2 \\ &= (dx)^2 + (dy)^2 + (\partial_x f(x, y)dx + \partial_y f(x, y)dy)^2\end{aligned} \tag{1.157}$$

Especially, in the absence of gravity, $S_M[x, y]$ is proportional to the path length

$$S_M[x, y] = \sqrt{2mE} \int_{(x_1,y_1)}^{(x_2,y_2)} ds \tag{1.158}$$

[8] It is said that Maupertuis described it in a more divine and profound manner.

Hence, in the absence of gravity, the trajectory of the particle is determined to minimize the length. We expect that the particle moves at a constant speed on the surface. In fact, Eq. (1.154) gives

$$t - t_1 = \sqrt{\frac{m}{2E}} \int_{(x_1,y_1)}^{(x,y)} ds \tag{1.159}$$

Since the velocity v is given by

$$v = \sqrt{\frac{2E}{m}}, \tag{1.160}$$

we get

$$v \cdot (t - t_1) = \int_{(x_1,y_1)}^{(x,y)} ds \tag{1.161}$$

as expected.

1.3.5 Application of the Principle of Least Action to Continuous Media

In Sect. 1.2.1, we have taken the continuum limit of the equations of motion to derive the wave equation. Here, we apply the principle of least action directly to the action of continuous media to derive the wave equation. This gives us a good exercise in functional differentiation.

In Sects. 1.1 and 1.2, we have dealt with a system consisting of $N-1$ mass points and N springs. The Lagrangian for the system is given by

$$L = \frac{m}{2}\sum_{i=1}^{N-1} \dot{u}_i^2 - \frac{m\omega^2}{2}\left(u_1^2 + \sum_{i=2}^{N-1}(u_i - u_{i-1})^2 + u_{N-1}^2\right) \tag{1.162}$$

From this, we have obtained the equations of motion:

$$\ddot{u}_i = -\omega^2 (u_i - u_{i-1} + u_i - u_{i+1}) \quad (i = 1, \cdots, N-1) \tag{1.163}$$

Then, by taking the continuum limit where the distance a between mass points goes to zero, we have derived the Lagrangian functional

$$L[u, \partial_t u] = \int_0^L dx \left[\frac{\rho}{2}(\partial_t u(x,t))^2 - \frac{\kappa}{2}(\partial_x u(x,t))^2\right] \tag{1.164}$$

from the Lagrangian (1.162) and obtained the wave equation

$$\rho \partial_x^2 u(x,t) - \kappa \partial_x^2 u(x,t) = 0 \tag{1.165}$$

from the equations of motion (1.163).

Now, we apply the principle of least action directly to the Lagrangian functional (1.164) to derive the wave Eq. (1.165). The action is given by the time integral of the Lagrangian functional:

$$S[u] \equiv \int_{t_0}^{t_1} dt\, L[u, \partial_t u] \tag{1.166}$$

To find the equations of motion for u, we consider the first-order change in S when u is varied infinitesimally by $\delta u(x,t)$. The equations of motion can then be derived, following the same principle as for the discrete system. The first-order change in $S[u]$ is calculated as follows:

$$\begin{aligned}\delta S &\equiv S[u+\delta u] - S[u] \\ &= \int_{t_0}^{t_1} dt\, \left(L[u+\delta u, \partial_t(u+\delta u)] - L[u, \partial_t u]\right) \\ &= \int_{t_0}^{t_1} dt \int_0^L dx\, \left(\rho \partial_t u(x,t) \partial_t \delta u(x,t) - \kappa \partial_x u(x,t) \partial_x \delta u(x,t)\right)\end{aligned} \tag{1.167}$$

Using the boundary conditions

$$u(0,t) = u(L,t) = 0, \tag{1.168}$$

we can integrate by parts to obtain

$$\int_0^L dx\, \partial_x u(x,t) \partial_x \delta u(x,t) = -\int_0^L dx\, \partial_x^2 u(x,t) \cdot \delta u(x,t) \tag{1.169}$$

Remembering that we fix $u(x,t)$ at $t = t_0$ and t_1 for $0 < x < L$, we integrate by parts to obtain

$$\int_{t_0}^{t_1} dt\, \partial_t u(x,t) \partial_t \delta u(x,t) = -\int_{t_0}^{t_1} dt\, \partial_t^2 u(x,t) \cdot \delta u(x,t) \tag{1.170}$$

Therefore, we obtain

$$\delta S = -\int_{t_0}^{t_1} dt \int_0^L dx\, \delta u(x,t) \left(\rho \partial_t^2 u(x,t) - \kappa \partial_x^2 u(x,t)\right) \tag{1.171}$$

For this to vanish for arbitrary $\delta u(x,t)$, we must find the wave equation:

$$\rho\partial_t^2 u(x,t) - \kappa\partial_x^2 u(x,t) = 0 \tag{1.172}$$

1.3.6 Adiabatic Invariants

When the parameters of a mechanical system are changed slowly compared with the period of motion, adiabatic invariants suffer no change. In early quantum mechanics, adiabatic invariants have played an important role in finding the quantization condition. We consider the simplest example of a harmonic oscillator.

The equation of motion of a simple harmonic oscillator is given by

$$\ddot{x} = -\omega^2(t)x \tag{1.173}$$

We assume that the angular frequency $\omega(t)$ is a function of time. We are interested in the case where the time variation of $\omega(t)$ is sufficiently small, i.e.,

$$\left|\frac{\dot{\omega}}{\omega}\right| \ll \omega \tag{1.174}$$

This amounts to an **adiabatic change**. For example, imagine changing very slowly the length of a pendulum's string.

Now, at time t, we express the total energy as

$$E(t) \equiv \frac{1}{2}\left(\dot{x}^2 + \omega(t)^2 x^2\right) \tag{1.175}$$

Then, we consider the ratio

$$S(t) \equiv \frac{E(t)}{\omega(t)} \tag{1.176}$$

Let us demonstrate that this quantity, $S(t)$, remains constant as long as $\omega(t)$ varies slowly. This $S(t)$ is referred to as the **action variable**. When $\omega(t)$ is strictly constant, $S(t)$ is also conserved due to the conservation of energy. The action variable can also be defined by the following integral over one period $\frac{2\pi}{\omega}$:

$$\frac{1}{2\pi}\oint dx\, \dot{x}(t) = \frac{1}{2\pi}\int_0^{\frac{2\pi}{\omega}} dt\, (\dot{x}(t))^2 = S \tag{1.177}$$

We are interested in a slow change of $\omega(t)$ in time. We can express $x(t)$ and $\dot{x}(t)$ in terms of the action variable and an angle variable $\varphi(t)$ as follows:

$$x(t) = \sqrt{\frac{2S(t)}{\omega(t)}} \sin\varphi(t) \tag{1.178a}$$

$$\dot{x}(t) = \sqrt{2\omega(t)S(t)} \cos\varphi(t) \tag{1.178b}$$

We check

$$\frac{\frac{1}{2}\left(\dot{x}^2 + \omega(t)^2 x^2\right)}{\omega(t)} = \omega(t)S(t)\frac{\cos^2\varphi(t) + \sin^2\varphi(t)}{\omega(t)} = S(t) \tag{1.179}$$

Using S and φ as independent mechanical variables instead of x and $\dot{x}$, we need to find the equations of motion for S and φ.

Differentiating (1.178a) with respect to time, we obtain

$$\dot{x} = \frac{1}{2}\left(\frac{\dot{S}}{S} - \frac{\dot{\omega}}{\omega}\right)\sqrt{\frac{2S}{\omega}} \sin\varphi + \dot{\varphi}\sqrt{\frac{2S}{\omega}} \cos\varphi \tag{1.180}$$

In order for this to match (1.178b), we must find

$$\frac{1}{2}\left(\frac{\dot{S}}{S} - \frac{\dot{\omega}}{\omega}\right) \sin\varphi + \dot{\varphi}\cos\varphi = \omega\cos\varphi \tag{1.181}$$

Differentiating (1.178b) with respect to time, we obtain

$$\ddot{x} = \frac{1}{2}\left(\frac{\dot{S}}{S} + \frac{\dot{\omega}}{\omega}\right)\sqrt{2\omega S} \cos\varphi - \dot{\varphi}\sqrt{2\omega S} \sin\varphi \tag{1.182}$$

For this to match the equation of motion (1.173), we must find

$$\frac{1}{2}\left(\frac{\dot{S}}{S} + \frac{\dot{\omega}}{\omega}\right) \cos\varphi - \dot{\varphi}\sin\varphi = -\omega\sin\varphi \tag{1.183}$$

Therefore, from (1.181) and (1.183), we obtain the following equations of motion for S and φ:

$$\frac{\dot{S}}{S} = -\frac{\dot{\omega}}{\omega}\cos 2\varphi \tag{1.184a}$$

$$\dot{\varphi} = \omega + \frac{\dot{\omega}}{2\omega}\sin 2\varphi \tag{1.184b}$$

If ω does not depend on time, we find

$$S = \text{constant}, \quad \varphi(t) = \omega t + \text{constant} \tag{1.185}$$

As long as ω changes adiabatically, it takes many multiples of the period $\frac{2\pi}{\omega}$ for ω to change even a small amount; meanwhile $\cos 2\varphi$ oscillates back and forth to average out. Hence, (1.184a) gives

$$\frac{\dot{S}}{S} \simeq -\frac{\dot{\omega}}{\omega}\langle\cos 2\varphi(t)\rangle \simeq 0 \tag{1.186}$$

where $\cos 2\varphi(t)$ is averaged over a period.

The adiabatic invariant S obtained here played an important role when Bohr was constructing quantum theory. Bohr's quantization condition is given as

$$S = n\frac{h}{2\pi} \quad (n = 1, 2, \ldots) \tag{1.187}$$

When the system is adiabatically changed, its quantum state should change smoothly. On the other hand, S takes quantized values; S must be an adiabatic invariant.

Chapter 2
Electromagnetism

Electricity and magnetism are almost always at the heart of physical phenomena.

2.1 Electromagnetism 1

Three Points to Remember

1. Understanding the relation between the SI and cgs-Gauss units requires physics.
2. Two of the four Maxwell's equations imply the existence of vector and scalar potentials; the remaining two imply the generation of electromagnetic fields by charge and current densities.
3. Electromagnetic waves propagate even in the absence of charge and current densities.

2.1.1 Understanding the Relation Between the SI and cgs-Gauss Units Requires Physics

International agreements dictate the use of the SI units for electromagnetism. Though the SI units are practical, they are not well suited for understanding electromagnetism. A system of units, where electric and magnetic fields have the same dimension and where the role of the speed of light is transparent, is preferable for understanding electromagnetism. Therefore, the cgs-Gauss units (or its rationalized version, the Heaviside units) is preferable even though it is out of fashion at present.

H. Sonoda, K. Ülker, *Basic Physics for Graduate Students*,
https://doi.org/10.1007/978-981-95-6607-5_2

SI Units

The SI (**I**nternational **S**ystem of units) underwent a significant revision on May 20, 2019. The base units the meter, the kilogram, the ampere, the mole, and the kelvin are now defined in terms of the fundamental physical constants such as the light velocity c, the Planck constant h, the elementary charge e, the Avogadro number N_A, and the Boltzmann constant k. Their numerical values are determined based on the most precise measurements up to the date of revision.

The units most relevant to electromagnetism are the following four (known as MKSA):

- Time 1s—Determined using atomic clocks based on ^{133}Cs. The frequency of radiation corresponding to the transition between the two energy levels ($F = 4, M = 0$) and ($F = 3, M = 0$) is defined as

$$\Delta\nu_{\mathrm{Cs}} \equiv 9{,}192{,}631{,}770\,\mathrm{Hz}$$

where F is the total spin of the nucleus and the valence electron. $\mathrm{Hz} = 1/\mathrm{s}$ is the unit of frequency. Therefore, 1s is defined as

$$1\mathrm{s} \equiv \frac{9{,}192{,}631{,}770}{\Delta\nu_{\mathrm{Cs}}}$$

- Length 1m—Defined as the distance light travels in $1/299{,}792{,}458$ s in vacuum. Hence, the speed of light is given exactly by

$$c \equiv 299{,}792{,}458\,\mathrm{m\,s^{-1}}$$

Therefore, 1m is defined in terms of c and $\Delta\nu_{\mathrm{Cs}}$ as

$$1\mathrm{m} \equiv \frac{c}{299{,}792{,}458}\,\mathrm{s} = \frac{9{,}192{,}631{,}770}{299{,}792{,}458}\frac{c}{\Delta\nu_{\mathrm{Cs}}}$$

- Mass 1kg—Before the revision the kilogram prototype made of a platinum-iridium cylinder was used. Now the kilogram is defined so that Planck's constant is given exactly as

$$h \equiv 6.62607015 \times 10^{-34}\,\mathrm{J\cdot s}$$

where $\mathrm{J} = \mathrm{kg\,m^2\,s^{-2}}$ is the unit of energy. Hence, 1kg is defined in terms of h, c, and $\Delta\nu_{\mathrm{Cs}}$ as

$$1\mathrm{kg} \equiv \frac{h}{6.62607015 \times 10^{-34}}\mathrm{m^{-2}s}$$

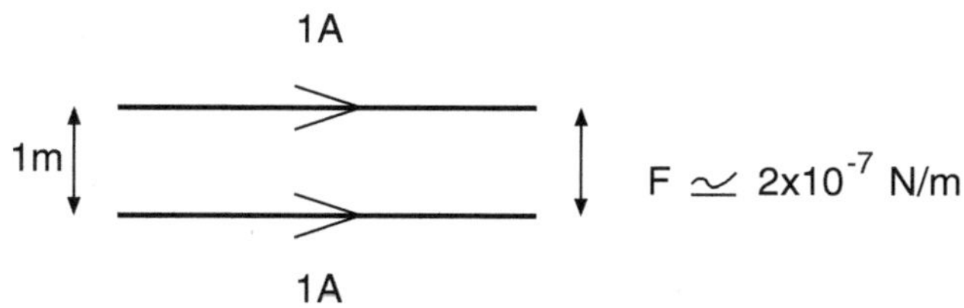

Fig. 2.1 The attractive force is 2×10^{-7} N/m to 9 decimal places

$$= \frac{(299{,}792{,}458)^2}{(6.62607015 \times 10^{-34})(9{,}192{,}631{,}770)} \frac{h\Delta\nu_{\text{Cs}}}{c^2}$$

- Current 1A—Before the revision, the unit of electric current was determined by the force between two parallel currents (Fig. 2.1). Now, it is defined by the elementary charge

$$e \equiv 1.602176634 \times 10^{-19}\,\text{C}$$

Then, the unit of electric current, Ampere, is defined by $\text{A} \equiv \text{C}\,\text{s}^{-1}$. Therefore, in terms of e and $\Delta\nu_{\text{Cs}}$,

$$\begin{aligned} 1\text{A} &\equiv \frac{e}{1.602176634 \times 10^{-19}} s^{-1} \\ &= \frac{1}{(1.602176634 \times 10^{-19})(9{,}192{,}631{,}770)} e\Delta\nu_{\text{Cs}} \end{aligned}$$

- The unit of the electric field[1]

$$\text{volt}\,\text{m}^{-1} \equiv \frac{\text{N}}{\text{C}} = \frac{\text{N}}{\text{A} \cdot \text{s}}$$

and that of the magnetic field

$$\text{Tesla} \equiv \frac{\text{volt}}{\text{m}} \cdot \frac{\text{s}}{\text{m}} = \frac{\text{N}}{\text{A} \cdot \text{m}}$$

are defined so that the force acting on the charge q is given by the Lorentz force law (Fig. 2.2):

$$\boxed{\vec{F} = q(\vec{E} + \vec{v} \times \vec{B})} \tag{2.1}$$

[1] Needless to say, the unit of force, Newton, can be written in terms of base units as $\text{N} = \text{kg}\cdot\text{m}\cdot\text{s}^{-2}$.

Fig. 2.2 Lorentz force

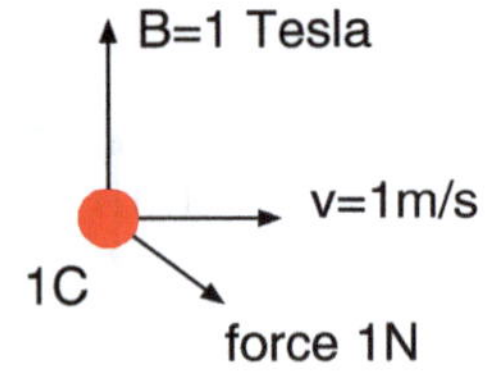

In the SI units, Maxwell's equations are written as follows:

$$\epsilon_0 \nabla \cdot \vec{E} = \rho \tag{2.2}$$

$$\frac{1}{\mu_0} \nabla \times \vec{B} - \epsilon_0 \partial_t \vec{E} = \vec{j} \tag{2.3}$$

$$\nabla \cdot \vec{B} = 0 \tag{2.4}$$

$$\nabla \times \vec{E} + \partial_t \vec{B} = 0 \tag{2.5}$$

Since the charge density has the dimension $\mathrm{C\,m^{-3}}$, the dimension of the vacuum permittivity ϵ_0 is given by

$$[\epsilon_0] = \frac{\mathrm{C}}{\mathrm{m}^3} \frac{\mathrm{m}}{\frac{\mathrm{N}}{\mathrm{C}}} = \frac{\mathrm{C}^2}{\mathrm{N} \cdot \mathrm{m}^2} = \frac{\mathrm{A}^2 \cdot \mathrm{s}^2}{\mathrm{N} \cdot \mathrm{m}^2} \tag{2.6}$$

The current density $\vec{j}$ has the dimension $\mathrm{C\,m^{-2}\,s^{-1}}$, and the vacuum permeability μ_0 has the dimension

$$[\mu_0] = \frac{\mathrm{Tesla}}{\mathrm{m}} \frac{\mathrm{m^2\,s}}{\mathrm{C}} = \frac{\mathrm{N}}{\mathrm{A\,m^2}} \frac{\mathrm{m}^2}{\mathrm{A}} = \frac{\mathrm{N}}{\mathrm{A}^2} \tag{2.7}$$

Before the revision, the numerical value of μ_0 was defined to be

$$\frac{\mu_0}{4\pi} \equiv 10^{-7} \frac{\mathrm{N}}{\mathrm{A}^2} \quad \text{(old SI)}$$

Not anymore.

Note that

$$\left[\frac{1}{\epsilon_0 \mu_0}\right] = \frac{\mathrm{m}^2}{\mathrm{s}^2} \tag{2.8}$$

has the dimension of a squared velocity, and this is consistent with the speed of light c given by

$$c^2 = \frac{1}{\epsilon_0 \mu_0} \tag{2.9}$$

Since the numerical value of c in the SI units is a defined quantity, either ϵ_0 or μ_0 must be measured. The numerical value of the elementary charge in the SI units is defined so that

$$\boxed{\frac{\mu_0}{4\pi} = 1 \times 10^{-7}\,\frac{\mathrm{N}}{\mathrm{A}^2}} \quad \textbf{approximate} \tag{2.10}$$

is valid up to 9 decimal places. This approximate value of μ_0 is a convenient physical constant to remember. From Eq. (2.9) we obtain approximately

$$\frac{1}{4\pi\epsilon_0} \simeq 1.00 \times 10^{-7}\,\frac{\mathrm{N}}{\mathrm{A}^2} \times c^2 \simeq 8.99 \times 10^9\,\frac{\mathrm{N}\cdot\mathrm{m}^2}{\mathrm{C}^2} \tag{2.11}$$

cgs-Gauss Units

We summarize the cgs-Gauss system of units.

- Time 1 s
- Length 1 cm $\equiv 10^{-2}$ m
- Mass 1 g $\equiv 10^{-3}$ kg
- Electric charge 1 **esu** (electrostatic unit)—Determined so that the force between two charges is given by

$$F = \frac{e_1 e_2}{r^2} \tag{2.12}$$

The unit of a force is dyn $\equiv$ g cm s^{-2}. Hence,

$$\mathrm{esu} \equiv \sqrt{\mathrm{dyn}}\cdot\mathrm{cm} = \mathrm{g}^{\frac{1}{2}}\,\mathrm{cm}^{\frac{3}{2}}\,\mathrm{s}^{-1} \tag{2.13}$$

The unit of electric current **statA** is defined by

$$\mathrm{statA} \equiv \mathrm{esu}\,\mathrm{s}^{-1}$$

Let q be an electric charge in the SI system, and let q_G be the same charge in the cgs-Gauss system. The Coulomb force is given by

$$F = \frac{q^2}{4\pi\epsilon_0 r^2} = \frac{q_G^2}{r^2}$$

Hence, the relation between the two charges is

$$q_G^2 = \frac{q^2}{4\pi\epsilon_0}$$

Therefore, for $q = 1\,\mathrm{C}$, the corresponding charge in the cgs-Gauss unit is (from Eq. (2.11))

$$q_G = \frac{1\,\mathrm{C}}{\sqrt{4\pi\epsilon_0}} \simeq \sqrt{8.99 \times 10^9 \times 10^5\,\mathrm{dyn}} \cdot 10^2\,\mathrm{cm} \simeq 3.00 \times 10^9\,\mathrm{esu}$$

This 3.00 comes from $\frac{\mu_0}{4\pi}\frac{1}{10^{-7}\mathrm{NA}^{-2}} \simeq 1.00$ and $\frac{c}{10^8\,\mathrm{ms}^{-1}} \simeq 3.00$. Hence, in the cgs-Gauss system, the elementary charge is

$$e_G \simeq 4.81 \times 10^{-10}\,\mathrm{esu}$$

- Electric field—Define $1\,\mathrm{statV/cm} \equiv 1\frac{\mathrm{esu}}{\mathrm{cm}^2}$ as the electric field created by a 1 esu charge at a distance of 1 cm. Note statV · esu = erg.
- Magnetic field—1 **Gauss** $\equiv$ 1 statV/cm is defined by the Lorentz force. For charge q_G, the Lorentz force is given by

$$\boxed{\vec{F} = q_G\left(\vec{E}_G + \frac{\vec{v}}{c} \times \vec{B}_G\right)} \tag{2.14}$$

  Since the velocity is divided by the speed of light, the magnetic field has the same dimension as the electric field. Furthermore, the dimension of magnetic charge is the same as that of electric charge. This is the convention introduced by Gauss. The unit of magnetic charge is usually referred to as **emu** (electromagnetic unit) rather than esu.

In the cgs-Gauss system of units, the speed of light c appears directly in Maxwell's equations. We use the subscript G for the electromagnetic fields to distinguish them from the electromagnetic fields in the SI units. We obtain the following Maxwell's equations:

$$\nabla \cdot \vec{E}_G = 4\pi \rho_G \tag{2.15}$$

$$\nabla \times \vec{B}_G - \frac{1}{c}\partial_t \vec{E}_G = \frac{4\pi}{c}\vec{j}_G \tag{2.16}$$

$$\nabla \cdot \vec{B}_G = 0 \tag{2.17}$$

$$\nabla \times \vec{E}_G + \frac{1}{c}\partial_t \vec{B}_G = 0 \tag{2.18}$$

The cgs-Gauss system of units has three main advantages:

1. Electric and magnetic fields have the same dimensions.
2. The vacuum permittivity ϵ_0 and permeability μ_0 are equal to 1.
3. The physical significance of the light velocity c is manifest.

Familiarity with both the SI and cgs-Gauss units can be helpful. The relation between the SI and cgs-Gauss units will be elaborated further in Sect. 2.4.1.

We can simplify the Maxwell equations in the cgs-Gauss units by removing the factors of 4π. This is called "rationalization," and the resulting system is due to **Heaviside**. The Gauss law is then given as

$$F = \frac{q_{H1} q_{H2}}{4\pi r^2} \tag{2.19}$$

In the Heaviside system of units, Maxwell's equations are given as

$$\nabla \cdot \vec{E}_H = \rho_H \tag{2.20}$$

$$\nabla \times \vec{B}_H - \frac{1}{c}\partial_t \vec{E}_H = \frac{1}{c}\vec{j}_H \tag{2.21}$$

$$\nabla \cdot \vec{B}_H = 0 \tag{2.22}$$

$$\nabla \times \vec{E}_H + \frac{1}{c}\partial_t \vec{B}_H = 0 \tag{2.23}$$

The Heaviside system of units is best suited for theoretical considerations and is a favorite choice of units in (theoretical) particle physics.

2.1.2 Two of the Four Maxwell's Equations Imply the Existence of Vector and Scalar Potentials—The Remaining Two Imply the Generation of Electromagnetic Fields by Charge and Current Densities

Maxwell's equations are usually written as four equations, but they are grouped into two. The first group gives the geometric properties inherent in the electromagnetic fields described by scalar and vector potentials. The second group gives the dynamical relation between the charge and current densities and the electromagnetic fields. The distinction between these two sets of equations is important.

In the cgs-Gauss system, Maxwell's equations are given as follows:[2]

Group (I)

$$\nabla \times \vec{E} = -\frac{1}{c}\frac{\partial \vec{B}}{\partial t} \tag{2.24}$$

$$\nabla \cdot \vec{B} = 0 \tag{2.25}$$

Group (II)

$$\nabla \cdot \vec{E} = 4\pi\rho \tag{2.26}$$

$$\nabla \times \vec{B} - \frac{1}{c}\frac{\partial \vec{E}}{\partial t} = \frac{4\pi}{c}\vec{j} \tag{2.27}$$

In this section we wish to explain that the two groups (I) and (II) have very different characteristics. Let us first consider group (I). Equation (2.25) implies that the magnetic field is given by a vector potential $\vec{A}$:

$$\vec{B} = \nabla \times \vec{A} \tag{2.28}$$

It is important to note that even if the magnetic field $\vec{B}$ is determined, the corresponding $\vec{A}$ is not uniquely determined. For any function $\chi(\vec{x}, t)$, we have

$$\nabla \times (\nabla\chi(\vec{x}, t)) = 0 \tag{2.29}$$

This implies

$$\nabla \times \left(\vec{A} + \nabla\chi\right) = \nabla \times \vec{A} \tag{2.30}$$

Both $\vec{A}$ and $\vec{A} + \nabla\chi$ give the same magnetic field. The transformation from the vector potential $\vec{A}$ to $\vec{A} + \nabla\chi$ is called a **gauge transformation**. Under a gauge transformation, the magnetic field $\vec{B}$ remains invariant.

[2] In the following, we will omit the subscript G that specifies the cgs-Gauss system.

Therefore, Eq. (2.24) that represents Faraday's law of electromagnetic induction can be written as

$$\nabla \times \vec{E} = -\frac{1}{c}\partial_t \left(\nabla \times \vec{A}\right) = -\frac{1}{c}\nabla \times \partial_t \vec{A}$$
$$\Longrightarrow \nabla \times \left(\vec{E} + \frac{1}{c}\partial_t \vec{A}\right) = 0 \tag{2.31}$$

Remembering (2.29), this implies the existence of a scalar potential $\phi(\vec{x}, t)$ so that

$$\vec{E} + \frac{1}{c}\partial_t \vec{A} = -\nabla\phi(\vec{x}, t) \tag{2.32}$$

The minus sign in front of $\nabla\phi$ is chosen so that in the static case the electric field points in the direction of the decreasing potential ϕ.

To summarize, we have found that the electric field can be expressed in terms of a vector potential $\vec{A}$ and a scalar potential ϕ as

$$\vec{E}(\vec{x}, t) = -\nabla\phi(\vec{x}, t) - \frac{1}{c}\partial_t \vec{A}(\vec{x}, t) \tag{2.33}$$

Let us now go back to the magnetic field. We have seen that under a gauge transformation

$$\vec{A}(\vec{x}, t) \longrightarrow \vec{A}(\vec{x}, t) + \nabla\chi(\vec{x}, t) \tag{2.34}$$

the magnetic field remains invariant. But what happens to the electric field? By performing a gauge transformation on the scalar potential as

$$\phi(\vec{x}, t) \longrightarrow \phi(\vec{x}, t) - \frac{1}{c}\partial_t \chi(\vec{x}, t) \tag{2.35}$$

we find that the electric field remains invariant:

$$-\nabla\left(\phi - \frac{1}{c}\partial_t \chi\right) - \frac{1}{c}\partial_t(\vec{A} + \nabla\chi) = -\nabla\phi - \frac{1}{c}\partial_t \vec{A} \tag{2.36}$$

The equations in Group (I) can be summarized as follows:

1. The electromagnetic fields $\vec{E}, \vec{B}$ can be expressed in terms of a scalar potential $\phi(\vec{x}, t)$ and a vector potential $\vec{A}(\vec{x}, t)$ as

$$\vec{B}(\vec{x}, t) = \nabla \times \vec{A}(\vec{x}, t) \tag{2.37}$$
$$\vec{E}(\vec{x}, t) = -\nabla\phi(\vec{x}, t) - \frac{1}{c}\partial_t \vec{A}(\vec{x}, t) \tag{2.38}$$

ϕ and $\vec{A}$ are yet to be determined.

2. Any potentials $\vec{A}(\vec{x}, t)$ and $\phi(\vec{x}, t)$ can be transformed via a **gauge transformation** using an arbitrary function $\chi(\vec{x}, t)$:

$$\vec{A}'(\vec{x}, t) = \vec{A}(\vec{x}, t) + \nabla\chi(\vec{x}, t) \tag{2.39}$$

$$\phi'(\vec{x}, t) = \phi(\vec{x}, t) - \frac{1}{c}\partial_t\chi(\vec{x}, t) \tag{2.40}$$

These new potentials give the same $\vec{E}$, $\vec{B}$ as the original potentials.

Next, let us consider the two Maxwell's equations in the second group (II). First, it is essential to note the consistency of these two equations. Differentiating Eq. (2.26) in time, we obtain

$$\partial_t(\nabla \cdot \vec{E}) = 4\pi\,\partial_t\rho \tag{2.41}$$

On the other hand, by taking the divergence of Eq. (2.27), and using

$$\nabla \cdot (\nabla \times \vec{B}) = 0, \tag{2.42}$$

we find

$$\nabla \cdot \partial_t\vec{E} = -4\pi\nabla \cdot \vec{j} \tag{2.43}$$

Since the left-hand sides of Eqs. (2.41) and (2.43) are equal, we must find

$$\boxed{\partial_t\rho + \nabla \cdot \vec{j} = 0} \tag{2.44}$$

This is an equation of continuity, representing the conservation of electric charge. In other words, the consistency of Maxwell's equations in the second group (II) requires charge conservation.

Now, using the results from the first group and rewriting the equations of the second group as equations for ϕ & $\vec{A}$, we obtain

$$\nabla \cdot \left(-\nabla\phi - \frac{1}{c}\partial_t\vec{A}\right) = 4\pi\rho \tag{2.45}$$

$$\nabla \times (\nabla \times \vec{A}) - \frac{1}{c}\partial_t\left(-\nabla\phi - \frac{1}{c}\partial_t A\right) = \frac{4\pi}{c}\vec{j} \tag{2.46}$$

Using the identity

$$\boxed{\nabla \times (\nabla \times \vec{A}) = -\nabla^2\vec{A} + \nabla(\nabla \cdot \vec{A})} \tag{2.47}$$

(where $\nabla^2 \equiv \partial_x^2 + \partial_y^2 + \partial_z^2$ is the Laplacian), we can rewrite the above two equations as

$$-\nabla^2\phi - \frac{1}{c}\partial_t \nabla \cdot \vec{A} = 4\pi\rho \tag{2.48}$$

$$-\nabla^2\vec{A} + \nabla(\nabla \cdot \vec{A}) + \frac{1}{c}\nabla\partial_t\phi + \frac{1}{c^2}\partial_t^2\vec{A} = \frac{4\pi}{c}\vec{j} \tag{2.49}$$

These equations describe how the charge density ρ and current density $\vec{j}$ generate the potentials ϕ and $\vec{A}$. Given specific ρ and $\vec{j}$ and ϕ, however, ϕ and $\vec{A}$ are not determined uniquely for two reasons. First, there is gauge invariance: transforming ϕ and $\vec{A}$ via a gauge transformation leaves the left-hand side of the equations unchanged. If ϕ and $\vec{A}$ satisfy these equations, so will the the potentials obtained by a gauge transformation. The second reason is that ϕ and $\vec{A}$ are not necessarily zero even in the absence of $\rho, \vec{j}$. This is because, as discussed in Sect. 2.1.3, electromagnetic waves can propagate without any sources.

Therefore, to solve the equations for the second group under given ρ and $\vec{j}$, it is necessary first to **fix the gauge** to eliminate the degrees of freedom of gauge transformations. For this purpose, in addition to Maxwell's equations, we impose a gauge-fixing condition. For calculations it is convenient to adopt the Lorenz gauge condition[3]

$$\nabla \cdot \vec{A} + \frac{1}{c}\partial_t\phi = 0 \tag{2.50}$$

If $\vec{A}$ and ϕ are given arbitrarily, they do not necessarily satisfy the Lorenz gauge condition. We look for a gauge transformation that makes the potentials satisfy the Lorenz condition:

$$\nabla \cdot (\vec{A} + \nabla\chi) + \frac{1}{c}\partial_t\left(\phi - \frac{1}{c}\partial_t\chi\right) = 0 \tag{2.51}$$

This implies

$$\left(\frac{1}{c^2}\partial_t^2 - \nabla^2\right)\chi = \nabla \cdot \vec{A} + \frac{1}{c}\partial_t\phi \tag{2.52}$$

As we explain in Sect. 2.2.1, this equation always has a solution. Therefore, we can always assume that $\vec{A}$ and ϕ satisfy the Lorenz gauge condition.

[3] Ludvig Lorenz (1829–91) was a Danish physicist, often confused with the famous Dutch physicist Hendrik Lorentz (1853–1928).

For potentials that satisfy the Lorenz condition (2.50), Eqs. (2.48) and (2.49) can be rewritten as simply as

$$\left(-\nabla^2 + \frac{1}{c^2}\partial_t^2\right)\phi = 4\pi\rho \tag{2.53}$$

$$\left(-\nabla^2 + \frac{1}{c^2}\partial_t^2\right)\vec{A} = \frac{4\pi}{c}\vec{j} \tag{2.54}$$

Clearly, the advantage of the Lorenz gauge is that the Maxwell equations for ϕ and $\vec{A}$ become two separate differential equations. The solution $\phi, \vec{A}$ to these equations satisfies the Lorenz condition (2.50) automatically thanks to the charge conservation (2.44).

The general solution to the above equations can be expressed as a sum of a particular solution ϕ_r, $\vec{A}_\mathrm{r}$ and a solution (free fields) for the case $\rho = \vec{j} = 0$:

$$\phi(\vec{x}, t) = \phi_\mathrm{r}(\vec{x}, t) + \phi_\mathrm{free}(\vec{x}, t) \tag{2.55}$$

$$\vec{A}(\vec{x}, t) = \vec{A}_\mathrm{r}(\vec{x}, t) + \vec{A}_\mathrm{free}(\vec{x}, t) \tag{2.56}$$

A particular solution is not uniquely determined. Adding any free field solution to a particular solution yields another particular solution. We will discuss the construction of ϕ_r, $\vec{A}_\mathrm{r}$ in Sect. 2.2.1. We now proceed to the final point of this section, the free field solution ϕ_free, $\vec{A}_\mathrm{free}$.

2.1.3 Electromagnetic Waves Propagate Even in the Absence of Charge and Current Densities

Electromagnetic waves are generated by electric charges and currents. Once generated, they propagate freely without any help from the charges and currents. They arise as homogeneous solutions to the wave equations.

When there is no electric charge or current densities, Eqs. (2.53) and (2.54) become wave equations for the waves propagating at the speed of light:

$$\left(-\nabla^2 + \frac{1}{c^2}\partial_t^2\right)\phi = 0 \tag{2.57}$$

$$\left(-\nabla^2 + \frac{1}{c^2}\partial_t^2\right)\vec{A} = 0 \tag{2.58}$$

Each component of the vector potential satisfies the same equation as the scalar potential. The general solution can be conveniently expressed as Fourier integrals:

$$\phi(\vec{x}, t) = \int \frac{d^3k}{(2\pi)^3} \left(\phi_{\vec{k}}\, e^{i\vec{k}\cdot\vec{x}-i\omega t} + \phi^*_{\vec{k}}\, e^{-i\vec{k}\cdot\vec{x}+i\omega t} \right) \tag{2.59}$$

$$\vec{A}(\vec{x}, t) = \int \frac{d^3k}{(2\pi)^3} \left(\vec{a}_{\vec{k}}\, e^{i\vec{k}\cdot\vec{x}-i\omega t} + \vec{a}^*_{\vec{k}}\, e^{-i\vec{k}\cdot\vec{x}+i\omega t} \right) \tag{2.60}$$

where $\vec{k}$ is the wave vector in the direction of propagation of the electromagnetic wave and

$$\omega \equiv |\vec{k}|c \tag{2.61}$$

is the angular frequency. The complex coefficients $\phi_{\vec{k}}$ and $\vec{a}_{\vec{k}} \equiv \vec{a}(\vec{k}, \omega) = (a_{\vec{k},1}, a_{\vec{k},2}, a_{\vec{k},3})$ are related by the Lorenz gauge condition (2.50):

$$\vec{k} \cdot \vec{a}_{\vec{k}} = \frac{\omega}{c}\, \phi_{\vec{k}} \tag{2.62}$$

It is clear from this relation that the longitudinal component of $\vec{a}_{\vec{k}}$, parallel to $\vec{k}$, is determined by $\phi_{\vec{k}}$.

Now, in general, a vector $\vec{a}$ can be decomposed into components parallel and perpendicular to $\vec{k}$ as

$$\vec{a} = \vec{a}_{\parallel} + \vec{a}_{\perp}\,, \tag{2.63}$$

where the longitudinal component is given by

$$\vec{a}_{\parallel} = \vec{k}\, \frac{(\vec{k} \cdot \vec{a})}{\vec{k}^2} \tag{2.64}$$

The part perpendicular to $\vec{k}$ is called the transverse part, which has two degrees of freedom corresponding to two unit vectors $\vec{\epsilon}_1$ and $\vec{\epsilon}_2 \equiv \frac{\vec{k}}{|\vec{k}|} \times \vec{\epsilon}_1$ (See Fig. 2.3). Therefore, any vector $\vec{a}$ can be decomposed as

$$\vec{a} = a_1\vec{\epsilon}_1 + a_2\vec{\epsilon}_2 + \vec{k}\frac{(\vec{k} \cdot \vec{a})}{\vec{k}^2} \tag{2.65}$$

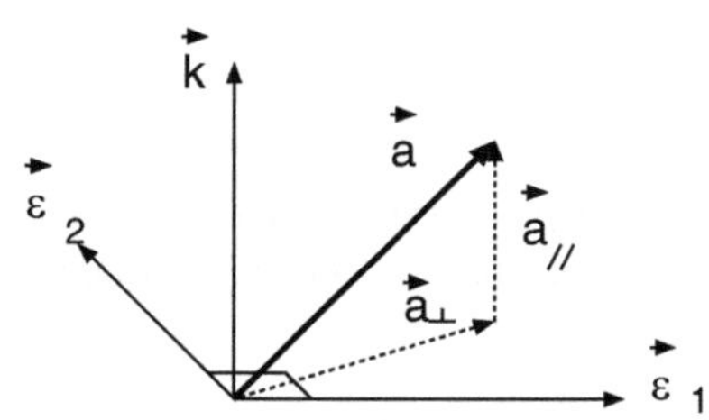

Fig. 2.3 Polarization vector orthogonal to the wave vector

Hence, in the Lorenz gauge, the free fields can be written as

$$\vec{A}(\vec{x}, t) = \int \frac{d^3k}{(2\pi)^3} \Big[\sum_{i=1,2} \vec{\epsilon}_{\vec{k},i} \left(a_{\vec{k},i} \mathrm{e}^{i\vec{k}\cdot\vec{x} - i\omega t} + a^*_{\vec{k},i} \mathrm{e}^{-i\vec{k}\cdot\vec{x} + i\omega t} \right)$$

$$+ \frac{c\vec{k}}{\omega} \left(\phi_{\vec{k}} \mathrm{e}^{i\vec{k}\cdot\vec{x} - i\omega t} + \phi^*_{\vec{k}} \mathrm{e}^{-i\vec{k}\cdot\vec{x} + i\omega t} \right) \Big] \tag{2.66}$$

$$\phi(\vec{x}, t) = \int \frac{d^3k}{(2\pi)^3} \left(\phi_{\vec{k}} \mathrm{e}^{i\vec{k}\cdot\vec{x} - i\omega t} + \phi^*_{\vec{k}} \mathrm{e}^{-i\vec{k}\cdot\vec{x} + i\omega t} \right) \tag{2.67}$$

Here, two independent transverse modes are chosen such that

$$\vec{k} \cdot \vec{\epsilon}_{\vec{k},i} = 0 \tag{2.68a}$$

$$\vec{\epsilon}_{\vec{k},i} \cdot \vec{\epsilon}_{\vec{k},j} = \delta_{ij} \tag{2.68b}$$

for $i, j = 1, 2$.

While keeping the Lorenz gauge condition, we can still change $\vec{A}$ and ϕ by a further gauge transformation with the gauge function

$$\chi(\vec{x}, t) = \int \frac{d^3k}{(2\pi)^3} \frac{c}{i\omega} \left(\phi_{\vec{k}} \mathrm{e}^{i\vec{k}\cdot\vec{x} - i\omega t} - \phi^*_{\vec{k}} \mathrm{e}^{-i\vec{k}\cdot\vec{x} + i\omega t} \right) \tag{2.69}$$

that satisfies

$$\frac{1}{c} \partial_t \chi = \phi \tag{2.70}$$

We obtain

$$\vec{A}_{\mathrm{tr}} \equiv \vec{A} + \nabla\chi = \int \frac{d^3k}{(2\pi)^3} \sum_{i=1,2} \vec{\epsilon}_{\vec{k},i} \left(a_{\vec{k},i} \mathrm{e}^{i\vec{k}\cdot\vec{x} - i\omega t} + a^*_{\vec{k},i} \mathrm{e}^{-i\vec{k}\cdot\vec{x} + i\omega t} \right)$$

$$\phi' \equiv \phi - \frac{1}{c} \partial_t \chi = 0 \tag{2.71}$$

This gives a vector potential only with transverse modes and no scalar potential. Since

$$\nabla \cdot \vec{A}_{\mathrm{tr}} = 0 \tag{2.72}$$

$\vec{A}_{\mathrm{tr}}$ and $\phi' = 0$ still satisfy the Lorenz gauge.

Therefore, the free electromagnetic fields are obtained as

$$\vec{E} = -\frac{1}{c} \partial_t \vec{A}_{\mathrm{tr}}$$

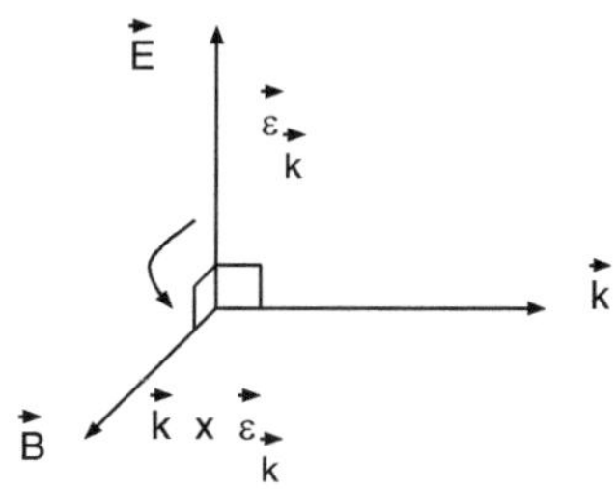

Fig. 2.4 Plane waves of linear polarization

$$= \int \frac{d^3k}{(2\pi)^3} \frac{i\omega}{c} \sum_{i=1,2} \vec{\epsilon}_{\vec{k},i} \left(a_{\vec{k},i} \mathrm{e}^{i\vec{k}\cdot\vec{x}-i\omega t} - a^*_{\vec{k},i} \mathrm{e}^{-i\vec{k}\cdot\vec{x}+i\omega t} \right) \tag{2.73}$$

$$\vec{B} = \nabla \times \vec{A}_{\mathrm{tr}}$$

$$= \int \frac{d^3k}{(2\pi)^3} i\vec{k} \times \sum_{i=1,2} \vec{\epsilon}_{\vec{k},i} \left(a_{\vec{k},i} \mathrm{e}^{i\vec{k}\cdot\vec{x}-i\omega t} - a^*_{\vec{k},i} \mathrm{e}^{-i\vec{k}\cdot\vec{x}+i\omega t} \right) \tag{2.74}$$

(Fig. 2.4) It is important to emphasize once more that for each wave vector $\vec{k}$, only two independent transverse modes are available.

Examples 3

1. **Relation between the cgs-Gauss units and the SI units**

   (a) Express an electric charge of 1 C in esu units.
   (b) Express a voltage of 1 volt in statV.
   (c) Express a magnetic field of 1 Tesla in gauss.

   Show not just the answers but also how to derive them.
2. **Vector analysis**—Derive the following formulas. ϕ is an arbitrary function, and $\vec{v}$ is an arbitrary vector field.

$$\nabla \cdot (\nabla\phi) = \nabla^2\phi$$

$$\nabla \cdot (\nabla \times \vec{v}) = 0$$

$$\nabla \times (\nabla\phi) = \vec{0}$$

3. **Maxwell's equations**—In the absence of charge and current, Maxwell's equations are given by

$$\nabla \cdot \vec{E} = 0 \tag{1}$$

$$\nabla \times \vec{B} - \frac{1}{c}\partial_t \vec{E} = 0 \tag{2}$$

$$\nabla \cdot \vec{B} = 0$$

$$\nabla \times \vec{E} + \frac{1}{c}\partial_t \vec{B} = 0 \tag{3}$$

(a) From the above, derive the wave equation for the electric field

$$\left(\nabla^2 - \frac{1}{c^2}\partial_t^2\right)\vec{E} = 0$$

Hint: You will need to use Eqs. (1–3). First, derive the formula

$$\nabla \times (\nabla \times \vec{v}) = -\nabla^2 \vec{v} + \nabla(\nabla \cdot \vec{v})$$

Then, take the curl (i.e., $\nabla\times$) of Eq. (3), and use the above formula as well as Eqs. (1, 2).

(b) Derive the general solution of the wave equation and Eq. (1) using the Fourier integral.

(c) Using the general solution of the electric field derived above and Eq. (3), derive the general solution of the magnetic field.

4. **Lagrangian of a charged particle**—The Lagrangian of a particle with mass m and charge e in an electromagnetic field is given by

$$L = \frac{m}{2}\left(\frac{d\vec{x}}{dt}\right)^2 + \frac{e}{c}\frac{d\vec{x}}{dt} \cdot \vec{A}(\vec{x}, t) - e\phi(\vec{x}, t)$$

where $\vec{A}$ and ϕ are vector and scalar potentials.

(a) Calculate the change in the Lagrangian under the gauge transformation

$$\vec{A} \to \vec{A} + \nabla\chi, \quad \phi \to \phi - \frac{1}{c}\partial_t\chi\,,$$

and show that it is a total time derivative.

(b) Explain that the equations of motion remain unchanged.

(c) Derive the equation of motion for the charged particle, and explain its physical significance.

(d) Derive the total energy E of the particle.

(e) Determine the temporal change of the total energy.

(f) Find the momentum $\vec{p} \equiv \frac{\partial L}{\partial \dot{\vec{x}}}$, conjugate to the coordinate $\vec{x}$.

(g) Express the total energy E using the momentum $\vec{p}$ without using velocity $\dot{\vec{x}}$. **Note**: This result is useful in quantum mechanics.

5. **Delta function**—The purpose of this problem is to derive the important formula for Fourier transformation

$$\int_{-\infty}^{\infty} \frac{dk}{2\pi}\, e^{ikx} = \delta(x)$$

(a) Assuming this formula, derive the inverse Fourier transformation

$$f(x) = \int_{-\infty}^{\infty} \frac{dk}{2\pi}\, \tilde{f}(k) e^{ikx}, \quad \text{where } \tilde{f}(k) \equiv \int_{-\infty}^{\infty} dx\, f(x) e^{-ikx}$$

(b) Compute the function directly.

$$\delta_\epsilon(x) \equiv \int_{-\infty}^{\infty} \frac{dk}{2\pi}\, e^{ikx-\frac{\epsilon}{2}k^2}$$

Hint: Complete the square.

(c) Plot $\delta_\epsilon(x)$ and observe the value, position, and width of its peak.

(d) Compute the integral $\int_{-\infty}^{\infty} dx\, \delta_\epsilon(x)$.

(e) Explain why $\delta(x) = \lim_{\epsilon \to 0+} \delta_\epsilon(x)$.

6. **Coulomb gauge**—Instead of the Lorenz gauge condition, we will solve Maxwell's Eqs. (2.48) and (2.49) using the Coulomb gauge condition

$$\nabla \cdot \vec{A} = 0$$

(a) Solve (2.48) in the Coulomb gauge to derive

$$\phi(\vec{x}, t) = \int d^3y\, \frac{\rho(\vec{y}, t)}{|\vec{x} - \vec{y}|}$$

(b) Substituting the above result into (2.49), determine the equation of motion for the vector potential $\vec{A}$.

(c) Explain why the charge conservation law

$$\partial_t \rho + \frac{1}{c} \nabla \cdot \vec{j} = 0$$

is necessary to solve the equation of motion derived above.

Solutions to Examples 3

1. You may use the conversion formula provided in the supplement:

$$q_G = \frac{q}{\sqrt{4\pi\epsilon_0}}, \quad V_G = \frac{qV}{q_G} = \sqrt{4\pi\epsilon_0}\, V, \quad B_G = \frac{q}{q_G} cB = \sqrt{\frac{4\pi}{\mu_0}}\, B$$

Alternatively, we can think in terms of physical quantities as follows.

(a) Let the charge q_G in the cgs-Gauss units correspond to $q = 1\,\mathrm{C}$. The Coulomb energy is given by

$$\frac{q^2}{4\pi\epsilon_0 r} = \frac{q_G^2}{r}$$

For $r = 1$ cm, we obtain

$$\frac{(1\,\mathrm{C})^2}{4\pi\epsilon_0\,(10^{-2}\,\mathrm{m})} \simeq 8.99 \times 10^{11}\,\mathrm{J} = 8.99 \times 10^{18}\,\mathrm{erg}$$

where erg $\equiv$ dyn cm $= 10^{-7}$ J. Since

$$\frac{\mathrm{esu}^2}{\mathrm{cm}} = \mathrm{erg}$$

we obtain

$$q_G \simeq \sqrt{8.99 \times 10^{18}}\,\mathrm{esu} \simeq \boxed{3.0 \times 10^9\,\mathrm{esu}}$$

(b) Let V in the SI correspond to the voltage $V_G = 1$ statV in the cgs-Gauss. For $q_G \simeq 3.0 \times 10^9$ esu obtained above, we have

$$q_G V_G \simeq 3.0 \times 10^9\,\mathrm{erg} = 3.0 \times 10^2\,\mathrm{J}$$

From $qV = q_G V_G$, we obtain

$$V = \frac{q_G V_G}{q} \simeq \frac{3.0 \times 10^2\,\mathrm{J}}{1\,\mathrm{C}} = \boxed{3.0 \times 10^2\,\mathrm{volt}}$$

(c) The Lorentz force acting on a charge moving at light velocity is

$$F = qcB = q_G B_G$$

For $q = 1\,\mathrm{C}$, $B = 1\,\mathrm{T}$, we find $qcB \simeq 3.0 \times 10^8$ N. Hence,

$$B_G = \frac{qcB}{q_G} = \frac{3.0 \times 10^{13}\,\mathrm{dyn}}{3.0 \times 10^9\,\mathrm{esu}} \simeq 1.0 \times \boxed{10^4\,\mathrm{gauss}}$$

In the old SI, this was exactly 10^4 gauss, but not anymore.

2. Omitted. (You can simply check them by calculations.)
3. (a) Taking the rotation (or curl) of Eq. (3), we obtain

$$\nabla \times (\nabla \times \vec{E}) + \frac{1}{c}\partial_t \nabla \times \vec{B} = 0$$

Hence,

$$-\nabla^2 \vec{E} + \nabla \underbrace{(\nabla \cdot \vec{E})}_{=0\,(1)} - \frac{1}{c}\partial_t \underbrace{\nabla \times \vec{B}}_{=\frac{1}{c}\partial_t \vec{E}\,(2)} = 0$$

We thus obtain

$$\left(-\nabla^2 + \frac{1}{c^2}\partial_t^2\right)\vec{E} = 0$$

Similarly, taking the rotation of (2) yields

$$\left(-\nabla^2 + \frac{1}{c^2}\partial_t^2\right)\vec{B} = 0$$

(b) The general solution to the wave equation is given by

$$\vec{E}(\vec{x}, t) = \int \frac{d^3k}{(2\pi)^3}\left(\vec{E}_{\vec{k}}\, e^{i\vec{k}\cdot\vec{x} - i\omega t} + \vec{E}^*_{\vec{k}}\, e^{-i\vec{k}\cdot\vec{x} + i\omega t}\right)$$

where $\omega \equiv |\vec{k}|c$, and $\vec{k} \cdot \vec{E}_{\vec{k}} = 0$.

(c)

$$\vec{B}(\vec{x}, t) = \int \frac{d^3k}{(2\pi)^3}\left(\vec{B}_{\vec{k}}\, e^{i\vec{k}\cdot\vec{x} - i\omega t} + \vec{B}^*_{\vec{k}}\, e^{-i\vec{k}\cdot\vec{x} + i\omega t}\right)$$

where

$$\frac{1}{|\vec{k}|}\vec{k} \times \vec{E}_{\vec{k}} = \vec{B}_{\vec{k}}$$

4. (a) Under the gauge transformation, the Lagrangian changes by

$$\Delta L = \frac{e}{c}\frac{d\vec{x}}{dt} \cdot \nabla\chi(\vec{x}(t), t) + \frac{e}{c}\partial_t \chi(\vec{x}(t), t) = \frac{e}{c}\frac{d}{dt}\chi(\vec{x}(t), t)$$

This is a total derivative with respect to time.

(b) Consequently, the action changes only by

$$\Delta \int_{t_1}^{t_2} dt\, L = \frac{e}{c}\left(\chi(\vec{x}_2, t_2) - \chi(\vec{x}_1, t_1)\right)$$

Under the variational principle, we minimize the action under an arbitrary variation of $\vec{x}(t)$ while fixing $\vec{x}(t_1), \vec{x}(t_2)$. Hence, the variation of the action does not depend on the gauge transformation; the equation of motion does not change.

(c) From

$$\frac{\partial L}{\partial \frac{d\vec{x}}{dt}} = m\frac{d\vec{x}}{dt} + \frac{e}{c}\vec{A}(\vec{x}(t), t), \quad \frac{\partial L}{\partial \vec{x}} = \frac{e}{c}\sum_{i=1}^{3}\frac{dx_i}{dt}\nabla A_i - e\nabla\phi$$

we obtain the equation of motion

$$\frac{d}{dt}\left(m\frac{d\vec{x}}{dt} + \frac{e}{c}\vec{A}(\vec{x}(t), t)\right) = \frac{e}{c}\sum_{i=1}^{3}\frac{dx_i}{dt}\nabla A_i - e\nabla\phi$$

Noting that

$$\frac{d}{dt}\vec{A}(\vec{x}(t), t) = \sum_{i=1}^{3}\frac{dx_i}{dt}\frac{\partial \vec{A}}{\partial x_i} + \partial_t\vec{A}$$

we get

$$-\frac{d}{dt}\vec{A} + \sum_i \frac{dx_i}{dt}\nabla A_i = \sum_i \frac{dx_i}{dt}\left(\nabla A_i - \frac{\partial \vec{A}}{\partial x_i}\right) - \partial_t\vec{A}$$
$$= \frac{d\vec{x}}{dt} \times \underbrace{\left(\nabla \times \vec{A}\right)}_{=\vec{B}} - \partial_t\vec{A}$$

Thus, we obtain

$$m\frac{d^2\vec{x}}{dt^2} = e\left(\vec{E} + \frac{1}{c}\frac{d\vec{x}}{dt} \times \vec{B}\right)$$

where $\vec{E} = -\nabla\phi - \frac{1}{c}\partial_t\vec{A}$. We have derived a familiar expression of the Lorentz force from the Lagrangian.

(d) The total energy E is

$$E \equiv \frac{d\vec{x}}{dt} \cdot \frac{\partial L}{\partial \frac{d\vec{x}}{dt}} - L = \frac{m}{2}\left(\frac{d\vec{x}}{dt}\right)^2 + e\phi(\vec{x}(t), t)$$

which is the sum of kinetic and potential energies. Under a gauge transformation, this changes as $E \to E + e\frac{1}{c}\partial_t\chi$. It is the kinetic energy $E - e\phi$ that is gauge invariant.

(e) The time evolution of energy is

$$\begin{aligned}\frac{d}{dt}E &= -\frac{\partial L}{\partial t}\\ &= -\frac{e}{c}\frac{d\vec{x}}{dt}\cdot\partial_t\vec{A} + e\partial_t\phi(\vec{x}(t),t)\\ &= e\frac{d\vec{x}}{dt}\cdot\left(-\frac{1}{c}\partial_t\vec{A} - \nabla\phi\right) + e\frac{d}{dt}\phi(\vec{x}(t),t)\\ &= e\frac{d\vec{x}}{dt}\cdot\vec{E} + e\frac{d}{dt}\phi(\vec{x}(t),t)\end{aligned}$$

On the other hand, the time evolution of kinetic energy is

$$\frac{d}{dt}(E - e\phi) = e\frac{d\vec{x}}{dt}\cdot\vec{E}$$

(f) The momentum is

$$\vec{p} \equiv \frac{\partial L}{\partial\frac{d\vec{x}}{dt}} = m\frac{d\vec{x}}{dt} + \frac{e}{c}\vec{A}(\vec{x}(t),t)$$

(g) In terms of momentum, we obtain

$$E = \frac{1}{2m}\left(\vec{p} - \frac{e}{c}\vec{A}\right)^2 + e\phi$$

5. (a)

$$\begin{aligned}f(x) &= \int dy\, f(y)\delta(x-y) = \int dy\, f(y)\int\frac{dk}{2\pi}\,\mathrm{e}^{ik(x-y)}\\ &= \int\frac{dk}{2\pi}\,\mathrm{e}^{ikx}\int dy\, f(y)\mathrm{e}^{-iky} = \int\frac{dk}{2\pi}\,\mathrm{e}^{ikx}\tilde{f}(k)\end{aligned}$$

(b)

$$\delta_\epsilon(x) = \mathrm{e}^{-\frac{x^2}{2\epsilon}}\int\frac{dk}{2\pi}\,\mathrm{e}^{-\frac{\epsilon}{2}\left(k-\frac{ix}{\epsilon}\right)^2} = \sqrt{\frac{2\pi}{\epsilon}}\mathrm{e}^{-\frac{x^2}{2\epsilon}}$$

(c)

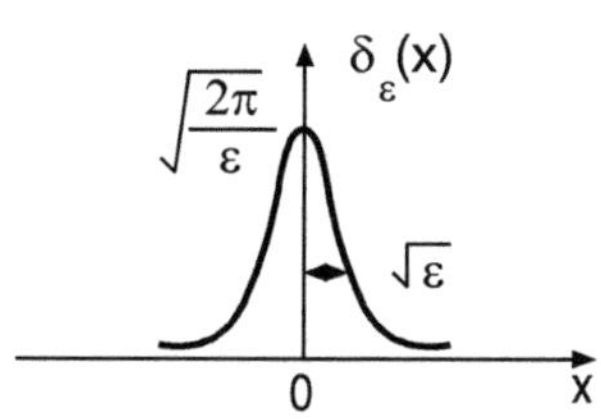

(d) $\int_{-\infty}^{\infty} dx\, \delta_\epsilon(x) = 1$
(e) As $\epsilon \to 0+$, $\int dx\, f(x)\delta_\epsilon(x) \to f(0)$.

6. (a) The Coulomb gauge condition reduces (2.48) to the Poisson equation

$$-\nabla^2\phi = 4\pi\rho$$

This is solved by

$$\phi(\vec{x}, t) = \int d^3y\, \frac{\rho(\vec{y}, t)}{|\vec{x} - \vec{y}|}$$

(b) Substituting the above result into (2.49), we obtain

$$\left(-\nabla^2 + \frac{1}{c^2}\partial_t^2\right)\vec{A} = \frac{4\pi}{c}\vec{j}_{\rm tr}$$

where

$$\vec{j}_{\rm tr}(\vec{x}, t) \equiv \vec{j}(\vec{x}, t) - \nabla \int d^3y\, \frac{\partial_t\rho(\vec{y}, t)}{|\vec{x} - \vec{y}|}$$

Note that in the Lorenz gauge, $\vec{j}_{\rm tr}$ is simply given by $\vec{j}$.

(c) $\nabla \cdot \vec{j}_{\rm tr}$ must vanish. Thus,

$$\nabla \cdot \vec{j}_{\rm tr} = \nabla \cdot \vec{j} - \frac{1}{c}\nabla^2 \int d^3y\, \frac{\partial_t\rho(\vec{y}, t)}{|\vec{x} - \vec{y}|} = \nabla \cdot \vec{j} + \partial_t\rho = 0$$

This is the equation for charge conservation. If this is satisfied, we obtain

$$\begin{aligned}
\vec{j}_{\rm tr}(\vec{x}, t) &= \vec{j}(\vec{x}, t) + \nabla_x \int d^3y\, \frac{\nabla_y \cdot \vec{j}(\vec{y}, t)}{|\vec{x} - \vec{y}|} \\
&= \vec{j}(\vec{x}, t) - \nabla_x \int d^3y\, \nabla_y \frac{1}{|\vec{x} - \vec{y}|} \cdot \vec{j}(\vec{y}, t) \\
&= \vec{j}(\vec{x}, t) + \nabla\left(\nabla \cdot \int d^3y\, \frac{\vec{j}(\vec{y}, t)}{|\vec{x} - \vec{y}|}\right)
\end{aligned}$$

Note that for any vector field $\vec{v}(\vec{x}, t)$ (that vanishes toward infinity),

$$\vec{v}_{\rm tr}(\vec{x}, t) \equiv \vec{v}(\vec{x}, t) + \nabla\left(\nabla \cdot \int d^3y\, \frac{\vec{v}(\vec{y}, t)}{|\vec{x} - \vec{y}|}\right)$$

satisfies $\nabla \cdot \vec{v}_{\rm tr} = 0$.

Exercises 3

1. **Electric dipole and magnetic moment** —For charge density $\rho(\vec{x})$, the electric dipole is defined as

$$\vec{d} \equiv \int d^3x\, \vec{x}\rho(\vec{x})$$

(a) Determine the dimension of $\vec{d}$.

Given that the radius of a hydrogen atom is approximately $a \equiv 10^{-8}$ cm, the natural magnitude of the electric dipole, with e as the elementary charge, is

$$d_0 \equiv ea$$

(b) Determine d_0 in units of $\mathrm{eV}/(\mathrm{statV}/\mathrm{cm})$.

The magnetic moment of a hydrogen atom, with $m \simeq 5 \times 10^5 \mathrm{eV}/c^2$ as the electron's mass, is of the order of the Bohr magneton

$$\mu_B \equiv \frac{e\hbar}{2mc}$$

(c) Determine μ_B in units of eV/gauss.

2. **Magnetic flux and electrical resistance**

   (a) Determine the elementary charge e in units of esu.
   (b) Given h as the Planck constant and c as the speed of light, show that $\frac{hc}{e}$ has the dimension of gauss $\cdot$ cm^2 = esu.
   (c) What is the unit for electrical resistance in the cgs-Gauss units?
   (d) Show that $\frac{h}{e^2}$ has the dimension of electrical resistance, and express its value in Ohms.

3. **Archimedes's principle**—As an application of Gauss's theorem, let us derive Archimedes's principle of buoyancy. In a liquid with mass density ρ, the pressure p satisfies

$$\nabla p = -\rho g \hat{z}$$

(where $\hat{z}$ is the unit vector in the z-direction). Prove that the buoyant force acting on an object enclosed by surface S with volume V is given by

$$\vec{F} = \hat{z}\rho g \underbrace{\int_V d^3x}_{\text{volume}}$$

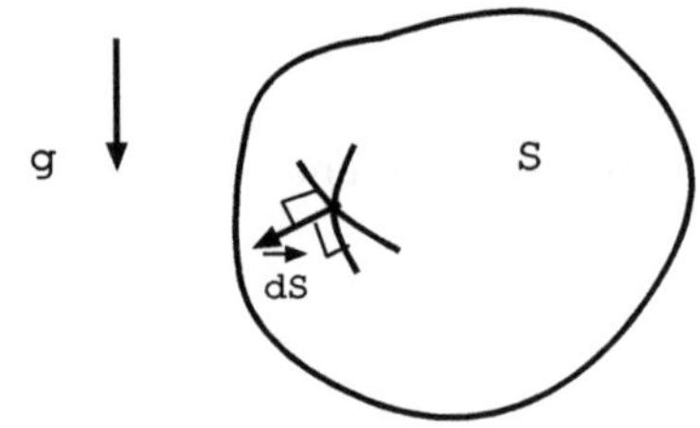

Hint: The buoyant force is given by

$$\vec{F} = -\int_S p d\vec{S}$$

where $d\vec{S}$ is the surface element (a vector perpendicular to S with the magnitude of the surface element). Proving

$$\int_S p\, d\vec{S} = \int_V \nabla p\, d^3x\,,$$

note that this is a vector equation. For the x component, this amounts to proving

$$\int_S (p\hat{x}) \cdot d\vec{S} = \int_V \partial_x p\, d^3x$$

using Gauss's theorem.

4. **Integration kernel**—We consider the product of two three-dimensional Fourier transforms given by

$$\tilde{F}(\vec{k}) = \frac{1}{k^2 + \kappa^2} \cdot \tilde{f}(\vec{k})$$

where

$$\tilde{f}(\vec{k}) = \int d^3x\, f(\vec{x}) e^{-i\vec{k}\cdot\vec{x}}$$

is the Fourier transform of a known function $f(\vec{x})$ and κ is a positive constant.

(a) Compute the inverse Fourier transform:

$$G(\vec{x}) \equiv \int \frac{d^3k}{(2\pi)^3} \frac{e^{i\vec{k}\cdot\vec{x}}}{k^2 + \kappa^2}$$

Hint: Use polar coordinates and the method of residues.

(b) Show that $G(\vec{x})$ satisfies the differential equation:

$$(-\nabla^2 + \kappa^2)G(\vec{x}) = \delta^{(3)}(\vec{x})$$

(c) Express the inverse Fourier transform $F(\vec{x})$ of $\tilde{F}(\vec{k})$ in the form of a spatial integral, using $G(\vec{x})$ and $f(\vec{x})$.
Answer: $F(\vec{x}) = \int d^3y\, G(\vec{x} - \vec{y}) f(\vec{y})$.
(d) Derive the differential equation satisfied by $F(\vec{x})$.
Answer: $\left(-\nabla^2 + \kappa^2\right) F(\vec{x}) = f(\vec{x})$.
(e) Determine the limit of $G(\vec{x})$ as $\kappa \to 0+$.

5. **Lagrangian density**—The Lagrangian density of the electromagnetic field, given a charge density ρ and current density $\vec{j}$, is

$$\mathcal{L} = \frac{1}{8\pi}\left(\vec{E}^2 - \vec{B}^2\right) - \rho\phi + \frac{1}{c}\vec{j}\cdot\vec{A}$$

where the electromagnetic field is given by a vector potential $\vec{A}$ and scalar potential ϕ as

$$\vec{E} = -\nabla\phi - \frac{1}{c}\partial_t\vec{A}, \quad \vec{B} = \nabla\times\vec{A}$$

The Lagrangian is the spatial integral of the Lagrangian density as

$$L = \int d^3x\, \mathcal{L}$$

which is a functional of ϕ and $\vec{A}$.

(a) For the Lagrangian L to change only by a total time derivative under a gauge transformation, show that the charge conservation law

$$\partial_t\rho + \nabla\cdot\vec{j} = 0$$

is necessary. (Refer to Example 3-4).
(b) Derive the equations of motion for ϕ and $\vec{A}$, and prove their equivalence to Maxwell's equations.
(c) In the absence of charge or current density, show that the energy density is given by

$$\boxed{\mathcal{E} \equiv \frac{1}{8\pi}\left(\vec{E}^2 + \vec{B}^2\right)}$$

Hint: Use the standard definition of energy $\sum_i \dot{q}_i \frac{\partial L}{\partial \dot{q}_i}$, where i should be replaced by the continuous variable $\vec{x}$.

(d) In the presence of charge and current, prove that the energy density $\mathcal{E}$ defined above satisfies

$$\partial_t \mathcal{E} + \nabla \cdot \vec{S} = -\vec{j} \cdot \vec{E} \,,$$

where the Poynting vector $\vec{S}$ is defined by

$$\boxed{\vec{S} \equiv \frac{c}{4\pi} \vec{E} \times \vec{B}}$$

(e) Explain the physical meaning of the above equation.

(f) Compute $\mathcal{E}$ and $\vec{S}$ for the plane wave with wave vector $\vec{k}$, and prove that $\frac{\vec{S}}{\mathcal{E}c}$ is a unit vector.

(g) Using the above result, normalize the plane wave solution such that $\mathcal{E} = \hbar|\vec{k}|c$. (One photon per unit volume.)

6. **Electromagnetic field of the sunlight**—The energy of the sunlight reaching the surface of the Earth (outside the atmosphere) is $1366\,\mathrm{W/m^2}$. From this, find the average values of the electric field (in units volt/m) and the magnetic field (in units Tesla). **Comment**: The average value calculated here is, strictly speaking, the square root of the mean square value.
7. **Lagrangian of a multi-particle system in the Coulomb gauge**—The Lagrangian of a system consisting of an electromagnetic field and N charged particles is

$$L = \int d^3x \, \frac{1}{8\pi} \left(\vec{E}^2 - \vec{B}^2 \right)$$
$$+ \sum_{i=1}^{N} \left[\frac{m_i}{2} \left(\frac{d\vec{x}_i}{dt} \right)^2 + \frac{e_i}{c} \frac{d\vec{x}_i}{dt} \cdot \vec{A}(\vec{x}_i, t) - e_i \phi(\vec{x}_i, t) \right]$$

where the electromagnetic field is given by

$$\vec{E} = -\nabla \phi - \frac{1}{c} \partial_t \vec{A}, \quad \vec{B} = \nabla \times \vec{A}.$$

(a) For $\vec{A}$ satisfying the Coulomb gauge condition

$$\nabla \cdot \vec{A} = 0,$$

show that ϕ is given by

$$\phi(\vec{x}, t) = \sum_{i=1}^{N} \frac{e_i}{|\vec{x} - \vec{x}_i(t)|}$$

Hint: The Euler-Lagrange equation for ϕ is

$$-\nabla^2 \phi(\vec{x}, t) = 4\pi \sum_i e_i \delta^{(3)}(\vec{x} - \vec{x}_i(t)).$$

(b) Derive Gauss's equation from the Coulomb gauge condition

$$\nabla \cdot \vec{E} = 4\pi \sum_i e_i \delta^{(3)}(\vec{x} - \vec{x}_i(t))$$

Hint: The time derivative of $\nabla \cdot \vec{A} = 0$ is zero.

(c) By substituting the ϕ obtained in (a) above into L, show that

$$L = \frac{1}{8\pi} \int d^3x \left[\frac{1}{c^2} \left(\partial_t \vec{A} \right)^2 - \vec{B}^2 \right]$$
$$+ \sum_{i=1}^{N} \left[\frac{m_i}{2} \left(\frac{d\vec{x}_i}{dt} \right)^2 + \frac{e_i}{c} \frac{d\vec{x}_i}{dt} \cdot \vec{A}(\vec{x}_i, t) \right] - \frac{1}{2} \sum_{i \neq j} \frac{e_i e_j}{|\vec{x}_i(t) - \vec{x}_j(t)|}$$

Ignore the infinite self-energy $\sum_{i=1}^{N} \frac{e_i^2}{0}$. **Comment**: Although the self-energy is infinite, it is merely a constant and therefore does not affect the equations of motion.

(d) Show that the total energy E is given by

$$\boxed{E = \frac{1}{8\pi} \int d^3x \left(\vec{E}^2 + \vec{B}^2 \right) + \sum_{i=1}^{N} \frac{m_i}{2} \left(\frac{d\vec{x}_i}{dt} \right)^2}$$

and check that it is conserved.

Comment: The integral includes the self-energy of the charged particles $\sum_i \frac{e_i^2}{0}$, which is infinite but can be treated as a time-independent constant. Recall that the energy defined in Example 3-4 is not conserved; it is because the energy of the electromagnetic field is not taken into account.

2.2 Electromagnetism 2

Two Points to Remember

1. The influence of charge and current densities propagates at the speed of light.
2. The amplitude of radiation from an accelerating charge decreases inversely as its distance.

2.2.1 The Influence of Charge and Current Densities Propagates at the Speed of Light

Electromagnetism is totally consistent with special relativity; Einstein invented relativity by studying Maxwell's equations. The influence of charge and current densities propagates at the maximum speed allowed by relativity, which is the speed of light. We can express this using a retarded Green function.

In the previous section, we have derived the wave equations in the Lorenz gauge as follows:

$$\left(-\nabla^2+\frac{1}{c^2}\partial_t^2\right)\phi=4\pi\rho \tag{2.75}$$

$$\left(-\nabla^2+\frac{1}{c^2}\partial_t^2\right)\vec{A}=\frac{4\pi}{c}\vec{j} \tag{2.76}$$

The purpose of this section is to find a particular solution to these equations. Not only the potentials ϕ and $\vec{A}$ but also the electromagnetic fields $\vec{E}$ and $\vec{B}$ are not uniquely determined, and this is due to the free field solutions, as explained in Sect. 2.1.3.

With this remark in mind, let us consider the wave equation satisfied by the scalar potential:

$$\left(-\nabla^2+\frac{1}{c^2}\partial_t^2\right)\phi=4\pi\rho \tag{2.77}$$

As a preparation, we introduce a Green function $G(\vec{x},t)$ by

$$\left(-\nabla^2+\frac{1}{c^2}\partial_t^2\right)G(\vec{x},t)=\delta(t)\delta^{(3)}(\vec{x}) \tag{2.78}$$

Using this, we can solve (2.77) as

$$\phi(\vec{x}, t) = \int d^3y dt' \, G(\vec{x} - \vec{y}, t - t') 4\pi\rho(\vec{y}, t') \tag{2.79}$$

The easiest way to find this Green function is by solving the equation for the Fourier transform $\tilde{G}$ of G:

$$\left(k^2 - \frac{\omega^2}{c^2}\right) \tilde{G}(\vec{k}, \omega) = 1 \tag{2.80}$$

From this, we obtain

$$\tilde{G}(\vec{k}, \omega) = \frac{1}{k^2 - \frac{\omega^2}{c^2}}, \tag{2.81}$$

and consequently

$$G(\vec{x}, t) = \int_{-\infty}^{\infty} \frac{d\omega}{2\pi} \, \mathrm{e}^{-i\omega t} \int \frac{d^3k}{(2\pi)^3} \, \mathrm{e}^{i\vec{k}\cdot\vec{x}} \, \frac{1}{k^2 - \frac{\omega^2}{c^2}} \tag{2.82}$$

We first integrate over ω and then over $\vec{k}$. Consider

$$\int_{-\infty}^{\infty} \frac{d\omega}{2\pi} \frac{\mathrm{e}^{-i\omega t}}{k^2 - \frac{\omega^2}{c^2}}$$

The integrand diverges at $\omega = \pm kc$ ($k \equiv |\vec{k}|$). There are various ways to handle this divergence. In general, the integrand can be interpreted as

$$\frac{1}{k^2 - \frac{\omega^2}{c^2}} \Rightarrow \frac{1}{k^2 - \frac{1}{c^2}(\omega + i\epsilon)^2} + a\,\delta(\omega - kc) + b\,\delta(\omega + kc) \tag{2.83}$$

where the constants a and b correspond to free field solutions. We set

$$a = b = 0 \tag{2.84}$$

so that G satisfies the **retarded condition**:

$$G(\vec{x}, t) = 0 \quad \text{for } t < 0 \tag{2.85}$$

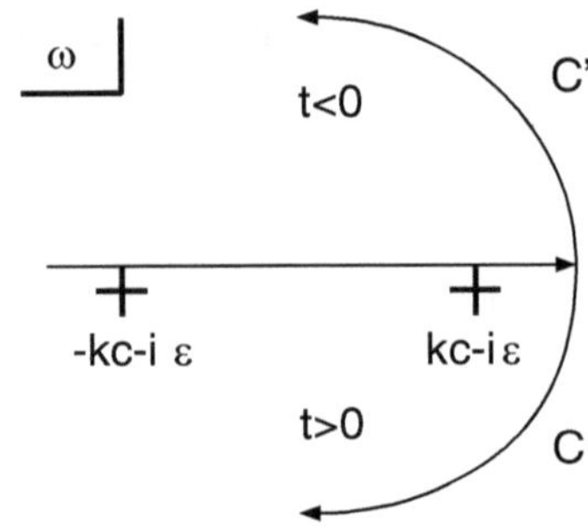

Fig. 2.5 Contours C, C' of integration on the complex plane

This is physically the correct choice. In fact, with $a = b = 0$, we find

$$
\begin{aligned}
&\int_{-\infty}^{\infty} \frac{d\omega}{2\pi} \frac{\mathrm{e}^{-i\omega t}}{k^2 - \frac{1}{c^2}(\omega + i\epsilon)^2} \\
&\overset{\text{if } t>0}{=} \oint_C \frac{d\omega}{2\pi} \frac{-\mathrm{e}^{i\omega t}}{\left(\frac{\omega}{c} - k + i\epsilon\right)\left(\frac{\omega}{c} + k + i\epsilon\right)} = ic^2 \left(\frac{\mathrm{e}^{ikct}}{-2kc} + \frac{\mathrm{e}^{-ikct}}{2kc} \right) \\
&= c^2 \frac{\sin kct}{kct} \qquad (2.86) \\
&\overset{\text{if } t<0}{=} \oint_{C'} \cdots = 0 \qquad (2.87)
\end{aligned}
$$

so that G satisfies the retarded condition (See Fig. 2.5).

Next, let us integrate over the wave vector $\vec{k}$.

$$
\begin{aligned}
G(\vec{x}, t) &= \int \frac{d^3k}{(2\pi)^3} \mathrm{e}^{i\vec{k}\cdot\vec{x}} \frac{\sin kct}{kc} c^2 \\
&= \int \frac{d\phi d\cos\theta k^2 dk}{(2\pi)^3} \mathrm{e}^{ikr\cos\theta} \frac{\sin kct}{kc} c^2 \\
&= \frac{1}{(2\pi)^2} \int_0^{\infty} k^2 dk \frac{2i \sin kr}{ikr} \frac{\sin kct}{kc} c^2 \\
&= \frac{1}{2\pi^2} \frac{c^2}{rc} \int_0^{\infty} dk \sin kr \sin kct \\
&= \frac{1}{2\pi^2} \frac{c^2}{rc} \int_0^{\infty} dk \frac{1}{2} (\cos k(r - ct) - \cos k(r + ct)) \\
&= \frac{1}{4\pi^2} \frac{c}{r} \pi \delta(r - ct) = \frac{1}{4\pi} \frac{c}{r} \delta(r - ct) \qquad (2.88)
\end{aligned}
$$

We have thus obtained the **retarded Green function** as

$$\boxed{G(\vec{x},t)=\theta(t)\frac{1}{4\pi}\frac{c}{r}\,\delta(r-ct)}\,, \tag{2.89}$$

where we have inserted the Heaviside function $\theta(t)$ so that $G(\vec{x},t)=0$ for $t<0$.[4]

Using this we obtain the scalar potential as

$$\begin{aligned}\phi_{\text{ret}}(\vec{x},t)&=\int d^3y\int_{-\infty}^{t}dt'\,\frac{c}{|\vec{x}-\vec{y}|}\,\delta(|\vec{x}-\vec{y}|-c(t-t'))\,\rho(\vec{y},t')\\&=\int d^3y\,\frac{1}{|\vec{x}-\vec{y}|}\,\underbrace{\rho\left(\vec{y},t-\frac{|\vec{x}-\vec{y}|}{c}\right)}_{\text{charge density in the past}}\end{aligned} \tag{2.90}$$

Similarly, we obtain the vector potential as

$$\begin{aligned}\vec{A}_{\text{ret}}(\vec{x},t)&=\int dt\int d^3y\,G(\vec{x}-\vec{y},t-t')\frac{4\pi}{c}\vec{j}(\vec{y},t')\\&=\int d^3y\,\frac{1}{|\vec{x}-\vec{y}|}\,\frac{1}{c}\,\underbrace{\vec{j}\left(\vec{y},t-\frac{|\vec{x}-\vec{y}|}{c}\right)}_{\text{current density in the past}}\end{aligned} \tag{2.91}$$

The above results have a clear physical meaning: the influence of charge and current densities propagates at the speed of light c. The past affects the future, but the future does not affect the past. This is known as **causality**.

Finally, let us check that the gauge potentials obtained above satisfy the Lorenz gauge condition. Noting that the Green function depends only on the difference in space and time, we find

$$\begin{aligned}&\nabla\cdot\vec{A}_{\text{ret}}(\vec{x},t)+\frac{1}{c}\frac{\partial}{\partial t}\phi_{\text{ret}}(\vec{x},t)\\&=\int dt'\int d^3y\left(\frac{\partial}{\partial\vec{x}}G(\vec{x}-\vec{y},t-t')\cdot\frac{1}{c}\vec{j}(\vec{y},t')\right.\\&\qquad\left.+\frac{1}{c}\frac{\partial}{\partial t}G(\vec{x}-\vec{y},t-t')\,\rho(\vec{y},t')\right)\\&=\frac{1}{c}\int dt'\int d^3y\left(-\frac{\partial}{\partial\vec{y}}G(\vec{x}-\vec{y},t-t')\cdot\vec{j}(\vec{y},t')\right.\\&\qquad\left.-\frac{\partial}{\partial t'}G(\vec{x}-\vec{y},t-t')\,\rho(\vec{y},t')\right)\end{aligned}$$

---

[4] We have introduced a retarded Green function in Sect. 1.3.1.

$$= -\frac{1}{c}\int dt' \int d^3y\, G(\vec{x}-\vec{y}, t-t')\left(\nabla\cdot\vec{j}(\vec{y},t') + \frac{\partial}{\partial t'}\rho(\vec{y},t')\right)$$
$$= 0 \qquad (2.92)$$

where we have used the charge conservation in the final step.

Liénard-Wiechert Potentials

Let us consider an arbitrary motion $\vec{x}(t)$ of a charged particle with charge e. The motion generates charge and current densities given by

$$\rho(\vec{x},t) = e\,\delta^{(3)}(\vec{x}-\vec{x}(t)) \qquad (2.93)$$
$$\vec{j}(\vec{x},t) = e\dot{\vec{x}}(t)\,\delta^{(3)}(\vec{x}-\vec{x}(t)) \qquad (2.94)$$

You can check in Example 4-3 that these ρ and $\vec{j}$ satisfy the charge conservation equation:

$$\partial_t\rho + \nabla\cdot\vec{j} = 0 \qquad (2.95)$$

Substituting these ρ and $\vec{j}$ into (2.90, 2.91), we obtain the Liénard-Wiechert potentials (see Example 4-3 for derivation):

$$\phi(\vec{x},t) = \frac{e}{|\vec{x}-\vec{x}(t')|\left(1-\frac{\dot{\vec{x}}(t')}{c}\cdot\frac{\vec{x}-\vec{x}(t')}{|\vec{x}-\vec{x}(t')|}\right)} \qquad (2.96)$$

$$\vec{A}(\vec{x},t) = \frac{e\dot{\vec{x}}(t')/c}{|\vec{x}-\vec{x}(t')|\left(1-\frac{\dot{\vec{x}}(t')}{c}\cdot\frac{\vec{x}-\vec{x}(t')}{|\vec{x}-\vec{x}(t')|}\right)} \qquad (2.97)$$

Here, the time t' is chosen so that the light emitted from the particle at t' reaches the point $\vec{x}$ at time t:

$$|\vec{x}-\vec{x}(t')| = c(t-t') \qquad (2.98)$$

As long as the charged particle moves at a speed less than c, this equation determines t' uniquely[5] (Fig. 2.6).

[5] Suppose two past times $t_1, t_2 > t_1$ satisfy $|\vec{x}-\vec{x}(t_1)| = c(t-t_1)$ and $|\vec{x}-\vec{x}(t_2)| = c(t-t_2)$. Then, using the triangle inequality, we find $|\vec{x}(t_2)-\vec{x}(t_1)| \geq c(t-t_1)-c(t-t_2) = c(t_2-t_1)$, which implies that the particle must have moved at a speed greater than c.

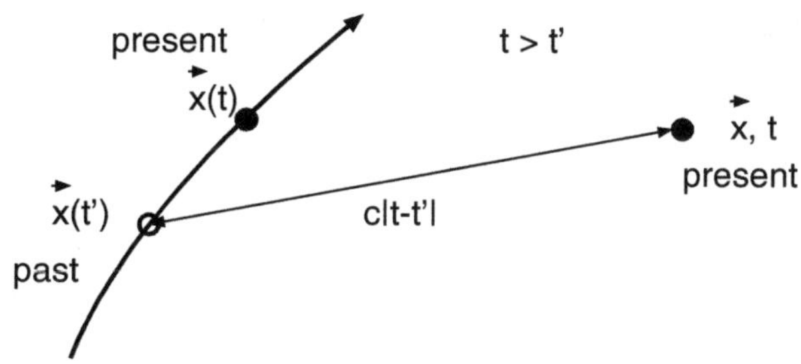

Fig. 2.6 Since the speed of light is finite, the direction of the emitted light is determined by the position of the charged particle in the past

Radiation from an Accelerating Charged Particle

Next, we will use the Liénard-Wiechert potentials to find the electromagnetic field generated by a charged particle. To do this, we simply substitute the Liénard-Wiechert potentials into the equations:

$$\vec{E} = -\nabla\phi - \frac{1}{c}\partial_t \vec{A}, \quad \vec{B} = \nabla \times \vec{A} \tag{2.99}$$

Calculating these without approximation is quite challenging; in the following we will make two assumptions and calculate the electromagnetic fields only approximately.[6]

Assumption 1 The charged particle is moving significantly slower than the speed of light (non-relativistic motion).

Assumption 2 We calculate the electromagnetic field at a location far from the charged particle compared with the wavelength of light.

To illustrate these assumptions, let us consider a typical example (Fig. 2.7).

Imagine an electric charge moving at speed $v \ll c$ (Assumption 1) in a region of size R. The period of its motion is of order $\frac{R}{v}$, and the angular frequency of the radiation it emits is of the order $\omega \sim \frac{v}{R}$. Hence, the typical wave length of the radiation is $\lambda = \frac{2\pi c}{\omega} \sim R\frac{c}{v} \gg R$. Let r be the distance from the charge to the observer where the electromagnetic fields are measured. Assumption 2 gives $r \gg \lambda$, i.e., $\frac{r}{R} \gg \frac{c}{v}$. Hence, $\frac{r}{R} \sim \frac{c^2}{v^2}$ or larger. Accordingly, the acceleration a of the charge is of order $a \sim \frac{v^2}{R}$, and $\frac{v^2}{a} \sim R \ll r$.

Let us now start calculating the electromagnetic fields, which are expressed by the gauge potentials (2.96, 2.97) through (2.99). Note that t' is a function of $\vec{x}$ and t defined by

$$c(t - t') = |\vec{x} - \vec{x}(t')| \tag{2.100}$$

[6] For the exact result without approximation, you may refer to M. Schwartz, *Principles of Electrodynamics* (Dover), for example.

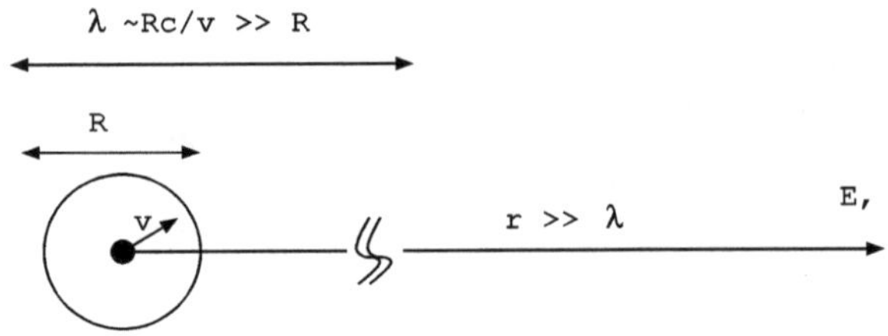

Fig. 2.7 A typical example satisfying the two assumptions

We first need to determine how this t' changes under small variations $\delta\vec{x}$ and δt of $\vec{x}$ and t. From the above equation, we obtain

$$c(\delta t - \delta t') = (\delta\vec{x} - \dot{\vec{x}}(t')\delta t') \cdot \frac{\vec{x} - \vec{x}(t')}{|\vec{x} - \vec{x}(t')|} \tag{2.101}$$

Note that $\vec{x}(t)$ is a given function of time so that

$$\delta\vec{x}(t') = \dot{\vec{x}}(t')\,\delta t'$$

By introducing a unit vector

$$\vec{\varepsilon} \equiv \frac{\vec{x} - \vec{x}(t')}{|\vec{x} - \vec{x}(t')|}, \tag{2.102}$$

we can rewrite the above equation as

$$(c - \vec{\varepsilon} \cdot \dot{\vec{x}}(t'))\delta t' = c\delta t - \vec{\varepsilon} \cdot \delta\vec{x} \tag{2.103}$$

We thus obtain

$$\frac{\partial t'}{\partial t} = \frac{c}{c - \vec{\varepsilon} \cdot \dot{\vec{x}}(t')}, \qquad \nabla t' = -\vec{\varepsilon}\,\frac{1}{c - \vec{\varepsilon} \cdot \dot{\vec{x}}(t')} \tag{2.104}$$

Please note that we have not yet introduced any approximations at this stage. Using the above results, we derive

$$\delta\vec{x}(t') = \dot{\vec{x}}(t')\left(\frac{c}{c - \vec{\varepsilon} \cdot \dot{\vec{x}}(t')}\delta t - \frac{\vec{\varepsilon} \cdot \delta\vec{x}}{c - \vec{\varepsilon} \cdot \dot{\vec{x}}(t')}\right) \tag{2.105}$$

$$\delta\dot{\vec{x}}(t') = \ddot{\vec{x}}(t')\left(\frac{c}{c - \vec{\varepsilon} \cdot \dot{\vec{x}}(t')}\delta t - \frac{\vec{\varepsilon} \cdot \delta\vec{x}}{c - \vec{\varepsilon} \cdot \dot{\vec{x}}(t')}\right) \tag{2.106}$$

Now, using Assumption 1, we approximate the Liénard-Wiechert potentials as

$$\phi(\vec{x}, t) \simeq \frac{e}{|\vec{x} - \vec{x}(t')|} \left(1 + \frac{\dot{\vec{x}}(t')}{c} \cdot \vec{\varepsilon}\right) \tag{2.107}$$

$$\vec{A}(\vec{x}, t) \simeq \frac{e}{|\vec{x} - \vec{x}(t')|} \frac{\dot{\vec{x}}(t')}{c} \tag{2.108}$$

where terms of quadratic order in velocity, $\frac{|\dot{\vec{x}}|}{c}$, are ignored. We then obtain

$$\begin{aligned} \delta\phi(\vec{x}, t) &= \frac{e}{|\vec{x} - \vec{x}(t')|} \frac{\vec{\varepsilon} \cdot \ddot{\vec{x}}(t')}{c} \delta t' \\ &+ e\delta \frac{1}{|\vec{x} - \vec{x}(t')|} \left(1 + \frac{\dot{\vec{x}}(t')}{c} \cdot \vec{\varepsilon}\right) + \frac{e}{|\vec{x} - \vec{x}(t')|} \frac{\dot{\vec{x}}(t')}{c} \cdot \delta\vec{\varepsilon} \end{aligned} \tag{2.109}$$

Since

$$\delta \frac{1}{|\vec{x} - \vec{x}(t')|} = -2 \frac{\vec{\varepsilon} \cdot \left(\delta\vec{x} - \delta\vec{x}(t')\right)}{|\vec{x} - \vec{x}(t')|^2}, \tag{2.110}$$

$$\delta\vec{\varepsilon} = \frac{1}{|\vec{x} - \vec{x}(t')|} \left[\delta\vec{x} - \delta\vec{x}(t') - 2\vec{\varepsilon} \left\{\vec{\varepsilon} \cdot \left(\delta\vec{x} - \delta\vec{x}(t')\right)\right\}\right] \tag{2.111}$$

the second line of (2.109) has an extra factor of $\frac{1}{|\vec{x} - \vec{x}(t')|}$ and can be ignored thanks to Assumption 2.

We thus obtain

$$\delta\phi(\vec{x}, t) \simeq \frac{e}{|\vec{x} - \vec{x}(t')|} \frac{\vec{\varepsilon} \cdot \ddot{\vec{x}}(t')}{c} \delta t' \tag{2.112}$$

$$\delta\vec{A}(\vec{x}, t) \simeq \frac{e}{|\vec{x} - \vec{x}(t')|} \frac{\ddot{\vec{x}}(t')}{c} \delta t' \tag{2.113}$$

Using Assumption 1 again, we find

$$\delta t' = \frac{\partial t'}{\partial t} \delta t + \nabla t' \cdot \delta\vec{x} \simeq \delta t - \vec{\varepsilon} \cdot \frac{\delta\vec{x}}{c} \tag{2.114}$$

and therefore

$$\delta\phi(\vec{x}, t) \simeq \frac{e}{|\vec{x} - \vec{x}(t')|} \frac{\vec{\varepsilon} \cdot \ddot{\vec{x}}(t')}{c} \left(\delta t - \vec{\varepsilon} \cdot \frac{\delta\vec{x}}{c}\right) \tag{2.115}$$

$$\delta\vec{A}(\vec{x}, t) \simeq \frac{e}{|\vec{x} - \vec{x}(t')|} \frac{\ddot{\vec{x}}(t')}{c} \left(\delta t - \vec{\varepsilon} \cdot \frac{\delta\vec{x}}{c}\right) \tag{2.116}$$

Hence, we obtain

$$-\nabla\phi(\vec{x},t) \simeq \frac{e}{|\vec{x}-\vec{x}(t')|}\frac{1}{c^2}\vec{\varepsilon}\left(\vec{\varepsilon}\cdot\ddot{\vec{x}}(t')\right) \tag{2.117}$$

$$-\frac{1}{c}\partial_t\vec{A}(\vec{x},t) \simeq -\frac{e}{|\vec{x}-\vec{x}(t')|}\frac{\ddot{\vec{x}}(t')}{c^2} \tag{2.118}$$

$$\nabla\times\vec{A}(\vec{x},t) \simeq -\frac{e}{|\vec{x}-\vec{x}(t')|}\frac{1}{c^2}\vec{\varepsilon}\times\ddot{\vec{x}}(t') \tag{2.119}$$

Therefore, the electromagnetic fields are given by

$$\vec{E}(\vec{x},t) \simeq \frac{1}{|\vec{x}-\vec{x}(t')|}\frac{-e}{c^2}\left\{\ddot{\vec{x}}(t')-\vec{\varepsilon}\left(\vec{\varepsilon}\cdot\ddot{\vec{x}}(t')\right)\right\} \tag{2.120}$$

$$\vec{B}(\vec{x},t) \simeq \frac{1}{|\vec{x}-\vec{x}(t')|}\frac{-e}{c^2}\vec{\varepsilon}\times\ddot{\vec{x}}(t') = \vec{\varepsilon}\times\vec{E}(\vec{x},t) \tag{2.121}$$

This result has three important characteristics, which can be summarized as follows. See Fig. 2.8.

1. The amplitude of the electromagnetic fields decreases inversely as the distance from the charged particle.
2. Away from the particle, the electromagnetic fields satisfy the equations of free fields. Hence, the electric and magnetic fields are orthogonal:

$$\vec{B} = \vec{\varepsilon}\times\vec{E} \tag{2.122}$$

3. The direction of the electric field is parallel to the vector $\ddot{\vec{x}}_\perp$, which is obtained by projecting the acceleration vector onto the plane perpendicular to $\vec{\varepsilon}$. (For a positive charge, $\vec{E}$ points opposite to $\ddot{\vec{x}}_\perp$.)

$$\vec{E}(\vec{x},t) \simeq \frac{1}{|\vec{x}-\vec{x}(t')|}\frac{-e}{c^2}\ddot{\vec{x}}_\perp \quad \text{where } \ddot{\vec{x}}_\perp \equiv \ddot{\vec{x}}-\vec{\varepsilon}\,(\vec{\varepsilon}\cdot\ddot{\vec{x}}) \tag{2.123}$$

2.2.2 *The Amplitude of Radiation from an Accelerating Charge Decreases Inversely as Its Distance*

It is important to remember the characteristics of light emitted by a non-relativistic charged particle. It is also helpful to memorize Larmor's formula for energy flow.

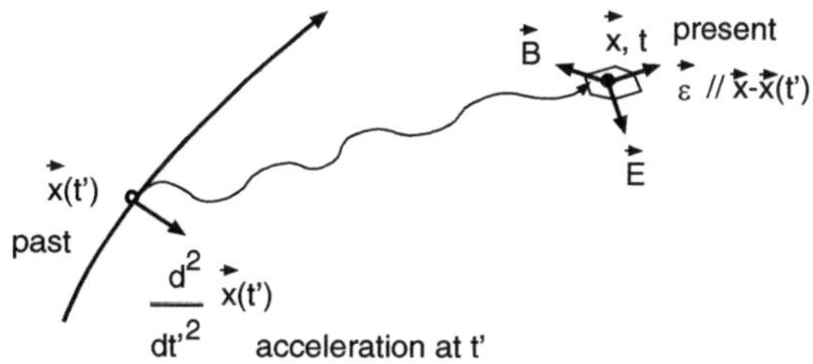

Fig. 2.8 Electromagnetic radiation from an accelerating charged particle

The first item in the summary of the previous section is so important that we make it the title of this section. In Exercise 3-5, we have derived the Poynting vector

$$\boxed{\vec{S} = \frac{c}{4\pi}\,\vec{E} \times \vec{B}} \tag{2.124}$$

to represent the flow of energy. We would like to calculate it for the radiation from a charged particle.

First, let us note some general properties. The magnitudes of the electric field and the magnetic field are equal and inversely proportional to the distance:

$$|\vec{E}| = |\vec{B}| \propto \frac{1}{r} \Longrightarrow S \propto \frac{1}{r^2} \tag{2.125}$$

Clearly the energy flow decreases with the square of the distance. Therefore, the total amount of energy passing through the entire spherical surface at a distance r remains independent of r, as required by the conservation law of energy.

Now, let us introduce spherical coordinates with $\vec{x}(t')$ as the origin and the acceleration $\ddot{\vec{x}}(t')$ taken along the z-axis (Fig. 2.9). In spherical coordinates, the results from the previous section (2.120, 2.121) can be expressed as follows:

$$\vec{E}(\vec{x}, t) = \frac{1}{r}\frac{-e}{c^2}\,a'\,\sin\theta(-\cos\theta\cos\phi, -\cos\theta\sin\phi, \sin\theta) \tag{2.126}$$

$$\vec{B}(\vec{x}, t) = \frac{1}{r}\frac{-e}{c^2}\,a'\,\sin\theta(\sin\phi, -\cos\phi, 0) \tag{2.127}$$

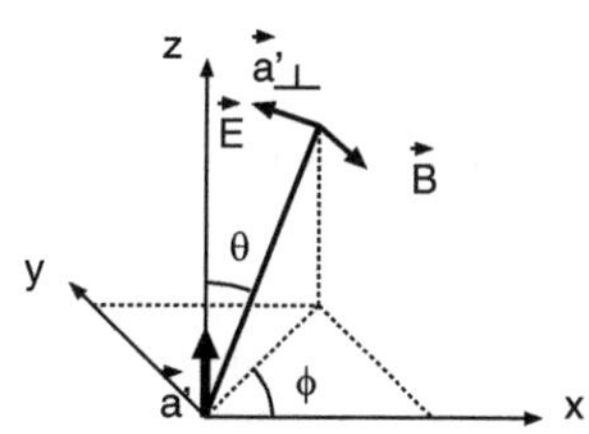

Fig. 2.9 Spherical coordinates

Here,

$$a' \equiv |\ddot{\vec{x}}(t')| \tag{2.128}$$

gives the magnitude of acceleration, and $a' \sin\theta$ is the projection onto the plane perpendicular to $\vec{x}$. Therefore, the energy flux is given by

$$\vec{S} = \frac{c}{4\pi r^2} \frac{e^2}{c^4} a'^2 \sin^2\theta \,(\sin\theta\cos\phi, \sin\theta\sin\phi, \cos\theta) \tag{2.129}$$

By considering a surface element $d\vec{A} = \frac{\vec{r}}{r} r^2 d\cos\theta d\phi$ at radius r, the energy flowing through it per unit time is given by

$$\vec{S} \cdot d\vec{A} = |\vec{S}|\, r^2 d\cos\theta d\phi = \frac{e^2 a'^2}{4\pi c^3} \sin^2\theta d\cos\theta d\phi \tag{2.130}$$

The angular distribution is $\sin^2\theta$, depending only on the angle θ with respect to the z-axis, and can be expressed as a function of the distance from the origin, as shown in Fig. 2.10.

Integrating this expression, we obtain the total energy flux:

$$\text{energyflux} = \frac{e^2 a'^2}{4\pi c^3} \int \sin^2\theta d\cos\theta d\phi = \boxed{\frac{2}{3}\frac{e^2 a'^2}{c^3}} \tag{2.131}$$

This result is known as **Larmor's formula**.

Larmor's formula is worth memorizing. You may not remember the factor of $\frac{2}{3}$, but remembering the rest is not too difficult. This is how you remember it. First, as the amplitude of the electromagnetic field is proportional to the acceleration a', the energy should depend on its square a'^2. Furthermore, since the amplitude is proportional to the charge e of the particle, the energy should also be proportional to e^2. Therefore,

$$\text{energy flux} = \text{const} \times e^2 a'^2 \tag{2.132}$$

is something you do not forget if you know physics. Since it should not depend on the particle's mass, the constant coefficient can depend only on the speed of light.

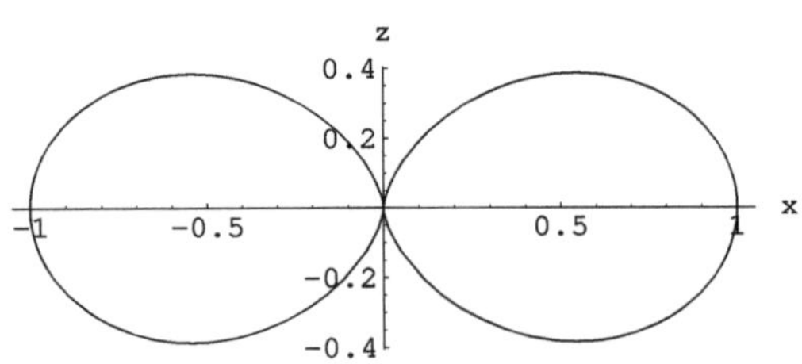

Fig. 2.10 Angular distribution of electromagnetic radiation. The distance from the origin indicates the intensity. Rotationally symmetric about the z-axis

Thus,

$$\text{energy flux} = \text{numerical const} \times \frac{e^2 a'^2}{c^n} \tag{2.133}$$

The exponent n can be determined through dimensional analysis. The dimensions of both sides are as follows:

$$[\text{energy flux}] = \frac{\text{erg}}{\text{s}}, \quad \left[\frac{e^2 a'^2}{c^n}\right] = \frac{\text{esu}^2\ \text{cm}^{2-n}}{\text{s}^{4-n}} = \frac{\text{erg cm}^{3-n}}{\text{s}^{4-n}} \tag{2.134}$$

Using $\text{esu}^2/\text{cm} = \text{erg}$, we obtain $n = 3$.

Since the SI charge q is related to e by

$$\frac{q^2}{4\pi\epsilon_0} = e^2$$

we can rewrite Larmor's formula in the SI units as

$$\text{energy flux} = \frac{2}{3}\frac{q^2 a'^2}{4\pi\epsilon_0 c^3} = \frac{q^2 a'^2}{6\pi\epsilon_0 c^3} \tag{2.135}$$

Dipole Radiation and Thomson Scattering

When unbound free electrons are exposed to light, they vibrate due to the electromagnetic field. Since electrons are charged particles, they should emit radiation due to this motion. Let us determine the relation between the electromagnetic field applied to the electron and the emitted radiation.

Assuming that the amplitude of the electron's vibration is much smaller than the wavelength of the incident light, we can consider the time variation of the electric and magnetic fields as uniform everywhere. The equation of motion for the electron is then given by

$$m\ddot{\vec{x}} = -e\left(\vec{E}(t) + \frac{1}{c}\dot{\vec{x}} \times \vec{B}(t)\right) \tag{2.136}$$

where the right-hand side is the Lorentz force. Recalling that the magnitude of the magnetic field is the same as that of the electric field, we see that the magnetic field is much less important if the motion of the electron is non-relativistic $|\dot{\vec{x}}| \ll c$. Hence, assuming that the electron's motion is non-relativistic, the equation of motion for the electron is simply

$$m\ddot{\vec{x}} = -e\vec{E}(t) \tag{2.137}$$

For the moment, let us assume that the angular frequency of the incident light is ω, so the electric field is given by

$$\vec{E}(t) = E \cos \omega t \, (0, 0, 1) \tag{2.138}$$

where we have chosen the direction of the electric field along the z-axis.

With the acceleration given by Eq. (2.137), the total energy radiated per unit time is given by Larmor's formula as

$$\text{energy flow} = \frac{2}{3} \frac{e^2}{c^3} \left(\frac{e}{m} \right)^2 E^2 \left\langle \cos^2 \omega t \right\rangle = \frac{1}{3} \frac{e^4}{m^2 c^3} E^2 \tag{2.139}$$

Here, we have used the time average of $\cos^2 \omega t$ as $\frac{1}{2}$.

On the other hand, the energy flow of the incident light is given per unit time and per unit area by

$$F_{in} = \frac{1}{8\pi} E^2 c \tag{2.140}$$

(This is obtained in Exercise 3-5.) To understand the relation between Eqs. (2.139 and 2.140), it is useful to introduce the concept of a **cross section**. Let us denote the "size" of the electron viewed from the light, or more precisely its cross section, as σ_T. In the presence of a uniform flow of energy given by Eq. (2.140), the energy that strikes the electron per unit time is given by

$$\text{energy flow} = \sigma_T \times F_{in} \tag{2.141}$$

This gives

$$\sigma_T = \frac{\text{energy flow}}{F_{in}} = \frac{8\pi}{3} \left(\frac{e^2}{mc^2} \right)^2 \simeq 0.665 \times 10^{-24} \text{cm}^2 \tag{2.142}$$

This is called the **Thomson cross section**. **The classical electron radius** is then defined as[7]

$$r_e \equiv \frac{e^2}{mc^2} \simeq 2.82 \times 10^{-13} \text{cm} \tag{2.143}$$

Using this, we can express σ_T as

$$\sigma_T = \frac{8\pi}{3} r_e^2 \tag{2.144}$$

[7] The classical electron radius r_e is defined such that the Coulomb energy $\frac{e^2}{r_e}$ equals the rest mass energy of the electron mc^2.

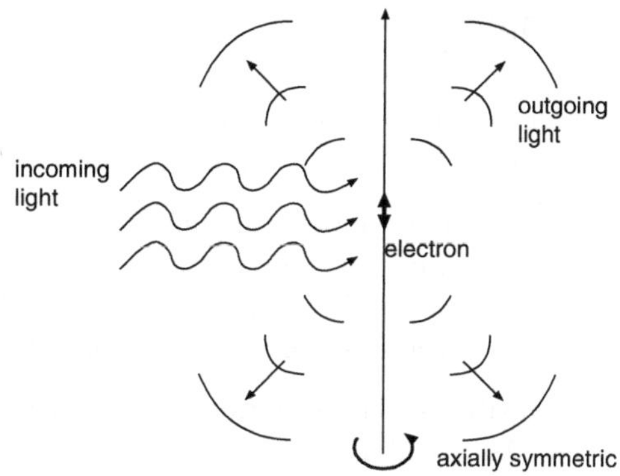

Fig. 2.11 Thomson scattering

The classical electron radius represents the electromagnetic "size" of the electron. It is important to note that the distribution of the scattered light is axially symmetric around the polarization vector of the incident light. Therefore, the amount of light scattered in the direction of the incident light and the amount scattered in the opposite direction are equal as in Fig. 2.11.

Justifying the Approximations Made to Obtain Thomson's Cross Section

To justify the approximations made to obtain the Thomson cross section, let us begin by considering the condition for the electron's motion to be non-relativistic. The velocity of the electron must be much smaller than the speed of light:

$$v = \frac{eE}{m\omega} \ll c \tag{2.145}$$

We can give a more physical interpretation of this condition by rewriting it as

$$eE\frac{c}{\omega} \ll mc^2 \tag{2.146}$$

The left-hand side represents the potential energy difference over a unit wavelength created by the electric field E, while the right-hand side represents the energy of the electron's rest mass. If the former is much smaller than the latter, the electron's motion can be considered non-relativistic. Another approximation involves the amplitude of the electron's motion being much smaller than the wavelength. This can be justified because the amplitude of motion is given by $\frac{eE}{m\omega^2}$ and the wavelength of light is $\frac{2\pi c}{\omega}$. If condition (2.145) is satisfied, we get automatically $\frac{eE}{m\omega^2} \ll \frac{2\pi c}{\omega}$.

Examples 4

1. **Cyclotron**—Consider a proton (mass 938 MeV$/c^2$) moving along a circle of radius 50 cm under a magnetic field $B = 3$ Tesla.

   (a) Find the cyclotron angular frequency in radians per second.
   (b) Determine the kinetic energy of the proton in eV.
   (c) Use Larmor's formula to calculate the electromagnetic energy radiated by this proton per second. Take the elementary charge as $e \simeq 4.8 \times 10^{-10}$ esu.

2. **Retarded Green function**[8]—Consider the equation of motion for a harmonic oscillator with friction:

$$\ddot{x} + \gamma\dot{x} + \Omega^2 x = f(t)$$

Here, $f(t)$ is an external force, and the friction coefficient is assumed to satisfy

$$0 < \gamma \ll \Omega$$

We first consider the case without external force. Since the equation of motion is a second-order differential equation, two linearly independent solutions exist. Let $x_1(t)$ be the solution that satisfies the initial condition $x_1(0) = 1$ and $\dot{x}_1(0) = 0$ at $t = 0$. Let $x_2(t)$ be the solution that satisfies $x_2(0) = 0$ and $\dot{x}_2(0) = 1$.

   (a) Find the homogeneous solutions $x_{1,2}(t)$.
   Hint: Search for solutions in the form of $x(t) = \exp(\alpha t)$. α satisfies a quadratic equation, and there are two solutions. Take a linear combination of these two solutions to satisfy the initial condition.

We next find the particular solution to the equation of motion in the form

$$x(t) = \int_{-\infty}^{\infty} dt'\, G(t - t') f(t'),$$

where the Green function $G(t)$ satisfies the retarded condition

$$G(t) = 0 \text{ for } t < 0$$

   (b) Determine the differential equation satisfied by $G(t)$.
   (c) Find the Fourier transform $\tilde{G}_\omega$ of the retarded Green function $G(t)$:

$$\tilde{G}_\omega \equiv \int_0^{\infty} dt\, \exp(i\omega t) G(t).$$

[8] The result of this problem has been used in Sect. 1.3.1.

(d) Obtain $G(t)$ by the inverse Fourier transform:

$$G(t) = \int_{-\infty}^{\infty} \frac{d\omega}{2\pi} \exp(-i\omega t)\tilde{G}_\omega$$

(e) Using the homogeneous solutions $x_{1,2}(t)$ and the particular solution $G(t)$, find the solution to the equation of motion that satisfies the initial condition $x(0) = x_0$ and $\dot{x}(0) = v_0$.

(f) How long does the influence of the initial condition persist?

3. **Liénard-Wiechert potential**—Consider the motion $\vec{x} = \vec{X}(t)$ of a charged particle with charge e.

(a) Explain why the charge density ρ and current density $\vec{j}$ are given by

$$\rho(\vec{x}, t) = e\,\delta^{(3)}(\vec{x} - \vec{X}(t))\,,$$

$$\vec{j}(\vec{x}, t) = e\,\dot{\vec{X}}(t)\,\delta^{(3)}(\vec{x} - \vec{X}(t))\,,$$

where $\delta^{(3)}(\vec{x}) \equiv \delta(x)\delta(y)\delta(z)$.

(b) Show that the above ρ, $\vec{j}$ satisfy the charge conservation equation:

$$\partial_t \rho + \nabla \cdot \vec{j} = 0$$

(c) In the Lorenz gauge, the scalar and vector potentials are given by

$$\phi(\vec{x}, t) = \int d^3y dt' \frac{c\rho(\vec{y}, t')}{|\vec{x} - \vec{y}|}\,\delta(c(t - t') - |\vec{x} - \vec{y}|)$$

$$\vec{A}(\vec{x}, t) = \int d^3y dt' \frac{\vec{j}(\vec{y}, t')}{|\vec{x} - \vec{y}|}\,\delta(c(t - t') - |\vec{x} - \vec{y}|)$$

Use this to show that the potential created by a charged particle is given by the following Liénard-Wiechert potentials:

$$\phi(\vec{x}, t) = \frac{e}{|\vec{x} - \vec{X}(t')|\left(1 - \frac{1}{c}\dot{\vec{X}}(t') \cdot \vec{\varepsilon}\right)}$$

$$\vec{A}(\vec{x}, t) = \frac{e\frac{1}{c}\dot{\vec{X}}(t')}{|\vec{x} - \vec{X}(t')|\left(1 - \frac{1}{c}\dot{\vec{X}}(t') \cdot \vec{\varepsilon}\right)}$$

where $t'(\vec{x}, t)$ is the solution to

$$|\vec{x} - \vec{X}(t')| = c(t - t')$$

and the unit vector $\vec{\varepsilon}(\vec{x}, t)$ is defined by

$$\vec{\varepsilon} \equiv \frac{\vec{x} - \vec{X}(t')}{|\vec{x} - \vec{X}(t')|}$$

Hint: It is easier to integrate over $\vec{y}$ first. Recall the formula $\delta(ax) = \frac{1}{|a|}\delta(x)$.

4. **Understanding the Liénard-Wiechert potentials**—We now understand geometrically the factor that appears in the denominator of the Liénard-Wiechert potentials, which is[9]

$$1 - \frac{1}{c}\dot{\vec{X}}(t') \cdot \vec{\varepsilon}$$

Consider an object of length L moving at a speed v toward the origin along the x-axis. Assume negligible spreading in the transverse directions (y, z). Inside the object, the electric charge is uniformly distributed, and the total charge is e.

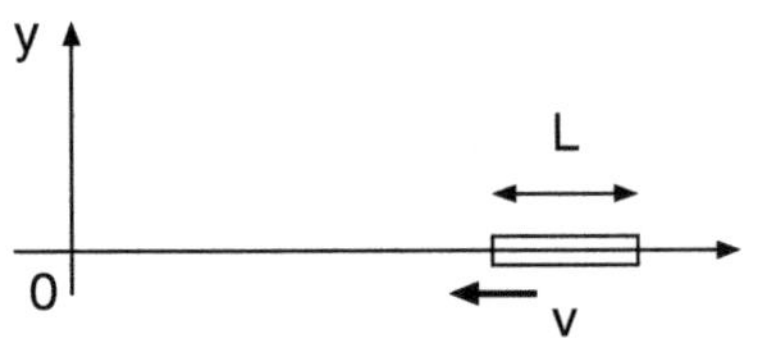

(a) If the object reaches $x = 0$ at time $t = T$, the charge density (charge per unit length) on the x-axis is given by

$$\rho(x, t) = \frac{e}{L} \cdot \begin{cases} 1 & T - \frac{x}{v} < t < T - \frac{x-L}{v} \\ 0 & \text{otherwise} \end{cases}$$

Show this.

(b) Calculate the retarded potential at $x = 0, t = 0$:

$$\phi = \int_{-\infty}^{\infty} dx \, \frac{\rho\left(x, -\frac{|x|}{c}\right)}{|x|}$$

assuming $r \equiv vT \gg L$. **Hint**: Determine the L independent factor.

5. **Retarded Green function in two dimensional space**—When the spatial dimension is 2, the retarded Green function is defined by the differential equation

$$\left(-\partial_x^2 - \partial_y^2 + \frac{1}{c^2}\partial_t^2\right) G_2(x, y, t) = \delta(x)\delta(y)\delta(t)$$

[9] M. Schwartz, *Principles of Electrodynamics* (Dover) Section 6-2.

and the retarded condition

$$G_2(x, y, t < 0) = 0$$

(a) Show that this retarded Green function can be expressed in terms of the retarded Green function $G_3(x, y, z, t)$ for three-dimensional space as

$$G_2(x, y, t) = \int_{-\infty}^{\infty} dz\, G_3(x, y, z, t)$$

Use

$$G_3(x, y, z, t) = \theta(t)\frac{c}{4\pi r}\delta(\sqrt{x^2 + y^2 + z^2} - ct)$$

to find $G_2(x, y, t)$.

(b) Explain the qualitative difference between the retarded Green functions in two- and three-dimensional spaces.
Hint: Consider which regions in space at $t < 0$ can influence $\vec{x} = \vec{0}$ at $t = 0$.

Solutions to Examples 4

1. (a) In the SI system, the cyclotron angular frequency is $\omega = \frac{eB}{m} = \frac{e}{mc^2}c^2 B$. Using $\frac{mc^2}{e} \simeq 938 \times 10^6$ V, we find

$$\omega \simeq \frac{1}{938 \times 10^6\,\mathrm{V}}(3.0 \times 10^8)^2\,\frac{\mathrm{m}^2}{\mathrm{s}^2} \cdot 3 \underbrace{\mathrm{T}}_{=\mathrm{V}\cdot\mathrm{s}/\mathrm{m}^2} \simeq 2.9 \times 10^8\,\mathrm{s}^{-1}$$

(b) The velocity of the proton is

$$\frac{v}{c} = \frac{r\omega}{c} = \frac{0.5\,\mathrm{m} \cdot 9.6 \times 10^7\,\mathrm{s}^{-1}}{3.0 \times 10^8\,\mathrm{m/s}} = 0.16$$

Therefore, the kinetic energy is

$$\frac{1}{2}mv^2 = \frac{1}{2}mc^2\left(\frac{v}{c}\right)^2 \simeq 12\,\mathrm{MeV}$$

(c) The energy of radiation per unit time is obtained from Larmor's formula as

$$P = \frac{2}{3}\frac{e^2 a^2}{c^3} = \frac{2}{3}\frac{e^2 r^2 \omega^4}{c^3} = \frac{2}{3}\frac{e^2}{r}\left(\frac{v}{c}\right)^3 \omega \simeq \frac{2}{3}\frac{e^2}{r_e}\frac{r_e}{r}\left(\frac{v}{c}\right)^3 \omega$$

in the cgs-Gauss system. Here, $r_e \simeq 2 \times 10^{-12}$ cm is the classical radius of the electron related to the mass energy of the electron as

$$m_e c^2 = \frac{e^2}{r_e} \simeq 5 \times 10^5 \text{ eV}$$

Thus,

$$P \simeq \frac{2}{3} \cdot 5 \times 10^5 \text{ eV} \cdot \frac{2 \times 10^{-12} \text{ cm}}{50 \text{ cm}} \cdot 0.16^3 \cdot 9.6 \times 10^7 \text{ s}^{-1} \simeq 5 \times 10^{-3} \text{ eV/s},$$

and the energy loss due to radiation during 1s is very small compared with the proton's mass energy 938 MeV $\simeq 1 \times 10^9$ eV.

2. (a) Substituting $x(t) = \mathrm{e}^{\alpha t}$ into $\ddot{x} + \gamma \dot{x} + \Omega^2 x = 0$, we obtain

$$\alpha^2 + \gamma\alpha + \Omega^2 = 0$$

Hence

$$\alpha = -\frac{\gamma}{2} \pm i\Omega_\gamma, \quad \text{where} \quad \Omega_\gamma \equiv \sqrt{\Omega^2 - \frac{\gamma^2}{4}}$$

Therefore,

$$x_1(t) = \mathrm{e}^{-\frac{\gamma}{2}t} \cos \Omega_\gamma t, \quad x_2(t) = \mathrm{e}^{-\frac{\gamma}{2}t} \frac{1}{\Omega_\gamma} \sin \Omega_\gamma t$$

(b) $\ddot{G}(t) + \gamma \dot{G}(t) + \Omega^2 G(t) = \delta(t)$

(c) $\tilde{G}_\omega$ satisfies

$$\left(-\omega^2 + i\omega\gamma + \Omega^2\right) \tilde{G}_\omega = 1$$

Hence,

$$\tilde{G}_\omega = \frac{1}{-\omega^2 + i\omega\gamma + \Omega^2} = \frac{1}{\left(\Omega_\gamma + \omega - \frac{i\gamma}{2}\right)\left(\Omega_\gamma - \omega + \frac{i\gamma}{2}\right)}$$

(d)

$$G(t) = \int_{-\infty}^{\infty} \frac{d\omega}{2\pi} \mathrm{e}^{-i\omega t} \frac{1}{\left(\Omega_\gamma + \omega - \frac{i\gamma}{2}\right)\left(\Omega_\gamma - \omega + \frac{i\gamma}{2}\right)}$$

$$= \theta(t) \cdot x_2(t)$$

(e) $\bar{x}(t) \equiv \int_0^t dt'\, G(t-t')f(t')$, with the lower limit of integration set to 0, is also a particular solution. It satisfies

$$\bar{x}(0) = 0, \quad \dot{\bar{x}}(0) = 0$$

So the solution to the motion equation is

$$x(t) = x_0 x_1(t) + v_0 x_2(t) + \bar{x}(t)$$

Note that $\bar{x}(t)$ depends on $f(t')$ only for $0 < t' < t$.

(f) About $1/\gamma$. For $t \gg 1/\gamma$, we find

$$x(t) \simeq \int_0^t dt'\, G(t-t')f(t'),$$

and the oscillation is determined only by the driving force, not by the initial condition.

3. (a) Omitted.

(b)

$$\begin{cases} \partial_t \rho(\vec{x}, t) &= -e\dot{\vec{X}}(t) \cdot \nabla\delta^{(3)}(\vec{x} - \vec{X}(t)) \\ \nabla \cdot \vec{j}(\vec{x}, t) &= e\dot{\vec{X}}(t) \cdot \nabla\delta^{(3)}(\vec{x} - \vec{X}(t)) \end{cases}$$

Therefore, $\partial_t \rho + \nabla \cdot \vec{j} = 0$.

(c) We calculate only ϕ.

$$\begin{aligned} \phi(\vec{x}, t) &= e \int d^3y dt' \frac{1}{|\vec{x} - \vec{y}|} c\delta^{(3)}(\vec{y} - \vec{X}(t'))\delta(c(t-t') - |\vec{x} - \vec{y}|) \\ &= e \int dt' \frac{1}{|\vec{x} - \vec{X}(t')|} c\delta(c(t-t') - |\vec{x} - \vec{X}(t')|) \end{aligned}$$

If the equation $f(t') = 0$ has only one solution $t' = t_0$, the identity

$$\delta(f(t')) = \frac{1}{|f'(t_0)|}\delta(t' - t_0)$$

holds in general. Now, let

$$f(t') \equiv c(t-t') - |\vec{x} - \vec{X}(t')|$$

so that

$$f'(t') = -c + \dot{\vec{X}}(t') \cdot \frac{\vec{x} - \vec{X}(t')}{|\vec{x} - \vec{X}(t')|}$$

We then obtain

$$\delta(f(t')) = \frac{1}{c}\frac{1}{1-\frac{1}{c}\dot{\vec{X}}(t')\cdot\vec{\varepsilon}}\delta(t'-t_0)$$

Thus,

$$\phi(\vec{x},t) = \frac{e}{|\vec{x}-\vec{X}(t')|}\frac{1}{1-\frac{1}{c}\dot{\vec{X}}(t')\cdot\vec{\varepsilon}}$$

4. (a) At time t, the object exists in the region $v(T-t) < x < v(T-t) + L$. Rewriting this condition in terms of t, we obtain

$$T - \frac{x}{v} < t < T - \frac{x-L}{v}$$

(b) At time $t = 0$, the left end of the object lies at $x = vT$. Let t' be the time when light emitted from the left end of the object reaches $x = 0$, and let t'' be that when light emitted from the right end reaches $x = 0$. Since the right end of the object is farther from $x = 0$ at t'' than at t', it is clear that $t'' < t'$. Furthermore, since the object is moving to the left, it is farther away from $x = 0$ at t'' than at t'. Therefore, the region of x whose emitted light reaches $x = 0$ at $t = 0$ will be longer than L. In fact, the length of this region is $\frac{L}{1-\frac{v}{c}}$, to be elaborated below.

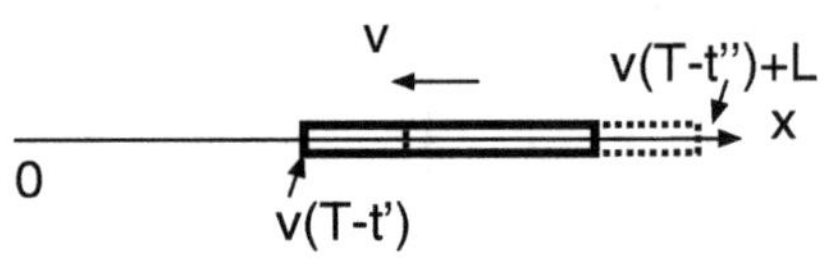

At t', the left end of the object is at $x = v(T-t')$. Thus,

$$-ct' = v(T-t') \Longrightarrow t' = -\frac{vT}{c-v}$$

At t'', the right end of the object is at $x = v(T-t'') + L$. Thus,

$$-ct'' = v(T-t'') + L \Longrightarrow t'' = -\frac{vT+L}{c-v}$$

Therefore, the apparent charge distribution at $t = 0$ is given by

$$v(T-t') < x < v(T-t'') + L \Longrightarrow X_1 \equiv \frac{vT}{1-\frac{v}{c}} < x < X_2 \equiv \frac{vT+L}{1-\frac{v}{c}}$$

The length of this region is $X_2 - X_1 = \frac{L}{1-\frac{v}{c}}$. Thus,

$$\phi = \frac{e}{L}\int_{X_1}^{X_2} dx\,\frac{1}{x} \simeq \frac{e}{L}\frac{1}{X_1}(X_2 - X_1) = \frac{e}{X_1}\frac{1}{1-\frac{v}{c}}\left(= \frac{e}{vT}\right)$$

5. (a) Integrating by parts we obtain

$$\left(-\partial_x^2 - \partial_y^2 + \frac{1}{c^2}\partial_t^2\right)\int_{-\infty}^{\infty} dz\, G_3(x, y, z, t)$$

$$= \int_{-\infty}^{\infty} dz\,\left(\partial_z^2 G_3 + \delta(t)\delta^{(3)}(\vec{x})\right) = \delta(t)\delta(x)\delta(y)$$

Therefore,

$$G_2(x, y, t) = \theta(t)\frac{c}{4\pi}\int_{-\infty}^{\infty} dz\,\frac{1}{\sqrt{x^2+y^2+z^2}}\delta\left(\sqrt{x^2+y^2+z^2} - ct\right)$$

Using

$$\frac{\partial}{\partial z}\sqrt{x^2+y^2+z^2} = \frac{z}{\sqrt{x^2+y^2+z^2}}$$

We obtain

$$G_2(x, y, t) = \theta(ct - \sqrt{x^2+y^2})\frac{c}{2\pi}\frac{1}{\sqrt{c^2t^2 - (x^2+y^2)}}$$

(b) Since G_2 is not proportional to $\delta(\sqrt{x^2+y^2} - ct)$, the influence of charge current density does not propagate at the speed of light, but at any speed less than or equal to the speed of light.

Exercises 4

1. **Relativistic charged particle**—Calculate the electromagnetic fields $\vec{E}$ and $\vec{B}$ created by a charged particle with charge e moving at constant velocity v along the x-axis. Illustrate the results.
2. **Polarization of electromagnetic waves**—Consider electromagnetic waves emitted by a charged particle undergoing non-relativistic circular motion of radius a in the xy-plane around the origin. Assume $R \gg a$. Determine the amplitude and polarization of the electromagnetic waves at the following points:

   (a) $\vec{x} = (R, 0, 0)$
   (b) $\vec{x} = (0, 0, R)$
   (c) $\vec{x} = \left(\frac{R}{\sqrt{2}}, 0, \frac{R}{\sqrt{2}}\right)$

3. **Circular current**—Explain briefly why a circular current does not radiate electromagnetic waves.
4. **Charged harmonic oscillator**—Consider a particle with charge e and mass m moving under the influence of a three-dimensional harmonic potential with angular frequency Ω. Given the time-dependent electric field

$$\vec{E}(t) = \mathrm{e}^{-i\omega t}\, E(0, 0, 1)$$

we consider the equation of motion

$$\ddot{\vec{x}} + \gamma\dot{\vec{x}} + \Omega^2\vec{x} = \frac{e}{m}\vec{E}(t)$$

where the friction coefficient $\gamma > 0$ is much smaller than Ω.

   (a) Solve the equation of motion to find a steady state.
   (b) Calculate the energy radiated by the charged particle in the steady state, averaged over time.
   (c) Find the scattering cross section of light by the electron.
   (d) Plot the scattering cross section as a function of ω, and highlight its characteristics.

5. **Calculation of friction coefficient** γ—Part of the origin of the friction coefficient γ in the previous problem can be traced to electromagnetic radiations. We first assume $\gamma = 0$.

   (a) Determine the energy $E(A)$ of a harmonic oscillator with amplitude A.
   (b) Calculate the rate of energy emission $P(A)$ for a harmonic oscillator with constant amplitude A and angular frequency Ω. (See Exercise 4-4b above.)
   (c) Find the value of γ defined by $P(A) = \gamma E(A)$.
   (d) Show that the condition $\gamma \ll \Omega$ can be expressed as

$$r_e \equiv \frac{e^2}{mc^2} \ll \frac{c}{\Omega}$$

   for an electron with charge e and mass m. In other words

   classical electron radius $\ll$ wavelength of emitted light

   (e) Explain that γ defined above is a friction coefficient.
   Hint: The time dependence of the energy of the oscillator with friction coefficient γ is given by $\mathrm{e}^{-\gamma t}$ (as long as $\gamma \ll \Omega$).

6. **Classical model of the hydrogen atom**—In the classical picture of the hydrogen atom, the electron revolves around the proton in a circular motion. However, since the electron carries an electric charge $-e$, it radiates electromagnetic waves as it revolves around the proton, gradually losing energy. Therefore, it is expected

that eventually the electron will lose all its energy and be absorbed by the proton. This implies the collapse of the atom. Let us estimate the time it takes for this collapse.[10]

(a) Determine the total energy $E(r)$ of the electron in a circular motion of radius r.
(b) Determine the energy $f(r)$ radiated by the electron in a circular motion of radius r per unit time. **Hint**: Use Larmor's formula.
(c) Show that in order for $\frac{|E(r)|}{f(r)}$ to be significantly larger than the period $T(r)$ of the circular motion of radius r, it is sufficient that

$$r \gg r_e \equiv \frac{e^2}{mc^2} \quad \text{electron's classical radius}$$

(d) Explain physically that when the above condition is met, the time dependence of the radius r can be approximately given by

$$\frac{d}{dt}E(r(t)) = -f(r(t))$$

(e) Solve the above equation to find the time dependence of $r(t)$.
(f) Starting with an initial electron orbit of radius $r(0) = 0.5 \times 10^{-8}$ cm (Bohr radius), calculate the time τ required for the atomic collapse by setting $r(\tau) = r_e$.
Hint: The answer does not change significantly by setting $r(\tau) = 0$.
(g) Approximately how many times does the electron circle around the proton before the atom collapses?
Hint: Calculate $\frac{\tau}{T(r(0))}$ for the estimate.

The lifetime of the atom obtained here is roughly the same as the lifetime of the excited states of the electron that can be calculated using quantum mechanics (refer to Sect. 5.2.2).

7. **Electric dipole approximation** —Consider charge and current densities oscillating at an angular frequency ω:

$$\rho(\vec{x}, t) = \rho(\vec{x})\mathrm{e}^{-i\omega t}, \quad \vec{j}(\vec{x}, t) = \vec{j}(\vec{x})\mathrm{e}^{-i\omega t}$$

(a) Show that the vector potential is given by

$$\vec{A}(\vec{x}, t) = \mathrm{e}^{-i\omega t} \int d^3y \, \frac{\vec{j}(\vec{y})}{c|\vec{x} - \vec{y}|} \, \mathrm{e}^{ik|\vec{x}-\vec{y}|}$$

where $k \equiv \frac{\omega}{c}$.

[10] The following calculations are not exact and are used as an approximation that is only valid for a slow decay.

Let us assume that $\rho,\ \vec{j}$ are localized in a region of length a near the origin. If $r \equiv |\vec{x}| \gg a$, you can use

$$|\vec{x}-\vec{y}| = \sqrt{r^2+\vec{y}^2-2\vec{x}\cdot\vec{y}} \simeq r - \frac{\vec{x}\cdot\vec{y}}{r},$$

and you obtain

$$\vec{A}(\vec{x},t) = \frac{\mathrm{e}^{i(kr-\omega t)}}{r}\int d^3y\, \frac{\vec{j}(\vec{y})}{c}\, \mathrm{e}^{-ik\hat{x}\cdot\vec{y}},$$

where $\hat{x} \equiv \frac{\vec{x}}{r}$.
Assuming further

$$\frac{2\pi}{k} \gg a$$

so that the wavelength of the radiated light is much longer than the size a of the region where charge and current densities are distributed, we can use

$$\mathrm{e}^{-ik\hat{x}\cdot\vec{y}} \simeq 1 \quad \textbf{electric dipole approximation}$$

Using this approximation, we obtain

$$\vec{A}(\vec{x},t) = \frac{\mathrm{e}^{i(kr-\omega t)}}{cr}\int d^3y\, \vec{j}(\vec{y})$$

(b) Define the electric dipole moment as

$$\vec{p} \equiv \int d^3y\, \vec{y}\rho(\vec{y})$$

Using the charge conservation law, derive

$$\vec{A}(\vec{x},t) = \frac{-ik\,\mathrm{e}^{i(kr-\omega t)}}{r}\,\vec{p}$$

(c) Retaining only the term proportional to $\frac{1}{r}$, derive

$$\vec{E} = \frac{k^2\mathrm{e}^{i(kr-\omega t)}}{r}\,\vec{p}_\perp \quad \text{where } \vec{p}_\perp \equiv \vec{p} - \hat{x}(\hat{x}\cdot\vec{p})$$

$$\vec{B} = \hat{x}\times\vec{E}$$

Hint: First, calculate $\vec{B} = \nabla \times \vec{A}$, and then use $\nabla \times \vec{B} = \frac{1}{c}\partial_t \vec{E}$ to calculate $\vec{E}$.

(d) Show that the energy radiated per unit time from the dipole is given by

$$\frac{1}{3}ck^4\vec{p}^{\,2}$$

(e) Calculate the energy radiated per unit time from the antenna placed along the z-axis, $z \in [-a/2, a/2]$, with a current

$$I(z,t) = I_0\left(1 - \frac{2|z|}{a}\right)\cos\omega t$$

Assume $a \ll \frac{2\pi c}{\omega}$.

(f) What is the value of I_0 for the antenna with $a = 5\,\text{cm}$ and $100\,\text{W}$ output at 1 MHz?

2.3 Electromagnetism 3

Three Points to Remember

1. The charge and current densities $\rho, \vec{j}$ induced in a material by the electromagnetic fields are given by the polarization and magnetization vectors $\vec{P}, \vec{M}$.
2. The relation between the electromagnetic fields $\vec{E}, \vec{B}$ and the polarization and magnetization vectors $\vec{P}, \vec{M}$ is given in terms of the dielectric constant ε and the magnetic permeability μ.
3. ε and μ determine the dispersion relation of light in a material.

2.3.1 *The Charge and Current Densities $\rho, \vec{j}$ Induced in a Material by the Electromagnetic Fields Are Given by the Polarization and Magnetization Vectors $\vec{P}, \vec{M}$*

The electromagnetic properties of a material are determined by the charge and current densities induced by the electromagnetic fields. The induced charge and current densities are expressed via the polarization and magnetization vectors. The former is electric and the latter magnetic. Let us understand the difference between these two.

The electromagnetic fields present within a material induce charge and current densities $\rho,\ \vec{j}$, which must satisfy the conservation law:

$$\partial_t \rho + \nabla \cdot \vec{j} = 0 \tag{2.147}$$

The most general form of $\rho,\ \vec{j}$ allowed by the conservation law is given by

$$\rho = -\nabla \cdot \vec{P} \tag{2.148}$$

$$\vec{j} = \partial_t \vec{P} + c\nabla \times \vec{M} \tag{2.149}$$

It is straightforward to verify that $\rho, \vec{j}$ given above satisfy the conservation law. Please note that $\vec{P}$ and $\vec{M}$ have the same dimensions, thanks to the light velocity c multiplied by $\vec{M}$.

In the following let us understand the physical meanings of $\vec{P}$ and $\vec{M}$.

Polarization Vector $\vec{P}$

Consider a medium made up of ions with a uniform charge density ρ_0 and electrons with $-\rho_0$. In equilibrium, they cancel, resulting in the zero total charge density. Since we are considering a classical model, both ions and electrons can be considered as continuous media.

Now, let us assume that the electron at coordinate $\vec{x}$ is displaced by $\vec{u}(\vec{x}, t)$. (Fig. 2.12) This induces a charge density

$$\rho(\vec{x}, t) = \rho_0 \nabla \cdot \vec{u}(\vec{x}, t) \tag{2.150}$$

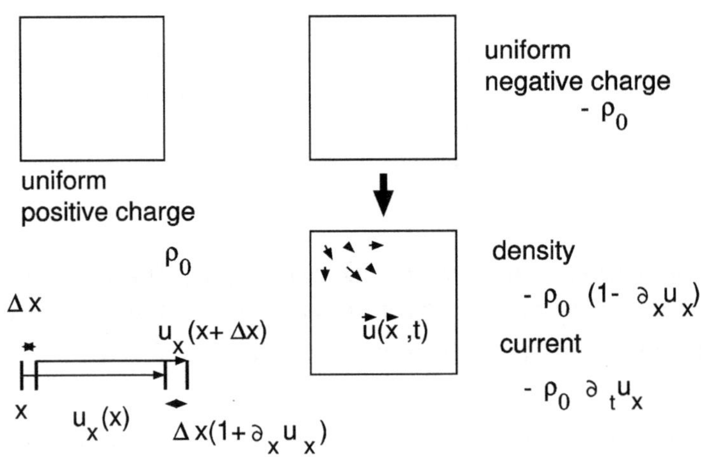

Fig. 2.12 Polarization vector

and also causes a current density

$$\vec{j}(\vec{x},t) = -\rho_0 \partial_t \vec{u}(\vec{x},t) \tag{2.151}$$

This implies that the polarization vector $\vec{P}(\vec{x},t)$ is given by

$$\vec{P}(\vec{x},t) = -\rho_0 \vec{u}(\vec{x},t) \tag{2.152}$$

Furthermore, if ions are displaced by $\vec{u}_i(\vec{x},t)$, the total polarization vector becomes

$$\vec{P}(\vec{x},t) = \rho_0 \left(\vec{u}_i(\vec{x},t) - \vec{u}(\vec{x},t)\right) \tag{2.153}$$

Magnetization Vector $\vec{M}$

The magnetization vector is a little more difficult to visualize. $\vec{M}$ represents the density of loop current. For intuitive understanding, let us approximate the continuous space as a collection of small cubes. Each side of a cube is to have a length a. Consider the point $\vec{x}$ at one vertex, and imagine a square defined by $\vec{x} + a\vec{e}_x$, $\vec{x} + a\vec{e}_y$, and $\vec{x} + a\left(\vec{e}_x + \vec{e}_y\right)$. This square is perpendicular to the z-axis (See the left side of Fig. 2.13). Let $I_z(\vec{x})$ be the magnitude of the counterclockwise loop current flowing along the edge of this square; the magnetic moment due to this current is given by

$$M_z(\vec{x})a^3 = \frac{1}{c} I_z(\vec{x})a^2 \tag{2.154}$$

Similarly, we can introduce counterclockwise loop currents $I_y(\vec{x})$, $I_z(\vec{x})$ and magnetic moments per unit volume $M_y(\vec{x})$, $M_z(\vec{x})$.

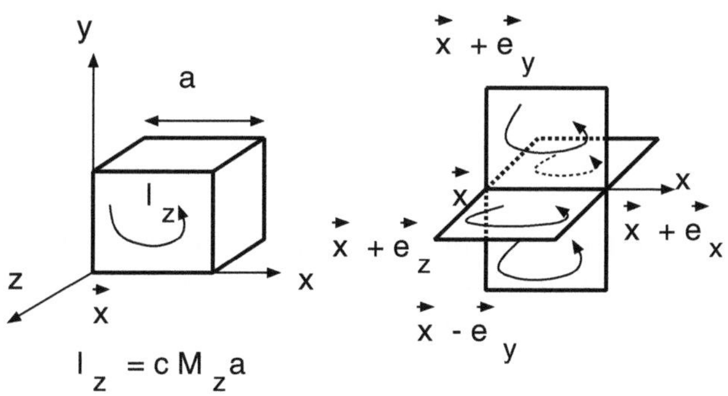

Fig. 2.13 A magnetization vector represents the magnetic moment due to loop currents

Next, let us calculate the total current flowing along the edge connecting point $\vec{x}$ and its adjacent point $\vec{x} + a\vec{e}_x$. As can be seen from the right side of Fig. 2.13, this edge is attached to four squares. Therefore, the current flowing through the area a^2 in the x-direction is found to be

$$\begin{aligned} j_x(\vec{x})a^2 &\equiv ac\left(M_z(\vec{x}) - M_z(\vec{x} - a\vec{e}_y) - M_y(\vec{x}) + M_y(\vec{x} - a\vec{e}_z)\right) \\ &\simeq a^2c\left(\partial_y M_z(\vec{x}) - \partial_z M_y(\vec{x})\right) \end{aligned} \tag{2.155}$$

Generalizing this, we obtain the current density given by

$$\vec{j}(\vec{x}) = c\nabla \times \vec{M}(\vec{x}) \tag{2.156}$$

Thus, the magnetization vector $\vec{M}$ represents the magnetic moment per unit volume generated by loop currents.

2.3.2 *The Relation Between the Electromagnetic Fields $\vec{E}$, $\vec{B}$ and the Polarization and Magnetization Vectors $\vec{P}$, $\vec{M}$ Is Given in Terms of the Dielectric Constant ε and the Magnetic Permeability μ*

Main objectives of condensed matter physics may well be the measurement of the dielectric constant and permeability and their theoretical understanding. Though we only deal with some classical models in Examples and Exercises, quantum mechanics is essential.

Given a uniform and isotropic continuous medium (which does not really exist in nature, but can be considered as an average over a certain range), upon exposition to weak electromagnetic fields, polarization and magnetization vectors are produced in proportion:[11]

$$4\pi\vec{P} = (\varepsilon - 1)\vec{E} \tag{2.157}$$

$$4\pi\vec{M} = \left(1 - \frac{1}{\mu}\right)\vec{B} \tag{2.158}$$

Here, both the **dielectric constant** ε and the **permeability** μ depend on the wave number k and angular frequency ω. Therefore, it is more accurate to express the above for the Fourier transforms as

[11] In a medium with preferred directions, ε and μ become second-order symmetric tensors: $4\pi P_i = \sum_j(\epsilon_{ij} - \delta_{ij})E_j$, $4\pi M_i = \sum_j(\delta_{ij} - (1/\mu)_{ij})B_j$.

$$4\pi \vec{P}(\vec{k}, \omega) = (\varepsilon(k, \omega) - 1)\vec{E}(k, \omega) \tag{2.159}$$

$$4\pi \vec{M}(k, \omega) = \left(1 - \frac{1}{\mu(k, \omega)}\right) \vec{B}(k, \omega) \tag{2.160}$$

It is natural to ask why E generates only P and B generates only M. The reason is somewhat advanced, based on the invariance of electromagnetism under parity transformation. Under the parity transformation, where the coordinate $\vec{x}$ transforms to $-\vec{x}$, the electromagnetic fields transform as

$$\vec{E}(\vec{x}, t) \longrightarrow -\vec{E}(-\vec{x}, t) \tag{2.161}$$

$$\vec{B}(\vec{x}, t) \longrightarrow \vec{B}(-\vec{x}, t) \tag{2.162}$$

If the charge and current densities transform simultaneously as

$$\rho(\vec{x}, t) \longrightarrow \rho(-\vec{x}, t) \tag{2.163}$$

$$\vec{j}(\vec{x}, t) \longrightarrow -\vec{j}(-\vec{x}, t) \tag{2.164}$$

then Maxwell's equations remain invariant (check this yourself). From the relations (2.148, 2.149) between charge and current densities and polarization and magnetization, the latter must transform as

$$\vec{P}(\vec{x}, t) \longrightarrow -\vec{P}(-\vec{x}, t) \tag{2.165}$$

$$\vec{M}(\vec{x}, t) \longrightarrow \vec{M}(-\vec{x}, t) \tag{2.166}$$

Therefore, in order for the relation between $(\vec{P}, \vec{M})$ and $(\vec{E}, \vec{B})$ to remain invariant under the parity transformation, $\vec{E}$ must only generate $\vec{P}$ and $\vec{B}$ only $\vec{M}$.[12]

Let us focus a little more on the Fourier transforms (2.159, 2.160). We define functions of the coordinate and time as

$$K_P(\vec{x}, t) \equiv \int \frac{d^3k}{(2\pi)^3} \int \frac{d\omega}{2\pi}\, \mathrm{e}^{i\vec{k}\cdot\vec{x} - i\omega t}\, (\varepsilon(k, \omega) - 1) \tag{2.167}$$

$$K_M(\vec{x}, t) \equiv \int \frac{d^3k}{(2\pi)^3} \int \frac{d\omega}{2\pi}\, \mathrm{e}^{i\vec{k}\cdot\vec{x} - i\omega t} \left(1 - \frac{1}{\mu(k, \omega)}\right) \tag{2.168}$$

Then, Eqs. (2.159, 2.160) can be expressed as

$$4\pi \vec{P}(\vec{x}, t) = \int d^3y dt'\, K_P(|\vec{x} - \vec{y}|, t - t')\vec{E}(\vec{y}, t') \tag{2.169}$$

[12] There are materials not invariant under parity. Then E can generate M, and B can generate P.

$$4\pi \vec{M}(\vec{x}, t) = \int d^3y dt' \, K_M(|\vec{x} - \vec{y}|, t - t')\vec{B}(\vec{y}, t') \tag{2.170}$$

The physical meaning of the integration kernels (or Green functions) $K_P(|\vec{x}|, t)$, $K_M(|\vec{x}|, t)$ is clear. The electric field $\vec{E}$ in $\vec{y}$ at time t' induces $\vec{P}$ in $\vec{x}$ at time t. Causality demands $t > t'$. But that is not all the constraint. We expect $K_P(|\vec{x} - \vec{y}|, t - t')$ to be non-vanishing only for a limited range of $|\vec{x} - \vec{y}|$ and $t - t'$. In the absence of correlations among the atoms, we expect $|\vec{x} - \vec{y}|$ to be of the order of inter atomic distance and $t - t'$ to be of the order of the lifetime of the atomic excited states.[13] When the wavelength of the electromagnetic field is much longer than the size of the atom, we can integrate over $\vec{y}$ first in the above equations to obtain

$$4\pi \vec{P}(t) = \int_{-\infty}^{t} dt' \, K_P(t - t') \, \vec{E}(t') \tag{2.171}$$

$$4\pi \vec{M}(t) = \int_{-\infty}^{t} dt' \, K_M(t - t') \, \vec{B}(t') \tag{2.172}$$

valid for any location $\vec{x}$. Both K_P and K_M vanish for $t < t'$; the consequences of this causality will be discussed at the end of Sect. 2.3.3.

Let us now rewrite the induced charge density in the material as ρ_{ind} and the current density as $\vec{j}_{ind}$. Furthermore, external charge and current densities other than these will be written as ρ_{ext}, $\vec{j}_{ext}$. Then, Maxwell's equations are

$$\begin{aligned} \nabla \cdot \vec{E} &= 4\pi(\rho_{ind} + \rho_{ext}) \\ &= 4\pi\left(-\nabla \cdot \vec{P} + \rho_{ext}\right) \end{aligned} \tag{2.173}$$

$$\begin{aligned} \nabla \times \vec{B} - \frac{1}{c}\frac{\partial}{\partial t}\vec{E} &= \frac{4\pi}{c}\left(\vec{j}_{ind} + \vec{j}_{ext}\right) \\ &= \frac{4\pi}{c}\left(\partial_t \vec{P} + c\nabla \times \vec{M} + \vec{j}_{ext}\right) \end{aligned} \tag{2.174}$$

Taking the induced charges and currents to the left side, we obtain

$$\nabla \cdot \left(\vec{E} + 4\pi \vec{P}\right) = 4\pi \rho_{ext} \tag{2.175}$$

$$\nabla \times \left(\vec{B} - 4\pi \vec{M}\right) - \frac{1}{c}\frac{\partial}{\partial t}\left(\vec{E} + 4\pi \vec{P}\right) = \frac{4\pi}{c}\vec{j}_{ext} \tag{2.176}$$

Define

$$\vec{D} \equiv \vec{E} + 4\pi \vec{P} \tag{2.177}$$

[13] The average time for an atom to return from an excited state to the ground state.

$$\vec{H} \equiv \vec{B} - 4\pi\vec{M} \tag{2.178}$$

so that the above equations can be rewritten as

$$\nabla \cdot \vec{D} = 4\pi\rho_{ext} \tag{2.179}$$

$$\nabla \times \vec{H} - \frac{1}{c}\frac{\partial}{\partial t}\vec{D} = \frac{4\pi}{c}\vec{j}_{ext} \tag{2.180}$$

In some textbooks, $\vec{E}$ is called the electric field, $\vec{D}$ the electric flux density, $\vec{B}$ the magnetic flux density, and $\vec{H}$ the magnetic field. We treat $\vec{D}$ and $\vec{H}$ as secondary quantities and choose not to assign them specific names. Therefore, we will continue to refer to $\vec{E}$ as the electric field and $\vec{B}$ as the magnetic field. In vacuum (regions without matter), $\vec{E} = \vec{D}$ and $\vec{H} = \vec{B}$.

For the Fourier transforms, we obtain from (2.159, 2.160)

$$\vec{D}(\vec{k}, \omega) = \varepsilon(k, \omega)\vec{E}(\vec{k}, \omega) \tag{2.181}$$

$$\vec{H}(\vec{k}, \omega) = \frac{1}{\mu(k, \omega)}\vec{B}(\vec{k}, \omega) \tag{2.182}$$

Hence, Eqs. (2.179, 2.180) are written as

$$\varepsilon(k, \omega)i\vec{k} \cdot \vec{E}(\vec{k}, \omega) = 4\pi\rho_{ext}(\vec{k}, \omega) \tag{2.183}$$

$$i\frac{1}{\mu(k, \omega)}\vec{k} \times \vec{B}(\vec{k}, \omega) + i\frac{\omega}{c}\varepsilon(k, \omega)\vec{E}(\vec{k}, \omega) = \frac{4\pi}{c}\vec{j}_{ext}(\vec{k}, \omega) \tag{2.184}$$

Example: Free Electron Plasma

Consider an electron gas with a uniform number density n. We treat the electrons as free. Electrons in the ionosphere can be treated as free electrons, and the same applies to the conduction electrons in metals. Additionally, in dielectrics where electrons are bound within atoms, if the frequency of the applied electric field is significantly higher than the electron oscillation frequency within the atoms, we can consider them as free electrons. The ions give a uniform charge density of $+ne$, resulting in the vanishing total charge density.

Under the influence of a uniform electric field $\vec{E}$ with angular frequency ω, an electron satisfies the equation of motion

$$m\ddot{\vec{x}} = -m\omega^2\vec{x} = -e\vec{E} \tag{2.185}$$

Solving this we obtain

$$\vec{x} = \frac{e}{m\omega^2}\vec{E} \tag{2.186}$$

Therefore, the polarization vector is

$$\vec{P} = -ne\vec{x} = -\frac{ne^2}{m\omega^2}\vec{E}\,, \tag{2.187}$$

and the dielectric constant ε is

$$\varepsilon(\omega) - 1 = -\frac{4\pi ne^2}{m\omega^2} \tag{2.188}$$

Defining the plasma frequency ω_{pl} as

$$\boxed{\omega_{pl}^2 \equiv \frac{4\pi ne^2}{m}} \tag{2.189}$$

we can rewrite $\varepsilon(\omega)$ as

$$\varepsilon(\omega) = 1 - \frac{\omega_{pl}^2}{\omega^2} \tag{2.190}$$

(Fig. 2.14)

For example, if we take $n = 10^4/\text{cm}^3$,[14] we find

$$\omega_{pl}^2 = \frac{4\pi \cdot 10^4\text{cm}^{-3} \cdot (4.8 \times 10^{-10}\text{esu})^2}{9.1 \times 10^{-28}\text{g}} \simeq 3.2 \times 10^{13}\,\text{s}^{-2} \tag{2.191}$$

which results in

$$\omega_{pl} \simeq 5.6 \times 10^6\,\text{s}^{-1} \tag{2.192}$$

If we take the number density as $n \sim \frac{1}{a^3} \sim 10^{24}/\text{cm}^3$ (where a is the Bohr radius), ω_{pl} becomes approximately the angular frequency of the electron in a hydrogen atom

$$\Omega_{\text{Bohr}} = \frac{Ry}{\hbar} = 2 \times 10^{16}\text{s}^{-1} \tag{2.193}$$

[14] The electron density in the ionosphere is typically in the range of 10^4 to $10^6\,\text{cm}^{-3}$.

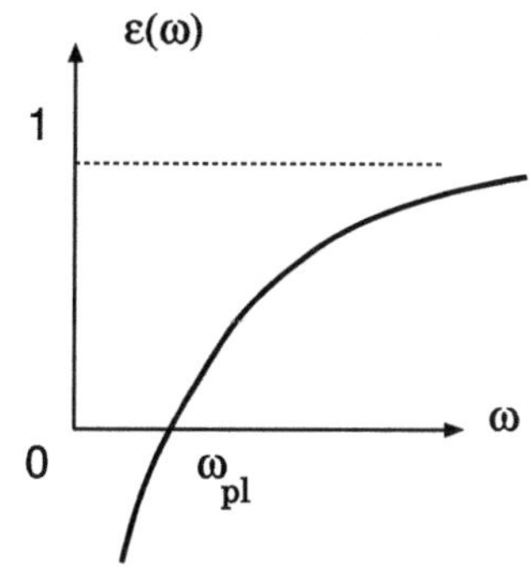

Fig. 2.14 Polarization of plasma

2.3.3 *ε and μ Determine the Dispersion Relation of Light in a Material*

In condensed matter physics, it is common to study the properties of matter by subjecting it to electromagnetic waves.

When there are no external charges or currents, the Maxwell equations valid in a material are given as follows:

$$\vec{k} \cdot \vec{B} = 0 \tag{2.194a}$$

$$\vec{k} \times \vec{E} - \frac{\omega}{c}\vec{B} = 0 \tag{2.194b}$$

$$\varepsilon\vec{k} \cdot \vec{E} = 0 \tag{2.194c}$$

$$\frac{1}{\mu}\vec{k} \times \vec{B} + \frac{\omega}{c}\varepsilon\vec{E} = 0 \tag{2.194d}$$

The first equation implies that $\vec{B}$ is a transverse wave orthogonal to the wave vector $\vec{k}$. The second equation gives

$$\vec{B} = \frac{c}{\omega}\vec{k} \times \vec{E} \tag{2.195}$$

so that the magnetic field is also orthogonal to the electric field. Taking the absolute value, we obtain

$$\frac{|\vec{B}|}{|\vec{E}|} = \frac{kc}{\omega} \tag{2.196}$$

The phase velocity is defined as

$$v_{ph} \equiv \frac{\omega}{k}, \tag{2.197}$$

and the index of refraction n is defined as

$$n \equiv \frac{c}{v_{ph}} \tag{2.198}$$

Hence, we obtain

$$\boxed{\frac{|\vec{B}|}{|\vec{E}|} = n(k, \omega)} \tag{2.199}$$

The ratio of the magnetic field to the electric field is equal to the refractive index.

If $\varepsilon \neq 0$, the third Maxwell equation shows that the electric field is also orthogonal to the wave vector $\vec{k}$. Furthermore, from the fourth equation, we obtain

$$\frac{1}{\mu} k^2 c^2 = \omega^2 \varepsilon \tag{2.200}$$

This leads to the important result

$$\frac{1}{n^2} = \frac{v_{ph}^2}{c^2} = \boxed{\frac{\omega^2}{k^2 c^2} = \frac{1}{\varepsilon(k, \omega)\mu(k, \omega)}} \tag{2.201}$$

This expression determines the **dispersion relation** between the wave number k and the angular frequency ω.

Example 1: Free Electron Plasma

In the previous section, when determining the dielectric constant in a free electron plasma, we assumed a wave number of zero. Similarly, when the amplitude of electron motion is much smaller than the wavelength, we can ignore the dependence of the dielectric constant on wave number:

$$\varepsilon(\omega) = 1 - \frac{\omega_{pl}^2}{\omega^2}, \quad \mu = 1 \tag{2.202}$$

This leads to the dispersion relation:

$$k^2 c^2 = \omega^2 - \omega_{pl}^2 \tag{2.203}$$

and

$$\omega = \sqrt{k^2 c^2 + \omega_{pl}^2} \tag{2.204}$$

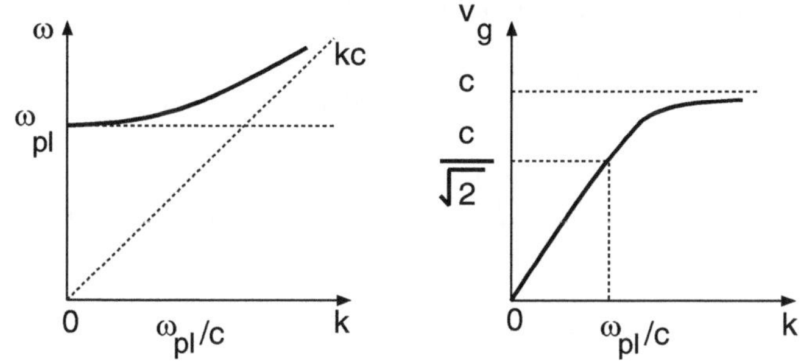

Fig. 2.15 Dispersion relation in free electron plasma

This gives the group velocity as

$$v_g \equiv \frac{d\omega}{dk} = \frac{c}{\sqrt{1 + \frac{\omega_{pl}^2}{k^2c^2}}} < c \tag{2.205}$$

It is worth noting that the group velocity is always less than the speed of light even though the phase velocity is always greater than the speed of light (Fig. 2.15).

Example 2: Superconductors

Type II superconductors satisfy

$$\varepsilon = 1, \quad \frac{1}{\mu} = 1 + \frac{1}{k^2\lambda_L^2} \tag{2.206}$$

where λ_L is the London penetration depth , which depends on the material but is roughly around 10^{-5} cm = 100 nm. This leads to the dispersion relation:

$$\frac{\omega^2}{k^2c^2} = 1 + \frac{1}{k^2\lambda_L^2} \tag{2.207}$$

Therefore, we find

$$\omega^2 = c^2\left(k^2 + \frac{1}{\lambda_L^2}\right) = k^2c^2 + \frac{c^2}{\lambda_L^2} \tag{2.208}$$

This has the same form as the dispersion relation in free electron plasma, with ω_{pl} replaced by $\frac{c}{\lambda_L}$.

Example 3: Quantum Electrodynamics (QED)

According to relativistic quantum electrodynamics, the vacuum can be considered as a dielectric, and the dielectric constant is given approximately by

$$\varepsilon(k,\omega) = \frac{1}{\mu(k,\omega)} = 1 - \frac{2}{\pi}\frac{e^2}{\hbar c}\int_0^1 dx\, x(1-x)\ln\left(1 + x(1-x)\lambda\!\!\!^{-}_e{}^2(k^2 - \omega^2/c^2)\right) \tag{2.209}$$

where

$$\lambda\!\!\!^{-}_e \equiv \frac{\hbar}{m_e c} \simeq 4\times 10^{-10}\mathrm{cm}$$

is the reduced Compton wavelength of the electron.[15] (Other charged particles also contribute, but the contribution of the electron field dominates due to its small mass.) For $\omega = 0$, $\varepsilon(k, 0)$ is a decreasing function of k, and $\varepsilon(0, 0) = 1$, so $\varepsilon(k, 0) \leq 1$. As shown in Example 5-2, the apparent charge at a distance $\frac{2\pi}{k}$ from the charge Q can be interpreted as $\frac{Q}{\varepsilon(k,0)}$. Therefore, $\frac{Q}{\varepsilon(k,0)}$ is an increasing function of k, indicating that the charge appears larger near the charge (Fig. 2.16). Thus, the vacuum acts as a dielectric, partially shielding the charge.[16]

The relativistic invariance of vacuum gives $\varepsilon = \frac{1}{\mu}$, which means the dispersion relation remains

$$\omega = k \tag{2.210}$$

Example 4: Higgs Phase (The Relativistic Version of Type II Superconductors with Massive Photons)

If photons have mass m_γ, quantum electrodynamics yields

$$\varepsilon = \frac{1}{\mu} = 1 + \frac{\frac{m_\gamma^2 c^4}{\hbar^2}}{k^2c^2 - \omega^2} \tag{2.211}$$

Similar to the case where $m_\gamma = 0$, $\varepsilon\mu = 1$ seems to imply that the dispersion relation is $\omega = kc$. However, it is important to note that ε diverges for $kc = \omega$. This makes it necessary to revisit the original Maxwell equations and reevaluate ε.

---

[15] **Reduced** means division by 2π.

[16] Equation (2.209) is an approximate formula for the case $k\lambda\!\!\!^{-}_e \ll e^{(\mathrm{positiveconst})\frac{\hbar c}{e^2}}$. We always find $\varepsilon(k, 0) > 0$.

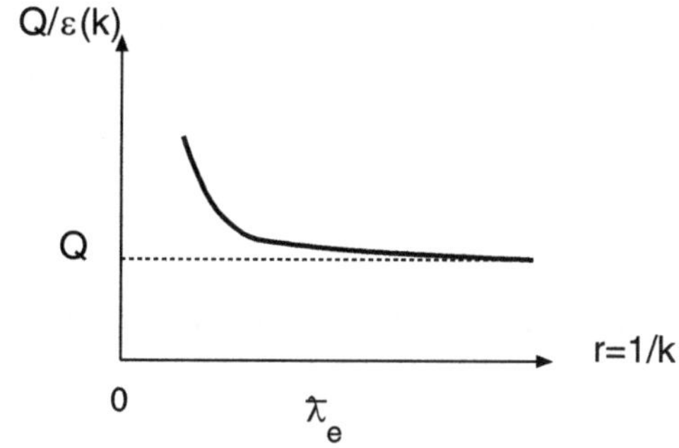

Fig. 2.16 Screening effect of charge by vacuum

Since $\varepsilon\mu = 1$, the Maxwell equations become as follows:

$$\vec{k} \cdot \vec{B} = 0 \tag{2.212a}$$

$$\vec{k} \times \vec{E} - \frac{\omega}{c}\vec{B} = 0 \tag{2.212b}$$

$$\varepsilon \vec{k} \cdot \vec{E} = 0 \tag{2.212c}$$

$$\varepsilon\left(\vec{k} \times \vec{B} + \frac{\omega}{c}\vec{E}\right) = 0 \tag{2.212d}$$

From Eq. (2.212b), we get

$$\vec{B} = \frac{c}{\omega}\vec{k} \times \vec{E} \tag{2.213}$$

Substituting this into Eq. (2.212d) and using Eq. (2.211), we obtain

$$\left(\omega^2 - k^2c^2 - \frac{m_\gamma^2 c^4}{\hbar^2}\right)\vec{E} = 0 \tag{2.214}$$

For $\vec{E} \neq 0$, the dispersion relation must be

$$\omega^2 = k^2c^2 + \frac{m_\gamma^2 c^4}{\hbar^2} \tag{2.215}$$

which has the same form as the dispersion relation in the free electron plasma.

Equation (2.211) gives

$$\varepsilon = \frac{1}{\mu} = 0 \tag{2.216}$$

which means that Eq. (2.212c) no longer implies that $\vec{E}$ is a transverse wave. Therefore, besides the two transverse modes, there is a longitudinal mode:

- Longitudinal mode (1 degree of freedom)

$$\vec{E} \parallel \vec{k}, \quad \vec{B} = 0$$

- Transverse modes (2 degrees of freedom)

$$\vec{E} \perp \vec{k}, \quad \vec{B} = \frac{c}{\omega}\vec{k} \times \vec{E}$$

Reflectance and Transmittance Depend on ε, μ

Let us consider a uniform medium with dielectric constant ε and magnetic permeability μ in the region $z > 0$, while vacuum is present for $z < 0$. Let us suppose that light enters the medium from the vacuum in the positive z direction. We wish to calculate how much of it is transmitted and how much is reflected (Fig. 2.17). Assume that the electric field is in the x direction so that the magnetic field is in the y direction. Let the amplitude of the incident wave be 1. We introduce the coefficients of transmission $t(\omega)$ and reflection $r(\omega)$ by

$$E_x(z) = \mathrm{e}^{-i\omega t} \begin{cases} t(\omega)\mathrm{e}^{ik'z} & \text{for} \quad z > 0 \\ \mathrm{e}^{ikz} + r(\omega)\mathrm{e}^{-ikz} & \text{for} \quad z < 0 \end{cases} \tag{2.217}$$

$$B_y(z) = \mathrm{e}^{-i\omega t} \begin{cases} \frac{k'c}{\omega} t(\omega)\mathrm{e}^{ik'z} & \text{for} \quad z > 0 \\ \mathrm{e}^{ikz} - r(\omega)\mathrm{e}^{-ikz} & \text{for} \quad z < 0 \end{cases} \tag{2.218}$$

The wave number k' in the medium is given by

$$k'^2 c^2 = \omega^2 \epsilon(\omega)\mu(\omega) \tag{2.219}$$

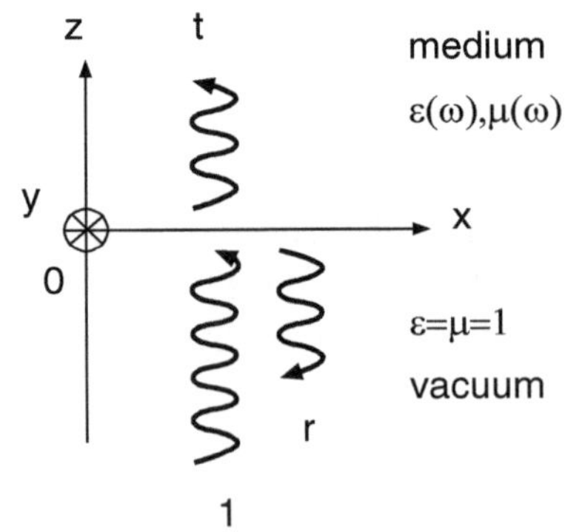

Fig. 2.17 Transmission and absorption of electromagnetic waves

The refractive index is defined by

$$n(\omega) \equiv \sqrt{\epsilon(\omega)\mu(\omega)} \tag{2.220}$$

Both E_x and $\frac{1}{\mu}B_y$ must be continuous at $z = 0$. (See Example 5-3.) Therefore, we find

$$E_x(z=0): \quad 1 + r(\omega) = t(\omega) \tag{2.221}$$

$$B_y(z=0): \quad 1 - r(\omega) = \frac{n(\omega)}{\mu(\omega)}\, t(\omega) \tag{2.222}$$

Hence, we obtain

$$t(\omega) = \frac{2}{1 + n/\mu(\omega)}, \quad r(\omega) = \frac{1 - n/\mu(\omega)}{1 + n/\mu(\omega)} \tag{2.223}$$

The proportion of energy that is reflected is given by

$$|r(\omega)|^2$$

The proportion of energy that enters the region $z > 0$ is expected to be $1 - |r(\omega)|^2$ from the law of energy conservation. Checking this for the general case is a bit of work, and we will do it in the appendix at the end of this section.

Kramers-Kronig Relations

The dielectric constant $\varepsilon(k, \omega)$ is a function of wave vector k and angular frequency ω. However, when the wavelength is much longer than the size of the atoms in the medium, we can disregard the dependence on the wave vector k. Therefore, we can consider the dielectric constant as a function of angular frequency alone, denoted as $\varepsilon(\omega)$.

In general, $\varepsilon(\omega)$ takes a complex value. There is a close relation between its real and imaginary parts, known as the **Kramers-Kronig relations**, named after their discoverers.

As we previously noted, the polarization vector can be written in the form of Eq. (2.171):

$$4\pi \vec{P}(t) = \int_{-\infty}^{t} dt'\, K_P(t - t')\vec{E}(t') \tag{2.224}$$

This equation indicates that the polarization vector at time t depends only on the electric field $\vec{E}(t')$ in the past $t' < t$. From the definition of the dielectric constant, we get

$$\varepsilon(\omega) - 1 = \int_0^{\infty} dt\, \mathrm{e}^{i\omega t}\, K_P(t) \tag{2.225}$$

Now, if we consider ω as a complex number with a non-negative imaginary part $\Im\omega \geq 0$, the exponential term $e^{i\omega t}$ tends to zero as $t \to \infty$, and the integral converges. Thus, $\varepsilon(\omega) - 1$ is well defined as a function of the complex variable ω and is analytic. However, for $\Im\omega < 0$, the integral may not necessarily converge. We then need to define the value of the integral through analytic continuation, and singularities are expected to occur.

In general, a function $f(\omega)$ that is analytic in the upper half complex plane can be expressed using the residue theorem as

$$f(\omega) = \int_{-\infty}^{\infty} \frac{d\omega'}{2\pi i} \frac{f(\omega')}{\omega' - \omega - i\epsilon} \tag{2.226}$$

For real ω, using the well-known relation

$$\frac{1}{\omega' - \omega - i\epsilon} = \mathbf{P}\frac{1}{\omega' - \omega} + i\pi\delta(\omega' - \omega) \tag{2.227}$$

where **P** denotes the principal part, we obtain

$$f(\omega) = \int_{-\infty}^{\infty} \frac{d\omega'}{2\pi i} f(\omega')\mathbf{P}\frac{1}{\omega' - \omega} + \frac{1}{2}f(\omega) \tag{2.228}$$

Taking the real and imaginary parts of the above, we obtain

$$\Re f(\omega) = \int_{-\infty}^{\infty} \frac{d\omega'}{\pi} \Im f(\omega')\mathbf{P}\frac{1}{\omega' - \omega} \tag{2.229}$$

$$\Im f(\omega) = \int_{-\infty}^{\infty} \frac{d\omega'}{\pi} (-\Re f(\omega'))\mathbf{P}\frac{1}{\omega' - \omega} \tag{2.230}$$

Applying this to $f(\omega) = \varepsilon(\omega) - 1$, we obtain the Kramers-Kronig dispersion relations as follows:

$$\Re(\varepsilon(\omega) - 1) = \int_{-\infty}^{\infty} \frac{d\omega'}{\pi} \Im\varepsilon(\omega')\mathbf{P}\frac{1}{\omega' - \omega} \tag{2.231}$$

$$\Im\varepsilon(\omega) = -\int_{-\infty}^{\infty} \frac{d\omega'}{\pi} \Re(\varepsilon(\omega') - 1)\mathbf{P}\frac{1}{\omega' - \omega} \tag{2.232}$$

Since $K_P(t)$ in Eq. (2.225) is a real-valued function, we find

$$\varepsilon(-\omega) = \varepsilon(\omega)^* \tag{2.233}$$

for real ω. Hence,

$$\Re\varepsilon(-\omega) = \Re\varepsilon(\omega), \quad \Im\varepsilon(-\omega) = -\Im\varepsilon(\omega) \tag{2.234}$$

Therefore, the Kramers-Kronig relations can be rewritten as follows:

$$\begin{aligned}\Re(\varepsilon(\omega)-1) &= \int_0^\infty \frac{d\omega'}{\pi}\,\Im\varepsilon(\omega')\mathbf{P}\left(\frac{1}{\omega'-\omega}+\frac{1}{\omega'+\omega}\right)\\ &= \int_0^\infty \frac{d\omega'}{\pi}\,\Im\varepsilon(\omega')\mathbf{P}\frac{2\omega'}{\omega'^2-\omega^2}\end{aligned} \tag{2.235}$$

$$\begin{aligned}\Im\varepsilon(\omega) &= -\int_0^\infty \frac{d\omega'}{\pi}\,\Re(\varepsilon(\omega')-1)\mathbf{P}\left(\frac{1}{\omega'-\omega}-\frac{1}{\omega'+\omega}\right)\\ &= -2\omega\int_0^\infty \frac{d\omega'}{\pi}\,\Re(\varepsilon(\omega')-1)\mathbf{P}\frac{1}{\omega'^2-\omega^2}\end{aligned} \tag{2.236}$$

As an application of the dispersion relations, we take $\omega \to \infty$ in Eq. (2.235), which leads to the asymptotic behavior:

$$\Re(\varepsilon(\omega)-1) \overset{\omega\to\infty}{\longrightarrow} -\frac{1}{\omega^2}\int_0^\infty \frac{d\omega'}{\pi}\,2\omega'\,\Im\varepsilon(\omega') \tag{2.237}$$

When ω is much larger than the atomic vibrational frequencies, the medium can be considered as a free electron plasma, and we obtain

$$\varepsilon(\omega)-1 \overset{\omega\to\infty}{\longrightarrow} -\frac{4\pi ne^2}{m\omega^2} \tag{2.238}$$

Hence, we obtain the sum rule:

$$\boxed{\int_0^\infty \frac{d\omega'}{\pi}\,2\omega'\,\Im\varepsilon(\omega') = \frac{4\pi ne^2}{m}} \tag{2.239}$$

Appendix: Energy Flow

We have shown that the proportion of energy reflected by the material is given by $|r(\omega)|^2$. In this appendix we verify that the proportion of energy entering the material is given by $1-|r(\omega)|^2$ as expected from the conservation of energy.

From Eqs. (2.217, 2.218), the electromagnetic wave propagating in the $+z$ direction into the region $z > 0$ can be expressed as follows:

$$E_x(z) = t(\omega)\mathrm{e}^{-i\omega t}\mathrm{e}^{ik'z} \tag{2.240}$$

$$B_y(z) = t(\omega)\frac{k'c}{\omega}\mathrm{e}^{-i\omega t}\mathrm{e}^{ik'z} \tag{2.241}$$

Here, the wave number k' is related to the angular frequency and refractive index by

$$\frac{k'^2 c^2}{\omega^2} = n(\omega)^2 = \varepsilon(\omega)\mu(\omega) \tag{2.242}$$

We split $n(\omega)$, which is complex, into its real and imaginary parts:

$$n(\omega) = n_1(\omega) + in_2(\omega) \tag{2.243}$$

To ensure that the amplitude of the electromagnetic wave decreases with increasing z, we must find

$$n_2(\omega) \geq 0 \tag{2.244}$$

The z component of the Poynting vector in the material is given by

$$S_z = \frac{c}{4\pi} E_x (B_y - 4\pi M_y) \tag{2.245}$$

Both E_x and $B_y - 4\pi M_y$ must be real, however, for this expression to be valid.[17] So, we must express the electromagnetic fields in matter without using the complex notation. We write $t = |t|e^{i\theta}$ and rewrite the electromagnetic fields as follows:

$$E_x(z) = |t|\, e^{-kn_2 z} \cos(kn_1 z - \omega t + \theta) \tag{2.246}$$

$$(B_y - 4\pi M_y)(z) = |t|\, |n/\mu|\, e^{-kn_2 z} \cos(kn_1 z - \omega t + \phi + \theta) \tag{2.247}$$

where we have introduced the phase $\phi(\omega)$ by

$$\frac{n(\omega)}{\mu(\omega)} = \left|\frac{n(\omega)}{\mu(\omega)}\right| e^{i\phi(\omega)} \tag{2.248}$$

The Poynting vector $S_z(z)$ can then be simplified as

$$\begin{aligned} S_z(z) &= \frac{c}{4\pi} |n/\mu||t|^2 e^{-2kn_2 z} \cos(kn_1 z - \omega t + \theta)\cos(kn_1 z - \omega t + \phi + \theta) \\ &\longrightarrow \frac{c}{8\pi} |n/\mu||t|^2 e^{-2kn_2 z} \cos\phi = \frac{c}{8\pi} \Re(n/\mu)|t|^2 e^{-2kn_2 z} \end{aligned} \tag{2.249}$$

where we have taken the time average.

[17] In Exercise 5-4, we derive that the Poynting vector in a magnetic material is given by $\vec{S} = \frac{c}{4\pi}\vec{E} \times (\vec{B} - 4\pi \vec{M})$, but this expression holds even in the presence of polarization.

Therefore, the flow of energy entering the material at $z = 0$ is given by

$$S_{tr} = \frac{c}{8\pi}\Re(n/\mu) \cdot |t|^2 \tag{2.250}$$

We compare this with the energy flow from $z < 0$:

$$S_{in} = \frac{c}{8\pi} \tag{2.251}$$

Using $|t|^2$ from (2.223), we obtain the transmittance of energy as

$$\frac{S_{tr}}{S_{in}} = \Re(n/\mu) \cdot |t|^2 = \frac{4\Re(n/\mu)}{|1 + n/\mu(\omega)|^2} \tag{2.252}$$

Comparing this with the reflectance of energy in Eq. (2.223)

$$|r|^2 = \frac{|1 - n/\mu(\omega)|^2}{|1 + n/\mu(\omega)|^2} = 1 - \frac{4\Re(n/\mu)}{|1 + n/\mu(\omega)|^2} \tag{2.253}$$

we obtain

$$\Re(n/\mu) \cdot |t|^2 = 1 - |r|^2 \tag{2.254}$$

as expected.

Examples 5

1. **Classical model of electric conduction**—We have studied the free electron plasma as a classical model for metals. To take resistance into account, we modify the equation of motion of an electron under the electric field $\vec{E}$ by introducing a friction coefficient:

$$\ddot{\vec{x}} + \gamma\dot{\vec{x}} = \frac{-e}{m}\vec{E}$$

   (a) Assuming that electrons are uniformly distributed with a number density n, calculate the dielectric constant $\varepsilon(\omega)$ of this medium at angular frequency ω.

   The electrical conductivity σ is defined so that the current density $\vec{j}$ under a time-independent electric field $\vec{E}$ is given by

$$\vec{j} = \sigma\vec{E}$$

(b) Using the result of (a), determine σ in terms of the plasma angular frequency $\omega_{pl} \equiv \sqrt{\frac{4\pi ne^2}{m}}$ and γ.
(c) In the cgs-Gauss system, what are the units of σ? Confirm that the result of (b) has the correct dimension.
(d) Look up for the values of n and σ for Na and Cu, and estimate the plasma angular frequency ω_{pl} and friction coefficient γ, using the free electron mass for m. **Hint**: For the conversion between the cgs-Gauss and SI units, refer to Sect. 2.1.1 and the supplementary note 2.4.1. The mass density is usually given in units of g cm^{-3}, atomic weight in g mol^{-1}, and resistivity $\frac{1}{\sigma}$ in $\Omega\cdot$m. You need to use $\Omega \simeq \frac{1}{4\pi\epsilon_0}\frac{1}{9.0\times 10^{11}}$ cm^{-1}s.

2. **Charge screening**—Consider a medium with a dielectric constant given as a function of wave number k:

$$\varepsilon(k) = 1 + \frac{\kappa^2}{k^2}$$

where κ is a positive constant.

(a) Under the condition of a time-independent charge density $\rho_{ext}(\vec{x})$, show that the scalar potential satisfies the equation

$$(-\nabla^2 + \kappa^2)\phi(\vec{x}) = 4\pi\rho_{ext}(\vec{x})$$

(b) For a point charge $\rho_{ext}(\vec{x}) = e\delta^{(3)}(\vec{x})$, solve the above differential equation, compute the electric field $\vec{E}$, and sketch the results.

3. **Boundary conditions** —Consider a uniform medium in $z < 0$. The polarization vector $\vec{P}$ and magnetization vector $\vec{M}$ are assumed to be discontinuous on the xy plane at $z = 0$ as follows:

$$\vec{P} = \begin{cases} \vec{P}_0 & z < 0 \\ \vec{0} & z > 0 \end{cases} \qquad \vec{M} = \begin{cases} \vec{M}_0 & z < 0 \\ \vec{0} & z > 0 \end{cases}$$

Here, $\vec{P}_0$ and $\vec{M}_0$ are assumed to be constant vectors.

(a) Determine the charge density (per unit area) present on the surface at $z = 0$. **Hint**: Calculate the charge density $\rho = -\nabla \cdot \vec{P}$.
(b) Using Maxwell's equations

$$\begin{cases} \nabla \cdot \vec{E} = 4\pi\rho \\ \nabla \times \vec{E} + \frac{1}{c}\frac{\partial}{\partial t}\vec{B} = \vec{0} \end{cases}$$

derive the relation (i.e., the discontinuity) between $\vec{E}(z = 0-)$ and $\vec{E}(z = 0+)$.

Hint: Use Green's theorem in the tangential direction and Gauss's theorem in the perpendicular direction to the surface $z = 0$.

(c) Determine the current density (per unit length) at the surface $z = 0$.
Hint: Calculate the current density $\vec{j} = c\nabla \times \vec{M}$.

(d) Using Maxwell's equations

$$\begin{cases} \nabla \cdot \vec{B} = 0 \\ \nabla \times \vec{B} - \frac{1}{c}\frac{\partial}{\partial t}\vec{E} = \frac{4\pi}{c}\vec{j} \end{cases}$$

derive the relation (i.e., the discontinuity) between $\vec{B}(z = 0-)$ and $\vec{B}(z = 0+)$.
Hint: Use Green's theorem in the tangential direction and Gauss's theorem in the perpendicular direction.

4. **Magnetization and current**—We consider time-independent magnetization $\vec{M}(\vec{x})$ and electric current $\vec{j}(\vec{x})$ in a material. The current density is given by

$$\vec{j}(\vec{x}) = c\nabla \times \vec{M}(\vec{x})$$

which satisfies

$$\nabla \cdot \vec{j}(\vec{x}) = 0$$

(a) Let S be a surface within the material. Let C be its boundary. If the induced current I_S flowing through S is non-vanishing:

$$I_S \equiv \int_S \vec{j} \cdot d\vec{S} \neq 0$$

show that the magnetization vector $\vec{M}$ cannot vanish along C.

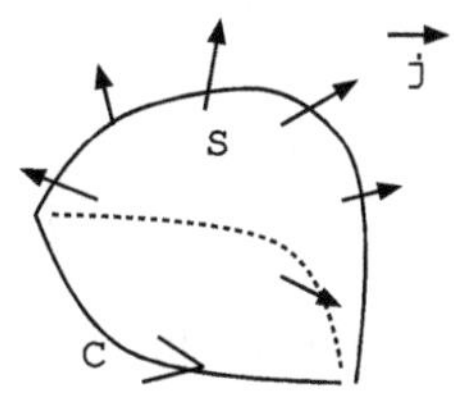

(b) Let $S = \{(x, y, 0) \mid x^2 + y^2 \leq a^2\}$ and $C = \{(x, y, 0) | x^2 + y^2 = a^2\}$. For $\vec{M}(x, y, 0) = \frac{f(r)}{r^2}(-y, x, 0)$, where $r \equiv \sqrt{x^2 + y^2}$, compute $\vec{j}$ and I_S.
Hint: You can use the result of (a) to calculate I_S.

5. **Meissner effect**—For type II superconductors, the reciprocal of the magnetic permeability is given by

$$\frac{1}{\mu(k)} = 1 + \frac{1}{k^2\lambda_L^2}$$

as a function of wave number k. Here, λ_L is known as **London's penetration depth**.
Assume that the superconductor occupies the half-space $x > 0$ separated from the vacuum at $x = 0$.

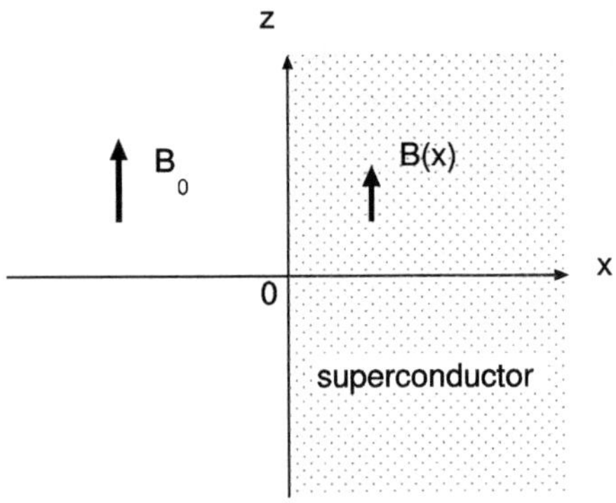

(a) Denoting the magnetization vector as $\vec{M}$, show that the given permeability implies

$$4\pi \nabla^2 \vec{M} = \frac{1}{\lambda_L^2} \vec{B}$$

(b) Assuming $\nabla \cdot \vec{M} = 0$, demonstrate that within the superconductor, the magnetic field satisfies London's differential equation:

$$\left(-\nabla^2 + \frac{1}{\lambda_L^2} \right) \vec{B} = \vec{0}$$

Hint: Take the curl of Ampère's law to obtain

$$\nabla \times \left(\nabla \times \vec{B} - \frac{4\pi}{c} \vec{j} \right) = \vec{0}$$

Use the result from (a) and the definition of the magnetization vector $\vec{j} = c\nabla \times \vec{M}$.

Now, consider a uniform magnetic field B_0 in the z direction in the region $x < 0$. Due to spatial symmetry, it is expected that a magnetic field $B(x)$ is induced in the z direction within the superconductor. Let us determine the x-dependence of this magnetic field.

(c) Assume that at the surface of the superconductor, $M_z(x = 0+) = 0$. Explain why the boundary condition $B(x = 0+) = B_0$ holds.
Hint: Use the results from Example 5-3.
(d) Use the above boundary condition to solve London's differential equation.
(e) Determine the current density and the magnetization M_z within the superconductor, and sketch the results.
Comment: From the above results, we conclude that even without applying an electric field, electric current flows in the superconductor. This implies zero electric resistance.

6. **Reflection from plasma**—Consider a free electron plasma with plasma angular frequency ω_{pl} occupying the half-space $z > 0$.[18]

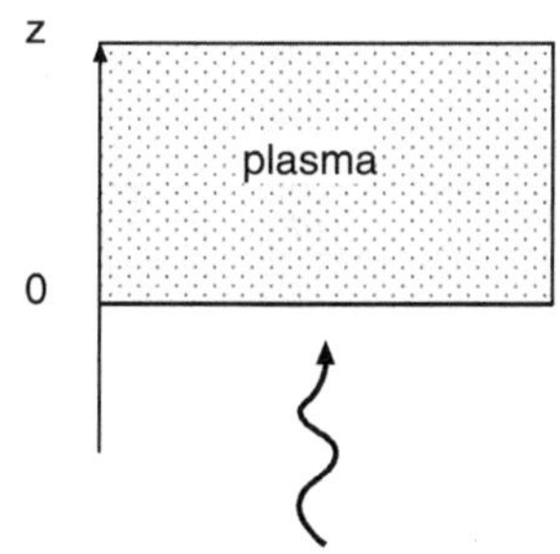

(a) Calculate the reflectivity of electromagnetic waves with angular frequency ω propagating in the $+z$ direction from the region $z < 0$.
Hint: Assume that the electric field of the electromagnetic wave is given as

$$E_x(z, t) = \mathrm{e}^{-i\omega t} \begin{cases} t(\omega)\mathrm{e}^{ik'z} & (z > 0) \\ \mathrm{e}^{ikz} + r(\omega)\mathrm{e}^{-ikz} & (z < 0) \end{cases}$$

Use the boundary conditions derived in Example 5-3 to determine $t(\omega)$ and $r(\omega)$.
(b) Use the above result to determine the maximum angular frequency at which total reflection occurs.
(c) For $n = 10^7\,\mathrm{cm}^{-3}$, calculate the maximum angular frequency obtained above.

[18] Think of the ionosphere at an altitude of 10 km above the Earth's surface.

Solutions to Examples 5

1. (a) For angular frequency ω, the equation of motion is

$$(-\omega^2 - i\omega\gamma)\vec{x} = -\frac{e}{m}\vec{E},$$

which gives

$$\vec{x} = \frac{e}{m}\frac{1}{\omega^2 + i\omega\gamma}\vec{E}$$

So, the polarization is

$$\vec{P} = -ne\vec{x} = -\frac{ne^2}{m}\frac{1}{\omega^2 + i\omega\gamma}\vec{E}$$

Therefore,

$$\varepsilon(\omega) - 1 = -\frac{4\pi ne^2}{m}\frac{1}{\omega^2 + i\omega\gamma} = -\frac{\omega_{pl}^2}{\omega^2 + i\omega\gamma}$$

$\varepsilon(\omega)$ has a pole at $\omega = -i\gamma$.

(b) At angular frequency ω,

$$\vec{j} = -i\omega\vec{P} = \frac{ne^2}{m}\frac{1}{-i\omega + \gamma}\vec{E}$$

In the limit $\omega \to 0$,

$$\vec{j} = \frac{\omega_{pl}^2}{4\pi\gamma}\vec{E} \Longrightarrow \boxed{\sigma = \frac{\omega_{pl}^2}{4\pi\gamma}}$$

(c) Since $[j] = \frac{\text{esu}}{\text{cm}^2\cdot\text{s}}$ and $[E] = \frac{\text{esu}}{\text{cm}^2}$, we obtain

$$[\sigma] = \frac{[j]}{[E]} = \text{s}^{-1},$$

which has the dimension of a frequency. ω_{pl} and γ also have the dimension s^{-1}. So, dimensionwise, the solution to (b) is correct.

(d) Assuming 1 conduction electron per atom for both Na and Cu, the electron number density n can be calculated from mass density ρ, atomic weight A, and Avogadro's number $N_A \simeq 6.0 \times 10^{23}$ /mol as follows:

$$\begin{cases} \rho_{\text{Na}} = 9.7 \times 10^2\,\text{kg/m}^3 \longrightarrow n_{\text{Na}} = 2.5 \times 10^{22}\,/\text{cm}^3 \\ \rho_{\text{Cu}} = 8.9 \times 10^3\,\text{kg/m}^3 \longrightarrow n_{\text{Cu}} = 8.3 \times 10^{22}\,/\text{cm}^3 \end{cases}$$

The resistance R_{SI} in the SI unit system is related to R in the cgs-Gauss system by

$$R_{\text{SI}} = \frac{1}{4\pi\epsilon_0} R$$

This gives

$$\Omega\,(\text{Ohm}) \simeq \frac{1}{4\pi\epsilon_0} \frac{1}{9.0 \times 10^{11}} \frac{\text{s}}{\text{cm}}$$

Hence,

$$\sigma_{\text{SI}} = 4\pi\epsilon_0\sigma$$

gives

$$\frac{1}{\Omega \cdot \text{m}} \simeq 4\pi\epsilon_0 \times \left(9.0 \times 10^9\right) \text{s}^{-1}$$

Converting to the cgs-Gauss system, we obtain

$$\begin{cases} \sigma_{\text{Na}} = 2.1 \times 10^7\,\Omega^{-1}\text{m}^{-1} \longrightarrow 1.9 \times 10^{17}\,\text{s}^{-1} \\ \sigma_{\text{Cu}} = 6.0 \times 10^7\,\Omega^{-1}\text{m}^{-1} \longrightarrow 5.4 \times 10^{17}\,\text{s}^{-1} \end{cases}$$

	$n[\text{cm}^{-3}]$	$\sigma[\text{s}^{-1}]$	$\omega_{pl}[\text{s}^{-1}]$	$\gamma[\text{s}^{-1}]$
Na	2.5×10^{22}	1.9×10^{17}	9.0×10^{15}	3.4×10^{13}
Cu	8.3×10^{22}	5.4×10^{17}	1.6×10^{16}	3.9×10^{13}

Clearly we find $\sigma \gg \omega_{pl} \gg \gamma$.

2. (a) Substituting $\vec{E}(\vec{k}) = -i\vec{k}\phi(\vec{k})$ into one of Maxwell's equations $i\vec{k} \cdot \varepsilon(k)\vec{E}(\vec{k}) = 4\pi\rho_{ext}(\vec{k})$, we obtain

$$k^2\varepsilon(k)\phi(\vec{k}) = 4\pi\rho_{ext}(\vec{k}) \Longrightarrow (k^2 + \kappa^2)\phi(\vec{k}) = 4\pi\rho_{ext}(\vec{k})$$

In real space, this gives

$$\left(-\nabla^2 + \kappa^2\right)\phi(\vec{x}) = 4\pi\rho_{ext}(\vec{x})$$

(b) Using

$$\phi(\vec{k}) = \frac{4\pi e}{k^2 + \kappa^2}$$

the inverse Fourier transformation gives

$$\phi(\vec{x}) = \int \frac{d^3k}{(2\pi)^3} \mathrm{e}^{i\vec{k}\cdot\vec{x}} \frac{4\pi e}{k^2 + \kappa^2} = e\frac{\mathrm{e}^{-\kappa r}}{r}$$

Therefore, the electric field is

$$\vec{E}(\vec{x}) = -\nabla\phi(\vec{x}) = \frac{e(1+\kappa r)\mathrm{e}^{-\kappa r}}{r^2} \longrightarrow \begin{cases} \frac{e}{r^2} & (kr \ll 1) \\ \frac{e\kappa}{r}\mathrm{e}^{-\kappa r} & (kr \gg 1) \end{cases}$$

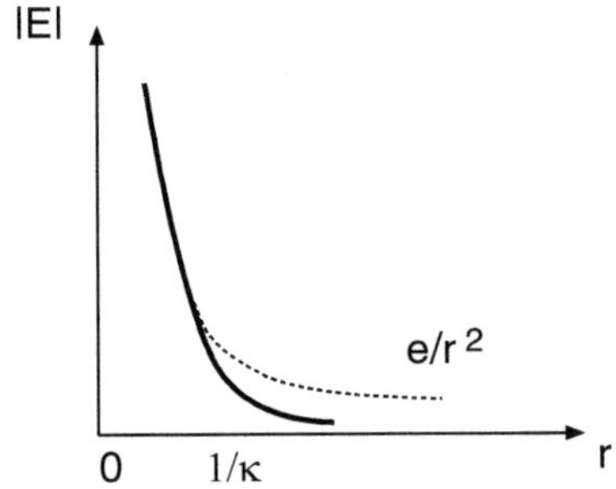

3. (a) From $\vec{P}(\vec{x}) = \vec{P}_0\theta(-z)$, we obtain

$$\rho(\vec{x}) = -\nabla \cdot \vec{P}(\vec{x}) = -P_{0,z}\frac{d}{dz}\theta(-z) = P_{0,z}\delta(z)$$

The surface charge density is $P_{0,z}$.

(b) $E_z(z = 0+) - E_z(z = 0-) = 4\pi P_{0,z}$. Therefore, $D_z = E_z + 4\pi P_z$ is continuous. On the other hand, $E_{x,y}$ is continuous at $z = 0$.

(c) From $\vec{M}(\vec{x}) = \vec{M}_0\theta(-z)$, we obtain

$$\vec{j}(\vec{x}) = \nabla \times \vec{M}(\vec{x}) = \begin{pmatrix} -\partial_z M_{0,y}\theta(-z) \\ \partial_z M_{0,x}\theta(-z) \\ 0 \end{pmatrix} = \delta(z)\begin{pmatrix} M_{0,y} \\ -M_{0,x} \\ 0 \end{pmatrix} = \delta(z)\vec{M}_0 \times \hat{z}$$

(d) B_z is continuous. But

$$B_x(z=0+) = B_x(z=0-) - 4\pi M_{0,x}$$
$$B_y(z=0+) = B_y(z=0-) - 4\pi M_{0,y}$$

Therefore, $H_x = B_x - 4\pi M_x$ and $H_y = B_y - 4\pi M_y$ are continuous.

4. (a) This is nothing more than the Ampère law, since $\vec{B} = 4\pi \vec{M}$ inside the material in the absence of an external current. We find

$$I_S = \int_S \vec{j} \cdot d\vec{S} = c \int_S \nabla \times \vec{M} \cdot d\vec{S}$$
$$= c \oint_C \vec{M} \cdot d\vec{x}$$

Hence, for $I_S \neq 0$, we must find $\vec{M}$ non-vanishing along C.

(b) We find

$$\vec{j} = c\nabla \times \vec{M} = c\left(0, 0, \frac{f'(r)}{r}\right)$$
$$I_S = c \oint_C \vec{M} \cdot d\vec{x} = 2\pi c f(a)$$

5. (a) The Fourier transform of $\vec{M}$ is given by

$$4\pi \tilde{\vec{M}}(\vec{k}) = \left(1 - \frac{1}{\mu(k)}\right) \tilde{\vec{B}}(\vec{k}) = -\frac{1}{k^2\lambda_L^2} \tilde{\vec{B}}(\vec{k})$$

Hence, we obtain

$$-k^2 4\pi \tilde{\vec{M}}(\vec{k}) = \frac{1}{\lambda_L^2} \tilde{\vec{B}}(\vec{k})$$

Taking the inverse transform, we obtain

$$\nabla^2 4\pi \vec{M}(\vec{x}) = \frac{1}{\lambda_L^2} \vec{B}(\vec{x})$$

(b)

$$\nabla \times \left(\nabla \times \vec{B}\right) = -\nabla^2 \vec{B}$$
$$\nabla \times \frac{4\pi}{c}\vec{j} = \nabla \times \left(\nabla \times 4\pi \vec{M}\right) = -4\pi \nabla^2 \vec{M} = -\frac{1}{\lambda_L^2} \vec{B}$$

From this, we obtain $(-\nabla^2 + 1/\lambda_L^2)\vec{B} = 0$.

(c) Since $B_z(x=0-) = B_z(x=0+) - 4\pi M_z(x=0+)$ and $M_z(x=0+) = 0$, we obtain $B_z(x=0+) = B_z(x=0-) = B_0$.

(d) $B_z(x)$ satisfies London's differential equation $\frac{d^2}{dx^2}B_z(x) = \frac{1}{\lambda_L^2}B_z(x)$ for $x > 0$. The solution is

$$B_z(x) = B_0 \mathrm{e}^{-\frac{x}{\lambda_L}}$$

(e) The current density is

$$j_y(x) = -\frac{c}{4\pi}\frac{d}{dx}B_z(x) = \frac{c}{4\pi\lambda_L}B_0\mathrm{e}^{-\frac{x}{\lambda_L}}$$

Near the surface, a current flows to ensure that there is no magnetic field inside the material. This is a characteristic of superconductors, where an electric current can flow in the absence of an electric field. On the other hand, a magnetization $M_z(x)$ can be obtained by integrating $j_y = -c\frac{d}{dx}M_z$ to get

$$M_z(x) = -\frac{1}{4\pi}B_0\left(1 - \mathrm{e}^{-\frac{x}{\lambda_L}}\right)$$

Hence, $B_z(x) - 4\pi M_z(x) = B_0$ does not depend on x.

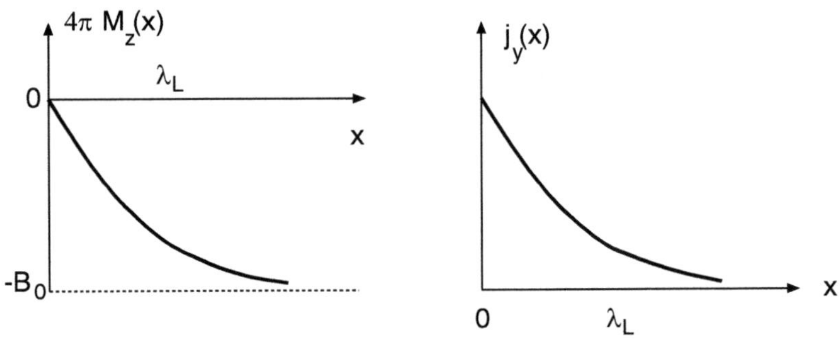

6. (a) The electric and magnetic fields are given in the form:

$$E_x(z,t) = \mathrm{e}^{-i\omega t}\begin{cases} t(\omega)\mathrm{e}^{ik'z} & (z>0) \\ \mathrm{e}^{ikz} + r(\omega)\mathrm{e}^{-ikz} & (z<0) \end{cases}$$

$$B_y(z,t) = \mathrm{e}^{-i\omega t}\begin{cases} t(\omega)n(\omega)\mathrm{e}^{ik'z} & (z>0) \\ \mathrm{e}^{ikz} - r(\omega)\mathrm{e}^{-ikz} & (z<0) \end{cases}$$

where

$$n(\omega) = \sqrt{\varepsilon(\omega)} = \sqrt{1 - \frac{\omega_{pl}^2}{\omega^2}}, \quad k'c = n(\omega)\omega$$

For $\omega < \omega_{pl}$, $\Im n(\omega) > 0$, i.e., the electromagnetic field decays exponentially for $z > 0$. Since E_x and B_y must be continuous at $z = 0$, we find

$$t(\omega) = 1 + r(\omega), \quad t(\omega)n(\omega) = 1 - r(\omega)$$

Therefore, we obtain

$$\begin{cases} t(\omega) &= \frac{2}{1+n(\omega)} = \frac{2}{1+\sqrt{1-\frac{\omega_{pl}^2}{\omega^2}}} \\ r(\omega) &= \frac{1-n(\omega)}{1+n(\omega)} = \frac{1-\sqrt{1-\frac{\omega_{pl}^2}{\omega^2}}}{1+\sqrt{1-\frac{\omega_{pl}^2}{\omega^2}}} \end{cases}$$

Hence, the reflectance is given by

$$|r(\omega)|^2 = \begin{cases} 1 & (\omega < \omega_{pl}) \\ \left| \frac{1-\sqrt{1-\frac{\omega_{pl}^2}{\omega^2}}}{1+\sqrt{1-\frac{\omega_{pl}^2}{\omega^2}}} \right|^2 & (\omega > \omega_{pl}) \end{cases}$$

For $\omega < \omega_{pl}$, electromagnetic waves are completely reflected because they cannot propagate in the plasma.

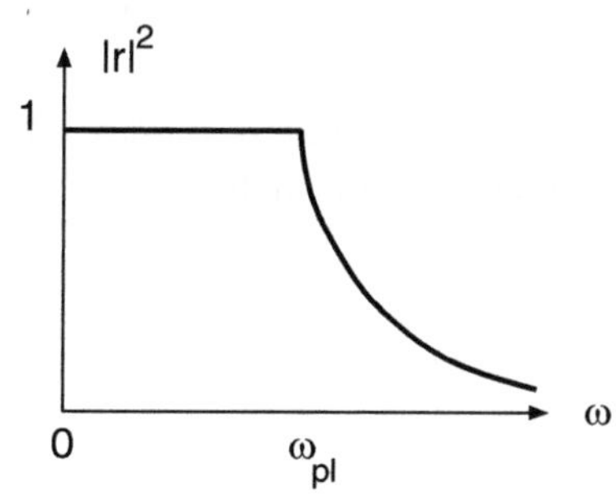

(b) ω_{pl}
(c) For $n = 10^7\,\text{cm}^{-3}$ we obtain

$$\omega_{pl}^2 = \frac{4\pi \cdot 10^7\,\text{cm}^{-3} \cdot (4.8 \times 10^{-10}\,\text{esu})^2}{9.1 \times 10^{-28}\,\text{g}} \simeq 3.2 \times 10^{16}\,\text{s}^{-2}$$

This gives

$$\omega_{pl} \simeq \boxed{1.8 \times 10^8 \text{ s}^{-1}}$$

Exercises 5

1. **Classical model of electric conductors**—Let us consider the dielectric constant obtained in Example 5-1:

$$\varepsilon(\omega) = 1 - \frac{\omega_{pl}^2}{\omega^2 + i\omega\gamma}$$

Here, ω_{pl} is a plasma angular frequency, and γ is a friction coefficient.

(a) Show that for $\omega \ll \gamma$ the dielectric constant above can be approximated as

$$\varepsilon(\omega) = 1 + i\frac{4\pi\sigma}{\omega}$$

where we define the conductivity as $\sigma \equiv \frac{\omega_{pl}^2}{4\pi\gamma}$.

(b) Using the above dielectric constant, determine the dispersion relation between the wave number k and angular frequency ω.

(c) Given this conductor occupying the region $z > 0$, send light with angular frequency ω in the $+z$ direction from the vacuum in $z < 0$. How deep does this light penetrate the conductor? Assume $\omega \ll \gamma$.

(d) Using the electric conductivity σ for Cu, plot the penetration length as a function of ω.

2. **Reflectance**—Show that the refractive index $n(\omega)$ and the square of the reflectance $|r(\omega)|^2$ for a material with dielectric constant $\varepsilon(\omega) = \varepsilon_1(\omega) + i\varepsilon_2(\omega)$ are given by

$$n(\omega) = \sqrt{\varepsilon_1^2 + \varepsilon_2^2}$$

$$|r(\omega)|^2 = \frac{1 + n - \sqrt{2(n + \varepsilon_1)}}{1 + n + \sqrt{2(n + \varepsilon_1)}}$$

Apply this result to the classical model of electric conductors discussed in 1 above to determine $|r(\omega)|^2$, and plot it.

Hint – Use $\frac{\omega}{\omega_{pl}}$ as the x-axis. You may consider using the value of $\frac{\gamma}{\omega_{pl}}$ for Cu.

3. **Classical model of dielectrics**—As a classical model of dielectrics (insulators), consider a medium with uniformly distributed electrons following the equation

of motion of a damping harmonic oscillator:

$$\ddot{\vec{x}} + \gamma\dot{\vec{x}} + \Omega^2\vec{x} = \frac{-e}{m}\vec{E}$$

The difference from Example 5-1 is that the electrons are bound in this model.

(a) Find the dielectric constant $\varepsilon(\omega)$ as a function of the angular frequency ω.
(b) Sketch the ω dependence of the real part ε_1 and the imaginary part ε_2 of $\varepsilon(\omega)$.
(c) Determine the constant A characterizing the asymptotic behavior of $\varepsilon_1(\omega)$ as $\omega \to \infty$:

$$\varepsilon_1(\omega) \stackrel{\omega\to\infty}{\longrightarrow} 1 - \frac{A}{\omega^2}$$

(d) Calculate the square of the reflectance $|r(\omega)|^2$ using the results from (b) above, and plot the results.
Hint – It is advisable to use $\frac{\omega}{\Omega}$ as the x-axis. It may be interesting to explore various values for $\frac{\omega_{pl}}{\Omega}$ and $\frac{\gamma}{\Omega}$.
(e) Show that the imaginary part $\varepsilon_2(\omega)$ obtained in (b) satisfies the sum rule:

$$\int_0^\infty \frac{d\omega}{\pi} 2\omega\varepsilon_2(\omega) = A$$

4. **Energy due to magnetization**—Consider a homogeneous magnetic material with the magnetization vector given by $\vec{M}(\vec{x}, t)$. Given an external current density $\vec{j}_{ext}$, Maxwell's equations are given as follows:

$$\nabla \cdot \vec{E} = 0$$

$$\nabla \times \vec{B} - \frac{1}{c}\partial_t \vec{E} = 4\pi \left(\nabla \times \vec{M} + \frac{1}{c}\vec{j}_{ext}\right)$$

$$\nabla \cdot \vec{B} = 0$$

$$\nabla \times \vec{E} + \frac{1}{c}\partial_t \vec{B} = \vec{0}$$

(a) Using the above Maxwell equations, derive

$$\frac{1}{4\pi}\partial_t \vec{E} \cdot \vec{E} + \frac{1}{4\pi}\partial_t \vec{B} \cdot (\vec{B} - 4\pi\vec{M}) = -\nabla \cdot \vec{S} - \vec{E} \cdot \vec{j}_{ext}$$

where

$$\vec{S} \equiv \frac{c}{4\pi}\vec{E} \times (\vec{B} - 4\pi\vec{M})$$

(b) Show that on the boundary of the magnetic material, only the normal component of $\vec{S}$ is continuous.
Hint: Use the result from Example 5-3.

We define the electric energy density as

$$\mathcal{E}_e \equiv \frac{1}{8\pi}\vec{E}^2$$

The magnetic energy density $\mathcal{E}_m$ is defined by its small change under a small change $\delta\vec{B}$ of the magnetic field as[19]

$$\boxed{\delta\mathcal{E}_m = \frac{1}{4\pi}\delta\vec{B} \cdot (\vec{B} - 4\pi\vec{M})}$$

(c) Assuming that the permeability μ is a constant, express $\mathcal{E}_m$ in terms of the magnetic field $\vec{B}$.

The equation from part (a) can be rewritten as

$$\partial_t(\mathcal{E}_e + \mathcal{E}_m) = -\nabla \cdot \vec{S} - \vec{E} \cdot \vec{j}_{ext}$$

(d) Interpret this result physically, and explain why the boundary conditions obtained in Example 5-3 are valid.

5. **Superconducting wire**—Consider a wire with radius a made of a superconductor. The permeability is given as a function of wave number k as

$$\frac{1}{\mu(k)} = 1 + \frac{1}{k^2\lambda_L^2}$$

where λ_L is the London penetration depth. Due to the axial symmetry, the current flows only in the direction of the wire (taken as the z-axis). The magnetic field vector can be expressed as a function of the radial coordinate $r \equiv \sqrt{x^2 + y^2}$:

$$\vec{B}(x, y, z) = \frac{B(r)}{r}(-y, x, 0)$$

[19] We will apply this result later in thermodynamics.

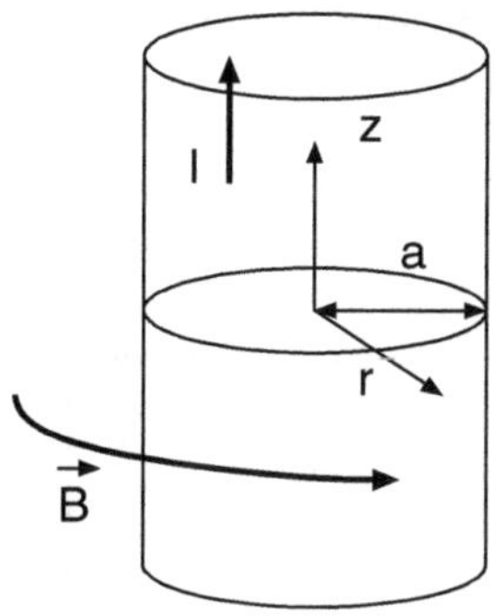

(a) Determine $B(a+)$ when the current flowing through the entire wire is denoted as I.
Hint: Use Ampère law.
(b) Solve for $B(r)$ by assuming continuity at $r = a$ and using the differential equation derived in Example 5-5.
Hint: You will obtain the differential equation for the modified Bessel functions $I_1(r/\lambda_L)$, $K_1(r/\lambda_L)$. The former vanishes at $r = 0$, and the latter diverges.
(c) Find the current density $\vec{j}$ within the wire.
Hint: Use Ampère's law to derive it from $\vec{B}$.
(d) Show that in the limit as $a \to \infty$, the magnetic field and current density obtained in Example 5-5 are recovered, where $B_0 = B(a)$.

Comment: Once a current flows through the wire and generates a magnetic field, a current is induced near the surface of the wire to prevent the formation of a magnetic field inside the wire (Meissner effect). This state is maintained. Therefore, no electric field is required to maintain the electric current.

(e) Find the magnetization vector $\vec{M}(\vec{r})$ for $r < a$ and $r > a$.
Hint: See Example 5-4. Inside the wire you have $\vec{B} = 4\pi\vec{M}$. Outside you must find $\vec{j} = c\nabla \times \vec{M} = 0$. You obtain

$$4\pi\vec{M}(\vec{r}) = \frac{B(r)}{r}(-y, x, 0) \qquad (r < a)$$

$$= \frac{aB(a)}{r^2}(-y, x, 0) \qquad (r > a)$$

The magnetization must be continuous at $r = a$, and as a consequence it becomes non-vanishing even at $r > a$.

6. **Cylindrical superconductor**—Similar to the previous problem, we consider an infinitely long superconductor with radius a. This time we apply a magnetic field B_0 parallel to the axis of the cylinder. Due to the axial symmetry, both the

magnetic field and magnetization vector inside the superconductor are assumed to have only the z-component.

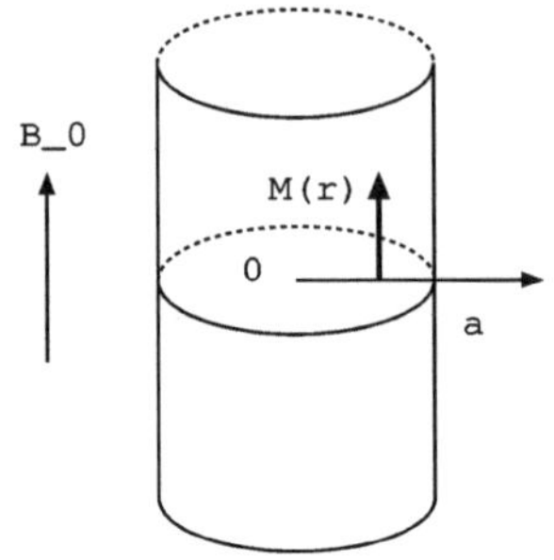

$$\vec{B} = (0, 0, B(r))$$

$$\vec{M} = (0, 0, M(r)) \qquad (0 < r \equiv \sqrt{x^2 + y^2} < a)$$

(a) Solve the differential equation for the magnetic field

$$\nabla^2 \vec{B}(\vec{r}) = \frac{1}{\lambda_L^2} \vec{B}(\vec{r})$$

under the boundary condition $B(r = a) = B_0$.

(b) Determine the current density within the superconductor.
Hint: $\vec{j} = \frac{c}{4\pi} \nabla \times \vec{B}$

(c) Find the magnetization $M(r)$ and sketch it.
Hint: Use $\nabla \times (\vec{B} - 4\pi \vec{M}) = \vec{0}$ to find it. The boundary condition $M(a) = 0$ gives $4\pi M(r) = B(r) - B_0$.

7. **Brewster's Angle**—Consider an infinite half-space $z < 0$ filled with a continuous medium A with refractive index n_1 and the remaining half-space $z > 0$ filled with a continuous medium B with refractive index n_2. We wish to consider the reflection and refraction of light with angular frequency ω sent from A to B. Assume that the incident light travels along the zx plane and the angle of incidence is θ_i. It is assumed that the y-component of the electric field is zero. Let θ_r be the angle of reflection and θ_t be the angle of refraction as shown in the figure below.

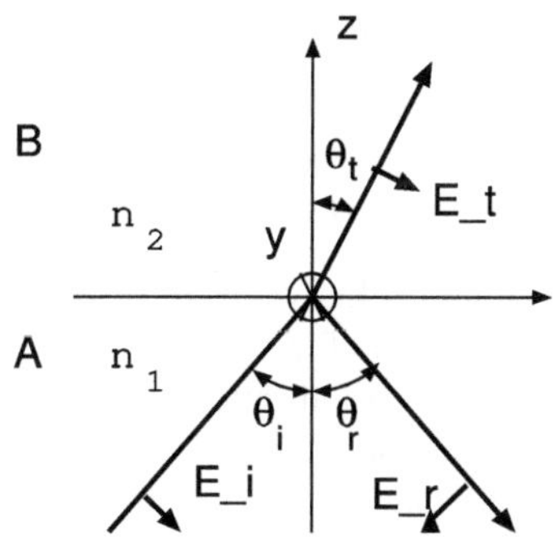

(a) Let E_i, E_r, E_t be the magnitudes of the electric fields for the incident, reflected, and refracted waves. Explain that the electric fields can be expressed as follows:

$$E_x(\vec{x}, t) = e^{-i\omega t} \times \begin{cases} E_i \cos\theta_i e^{in_1 k\hat{i}\cdot\vec{x}} & (z < 0) \\ -E_r \cos\theta_r e^{in_1 k\hat{r}\cdot\vec{x}} & (z < 0) \\ E_t \cos\theta_t e^{in_2 k\hat{t}\cdot\vec{x}} & (z > 0) \end{cases}$$

$$E_z(\vec{x}, t) = e^{-i\omega t} \times \begin{cases} -E_i \sin\theta_i e^{in_1 k\hat{i}\cdot\vec{x}} & (z < 0) \\ -E_r \sin\theta_r e^{in_1 k\hat{r}\cdot\vec{x}} & (z < 0) \\ -E_t \sin\theta_t e^{in_2 k\hat{t}\cdot\vec{x}} & (z > 0) \end{cases}$$

where $k \equiv \frac{\omega}{c}$, and

$$\hat{i} \equiv \sin\theta_i \hat{x} + \cos\theta_i \hat{z}$$

$$\hat{r} \equiv \sin\theta_r \hat{x} - \cos\theta_r \hat{z}$$

$$\hat{t} \equiv \sin\theta_t \hat{x} + \cos\theta_t \hat{z}$$

(b) Express the magnetic field in a similar manner.

(c) Using the boundary conditions at $z = 0$ (see Example 5-3), derive Snell's law:

$$\theta_i = \theta_r, \quad n_1 \sin\theta_i = n_2 \sin\theta_t$$

(d) Similarly, derive Snell's relations:

$$E_r = \frac{n_2 \cos\theta_i - n_1 \cos\theta_t}{n_1 \cos\theta_t + n_2 \cos\theta_i} E_i$$

$$E_t = \frac{2n_1 \cos\theta_i}{n_1 \cos\theta_t + n_2 \cos\theta_i} E_i$$

(e) Using Snell's law and relations, demonstrate that for $E_r = 0$

$$\boxed{\theta_i + \theta_t = \frac{\pi}{2}}$$

has to be satisfied. This θ_i is known as the **Brewster angle**.

(f) Determine the Brewster angle for $n_1 = 1,\ n_2 = 1.3$(water).

(g) Determine the condition for total reflection.
Hint: Find the condition for the relative size of n_1 and n_2, as well as the range of θ_i where Snell's law has no solution.
Note: Total reflection does not imply the complete absence of electromagnetic fields in $z > 0$; instead the fields decay exponentially as z increases.

2.4 Supplementary Notes

2.4.1 *Relation Between the SI and cgs-Gauss Units for Electromagnetism*

We start with tables of the SI and cgs-Gauss units (Tables 2.1 and 2.2)

To figure out the relation between the two systems of electromagnetic units, we recall the following three laws:

1. Coulomb's law

$$F = \frac{q^2}{4\pi\epsilon_0 r^2} = \frac{q_G^2}{r^2} \tag{2.255}$$

Table 2.1 SI electromagnetic units; note that $\frac{\mu_0}{4\pi}$ is a measured quantity

Physical quantity	Symbol	Unit
Electric field	E	$\frac{\text{volt}}{\text{m}} = \frac{\text{N}}{\text{A·s}}$
Magnetic field	B	$\text{Tesla} = \frac{\text{N}}{\text{A·m}}$
Electric charge	Q	$\text{C} = \text{A} \cdot \text{s}$
Electric current	I	A
Voltage	V	$\text{volt} = \frac{\text{J}}{\text{A·s}}$
Inductance	L	$\text{Henry} = \frac{\text{J}}{\text{A}^2}$
Resistance	R	$\text{Ohm} = \frac{\text{J}}{\text{A}^2\text{·s}}$
Capacitance	C	$\text{Farad} = \frac{\text{A}^2\text{·s}^2}{\text{J}}$
Vacuum permeability	$\frac{\mu_0}{4\pi}$	$\simeq 1.000000000 \times 10^{-7}\ \frac{\text{N}}{\text{A}^2}$
Vacuum permittivity	$\frac{1}{4\pi\epsilon_0}$	$= \frac{\mu_0}{4\pi}c^2 \simeq 9.0 \times 10^9\ \frac{\text{N·m}^2}{(\text{A·s})^2}$

Table 2.2 cgs-Gauss electromagnetic units

Physical quantity	Symbol	Unit
Electric field	E_G	$\frac{\text{esu}}{\text{cm}^2} = \frac{\text{statV}}{\text{cm}}$
Magnetic field	B_G	$\text{gauss} = \frac{\text{esu}}{\text{cm}^2}$
Electric charge	Q_G	$\text{esu} \equiv \text{dyn}^{\frac{1}{2}}\ \text{cm}$
Electric current	I_G	$\text{statA} \equiv \frac{\text{esu}}{\text{s}}$
Voltage	V_G	$\text{statV} \equiv \frac{\text{esu}}{\text{cm}}$
Inductance	L_G	$\frac{\text{s}^2}{\text{cm}}$
Resistance	R_G	$\frac{s}{\text{cm}}$
Capacitance	C_G	cm

2. Lorentz force law

$$F = q\,(E + vB) = q_G\left(E_G + \frac{v}{c}B_G\right) \tag{2.256}$$

3. Kirchhoff's law

$$V = L\frac{dI}{dt} + RI + \frac{1}{C}\int^t dt\, I\,, \tag{2.257a}$$

$$V_G = L_G\frac{dI_G}{dt} + R_G I_G + \frac{1}{C_G}\int^t dt\, I_G\,. \tag{2.257b}$$

We find the following relations:

$$q_G = \frac{q}{\sqrt{4\pi\epsilon_0}} \tag{2.258a}$$

$$I_G = \frac{I}{\sqrt{4\pi\epsilon_0}} \tag{2.258b}$$

$$E_G = \sqrt{4\pi\epsilon_0}\,E \tag{2.258c}$$

$$B_G = \sqrt{\frac{4\pi}{\mu_0}}\,B \tag{2.258d}$$

$$V_G = \sqrt{4\pi\epsilon_0}\,V \tag{2.258e}$$

$$L_G = 4\pi\epsilon_0 L \tag{2.258f}$$

$$R_G = 4\pi\epsilon_0 R \tag{2.258g}$$

$$C_G = \frac{1}{4\pi\epsilon_0}C \tag{2.258h}$$

These translate into the relation among the units:

$$\text{esu} = \frac{1}{\sqrt{4\pi\epsilon_0}} \frac{1}{3.0 \times 10^9} \text{C} \tag{2.259a}$$

$$\text{statA} = \frac{1}{\sqrt{4\pi\epsilon_0}} \frac{1}{3.0 \times 10^9} \text{A} \tag{2.259b}$$

$$\frac{\text{statV}}{\text{cm}} = \sqrt{4\pi\epsilon_0}(3.0 \times 10^4) \frac{\text{volt}}{\text{m}} \tag{2.259c}$$

$$\text{gauss} = \sqrt{\frac{4\pi}{\mu_0}}(1.0 \times 10^{-4})\,\text{T} \tag{2.259d}$$

$$\text{statV} = \sqrt{4\pi\epsilon_0}(3.0 \times 10^2)\,\text{volt} \tag{2.259e}$$

$$\frac{\text{s}}{\text{cm}} = 4\pi\epsilon_0(9.0 \times 10^{11})\,\Omega \tag{2.259f}$$

$$\frac{\text{s}^2}{\text{cm}} = 4\pi\epsilon_0(9.0 \times 10^{11})\,\text{H} \tag{2.259g}$$

$$\text{cm} = \frac{1}{4\pi\epsilon_0} \frac{1}{9.0 \times 10^{11}} \text{F} \tag{2.259h}$$

Note that 3.0 in the above is approximate. It actually equals

$$\frac{\frac{\mu_0}{4\pi}}{10^{-7}\,\text{N}\,\text{A}^{-2}} \times \frac{c}{\text{m}\,\text{s}^{-1}} = 1.000000000 \times 2.99792458 \simeq 3.0$$

Since μ_0 is measured only up to 9 digits, the value is only approximate. The same goes with 1.0 and 9.0; they are approximate.

It is common to come across with such a relation as[20]

$$1\,\text{esu} = \frac{1}{3.0 \times 10^9}\,\text{C}$$

Strictly speaking, esu and C have different dimensions, and it is more correct to write (2.259a) using $\sqrt{4\pi\epsilon_0}$.

All the equations in (2.259) give $4\pi\epsilon_0$ (or $\mu_0/(4\pi)$) in various units. For example, the first equation gives

$$\sqrt{4\pi\epsilon_0} = \frac{1}{3.0 \times 10^9} \frac{\text{C}}{\text{esu}}$$

[20] For example, you find this on p. 339 of M. Schwartz, *Principles of Electrodynamics* (Dover).

and the last one gives

$$4\pi\epsilon_0 = \frac{1}{9.0 \times 10^{11}} \frac{\text{F}}{\text{cm}}$$

2.4.2 *Line Integrals and Surface Integrals*

The concepts of line integrals and surface integrals are well-known in electromagnetism, but their abstract notations, though giving a vague sense of understanding, often fail to tell us how to compute. Here, we define the integrals concretely using parametric representations of curves and surfaces.

Let us consider integrals over curves and surfaces in three-dimensional Euclidean space. We will use Cartesian coordinates x, y, and z for the three-dimensional space.

Curves

A curve is a one-dimensional subspace and can be described by a single parameter. Let us denote this parameter as t. Therefore, a curve can be determined by specifying the coordinates using three functions of t as follows:

$$\vec{x}(t) = (x(t), y(t), z(t)) \qquad 0 \le t \le t_{\max} \tag{2.260}$$

(Fig. 2.18) We will refer to this curve as C. It is oriented toward increasing t. The line element is given by

$$d\vec{x}(t) = dt \left(\frac{dx(t)}{dt}, \frac{dy(t)}{dt}, \frac{dz(t)}{dt} \right) \tag{2.261}$$

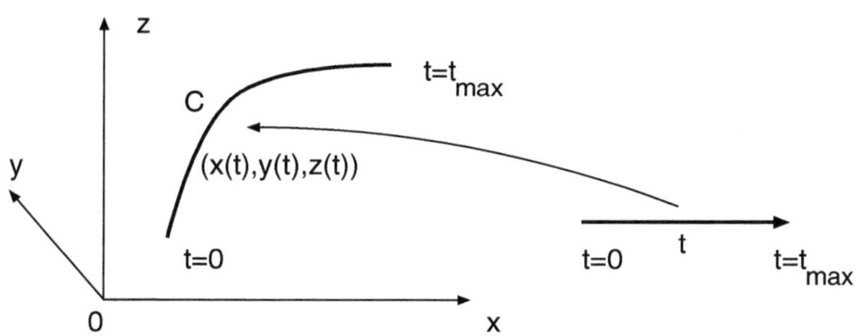

Fig. 2.18 Parametric representation of a curve

Hence, the line integral of a vector field $\vec{v}(\vec{x})$ along the curve C is defined as

$$\int_C d\vec{x} \cdot \vec{v} \equiv \int_0^{t_{\max}} dt \frac{d\vec{x}}{dt} \cdot \vec{v}(\vec{x}(t))$$
$$= \int_0^{t_{\max}} dt \left\{ \frac{dx(t)}{dt} v_1(x(t), y(t), z(t)) + \frac{dy(t)}{dt} v_2(x(t), y(t), z(t)) \right.$$
$$\left. + \frac{dz(t)}{dt} v_3(x(t), y(t), z(t)) \right\} \tag{2.262}$$

There are infinitely many ways of parametrize a curve. Let us show that the value of the integral is independent of the choice of parametrization. Consider using a different parameter $\tilde{t}(t)$ to define the integral:

$$\int_C \vec{v} \cdot d\vec{x} \equiv \int_0^{\tilde{t}_{\max}} d\tilde{t} \frac{d\vec{x}(\tilde{t})}{d\tilde{t}} \cdot \vec{v}(\vec{x}(\tilde{t})) \tag{2.263}$$

Since $\tilde{t}$ is an increasing function of the original parameter t, the integral becomes

$$= \int_0^{t_{\max}} \left(\frac{d\tilde{t}(t)}{dt} dt \right) \left(\frac{d\vec{x}}{dt} \frac{1}{\frac{d\tilde{t}}{dt}} \right) \cdot \vec{v}(\vec{x}(\tilde{t}(t))) = \int_0^{t_{\max}} dt \frac{d\vec{x}}{dt} \cdot \vec{v}(\vec{x}(t)) \tag{2.264}$$

Thus, the integral is independent of the choice of parametrization.

Example: Circle

Let us parameterize a circle C of radius r with a parameter $0 \le t \le 1$:

$$x(t) = r \cos 2\pi t, \quad y(t) = r \sin 2\pi t \tag{2.265}$$

Given a vector field in the xy plane

$$v_x(x, y) = y, \quad v_y(x, y) = -x \tag{2.266}$$

we obtain the line integral along C as

$$\oint_C \vec{v} \cdot d\vec{x} \equiv \int_0^1 dt \left(\frac{dx(t)}{dt} v_x(x(t), y(t)) + \frac{dy(t)}{dt} v_y(x(t), y(t)) \right)$$
$$= \int_0^1 dt \, (-2\pi r \sin 2\pi t \cdot r \sin 2\pi t + 2\pi r \cos 2\pi t \cdot (-r \cos 2\pi t))$$
$$= -2\pi r^2 \int_0^1 dt = -2\pi r^2$$

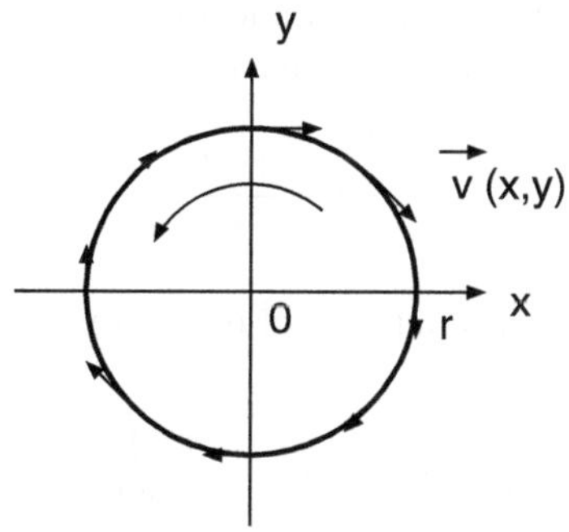

Fig. 2.19 Integral of $\vec{v}(x, y)$ along a counterclockwise circle of radius r

See Fig. 2.19.

Surfaces

A surface is a two-dimensional subspace and can be described by two parameters. Let us denote these parameters as s and t. Then a surface can be determined by specifying the coordinates using three functions of s and t as follows:

$$\vec{x}(s,t) = (x(s,t), y(s,t), z(s,t)) \qquad 0 \le s \le s_{\max}, \quad 0 \le t \le t_{\max} \tag{2.267}$$

We will refer to this surface as S (Fig. 2.20). The area element is given by the vector product:

$$d\vec{S} \equiv \frac{\partial \vec{x}(s,t)}{\partial s} \times \frac{\partial \vec{x}(s,t)}{\partial t} dsdt \tag{2.268}$$

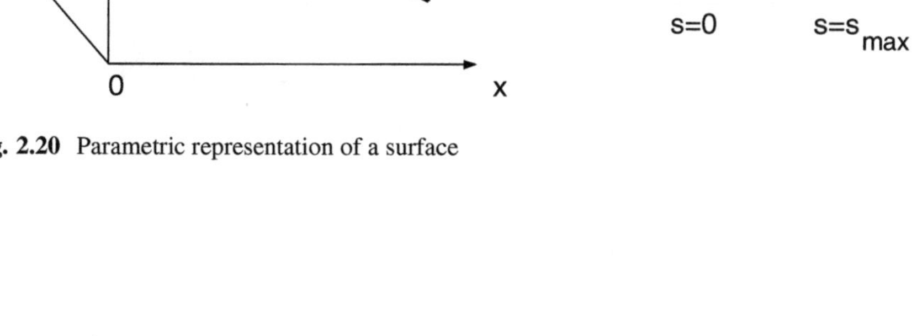

Fig. 2.20 Parametric representation of a surface

This is an infinitesimal vector orthogonal to the surface. For example, its x component is given by

$$\left(\frac{\partial y(s,t)}{\partial s}\cdot\frac{\partial z(s,t)}{\partial t}-\frac{\partial y(s,t)}{\partial t}\cdot\frac{\partial z(s,t)}{\partial s}\right)dsdt \tag{2.269}$$

The integral of the vector field $\vec{v}(\vec{x})$ over the surface S is defined as

$$\begin{aligned}\int_S \vec{v}\cdot d\vec{S} &\equiv \int_0^{s_{max}} ds \int_0^{t_{max}} dt\, \vec{v}(\vec{x})\cdot\left(\frac{\partial\vec{x}(s,t)}{\partial s}\times\frac{\partial\vec{x}(s,t)}{\partial t}\right)\\ &= \int_0^{s_{max}} ds \int_0^{t_{max}} dt\\ &\cdot\left[v_1(x(s,t),y(s,t),z(s,t))\left\{\frac{\partial y(s,t)}{\partial s}\frac{\partial z(s,t)}{\partial t}-\frac{\partial y(s,t)}{\partial t}\frac{\partial z(s,t)}{\partial s}\right\}\right.\\ &\left.\quad+\cdots\right]\end{aligned} \tag{2.270}$$

Like line integrals, we can show that this surface integral is independent of the choice of parameters. We omit a proof.

Example: Torus

A two-dimensional torus is represented by two parameters s and t ($0 \le s \le 1$, $0 \le t \le 1$), where we identify the edges at $s = 0$ and $s = 1$ and also the edges at $t = 0$ and $t = 1$. (See the left side of Fig. 2.21.) A two-dimensional torus embedded in three-dimensional space can be given as follows:

$$\begin{cases} x(s,t) &= (a + r\cos 2\pi t)\cos 2\pi s\\ y(s,t) &= (a + r\cos 2\pi t)\sin 2\pi s\\ z(s,t) &= r\sin 2\pi t\end{cases} \tag{2.271}$$

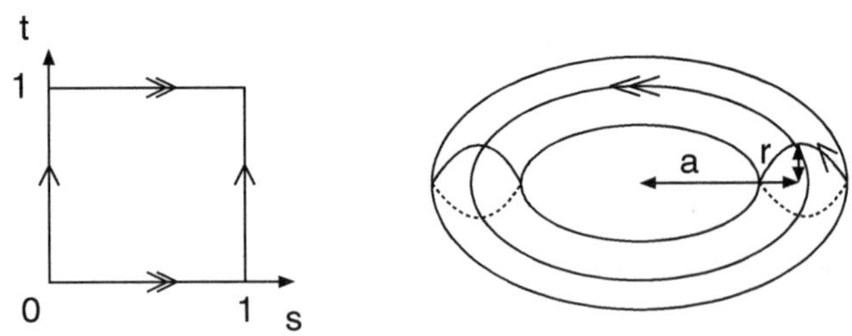

Fig. 2.21 Integration over the torus

2.4.3 *Vector Analysis in a Nutshell*

The key to mastering vector analysis is to think concretely without being deceived by abstract notations.

Notation

1. **Gradient**: A vector field is built from a scalar function $f(\vec{x})$.

$$\text{grad}\, f = \nabla f(\vec{x}) \equiv (\partial_x f(\vec{x}), \partial_y f(\vec{x}), \partial_z f(\vec{x})) \tag{2.272}$$

2. **Divergence**: A scalar function is built from a vector field $\vec{v}(\vec{x})$.

$$\text{div}\, \vec{v} = \nabla \cdot \vec{v}(\vec{x}) \equiv \partial_x v_x(\vec{x}) + \partial_y v_y(\vec{x}) + \partial_z v_z(\vec{x}) \tag{2.273}$$

3. **Rotation** or **Curl**: A vector field is built from a vector field $\vec{v}(\vec{x})$.

$$\text{rot}\, \vec{v} = \text{curl}\, \vec{v} = \nabla \times \vec{v}(\vec{x}) \equiv (\partial_y v_z - \partial_z v_y, \partial_z v_x - \partial_x v_z, \partial_x v_y - \partial_y v_x) \tag{2.274}$$

Geometrical Interpretation

1. $\nabla f(\vec{x})$: The vector field $\nabla f(\vec{x})$ is orthogonal to contours of the function $f(\vec{x})$ along which the function is constant. Consider a contour of f passing through the point $\vec{x}$. For a point $\vec{x} + \Delta\vec{x}$ to be on the same contour, it must satisfy

$$f(\vec{x} + \Delta\vec{x}) = f(\vec{x}) \tag{2.275}$$

Expanding the left-hand side with respect to $\Delta\vec{x}$, we obtain

$$\Delta\vec{x} \cdot \nabla f(\vec{x}) = 0 \tag{2.276}$$

This implies that the tangent of the contour at point $\vec{x}$ is orthogonal to the vector $\nabla f(\vec{x})$ (Fig. 2.22).

Fig. 2.22 Gradient ∇f

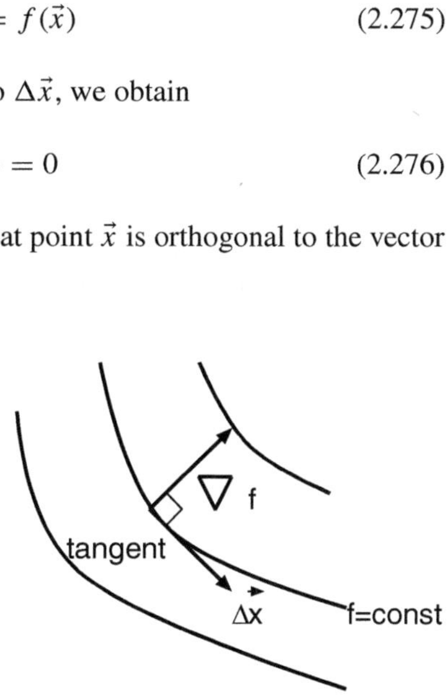

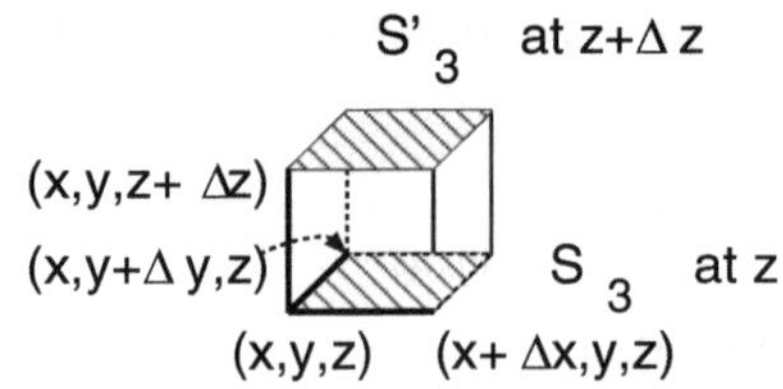

Fig. 2.23 Small cube

2. $\nabla \cdot \vec{v}(\vec{x})$: Interpreting the vector field $\vec{v}(\vec{x})$ as a fluid velocity at each point in space, $\nabla \cdot \vec{v}$ represents the fluid's source or sink. Consider a small cube defined by $\vec{x} = (x, y, z)$ as one vertex, and three intervals $\Delta x\ \vec{e}_1 = (\Delta x, 0, 0)$, $\Delta y\ \vec{e}_2 = (0, \Delta y, 0, 0)$, and $\Delta z\ \vec{e}_3 = (0, 0, \Delta z)$ (See Fig. 2.23). The amount of fluid flowing out of the faces S_3 and S_3' in the $-z$ and $+z$ directions, respectively, is given by

$$(v_z(x, y, z + \Delta z) - v_z(x, y, z))\,\Delta x \Delta y \simeq \partial_z v_z(x, y, z)\Delta x \Delta y \Delta z \tag{2.277}$$

Similarly, the amount flowing out of the two faces perpendicular to the x-axis is

$$(v_x(x + \Delta x, y, z) - v_x(x, y, z))\,\Delta y \Delta z \simeq \partial_x v_x(x, y, z)\Delta x \Delta y \Delta z\,, \tag{2.278}$$

and out of the two faces perpendicular to the y-axis

$$\left(v_y(x, y + \Delta y, z) - v_y(x, y, z)\right)\Delta x \Delta z \simeq \partial_x v_x(x, y, z)\Delta x \Delta y \Delta z \tag{2.279}$$

Thus, the total amount of fluid flowing out of this small cube is given by

$$\left(\partial_x v_x + \partial_y v_y + \partial_z v_z\right)\Delta x \Delta y \Delta z = (\nabla \cdot \vec{v}) \cdot \Delta x \Delta y \Delta z \tag{2.280}$$

This shows that $\nabla \cdot \vec{v}$ represents the density of sources (if positive) or sinks (if negative).

3. $\nabla \times \vec{v}$: Similarly, interpreting $\vec{v}$ as the velocity field of a fluid, $\nabla \times \vec{v}$ represents "vorticity" or "circulation." To find the circulation around the z-axis at point $\vec{x}$, consider a small area defined by points $\vec{x}$, $\vec{x} + \Delta x\ \vec{e}_1$, and $\vec{x} + \Delta y\ \vec{e}_2$ (See Fig. 2.24). The circulation around the z-axis can be obtained by integrating the flow vectors in the direction indicated by the arrows in the figure. Therefore, the circulation in the z direction is given by

$$\begin{aligned} &v_x(x, y, z)\Delta x + v_y(x + \Delta x, y, z)\Delta y \\ &\quad - v_x(x, y + \Delta y, z)\Delta x - v_y(x, y, z)\Delta y \\ &\simeq \left(\partial_x v_y(x, y, z) - \partial_y v_x(x, y, z)\right)\Delta x \Delta y \end{aligned} \tag{2.281}$$

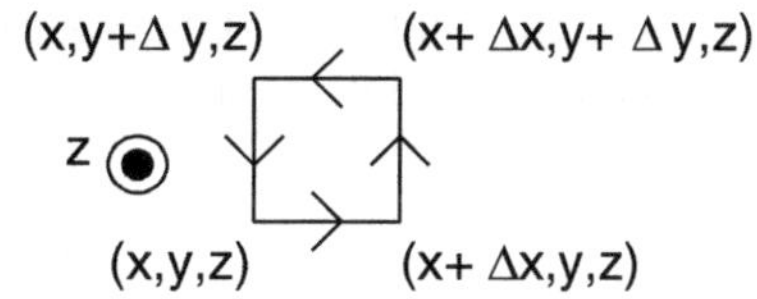

Fig. 2.24 $\nabla \times \vec{v}$ represents circulation

This shows that $\nabla \times \vec{v}$ is a vector field representing circulation per unit area.

Stokes's Theorem

Stokes's theorem holds generally in n-dimensional space, but here we will limit our discussion to $n = 1, 2, 3$ (See Fig. 2.25).[21]

- 1-dimensional case (Newton and Leibniz's Theorem)—Given any function $f(\vec{x})$, for any arbitrary curve C connecting two points P and Q, we find

$$\int_C \nabla f \cdot d\vec{x} = f(Q) - f(P) \tag{2.282}$$

  where the line integral is taken from P to Q.
- 2-dimensional case (Green's Theorem)—Given any vector field $\vec{v}(\vec{x})$, for any surface S and its boundary $\partial S = C$, we find

$$\int_S (\nabla \times \vec{v}) \cdot d\vec{S} = \oint_C \vec{v} \cdot d\vec{x} \tag{2.283}$$

  Here, the surface S has an orientation, and the surface integral element $d\vec{S}$ is facing outward. The closed curve C is oriented counterclockwise when viewed from the outside.

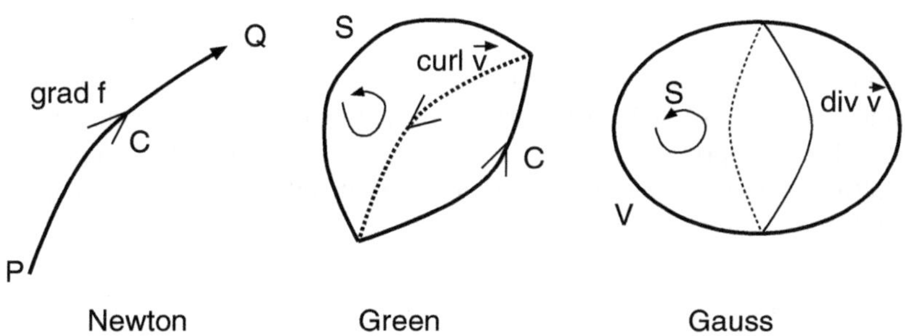

Fig. 2.25 All of these are Stokes's theorem

[21] For line and surface integrals, see the preceding Sect. 2.4.2.

- 3-dimensional case (Gauss's Theorem)—Given any vector field $\vec{v}(\vec{x})$, for any volume V and its boundary $\partial V = S$, we find

$$\int_V \nabla \cdot \vec{v}\, dV = \int_S \vec{v} \cdot d\vec{S} \tag{2.284}$$

Here, the surface integral element $d\vec{S}$ is facing outward.

Stokes's theorem, unlike Poincaré's Lemma explained below, is much simpler and can be derived straightforwardly by following the definition of integrals. Its essence is nothing more than the fundamental theorem of calculus, written explicitly for 1, 2, 3 dimensions as follows:

$$\int_0^1 ds\, \frac{df(s)}{ds} = f(1) - f(0)\,, \tag{2.285}$$

$$\int_0^1 ds \int_0^1 dt \left(\frac{\partial f(s,t)}{\partial s} - \frac{\partial g(s,t)}{\partial t} \right)$$
$$= \int_0^1 dt\, (f(1,t) - f(0,t)) - \int_0^1 ds\, (g(s,1) - g(s,0))\,, \tag{2.286}$$

$$\int_0^1 ds \int_0^1 dt \int_0^1 du \left(\frac{\partial f(s,t,u)}{\partial s} + \frac{\partial g(s,t,u)}{\partial t} + \frac{\partial h(s,t,u)}{\partial u} \right)$$
$$= \int_0^1 dt \int_0^1 du\, (f(1,t,u) - f(0,t,u)) + \int_0^1 ds \int_0^1 du\, (g(s,1,u)$$
$$- g(s,0,u))$$
$$+ \int_0^1 ds \int_0^1 dt\, (h(s,t,1) - h(s,t,0)) \tag{2.287}$$

In the following, we prove only Green's theorem. First, we need to understand the left-hand side of Eq. (2.283). Any surface S is defined by two parameters s and t as $\vec{x}(s,t)$.[22] By appropriately scaling the parameters, we can set their range to [0, 1]. (Fig. 2.26) Therefore, using the definition of surface integrals,[23] we can express the left-hand side of Green's theorem as follows:

$$\int_S (\nabla \times \vec{v}) \cdot d\vec{S} = \int_0^1 ds \int_0^1 dt \left[(\partial_y v_z - \partial_z v_y) \left(\frac{\partial y}{\partial s}\frac{\partial z}{\partial t} - \frac{\partial y}{\partial t}\frac{\partial z}{\partial s} \right) \right.$$

[22] Even non-trivial surfaces like tori can be constructed by gluing simple surfaces as we consider here. Similarly, for Gauss's theorem, a closed surface can be constructed by gluing two or more open surfaces.

[23] See Sect. 2.4.2 on surface integrals.

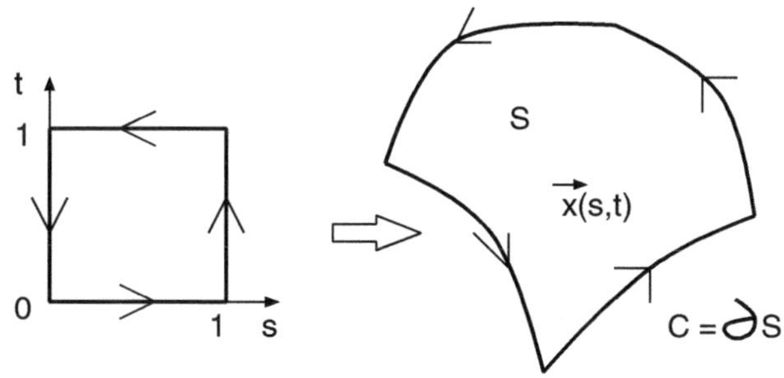

Fig. 2.26 Parametric representation of a surface and its boundary

$$+ (\partial_z v_x - \partial_x v_z)\left(\frac{\partial z}{\partial s}\frac{\partial x}{\partial t} - \frac{\partial z}{\partial t}\frac{\partial x}{\partial s}\right) + (\partial_x v_y - \partial_y v_x)\left(\frac{\partial x}{\partial s}\frac{\partial y}{\partial t} - \frac{\partial x}{\partial t}\frac{\partial y}{\partial s}\right)\Bigg]$$

$$= \int_0^1 ds \int_0^1 dt \Bigg[$$

$$\frac{\partial}{\partial s}\left(v_x(\vec{x}(s,t))\frac{\partial x(s,t)}{\partial t} + v_y(\vec{x}(s,t))\frac{\partial y(s,t)}{\partial t} + v_z(\vec{x}(s,t))\frac{\partial z(s,t)}{\partial t}\right)$$

$$- \frac{\partial}{\partial t}\left(v_x(\vec{x}(s,t))\frac{\partial x(s,t)}{\partial s} + v_y(\vec{x}(s,t))\frac{\partial y(s,t)}{\partial s} + v_z(\vec{x}(s,t))\frac{\partial z(s,t)}{\partial s}\right)\Bigg]$$

$$= \int_0^1 ds \int_0^1 dt\ (\partial_s f(s,t) - \partial_t g(s,t)) \tag{2.288}$$

where we have defined f, g by

$$f(s,t) \equiv v_x(\vec{x}(s,t))\frac{\partial x(s,t)}{\partial t} + v_y(\vec{x}(s,t))\frac{\partial y(s,t)}{\partial t} + v_z(\vec{x}(s,t))\frac{\partial z(s,t)}{\partial t}$$

$$= \vec{v}(\vec{x}(s,t)) \cdot \frac{\partial \vec{x}(s,t)}{\partial t} \tag{2.289a}$$

$$g(s,t) \equiv v_x(\vec{x}(s,t))\frac{\partial x(s,t)}{\partial s} + v_y(\vec{x}(s,t))\frac{\partial y(s,t)}{\partial s} + v_z(\vec{x}(s,t))\frac{\partial z(s,t)}{\partial s}$$

$$= \vec{v}(\vec{x}(s,t)) \cdot \frac{\partial \vec{x}(s,t)}{\partial s} \tag{2.289b}$$

This has the same form as the left-hand side of Eq. (2.286). Therefore, using Eq. (2.286), we obtain

$$\int_S (\nabla \times \vec{v}) \cdot d\vec{S} = \int_0^1 dt\ (f(1,t) - f(0,t)) - \int_0^1 ds\ (g(s,1) - g(s,0))$$

$$= \int_0^1 dt\, \vec{v}(\vec{x}(1,t)) \cdot \frac{\partial \vec{x}(1,t)}{\partial t} - \int_0^1 dt\, \vec{v}(\vec{x}(0,t)) \cdot \frac{\partial \vec{x}(0,t)}{\partial t}$$

$$- \int_0^1 ds\, \vec{v}(\vec{x}(s,1)) \cdot \frac{\partial \vec{x}(s,1)}{\partial s} + \int_0^1 ds\, \vec{v}(\vec{x}(s,0)) \cdot \frac{\partial \vec{x}(s,0)}{\partial s}$$

$$= \oint_{\partial S} \vec{v}(\vec{x}) \cdot d\vec{x} \tag{2.290}$$

which is a line integral along the closed boundary ∂S of the surface S.

Poincaré's Lemma

There are two important properties regarding the gradient, curl, and divergence:

$$\nabla \times (\nabla f) = 0 \tag{2.291}$$

$$\nabla \cdot (\nabla \times \vec{v}) = 0 \tag{2.292}$$

where f is any function and $\vec{v}$ is any vector field. The above equations can be easily derived following the definitions. Poincaré's Lemma states the converse:

1. If $\nabla \times \vec{v} = 0$, then there exists a function f such that $\vec{v}$ can be obtained as its gradient:

$$\vec{v} = \nabla f \tag{2.293}$$

   In this case, f is not unique, and $f + \text{const}$ would yield the same result.
2. If $\nabla \cdot \vec{v} = 0$, then there exists a vector field $\vec{A}$ such that $\vec{v}$ can be obtained as its curl:

$$\vec{v} = \nabla \times \vec{A} \tag{2.294}$$

   Here, $\vec{A}$ is not unique; for any function f, $\vec{A} + \nabla f$ would also yield the same result because $\nabla \times \nabla f = 0$.

   Let us prove Poincaré's lemma below. We will prove it by constructing f, $\vec{A}$.

1. Using $\vec{v}$ that satisfies $\nabla \times \vec{v} = 0$, the function f is constructed as follows. First, choose an arbitrary point, say the origin, and set the function's value there to zero (any constant would do). The function's value at any other point $\vec{x}$ is defined by any curve C connecting the origin $\vec{x} = 0$ to this $\vec{x}$ as

$$f(\vec{x}) \equiv \int_C \vec{v} \cdot d\vec{x} \tag{2.295}$$

   The integral does not depend on the choice of the curve C (Fig. 2.27). For example, if another curve C' is taken, then for $C - C'$ (first trace C to reach $\vec{x}$, then trace C' in reverse to return to the origin), using Green's theorem and the

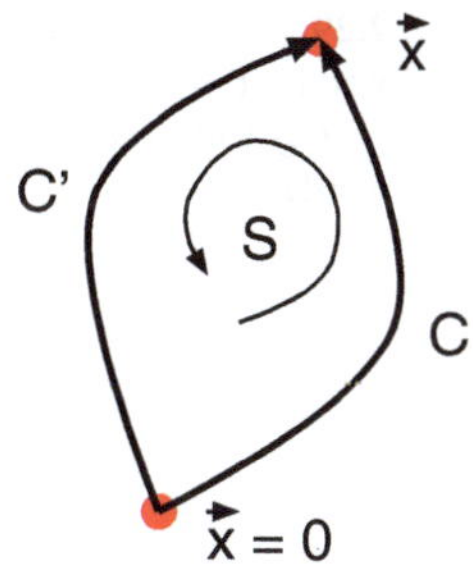

Fig. 2.27 Both curves C and C' yield the same integral value

assumption $\nabla \times \vec{v} = 0$, we find

$$\int_C \vec{v} \cdot d\vec{x} - \int_{C'} \vec{v} \cdot d\vec{x} = \int_{C-C'} \vec{v} \cdot d\vec{x} = \int_S (\nabla \times \vec{v}) \cdot d\vec{S} = 0 \qquad (2.296)$$

Here, S is any surface whose boundary is the closed curve $C - C'$. Obviously, $f(\vec{x})$ defined by Eq. (2.295) satisfies $\nabla f = \vec{v}$.

2. It is a little more involved to construct $\vec{A}$ such that $\vec{v} = \nabla \times \vec{A}$. We give some details below. First note that a vector field is uniquely determined by specifying the value of its line integral along any curve. For example, to specify the z component of the vector field $\vec{A}(\vec{x})$ at point $\vec{x}$, it is sufficient to specify the integral of $A_z(\vec{x})$ over the infinitesimal line segment dz along the z axis. Now, we construct the vector field $\vec{A}(\vec{x})$ as follows. First, choose a standard curve $C_{\vec{x}}$ to go from the origin to each point $\vec{x}$ in space. Any choice will do. It could be a straight line connecting the origin to $\vec{x}$, or a broken line that first goes along the x-axis to $(x, 0, 0)$, then along the y-axis to $(x, y, 0)$, and finally along the z-axis to $\vec{x} = (x, y, z)$ (Fig. 2.28).

   Once a standard curve is chosen for all points in space, the integral of the vector field $\vec{A}$ along that line is defined to be zero.[24]

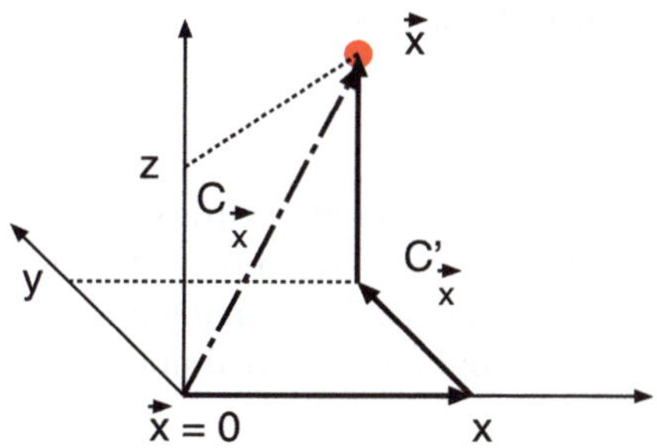

Fig. 2.28 There are multiple ways to choose the standard curve

[24] The choice of the standard curves $C'_{\vec{x}}$ as given in Fig. 2.28 amounts to choosing $A_z(x, y, z) = A_y(x, y, 0) = A_x(x, 0, 0) = 0$. In theoretical physics, this choice of the standard curve is called the **axial gauge**.

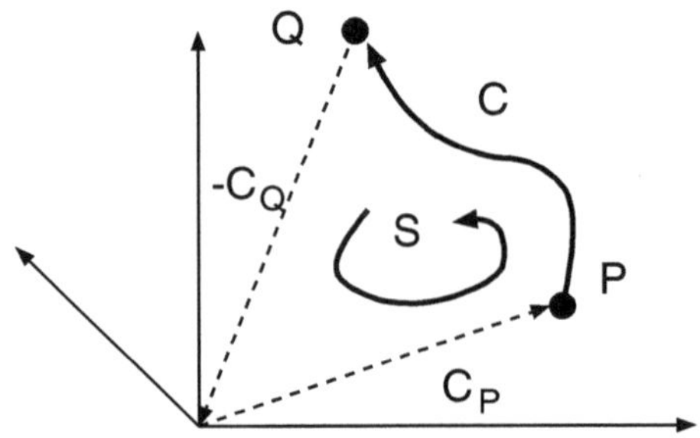

Fig. 2.29 S is a surface enclosed by C_P, C, and $-C_Q$ (C_Q followed backward)

$$\int_{C_P} \vec{A} \cdot d\vec{x} \equiv 0 \quad \text{any point } P \tag{2.297}$$

Now, consider an arbitrary curve C that connects point P to point Q. The line integrals of $\vec{A}$ along C_P and C_Q are zero by definition. Using the given vector field $\vec{v}$, we define the integral of $\vec{A}$ along C as follows (Fig. 2.29):

$$\int_C \vec{A} \cdot d\vec{x} = \int_{C_P + C - C_Q} \vec{A} \cdot d\vec{x} \equiv \int_S \vec{v} \cdot d\vec{S} \tag{2.298}$$

where S is an arbitrary surface enclosed by C_P, C, and C_Q.

Note that this definition does not depend on the choice of surface S. If another surface S' is chosen, then, thanks to the assumption $\nabla \cdot \vec{v} = 0$, we find

$$\int_{S'} \vec{v} \cdot d\vec{S} - \int_S \vec{v} \cdot d\vec{S} = \int_V \nabla \cdot \vec{v} \, dV = 0 \tag{2.299}$$

where V represents the region enclosed by S' and S.

We have thus defined the vector field $\vec{A}(\vec{x})$ by giving its line integral along any curve. Now, we must explain that the vector field $\vec{A}$ constructed by (2.298) indeed satisfies

$$\nabla \times \vec{A} = \vec{v} \tag{2.300}$$

Let us consider an arbitrary surface S with boundary C. Take any point P on C, which can be thought of as starting and ending at P. Let S' be a cone-like surface formed by the origin and the boundary C (Fig. 2.30).

According to the definition (2.298), we find

$$\oint_C \vec{A} \cdot d\vec{x} \equiv \int_{S'} \vec{v} \cdot d\vec{S} \tag{2.301}$$

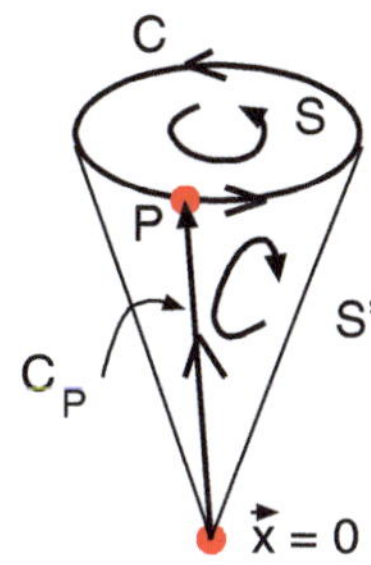

Fig. 2.30 Checking $\nabla \times \vec{A} = \vec{v}$

Since $\nabla \cdot \vec{v} = 0$, however, the integral over S' is the same as over S:

$$\int_{S'} \vec{v} \cdot d\vec{S} = \int_S \vec{v} \cdot d\vec{S} \tag{2.302}$$

Therefore, we find

$$\oint_C \vec{A} \cdot d\vec{x} = \int_S \vec{v} \cdot d\vec{S} \tag{2.303}$$

On the other hand, according to Green's theorem, we find

$$\oint_C \vec{A} \cdot d\vec{x} = \int_S (\nabla \times \vec{A}) \cdot d\vec{S} \tag{2.304}$$

Hence,

$$\int_S \vec{v} \cdot d\vec{S} = \int_S (\nabla \times \vec{A}) \cdot d\vec{S} \tag{2.305}$$

For this to hold for any surface S, we must find

$$\vec{v} = \nabla \times \vec{A} \tag{2.306}$$

This is what we wanted to show.

Changing the choice of standard curves may change the vector field $\vec{A}$, which also satisfies $\vec{v} = \nabla \times \vec{A}$. We can show that the change of $\vec{A}$ can be written as ∇f. We omit a proof.

Chapter 3
Thermodynamics and Statistical Mechanics

There is no field in physics as well developed as thermodynamics. One who conquers thermodynamics conquers physics; Einstein was a master of thermodynamics. Statistical mechanics, together with quantum mechanics, forms the two pillars of modern physics.

3.1 Thermodynamics

Two Points to Remember

1. All thermodynamic quantities can be derived from a free energy.
   - Helmholtz's free energy $F(T, V)$ depends on temperature and volume (or magnetization M).
   - Gibbs's free energy $G(T, p)$ depends on temperature and pressure (or external magnetic field H_e).
2. Under given conditions, thermal equilibrium is determined such that the free energy is minimized.

3.1.1 *All Thermodynamic Quantities Can Be Derived from a Free Energy*

Once you know the free energy of a system, you can derive all its thermodynamic quantities. In a typical undergraduate course on thermodynamics, much time is spent on entropy, and not enough time is given to learning about free energies. It is the goal of this section to become comfortable with free energies.

H. Sonoda, K. Ülker, *Basic Physics for Graduate Students*,
https://doi.org/10.1007/978-981-95-6607-5_3

Let us consider a homogeneous, i.e., both uniform and isotropic, system of volume V at absolute temperature T (such as a gas, a liquid, and even a solid as long as homogeneity is a good approximation). For a gas, the first law of thermodynamics states

$$Q = dU + p dV \tag{3.1}$$

where Q is the infinitesimal heat given to the gas, U is the energy,[1] p is the pressure, and V is the volume of the gas. This is nothing but the conservation of energy, and we recognize heat as a form of transferring energy. The second law of thermodynamics, usually formulated either as Thomson's principle or as Clausius's principle, boils down to the integrability of

$$\frac{Q}{T} = dS \tag{3.2}$$

Namely, for a gas, there is a state function $S(T, V)$ that satisfies

$$dS = \frac{1}{T} dU + \frac{p}{T} dV \tag{3.3}$$

The state function S is the entropy of the gas. We refer the reader to a textbook on thermodynamics[2] for the original formulations of the second law by Thomson (Lord Kelvin) and by Clausius and their equivalence. Using the entropy, the second law can be stated that the total entropy never decreases.

In Example 6-1 we review that Helmholtz's free energy

$$\boxed{F(T, V) \equiv U(T, V) - TS(T, V)} \tag{3.4}$$

satisfies

$$\frac{\partial F(T, V)}{\partial T} = -S(T, V) \tag{3.5}$$

$$\frac{\partial F(T, V)}{\partial V} = -p(T, V) \tag{3.6}$$

Both the entropy and the pressure are obtained from the free energy. By differentiating the entropy with respect to T, we obtain the heat capacity at constant volume:

$$C_V = T \frac{\partial S(T, V)}{\partial T} \tag{3.7}$$

Equation (3.6) is an **equation of state**, expressing pressure in terms of temperature and volume.

[1] In thermodynamics U is often called the internal energy.

[2] For example, E. Fermi, *Thermodynamics* (Dover).

The entropy and pressure are not completely independent since they are obtained from a single function $F(T, V)$. In fact, these two quantities are connected by *Maxwell's relation*:

$$-\frac{\partial}{\partial V}\frac{\partial}{\partial T}F(T,V)=\boxed{\left(\frac{\partial S}{\partial V}\right)_T=\left(\frac{\partial p}{\partial T}\right)_V}=-\frac{\partial}{\partial T}\frac{\partial}{\partial V}F(T,V) \tag{3.8}$$

Example: Boyle's Law

Let us consider a gas that satisfies Boyle's law:

$$pV = f(T) \tag{3.9}$$

Taking the pressure p as a function $p(T, V)$ of T, V, we obtain

$$\left(\frac{\partial p}{\partial T}\right)_V = \frac{1}{V}f'(T) \tag{3.10}$$

where $f'(T) = \frac{df(T)}{dT}$. Maxwell's relation gives

$$\left(\frac{\partial S}{\partial V}\right)_T = \frac{1}{V}f'(T) \tag{3.11}$$

Integrating this, we get

$$S(T, V) = f'(T)\ln V + s(T) \tag{3.12}$$

where $s(T)$, a function of only T, is to be determined. By differentiating $S(T, V)$ with respect to T, we obtain the heat capacity at constant volume as

$$C_V = T\left(\frac{\partial S}{\partial T}\right)_V = T\left(f''(T)\ln V + s'(T)\right) \tag{3.13}$$

In particular, for an ideal gas, we have

$$f(T) = Nk_B T \tag{3.14}$$

Hence, we obtain

$$f''(T) = 0 \Longrightarrow C_V(T) = Ts'(T) \tag{3.15}$$

This shows that the heat capacity of an ideal gas is independent of volume.

In general if entropy and pressure are given as functions of T and V, we can obtain the free energy by integration. In short, this can be expressed as

$$\boxed{\text{entropy } S(T, V) \ \& \ \text{equation of state } p(T, V) \implies F(T, V)}$$

Let us rewrite this relation more quantitatively. Integrating Eq. (3.5) with respect to temperature, we get

$$F(T, V) = -\int_{T_0}^{T} dT' \, S(T', V) + f(V) \tag{3.16}$$

Here, T_0 is an arbitrary constant temperature, and $f(V)$ is a function of volume, independent of temperature. Differentiating this expression with respect to volume and using Maxwell's relation (3.8), we obtain

$$\begin{aligned} \frac{\partial F(T, V)}{\partial V} &= -\int_{T_0}^{T} dT' \, \frac{\partial S(T', V)}{\partial V} + f'(V) \\ &= -\int_{T_0}^{T} dT' \, \frac{\partial p(T', V)}{\partial T'} + f'(V) \\ &= p(T_0, V) - p(T, V) + f'(V) \end{aligned} \tag{3.17}$$

where $f'(V) = \frac{df(V)}{dV}$. From Eq. (3.6), however, this should be equal to $-p(T, V)$, so we find

$$f'(V) = -p(T_0, V) \tag{3.18}$$

Thus, we get

$$f(V) = -\int_{V_0}^{V} dV' \, p(T_0, V') + f_0 \, , \tag{3.19}$$

where f_0 is a constant of integration. Therefore, combining all of the above, we obtain

$$F(T, V) = -\int_{T_0}^{T} dT' \, S(T', V) - \int_{V_0}^{V} dV' \, p(T_0, V') + f_0 \tag{3.20}$$

Clearly, the constant of integration equals the free energy at T_0, V_0: $f_0 = F(T_0, V_0)$.

Thus, the free energy is determined from $S(T, V)$ and $p(T, V)$. The undetermined constant f_0 reflects the ambiguity of the absolute value of energy by an additive constant. By changing the order of integration, we can obtain similarly

$$F(T, V) = -\int_{V_0}^{V} dV' \, p(T, V') - \int_{T_0}^{T} dT' \, S(T', V_0) + f_0 \tag{3.21}$$

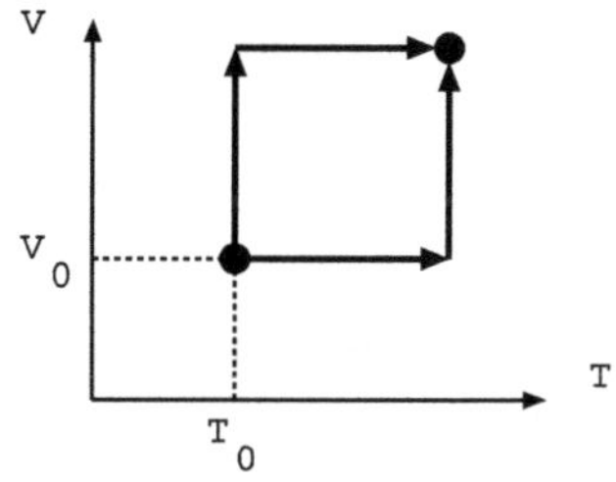

Fig. 3.1 The difference between $F(T, V)$ and $F(T_0, V_0) = f_0$ does not depend on the contour of integration as guaranteed by Maxwell's relation

If we know the specific heat at constant volume instead of entropy, we obtain the free energy in a similar manner (Fig. 3.1).

Example: Free Energy of an Ideal Gas

An ideal gas, composed of randomly moving non-interacting point particles, satisfies the equation of state[3]

$$p = \frac{Nk_BT}{V} \tag{3.22}$$

where N is the number of particles (atoms or molecules) and $k_B \simeq 1.38 \times 10^{-23}\,\mathrm{J\,K^{-1}}$ is the Boltzmann constant. The heat capacity must be a function of temperature alone, as we have shown in Eq. (3.15). Let us consider the simplest case where the heat capacity is a constant:

$$T\frac{\partial S(T, V)}{\partial T} = Nk_Bc_V \tag{3.23}$$

Integrating this with respect to temperature, we obtain

$$S(T, V) = N\,(k_Bc_V \ln T + s(V)) \tag{3.24}$$

The function of volume, $s(V)$, can be found from Maxwell's relation. From

$$\frac{\partial S(T, V)}{\partial V} = Ns'(V), \quad \frac{\partial p(T, V)}{\partial T} = \frac{Nk_B}{V} \Longrightarrow \frac{ds(V)}{dV} = \frac{k_B}{V} \tag{3.25}$$

we obtain

$$s(V) = k_B \ln \frac{V}{N} + s_0\,, \tag{3.26}$$

[3] The ideal gas is an approximation that works well at high temperatures and low pressures. At low temperatures, the interactions between particles cannot be neglected, and at high pressures the volume of the particles cannot be neglected.

where s_0 is a constant.[4] Therefore, we obtain

$$S(T, V) = N\left\{k_B\left(c_V \ln T + \ln\frac{V}{N}\right) + s_0\right\}. \tag{3.27}$$

From this, using formula (3.20), we can determine the free energy of an ideal gas as follows:

$$F(T, V) = N\left[-k_B T\left(c_V(\ln T - 1) + \ln\frac{V}{N}\right) - s_0 T + \varepsilon_0\right] \tag{3.28}$$

The constants s_0 and ε_0 can be expressed in terms of the mass of an atom or molecule using quantum mechanics (See Sect. 3.2.2).

Energy U

The energy U is given in terms of the free energy F by

$$U = F + TS = F - T\frac{\partial F(T, V)}{\partial T} = -T^2\frac{\partial}{\partial T}\frac{F(T, V)}{T} \tag{3.29}$$

The derivative of energy with respect to temperature under constant volume is given by

$$\frac{\partial U(T, V)}{\partial T} = \frac{\partial F(T, V)}{\partial T} + S + T\frac{\partial S(T, V)}{\partial T} = T\frac{\partial S(T, V)}{\partial T} = C_V(T, V) \tag{3.30}$$

which is the heat capacity at constant volume.

Relation to Statistical Mechanics

Equation (3.4) gives the free energy in terms of energy and entropy. This particular combination of energy and entropy can be understood clearly from the viewpoint of statistical mechanics. According to statistical mechanics, given a macroscopic system, the probability of realizing a microscopic state with energy E is proportional to the Boltzmann weight $e^{-\frac{E}{k_B T}}$. Let us assume that the number of microscopic states between energies E and $E + \Delta E$ is given by the function of energy $S(E)$ as

$$\frac{\Delta E}{\epsilon_0} e^{\frac{S(E)}{k_B}} \tag{3.31}$$

[4] As the argument of the logarithm we have chosen V/N instead of V so that $S(T, V) = NS(T, V/N)$.

where ϵ_0 has the unit of energy. Then, the probability of the energy of this system being between E and $E + \Delta E$ is given by

$$\frac{1}{Z}\frac{\Delta E}{\epsilon_0}\,\mathrm{e}^{-\frac{E}{k_BT}+\frac{S(E)}{k_B}} \tag{3.32}$$

where Z is a normalization factor that ensures the total probability to be 1.

If we define a function of energy by

$$F_{\mathrm{SM}}(E,T) \equiv E - TS(E)\,, \tag{3.33}$$

then the above probability can be rewritten as

$$\mathrm{const} \times \mathrm{e}^{-\frac{F_{\mathrm{SM}}(E)}{k_BT}} \tag{3.34}$$

Hence, the most probable energy at temperature T is the one that minimizes $F_{\mathrm{SM}}(E,T)$. We denote this energy as $E^*(T)$, and in the vicinity of $E = E^*(T)$ we can approximate

$$\frac{F_{\mathrm{SM}}(E)}{k_BT} \simeq \mathrm{const} + \frac{1}{2\sigma}(E - E^*(T))^2 \tag{3.35}$$

(See Fig. 3.2.) Now, if the number of degrees of freedom of the system is $N \gg 1$, $E^*(T)$ is also of order N. As shown later in Sect. 3.2.1, σ is also of order N. Hence, the fluctuations of E are of order $\sqrt{N}$ and are much smaller compared with the mean value $E^*(T)$:

$$\langle (E - E^*(T))^2 \rangle = \sigma \propto N \tag{3.36}$$

Thus, identifying the most probable energy $E^*(T)$ as the energy U in thermodynamics, and $S(E^*(T))$ as the entropy $S(T)$, we obtain

$$F_{\mathrm{SM}}(E^*(T),T) = E^*(T) - TS(E^*(T)) = U - TS(T) \tag{3.37}$$

This is Helmholtz's free energy. The particular combination of energy and entropy arises naturally from statistical mechanics.

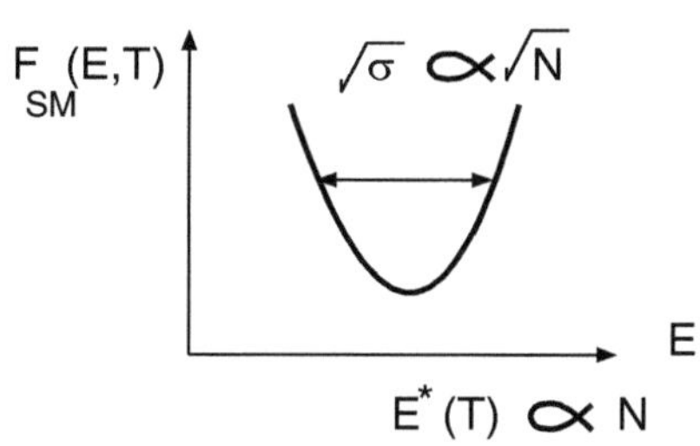

Fig. 3.2 The relative fluctuations around the most probable value are inversely proportional to $\sqrt{N}$

Gibbs Free Energy

Until now, we have considered temperature T and volume V as independent variables, but in a laboratory environment it is often more convenient to use pressure p instead of V. The free energy corresponding to the choice of independent variables T and p is called the Gibbs free energy, and it is defined by

$$G(T, p) \equiv F(T, V) + pV \tag{3.38}$$

Here, V should be considered as a function of T and p. An infinitesimal change of the Gibbs energy is given by

$$\begin{aligned} dG &= \underbrace{\frac{\partial F(T, V)}{\partial T}}_{-S} dT + \underbrace{\frac{\partial F(T, V)}{\partial V}}_{-p} dV + p\,dV + V\,dp \\ &= -SdT + Vdp \end{aligned} \tag{3.39}$$

Hence,

$$\frac{\partial G(T, p)}{\partial T} = -S, \quad \frac{\partial G(T, p)}{\partial p} = V \tag{3.40}$$

This gives Maxwell's relation

$$\boxed{\left(\frac{\partial S}{\partial p}\right)_T = -\left(\frac{\partial V}{\partial T}\right)_p} \tag{3.41}$$

Example: Ideal Gas

We have already obtained the Helmholtz free energy for an ideal gas:

$$F(T, V) = N\left[-k_B T\left(c_V(\ln T - 1) + \ln\frac{V}{N}\right) - s_0 T + \varepsilon_0\right] \tag{3.42}$$

From this, we derive

$$\begin{aligned} G(T, p) &= F(T, V) + pV \\ &= N\left[-k_B T\left(c_V(\ln T - 1) + \ln\frac{k_B T}{p}\right) - s_0 T + \varepsilon_0 + k_B T\right] \\ &= Nk_B T\left[(c_V + 1)(1 - \ln T) + \ln p - \left(\frac{s_0}{k_B} + \ln k_B\right) + \frac{\varepsilon_0}{k_B T}\right] \end{aligned} \tag{3.43}$$

where we have used the equation of state. Therefore, the entropy is given by

$$S(T, p) = -\frac{\partial G(T, p)}{\partial T} = N \left[k_B(c_V + 1) \ln T + s_0 + k_B \ln \frac{k_B}{p} \right] \quad (3.44)$$

Furthermore, the specific heat at constant pressure is given by

$$C_p(T, p) \equiv T \frac{\partial S(T, p)}{\partial T} = N k_B(c_V + 1), \quad (3.45)$$

which is a constant related to the specific heat at constant volume by Mayer's relation:

$$C_p(T, p) = C_V(T, V) + N k_B \quad (3.46)$$

3.1.2 *Under Given Conditions, Thermal Equilibrium Is Determined Such that the Free Energy Is Minimized*

The thermal equilibrium of an isolated system is determined by the condition of the maximum entropy. The thermal equilibrium of a system at a given temperature (or a system in a heat bath) is determined instead by the condition of the minimum free energy.

To consider the thermal equilibrium of a system, let us use the viewpoint of statistical mechanics, introduced in the previous section. We consider a system at a fixed volume V in thermal equilibrium with a heat bath of temperature T. Let $p(E)$ be the probability that the system has energy E at temperature T.[5] Then, we have

$$-\ln p(E) \approx F_{\text{SM}}(E, T) \equiv E - T \cdot S(E) \quad (3.47)$$

Here, $\exp\left(\frac{S(E)}{k_B}\right)$ is proportional to the number of microscopic states of the system with energy E. It is the energy E minimizing $F_{\text{SM}}(E, T)$ that is realized in thermal equilibrium.

Phase Transitions

Now, let us consider a scenario where the statistical mechanical free energy $F_{\text{SM}}(T, E)$ has two local minima as shown in Fig. 3.3.[6] We first assume $F_{\text{SM}}(E, T)$ as shown on the left side of the figure. The state with higher entropy, which is the

[5] In the following discussion, we assume that the volume of the system V is fixed.

[6] As we will see shortly, the gas state has higher entropy than the liquid state, justifying our terminology. Instead of "liquid" and "gas," we could use "low entropy state" and "high entropy state," but "liquid" and "gas" sound more familiar.

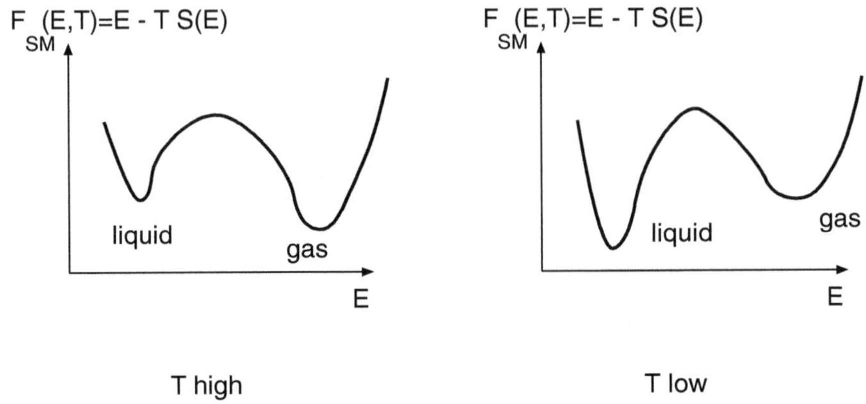

Fig. 3.3 At high temperatures, the system is in a gas phase; at low temperatures, in a liquid phase

state of gas, is realized even though its energy is higher than that of liquid. This takes place at high temperatures, where the second term TS of F_{SM} is more important than E. Conversely, at low temperatures, the energy becomes more important, and the state with lower energy, which is the state of liquid, is realized. This is illustrated on the right side of the figure.

Now, out of the two local minima of F_{SM}, we denote the one with lower energy (i.e., the liquid) as $F_{liq}(T)$ and that with higher energy (i.e., the gas) as $F_{gas}(T)$. Rewriting what we have stated above, we find

$$F_{liq}(T) > F_{gas}(T) \tag{3.48}$$

at high temperatures, and the gas state is realized. At low temperatures, we find

$$F_{liq}(T) < F_{gas}(T)\,, \tag{3.49}$$

and the liquid state is realized.

There should be an intermediate temperature, denoted as T_{tr}, where the free energies of the liquid and gas states become equal:

$$F_{gas}(T_{tr}) = F_{liq}(T_{tr}) \tag{3.50}$$

This temperature T_{tr} is called the **transition temperature**. At this temperature, the liquid and gas are equally likely, and they coexist. Therefore, when we plot the temperature dependence of the free energies of the liquid and gas states, it looks

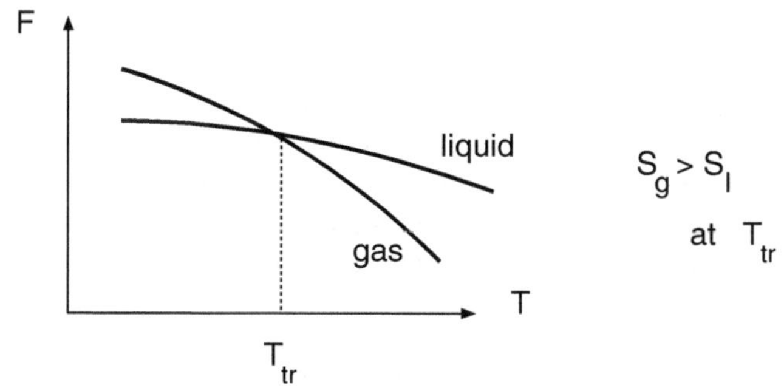

Fig. 3.4 First-order phase transition

like Fig. 3.4. It is worth noting that the free energy is concave (convex upward) as a function of temperature:[7]

$$\frac{d^2F(T)}{dT^2} = -\frac{dS}{dT} = -\frac{1}{T}C_V < 0 \tag{3.51}$$

As is evident from Fig. 3.4, at the transition temperature T_{tr} we find

$$\underbrace{\frac{dF_{\text{liq}}}{dT}}_{=-S_{\text{liq}}} > \underbrace{\frac{dF_{\text{gas}}}{dT}}_{=-S_{\text{gas}}} \tag{3.52}$$

This implies that the entropy difference between the gas and liquid states is always positive:

$$S_{\text{gas}}(T_{tr}) - S_{\text{liq}}(T_{tr}) = \frac{L}{T_{tr}} > 0 \tag{3.53}$$

where L is called the **latent heat**. The latent heat is the heat required to convert the liquid into the gas at the transition temperature.[8] The state (phase) realized on the high-temperature side has inevitably higher entropy. Since the free energies of the two phases are equal at the transition temperature, we find

$$\begin{cases} F_{\text{liq}} = E_{\text{liq}} - T_{tr}\, S_{\text{liq}} \\ F_{\text{gas}} = E_{\text{gas}} - T_{tr}\, S_{\text{gas}} \end{cases} \overset{F_{\text{liq}}=F_{\text{gas}}}{\Longrightarrow} E_{\text{gas}} - E_{\text{liq}} = L \tag{3.54}$$

[7] The specific heat C_V is always positive.

[8] We all know in our kitchen experiments it is impossible to raise the temperature of a boiling water. The energy supplied (the heat added) to the water is used entirely for changing the phase of the water rather than its temperature.

Hence, the latent heat is the difference in energy.[9]

Now, let us revisit what we have just observed, as it is very important to grasp the essence. The liquid state has lower energy and entropy, while the gas state has higher energy and entropy. For simplicity let us assume that both energy and entropy are constant, independent of temperature. We denote the energy of the liquid as ε_l, the energy of the gas as $\varepsilon_g > \varepsilon_l$, the entropy of the liquid as σ_l, and the entropy of the gas as $\sigma_g > \sigma_l$. At a given temperature, the phase that has a lower free energy is chosen. When the temperature is low, energy is more crucial, and the state with lower energy, i.e., the liquid state, is chosen. On the other hand, at high temperatures, entropy becomes more important, and the state with lower free energy is the gas state, which has a larger entropy. We can easily calculate the transition temperature. The free energy of the liquid is

$$f_l \equiv \varepsilon_l - T\sigma_l,$$

and that of the gas is

$$f_g \equiv \varepsilon_g - T\sigma_g$$

Therefore, the transition temperature is given by

$$f_l = f_g \Longleftrightarrow T_{tr} = \frac{\varepsilon_g - \varepsilon_l}{\sigma_g - \sigma_l} \tag{3.55}$$

So far we have assumed that the volume of the system is fixed. Similarly, we can discuss the equilibrium state of the system by assuming that the pressure is fixed. In this case, what needs to be minimized is the Gibbs free energy, denoted by $G(T, p)$. The left side of Fig. 3.5 shows the temperature dependence of the free energy with fixed pressure, while the right figure shows the pressure dependence at a fixed temperature. Here are two important points to note:

1. Entropy is discontinuous at $T = T_{tr}$ (latent heat).
2. Volume is discontinuous at $p = p_{tr}$ (density discontinuity).

At the transition between liquid and gas, the free energy G is continuous, but its first derivatives, entropy and pressure, become discontinuous. Such a transition from one phase to another is called a **first-order phase transition**.[10]

---

[9] Note that we are assuming constant volume here. Since the densities are different for the liquid and gas, in order to keep the volume constant, you would need to change the pressure gradually at the transition temperature. In the case of transition at constant pressure, the required latent heat would be a difference in enthalpy $H \equiv E + pV$, not just energy.

[10] Phase transitions where both the free energy and its first-order derivatives are continuous are called **continuous phase transitions**.

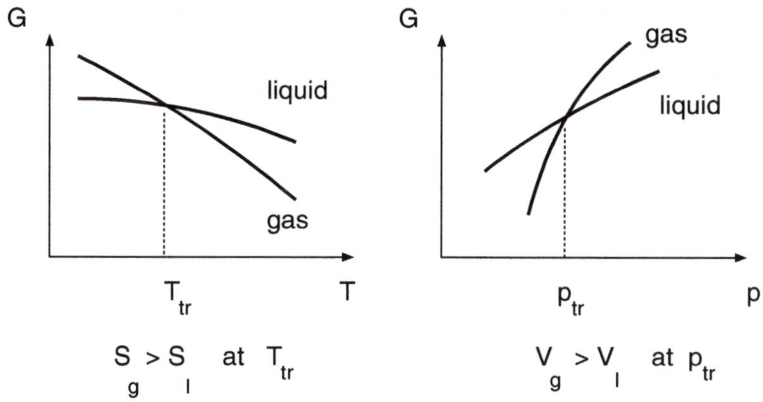

Fig. 3.5 First-order phase transition can also be understood from Gibbs energy

Fig. 3.6 Phase diagram on the Tp diagram

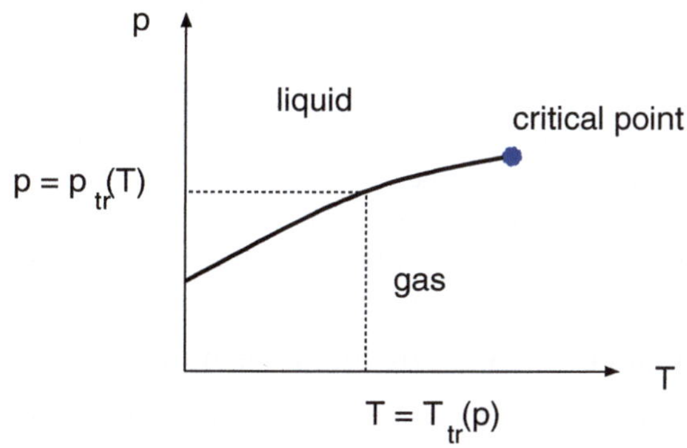

When we plot the phases in the (T, p) plane, it looks like the diagram in Fig. 3.6. As the temperature is increased at a fixed pressure, the liquid becomes a gas at the transition temperature $T_{tr}(p)$. Conversely, as the pressure is increased at a fixed temperature, the gas becomes a liquid at the transition pressure $p_{tr}(T)$. The point where the boundary (or line of coexistence) between the liquid and gas phases ends is called a **critical point**—the latent heat becomes zero, and the liquid and gas have the same density.

Clausius-Clapeyron Equation

The transition pressure $p_{tr}(T)$ from liquid to gas is a function of temperature. For example, in the case of water, the boiling temperature defined by $p_{tr}(T) = 1\,\text{atm}$ is given by $T \simeq 373\,\text{K}$ (Kelvin) or $100\,^\circ\text{C}$ (degree Celsius).

Now, let us determine the temperature dependence of the transition pressure $p_{tr}(T)$ in general. By definition of $p_{tr}(T)$, the Gibbs free energies of the gas and liquid phases are equal:

$$G_{\text{gas}}(T, p_{tr}(T)) = G_{\text{liq}}(T, p_{tr}(T)) \tag{3.56}$$

Differentiating this with respect to temperature, we get

$$\begin{aligned} &- S_{\text{gas}}(T, p_{tr}(T)) + V_{\text{gas}}(T, p_{tr}(T))\frac{dp_{tr}(T)}{dT} \\ &= -S_{\text{liq}}(t, p_{tr}(T)) + V_{\text{liq}}(T, p_{tr}(T))\frac{dp_{tr}(T)}{dT} \end{aligned} \tag{3.57}$$

From this, we obtain the **Clausius-Clapeyron equation**:

$$\frac{dp_{tr}(T)}{dT} = \frac{S_{\text{gas}} - S_{\text{liq}}}{V_{\text{gas}} - V_{\text{liq}}} = \frac{1}{T}\frac{L(T)}{\Delta V(T)} \tag{3.58}$$

where $L(T) \equiv T\left(S_{\text{gas}} - S_{\text{liq}}\right)$ is the latent heat and $\Delta V \equiv V_{\text{gas}} - V_{\text{liq}}$ is the difference in volume.

We usually find $\Delta V > 0$ so that $\frac{dp_{tr}(T)}{dT} > 0$. With an increase in temperature, the transition pressure becomes higher. This is the case with water. The boiling point of water is lower at high altitudes where the atmospheric pressure is lower.

Definition of the Absolute Temperature

Theoretically it is the best to use units of energy, such as J or erg, for the absolute temperature. Before the 2019 revision of the SI, the unit K (Kelvin) of the absolute temperature was defined so that the triple point of water (the temperature at which liquid, solid, and gas phases coexist) has $T = 273.16\,\text{K}$ exactly (Fig. 3.7). Currently, the Kelvin is defined by the exact value for the Boltzmann constant[11]

$$k_B \equiv 1.380\,649 \times 10^{-23}\,\text{J}\,\text{K}^{-1} \tag{3.59}$$

Hence,

$$\text{K} \equiv \frac{1.380\,649 \times 10^{-23}\,\text{J}}{k_B}$$

The triple point of water remains approximately at $T = 273.16\,\text{K}$. The temperature is a measured quantity now.

[11] It is more common to write k for the Boltzmann constant, but in this book we use k_B to distinguish it from k we often use for a wave number.

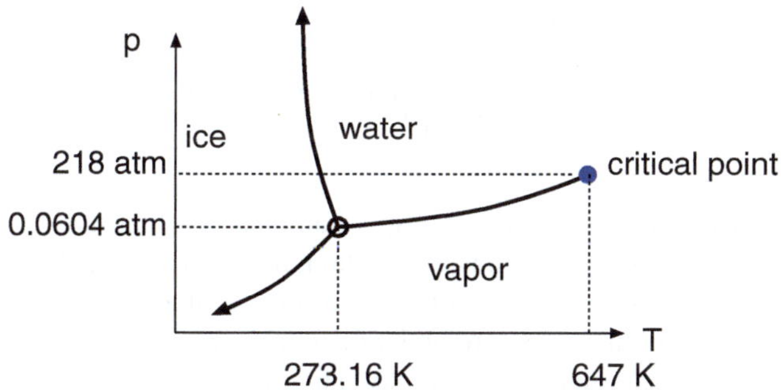

Fig. 3.7 The absolute temperature was defined by the temperature of the triple point of water before the 2019 revision of the SI

Examples 6

1 **Helmholtz's free energy**—Let $U(T, V)$ and $p(T, V)$ be the energy and pressure of a gas at temperature T and volume V. The second law of thermodynamics implies that there is a state function $S(T, V)$, called the entropy, that satisfies

$$dS(T, V) = \frac{1}{T}dU + \frac{p}{T}dV$$

This gives

$$\frac{\partial S(T, V)}{\partial T} = \frac{1}{T}\frac{\partial U(T, V)}{\partial T}$$
$$\frac{\partial S(T, V)}{\partial V} = \frac{1}{T}\frac{\partial U(T, V)}{\partial V} + \frac{p}{T}$$

(a) Show that $F(T, V) \equiv U(T, V) - TS(T, V)$ satisfies

$$dF = -SdT - pdV$$

(b) From the integrability of $S(T, V)$, i.e.,

$$\frac{\partial}{\partial T}\frac{\partial}{\partial V}S = \frac{\partial}{\partial V}\frac{\partial}{\partial T}S,$$

show

$$\frac{\partial U(T, V)}{\partial V} = T\frac{\partial p(T, V)}{\partial T} - p(T, V)$$

Hint: Use the integrability of $U(T, V)$ as well.

(c) Derive Maxwell's relation

$$\frac{\partial S(T,V)}{\partial V}=\frac{\partial p(T,V)}{\partial T}$$

2 **Ideal gas**—Determine the Gibbs energy of an ideal gas that satisfies the equation of state $pV = Nk_BT$ and has a constant heat capacity Nk_Bc_p at constant pressure, as a function of absolute temperature T and pressure p. **Hint**: You can use $F(T, V)$ obtained before and calculate $G(T, p) = F(T, V) + pV$ as a function of T, p. Alternatively, you may integrate

$$\frac{\partial G(T,p)}{\partial T}=-S(T,p),\qquad \frac{\partial G(T,p)}{\partial p}=V(T,p)$$

where $S(T, p)$ is obtained from $Nk_Bc_p = T\frac{\partial S(T,p)}{\partial T}$.

3 **Thermodynamic relations**—Given the Helmholtz free energy $F(T, V)$ as a function of temperature T and volume V, the internal energy is defined as

$$U \equiv F(T,V)-T\frac{\partial F(T,V)}{\partial T}$$

We define the entropy by $S(T, V) \equiv -\frac{\partial F(T,V)}{\partial T}$ and pressure by $p \equiv -\frac{\partial F(T,V)}{\partial V}$.

(a) Derive the first law of thermodynamics

$$dU = TdS - pdV$$

(b) Derive

$$\left(\frac{\partial U}{\partial V}\right)_T = T\left(\frac{\partial p}{\partial T}\right)_V - p$$

(c) We assume that the equation of state is given in the form:

$$p = Tf(V)$$

Show that the energy of the system depends only on T and does not depend on V.

Note: Therefore, for a gas that satisfies the equation of state of an ideal gas, $p = T\frac{Nk_B}{V}$, the energy does not depend on V.

4 **Change of variables**—Let $Y(x, y)$ be a function of two independent variables x and y. Treating x and Y as independent variables, show that the partial derivatives with respect to x and Y are given by those with respect to x and y as follows:

$$\left(\frac{\partial f}{\partial x}\right)_Y = \frac{\partial f}{\partial x} - \frac{\partial f}{\partial y}\frac{\frac{\partial Y}{\partial x}}{\frac{\partial Y}{\partial y}}$$

$$\left(\frac{\partial f}{\partial Y}\right)_x = \frac{\frac{\partial f}{\partial y}}{\frac{\partial Y}{\partial y}}$$

Hint:

$$dY = \frac{\partial Y}{\partial x}dx + \frac{\partial Y}{\partial y}dy$$

Solve for dy in terms of dx and dY, and then substitute the result into $df(x, y)$ to obtain the desired expressions. This result has wide-ranging applications, as will be seen in Exercise 6-3.

5 **Landau theory of ferromagnets**—Consider a uniform isotropic magnetic material at temperature T. Let M be the magnetization of the magnet under a uniform external magnetic field H_e applied in the z direction.

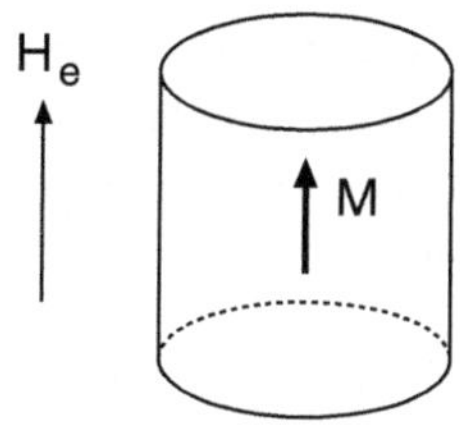

The energy U of the magnet satisfies[12]

$$dU = TdS + H_e dM$$

Following Landau, we assume that, given T and M, the Helmholtz free energy of the magnet is given in the following form:

$$F(T, M) = a(T)M^2 + bM^4$$

where $b > 0$ is a positive constant and $a(T)$ is a function of temperature

$$a(T) = c \cdot (T - T_c)$$

with positive constants c and T_c.

[12] See Sect. 3.3.4.

(a) From the definition $F(T, M) \equiv U - TS$, derive the general expression $dF(T, M) = -SdT + H_e dM$.
(b) Draw graphs showing the M dependence of $F(T, M)$ for the cases $T > T_c$ and $T < T_c$.

In thermal equilibrium at T and H_e,

$$F(T, M) - H_e M$$

is minimized.[13]

(c) For both $T > T_c$ and $T < T_c$, determine the magnetization M as a function of temperature at $H_e = 0$.
(d) Based on the above results, plot the phase diagram of the system in the (T, M) plane. **Comment**: This is similar to the (T, V) phase diagram of a liquid-gas system.
(e) Draw the corresponding phase diagram in the (T, H_e) plane. Show where M is discontinuous. **Comment**: This is similar to the (T, p) phase diagram of a liquid-gas system.
(f) Assuming that the external magnetic field H_e is small, determine M as a function of H_e for the cases $T < T_c$, $T > T_c$, and $T = T_c$. **Comment**: At the critical point $T = T_c$, $H_e = 0$, $\left(\frac{\partial M}{\partial H_e}\right)_T$ diverges.

6 **Landau theory of a liquid-gas phase transition**—Consider a liquid-gas system near the critical point. Following Landau, we assume that the Helmholtz free energy of the system near the critical temperature T_c can be expressed as a function of temperature T and volume V in the following form:

$$F(T, V) = (-p_c - a(T))(V - V_c) + b(T)(V - V_c)^2 + c(T)(V - V_c)^3 + d\,(V - V_c)^4$$

where p_c, V_c, d are positive constants and $a(T), b(T), c(T)$ are functions of temperature only, given by

$$a(T) = a \cdot (T - T_c), \quad b(T) = b \cdot (T - T_c), \quad c(T) = c \cdot (T - T_c)$$

(a) By shifting the volume as $V' \equiv V + \frac{c}{4d} \cdot (T - T_c)$, show that the dependence on V' can be rewritten as

$$F(T, V) \simeq (-p_c - a(T))(V' - V_c) + b(T)(V' - V_c)^2 + d\,(V' - V_c)^4$$

You can ignore the additive terms dependent only on temperature.

[13] This will be explained in the next section on statistical mechanics.

In thermal equilibrium at temperature T and pressure p, the volume V is determined such that

$$\begin{aligned} &F(T, V) + p(V' - V_c) \\ &= F(T, V) + (p - p_c - a(T))(V' - V_c) + (p_c + a(T))(V' - V_c) \end{aligned}$$

is minimized.

(b) For the case $T < T_c$, determine the vapor pressure $p_{tr}(T)$. **Hint**: At $p = p_{tr}(T)$, the gas and liquid coexist, so there are two solutions $V_l(T) < V_g(T)$ for V that minimizes $F(T, V) + p(V' - V_c)$.
(c) For $T < T_c$, determine $V_g(T) - V_l(T)$.
(d) Determine V as a function of p for the cases $T < T_c$, $T = T_c$, and $T > T_c$.

Comment: Is the similarity between this and the preceding problems clear? In Exercise 6-8, we demonstrate that a system satisfying the van der Waals equation of state behaves as predicted by Landau's theory.

Ferromagnet	Liquid-Gas
M	$V_c - V'$
H_e	$p - p_c - a(T)$

7 **Ideal rechargeable battery**—The thermodynamic state of a rechargeable battery is given by a free energy $F(q, T)$ as a function of the charge q stored in the battery and temperature T. We assume that the free energy is given in the form

$$F(q, T) = F_0(T) + qv(T)$$

where $v(T)$ is the voltage of the battery independent of q. At constant temperature, the change of the free energy is

$$\Delta F(q, T) = v(T)\Delta q$$

As the battery releases charge $-\Delta q > 0$, its free energy decreases, and it does positive amount of work $v(T)(-\Delta q)$. The energy $U(q, T)$ is defined by

$$U(q, T) \equiv F(q, T) - T\frac{\partial F(q, T)}{\partial T}$$

(a) Derive Helmholtz's relation

$$\frac{\partial U(q, T)}{\partial q} = v(T) - Tv'(T)$$

(b) Relate the sign of $\frac{\partial S(T,q)}{\partial q}$ to that of $v'(T)$, where $S(T, q) \equiv -\frac{\partial F(q,T)}{\partial T}$ is the entropy.
(c) Typically we find $v'(T) > 0$, i.e., the voltage increases as the temperature rises, since the chemical reactions become more active. Does the battery generate or absorb heat as we charge it ($\Delta q > 0$)?

Solutions to Examples 6

1 (a) By definition of F we obtain

$$dF = dU - TdS - SdT = -pdV - TdS$$

(b) From $dS = \frac{1}{T}dU + \frac{p}{T}dV$, we find

$$\frac{\partial S(T,V)}{\partial V} = \frac{1}{T}\frac{\partial U(T,V)}{\partial V} + \frac{p(T,V)}{T}$$
$$\frac{\partial S(T,V)}{\partial T} = \frac{1}{T}\frac{\partial U(T,V)}{\partial T}$$

Hence,

$$\frac{\partial}{\partial T}\frac{\partial S(T,V)}{\partial V} = -\frac{1}{T^2}\frac{\partial U}{\partial V} + \frac{1}{T}\frac{\partial}{\partial T}\frac{\partial U}{\partial V} + \frac{1}{T}\frac{\partial p}{\partial T} - \frac{1}{T^2}p$$
$$\frac{\partial}{\partial V}\frac{\partial S(T,V)}{\partial T} = \frac{1}{T}\frac{\partial}{\partial V}\frac{\partial U}{\partial T}$$

Using the integrability of S and U, we obtain

$$\frac{\partial U(T,V)}{\partial V} = T\frac{\partial p(T,V)}{\partial T} - p(T,V)$$

(c) We obtain

$$\frac{\partial S(T,V)}{\partial V} = \frac{1}{T}\frac{\partial U}{\partial V} + \frac{p}{T} = \frac{\partial p(T,V)}{\partial T}$$

2

$$\begin{cases} \frac{\partial S}{\partial T} &= Nk_B\frac{c_p}{T} \\ \frac{\partial S}{\partial p} &= -\frac{\partial V}{\partial T} = -Nk_B\frac{1}{p} \end{cases}$$

We integrate the above to obtain

$$S(T,p) = Nk_B\left(c_p \ln T - \ln p + s_0\right)$$

where s_0 is a constant of integration. Therefore,

$$\begin{cases} \frac{\partial G}{\partial T} &= Nk_B\left(-c_p \ln T + \ln p - s_0\right) \\ \frac{\partial G}{\partial p} &= N\frac{k_B T}{p} \end{cases}$$

Integrate the above to obtain

$$\boxed{G(T,p) = N\Big[k_B T\left(-c_p(\ln T - 1) + \ln p - s_0\right) + \epsilon_0\Big]}$$

where ϵ_0 is a constant of integration.

3 (a)

$$\begin{aligned} dU &= d\left(F - T\frac{\partial F}{\partial T}\right) = d\left(F + TS\right) \\ &= -SdT - pdV + dT \cdot S + TdS = TdS - pdV \end{aligned}$$

(b)

$$\left(\frac{\partial U}{\partial V}\right)_T = T\left(\frac{\partial S}{\partial V}\right)_T - p = T\left(\frac{\partial p}{\partial T}\right)_V - p$$

(c) Since $\left(\frac{\partial p}{\partial T}\right)_V = f(V)$, we find $\left(\frac{\partial U}{\partial V}\right)_T = 0$.

4

$$dY = \frac{\partial Y}{\partial x}dx + \frac{\partial Y}{\partial y}dy$$

Hence,

$$dy = \frac{1}{\frac{\partial Y}{\partial y}}\left(dY - \frac{\partial Y}{\partial x}dx\right)$$

We then obtain

$$\begin{aligned} df(x,y) &= \frac{\partial f}{\partial x}dx + \frac{\partial f}{\partial y}dy \\ &= \left(\frac{\partial f}{\partial x} - \frac{\frac{\partial Y}{\partial x}}{\frac{\partial Y}{\partial y}}\frac{\partial f}{\partial y}\right)dx + \frac{\frac{\partial f}{\partial y}}{\frac{\partial Y}{\partial y}}dY \end{aligned}$$

5 (a)

$$dF = dU - dT \cdot S - TdS = TdS + H_e dM - SdT - TdS = H_e dM - SdT$$

(b) $F(T, M) = c(T - T_c)M^2 + bM^4$

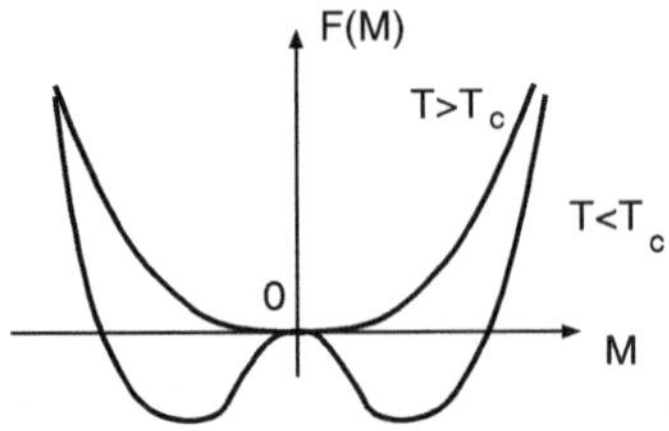

The minimization of $F(T, M) - H_e M$ under given T and H_e can be obtained graphically as follows. It is minimized at M where the tangent to the curve $F(T, M)$ equals H_e: $\frac{\partial}{\partial M} F(T, M) = H_e$. (For more details, see Sect. 3.3.2 on the Legendre transformations.)

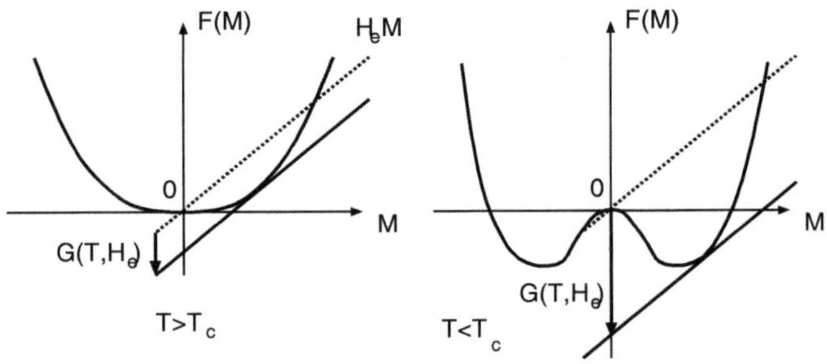

(c) $M(T)$ that minimizes $F(T, M)$ is found as

$$M(T) = \begin{cases} \pm\sqrt{\frac{c}{2b}(T_c - T)} & (T < T_c) \\ 0 & (T > T_c) \end{cases}$$

(d) For $T < T_c$, the region given by $|M| < \sqrt{\frac{c}{2b}(T_c - T)}$ is excluded (left-hand side of the figure below).

(e) For $T > T_c$, M is a continuous function of H_e. For $T < T_c$, M is discontinuous at $H_e = 0$ (right-hand side of the figure below).

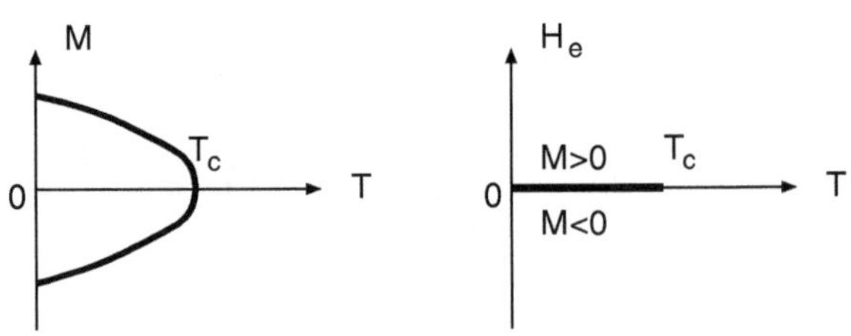

(f)

$$F(T, M) - H_e M = c(T - T_c)M^2 + bM^4 - H_e M$$

This is minimized with respect to M. Hence, the derivative with respect to M must be zero:

$$\frac{\partial}{\partial M} F(T, M) = H_e \Longrightarrow 2c(T - T_c)M + 4bM^3 = H_e$$

For $T > T_c$ and small H_e, we can neglect the b term, and

$$M \simeq \frac{H_e}{2c(T - T_c)} \Longrightarrow \left.\frac{\partial M}{\partial H_e}\right|_{H_e=0} = \frac{1}{2c(T - T_c)}$$

For $T < T_c$,

$$M \simeq M(T) + \frac{H_e}{4c(T_c - T)} \Longrightarrow \left.\frac{\partial M}{\partial H_e}\right|_{H_e=0} = \frac{1}{4c(T_c - T)}$$

For $T = T_c$,

$$M = \left(\frac{H_e}{4b}\right)^{\frac{1}{3}} \Longrightarrow \frac{\partial M}{\partial H_e} \propto H_e^{-\frac{2}{3}}$$

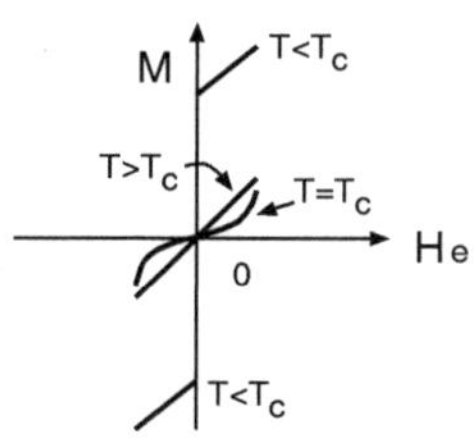

6 (a) Ignoring terms of second order or higher in $T - T_c$, we find

$$d(V' - V_c)^4 \simeq d(V - V_c)^4 + c(T)(V - V_c)^3,$$
$$b(T)(V' - V_c)^2 \simeq b(T)(V - V_c)^2$$

We thus obtain

$$F(T, V) \simeq (-p_c - a(T))(V' - V_c) + b(T)(V' - V_c)^2 + d(V' - V_c)^4$$

Hence,

$$F(T, V) + p(V' - V_c)$$
$$= b(T)(V' - V_c)^2 + d(V' - V_c)^4 + (p - p_c - a(T))(V' - V_c)$$

has the same form as $F(T, M) - H_e M$ in Example 6-5 (the preceding problem).

(b) Rewrite the condition $H_e = 0$ in Example 6-5 to get $p_{tr}(T) = p_c + a(T)$.

(c) Rewrite the solution $2|M(T)| = \sqrt{\frac{2c}{b}(T_c - T)}$ from 6-5(c) to obtain

$$V_g(T) - V_l(T) = \sqrt{\frac{2b}{d}(T_c - T)}$$

(d) We can rewrite the results from Example 6-5(f). For $T > T_c$ and for small $|p - p_{tr}(T)|$,

$$V_c - V'(T) \simeq \frac{p - p_{tr}(T)}{2b(T - T_c)}$$

For $T < T_c$,

$$\begin{cases} V_c - V_l'(T) \simeq \frac{p - p_{tr}(T)}{4b(T_c - T)} & (p > p_{tr}(T)) \\ V_c - V_g'(T) \simeq \frac{p - p_{tr}(T)}{4b(T_c - T)} & (p < p_{tr}(T)) \end{cases}$$

For $T = T_c$,

$$V_c - V \simeq \left(\frac{p - p_c}{4d}\right)^{\frac{1}{3}} \Longrightarrow p - p_c \simeq 4d\,(V_c - V)^3$$

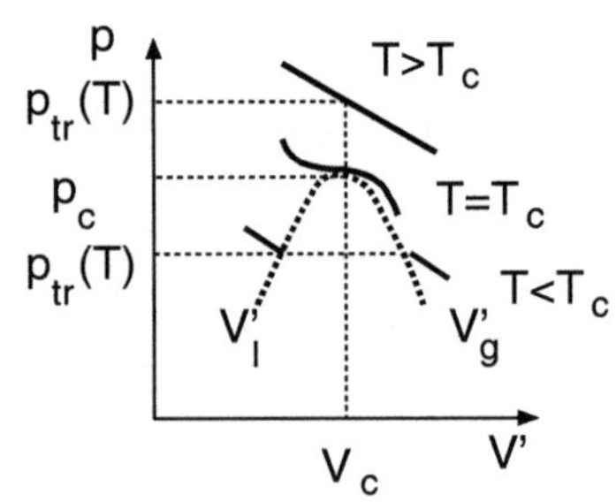

7 (a) By definition we obtain

$$U(q, T) = F_0(T) + qv(T) - T\left(F_0'(T) + qv'(T)\right)$$

Hence, we obtain the Helmholtz relation

$$\frac{\partial U(q,T)}{\partial q} = v(T) - Tv'(T)$$

(b) The entropy is

$$S(q,T) = -\frac{\partial F(q,T)}{\partial T} = -F_0'(T) - qv'(T)$$

Hence,

$$\frac{\partial S(q,T)}{\partial q} = -v'(T)$$

(c) $T\frac{\partial S(q,T)}{\partial q}dq = -Tv'(T)dq$ is the heat given to the battery as we store charge dq. If $v'(T) > 0$, the battery releases heat ($TdS < 0$) as it is charged ($dq > 0$). Conversely, the battery absorbs heat ($TdS > 0$) as it discharges ($dq < 0$). We can think of this more physically as follows. Since $F = U - TS$, a small change at constant temperature satisfies

$$\Delta F = \Delta U - T\Delta S = \Delta U - Q$$

where Q is the heat absorbed by the battery. Since $\Delta F = v(T)\Delta q$ and $\Delta U = u(T)\Delta q$, we find

$$Q = (u(T) - v(T))\Delta q = -Tv'(T)\Delta q$$

from the Helmholtz relation. We have assumed $v'(T) > 0$. Hence, $\Delta q > 0$ (charging) is exothermic ($Q < 0$), and $\Delta q < 0$ (discharging) is endothermic ($Q > 0$).

Exercises 6

1 **Specific heat of a solid**—Suppose that the equation of state for a solid is empirically given as

$$\left(\frac{\partial V}{\partial T}\right)_p = a + b\,p + c\,p^2$$

Determine the change in entropy when the pressure is changed from p_A to p_B at temperature T.
Hint: Use Maxwell's relation derived from Gibbs energy $G(T, p)$.

2 **Adiabatic demagnetization**—Consider a magnetic substance whose magnetization at an external magnetic field H_e follows Curie's law:

$$M = \frac{CH_e}{T}$$

where $C > 0$. The energy of the magnet at $M = 0$ is given by

$$U(T, M = 0) = aT^2$$

where $a > 0$.

(a) Show generally that $U(T, M)$ depends only on T if $H_e = Tf(M)$. **Hint**: Use $dU = TdS + H_e dM$. Determine $\frac{\partial S(T,M)}{\partial M}$ from Maxwell's relation. See Example 6-3.

Thus, $U(T) = aT^2$ independent of M.

(b) Determine the specific heat C_{H_e} at constant external magnetic field.
(c) Determine the change in temperature when the external magnetic field is changed from H_e to 0 adiabatically $\Delta S = 0$.

3 **Thermodynamic inequalities**

(a) Show that the thermal expansion coefficient at constant pressure is

$$\left(\frac{\partial V}{\partial T}\right)_p = -\frac{\left(\frac{\partial p}{\partial T}\right)_V}{\left(\frac{\partial p}{\partial V}\right)_T}$$

Hint: Compute $dp(T, V)$ and set it to zero.
(b) Derive the following relation:

$$\left(\frac{\partial S}{\partial T}\right)_p = \left(\frac{\partial S}{\partial T}\right)_V - \frac{\left\{\left(\frac{\partial p}{\partial T}\right)_V\right\}^2}{\left(\frac{\partial p}{\partial V}\right)_T}$$

Hint: Use the results from Example 6-4 and Maxwell's relation.
(c) The stability of thermal equilibrium requires

$$\left(\frac{\partial p(T, V)}{\partial V}\right)_T < 0$$

Using this, show that in general the specific heat at constant pressure is greater than the specific heat at constant volume.

$$C_p \equiv T\left(\frac{\partial S}{\partial T}\right)_p > C_V \equiv T\left(\frac{\partial S}{\partial T}\right)_V$$

(d) For an ideal gas, determine $C_p - C_V$.

(e) Derive

$$\left(\frac{\partial p}{\partial V}\right)_S = \left(\frac{\partial p}{\partial V}\right)_T - \frac{\left\{\left(\frac{\partial p}{\partial T}\right)_V\right\}^2}{\left(\frac{\partial S}{\partial T}\right)_V}$$

(f) Using $C_V > 0$, show that the adiabatic compressibility is greater than the isothermal compressibility:

$$-\frac{V}{p}\left(\frac{\partial p}{\partial V}\right)_S > -\frac{V}{p}\left(\frac{\partial p}{\partial V}\right)_T$$

(g) For an ideal gas, verify the above.
(h) Show in general

$$\frac{C_P}{C_V} = \frac{\left(\frac{\partial p}{\partial V}\right)_S}{\left(\frac{\partial p}{\partial V}\right)_T}$$

4 **Clausius-Clapeyron equation**—Search for experimental data for water to verify the Clausius-Clapeyron equation

$$\frac{dp_{tr}(T)}{dT} = \frac{1}{T}\frac{L(T)}{\Delta V}$$

5 **Van der Waals equation of state** —Consider a gas that satisfies the van der Waals equation of state

$$\left(p + a\frac{N^2}{V^2}\right)(V - Nb) = Nk_BT$$

where a, b are positive constants.

(a) What is the volume dependence of the heat capacity at constant volume C_V?

Let us take the heat capacity at constant volume as a positive constant Nk_Bc_V. (This is consistent with what we find above.)

(b) Determine the entropy S as a function of temperature T and volume V.
(c) Determine the Helmholtz free energy $F(T, V)$.
(d) How much does the temperature change when a gas expands freely by a volume of ΔV under zero external pressure?
Hint: You need $\left(\frac{\partial V}{\partial T}\right)_U$. This can be derived from $dU = TdS - pdV = 0$. For an ideal gas, the temperature change is zero, but it is non-zero in this case.
(e) Please explain the physical reason for the sign of the result obtained above.

6 **Superconducting phase transition**—Consider a cylindrical material of volume V under an external magnetic field H_e. This material becomes superconducting at temperatures $T < T_c$ as long as the external magnetic field is small $|H_e| < H_c(T)$.[14] In the superconducting phase, the total magnetic flux inside the material vanishes, and the total magnetization M becomes[15]

$$4\pi M = -H_e V \qquad \text{superconducting phase}$$

On the other hand, in the normal phase, there is no magnetization:

$$4\pi M = 0 \qquad \text{normal phase}$$

The energy U satisfies

$$dU = T dS + H_e dM ,$$

and the Gibbs free energy G is defined as

$$G(T, H_e) \equiv U - TS - H_e M$$

We write the Gibbs energy as G_s in the superconducting phase, G_n in the normal phase.

(a) Express dG in terms of dT and dH_e.
(b) Determine the H_e dependence of $G_s(T, H_e)$.
(c) Determine the H_e dependence of $G_n(T, H_e)$.

At a temperature $T < T_c$, G_s and G_n become equal at the critical magnetic field $H_c(T)$:

$$G_s(T, H_c(T)) = G_n(T, H_c(T))$$

Both phases coexist.

(c) Express $G_n(T, 0) - G_s(T, 0)$ in terms of $H_c(T)$.
(d) At $T < T_c$ as we increase the magnetic field past $H_c(T)$, the system goes from the superconductive phase to the normal phase. Determine the heat (latent heat) that must be given to the system at the critical magnetic field $H_c(T)$.
(e) Show that the entropy is continuous but the heat capacity is discontinuous between the two phases at $T = T_c$ and $H_e = 0$:

$$S_n(T_c + 0, 0) = S_s(T_c - 0, 0), \quad C_n(T_c + 0, 0) \neq C_s(T_c - 0, 0)$$

---

[14] We call $H_c(T)$ a critical magnetic field though the transition is of first order.

[15] Here, V is assumed to be macroscopic, and we disregard the London penetration length $\lambda_L \simeq 10^{-5}$ cm.

where the heat capacities are given by

$$C_n(T,0) \equiv T\frac{\partial S_n(T,0)}{\partial T}, \quad C_s(T,0) \equiv T\frac{\partial S_s(T,0)}{\partial T}$$

(f) Suppose that the critical magnetic field at $T < T_c$ is given by

$$H_c(T) = H_c(0)\left(1 - \frac{T^2}{T_c^2}\right)$$

Calculate $C_n(T_c, 0) - C_s(T_c, 0)$.

7 **Law of mass action**—Consider 1 mole of gas that obeys the ideal equation of state

$$pV = RT$$

where $R \equiv N_A k_B \simeq 8.3\,\mathrm{J\,mol^{-1}\,K^{-1}}$ is the gas constant.

(a) Let $g(T, p)$ be the Gibbs energy. We define

$$\phi(T) \equiv g(T, p) - RT \ln \frac{p}{RT}$$

Show that $\phi(T)$ depends only on the temperature T but not on the pressure p.

(b) Show that $\phi(T)$ satisfies

$$\frac{d}{dT}\frac{\phi(T)}{RT} = -\frac{u(T)}{RT^2}$$

where $u(T)$ is energy. **Note**: From Example 6-3, $u(T)$ is a function of temperature only.

We next consider the equilibrium of gas under a chemical reaction:

$$i_1 A_1 + \cdots + i_I A_I \longleftrightarrow j_1 B_1 + \cdots + j_J B_J$$

For instance, i_1 represents the number of moles of gas A_1.

(c) In an equilibrium state at T and p, demonstrate

$$\begin{aligned} &i_1 g_{A_1}(T, p_{A_1}) + \cdots + i_I g_{A_I}(T, p_{A_I}) \\ &\quad = j_1 g_{B_1}(T, p_{B_1}) + \cdots + j_J g_{B_J}(T, p_{B_J}) \end{aligned}$$

where p_{A_1} is the partial pressure of gas A_1 and $g_{A_1}(T, p_{A_1})$ represents the Gibbs energy of 1 mole of gas A_1 at partial pressure p_{A_1}.

Denoting the number of moles of gas A_1 per unit volume by $[A_1]$, the following relation holds in equilibrium:

$$\frac{[B_1]^{j_1}\cdots[B_J]^{j_J}}{[A_1]^{i_1}\cdots[A_I]^{i_I}} = K(T)$$

(d) Show that the above $K(T)$ is given by

$$K(T) = \exp\left[\frac{1}{RT}\big(i_1\phi_{A_1}(T) + \cdots + i_I\phi_{A_I}(T) - j_1\phi_{B_1}(T) - \cdots - j_J\phi_{B_J}(T)\big)\right]$$

(e) Calculate $\frac{d\ln K(T)}{dT}$ and interpret the result physically.

8 **Van der Waals equation of state and Landau theory**—We would like to follow up on the van der Waals equation of state considered in Exercise 6-5 above. The equation of state is

$$\left(p + a\frac{N^2}{V^2}\right)(V - Nb) = Nk_BT \qquad (a, b > 0)$$

This determines the volume V in terms of p, T.

(a) For particular pressure and temperature p_c, T_c, the equation of state has a triple root for volume V_c. Show

$$V_c = 3Nb, \quad p_c = \frac{a}{27b^2}, \quad k_BT_c = \frac{8a}{27b}$$

Hint: For $p = p_c$ and $T = T_c$, the equation of state reduces to $(V - V_c)^3 = 0$.

(b) Introducing dimensionless variables by

$$p^* \equiv \frac{p}{p_c}, \quad T^* \equiv \frac{T}{T_c}, \quad V^* \equiv \frac{V}{V_c}$$

show that the equation of state can be expressed as

$$\left(p^* + \frac{3}{V^{*2}}\right)\left(V^* - \frac{1}{3}\right) = \frac{8}{3}T^*$$

(c) What is the physical significance of this rewriting?

(d) Define $\Delta p^* \equiv p^* - 1$, $\Delta T^* \equiv T^* - 1$, and $\Delta V^* \equiv V^* - 1$. Show that up to the third order in ΔV^* and the first order in ΔT^* the equation of state can be approximated as

$$\Delta p^* = 4\Delta T^* - 6\Delta T^* \cdot \Delta V^* + 9\Delta T^* \cdot \left(\Delta V^*\right)^2 - \frac{3}{2}\left(\Delta V^*\right)^3$$

(The coefficient of $(\Delta V^*)^3$ has a correction first order in ΔT^*; we have ignored it.)

(e) Integrate the approximate equation of state obtained above with respect to ΔV^* to determine the Helmholtz free energy $F(T^*, V^*)$:

$$\frac{\partial F(T^*, V^*)}{\partial V^*} = -p^* = -1 - \Delta p^*$$

Comment: Therefore, the Landau theory discussed in Example 6-6 is valid here.

9 **Joule-Thomson effect**—We consider passing a gas through a porous material from a region of high pressure (A) to another of low pressure (B). Let p_A be the pressure of A and p_B be that of B. Let V_A be the volume of gas in A. After passing, its volume changes to V_B.

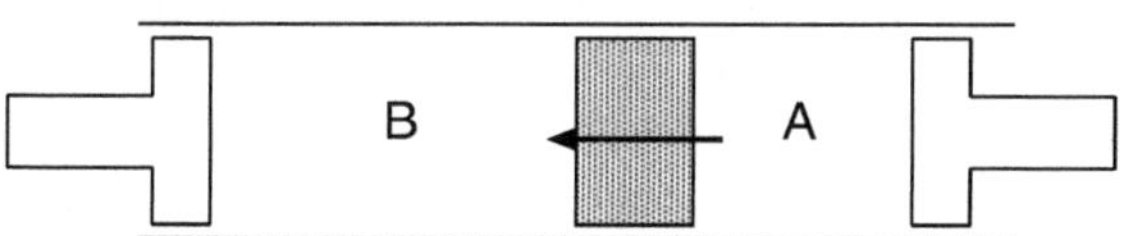

(a) Show that the enthalpy $H \equiv U + pV$ is unchanged after passing.

(b) Show that the Joule-Thomson coefficient $\left(\frac{\partial T}{\partial p}\right)_H$ (H is fixed) is given by

$$\left(\frac{\partial T}{\partial p}\right)_H = \frac{1}{C_p}\left(T\left(\frac{\partial V}{\partial T}\right)_p - V\right)$$

where C_p is the heat capacity at constant pressure. **Hint**: $C_p = T\left(\frac{\partial S}{\partial T}\right)_p = \frac{\partial H(T,p)}{\partial T}$

(c) Find the Joule-Thomson coefficient for the van der Waals gas considered in Exercise 6-5. Let C_V be the constant heat capacity at constant volume. What is the constraint on the parameters a, b so that the Joule-Thomson coefficient is positive?

3.2 Statistical Mechanics

Three Points to Remember

1. The canonical distribution is characterized by temperature T.
2. The grand canonical distribution is characterized by temperature T and chemical potential μ.
3. To understand cooperative phenomena, the mean field approximation is the first step.

3.2.1 The Canonical Distribution Is Characterized by Temperature T

The canonical distribution is the most important concept in statistical mechanics, and we must make sure to understand it well. The important relation between the specific heat and the fluctuation of energy often escapes our attention.

Consider a system in volume V in thermal equilibrium with a heat bath (or reservoir) at absolute temperature T.[16] The size of this system itself may be either macroscopic or microscopic. The heat bath, however, must be macroscopic and must be sufficiently larger than the system of interest. The probability that our system is in the microscopic state n with energy E_n is proportional to the Boltzmann factor

$$\mathrm{e}^{-\frac{E_n}{k_B T}}$$

where k_B is the Boltzmann constant and is given by[17]

$$k_B \simeq 1.38 \times 10^{-23}\,\mathrm{J/K} \tag{3.60}$$

By normalizing the Boltzmann factor, we obtain the probability

$$p_n = \frac{1}{Z(T)}\,\mathrm{e}^{-\frac{E_n}{k_B T}}, \tag{3.61}$$

where the normalization constant is the sum of the Boltzmann factors for all states

$$Z(T) \equiv \sum_n \mathrm{e}^{-\frac{E_n}{k_B T}} \tag{3.62}$$

[16] If the system consists of identical particles, their number N is fixed throughout Sect. 3.2.1.

[17] See the last subsection of Sect. 3.1.2 for the SI definition of k_B.

so that $\sum_n p_n = 1$. We first derive the above results and then proceed with their applications.

Derivation of the Boltzmann Factor

Let us recall the first law of thermodynamics:

$$dU = TdS - pdV \tag{3.63}$$

It follows that

$$\boxed{\left(\frac{\partial S}{\partial U}\right)_V = \frac{1}{T}} \tag{3.64}$$

Let us apply this expression to the entropy of a heat bath at temperature T. If the total energy of the heat bath and our system is U_0, and the energy of the system is E_n, the energy of the heat bath becomes $U_0 - E_n$. Therefore, the entropy of the heat bath is

$$S(U_0 - E_n) = S(U_0) - \frac{E_n}{T} \tag{3.65}$$

Since $U_0 \gg E_n$, it is acceptable to assume that this formula has no corrections. The temperature of a sufficiently large heat bath does not change even if it loses a finite energy E_n. Given a specific state n of the system, the entropy of the whole system (heat bath and the system) is equal to the entropy of the heat bath:

$$S_{\text{total}}(U_0 - E_n, E_n) = S(U_0 - E_n) = S(U_0) - \frac{E_n}{T} \tag{3.66}$$

We interpret $e^{\frac{1}{k_B}S(U_0-E_n)}$ as the number of microscopic states of the heat bath with energy $U_0 - E_n$.

We now adopt the **principle of equal weight**, i.e., as long as the total energy of the combined system, consisting of a heat bath and the system of our interest, remains constant, all microscopic states of the combined system have the same probability. Hence, the relative probability that the state of the system is n is determined solely by its energy E_n:

$$\mathrm{e}^{\frac{S_{\text{total}}}{k_B}} = \underbrace{\mathrm{e}^{\frac{S(U_0)}{k_B}}}_{E_n \text{ independent}} \cdot \underbrace{\mathrm{e}^{-\frac{E_n}{k_B T}}}_{\text{Boltzmann weight}} \tag{3.67}$$

Hence, the probability of the microscopic state n with energy E_n is proportional to the Boltzmann factor. This probability distribution is called the **canonical distribution** (Fig. 3.8).

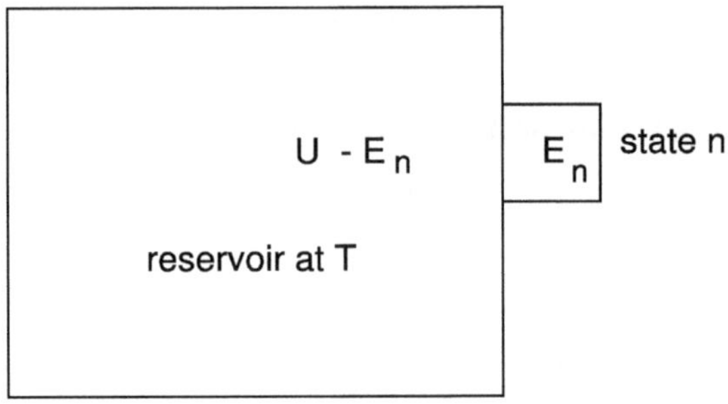

Fig. 3.8 The canonical distribution is determined only by the absolute temperature T

The Partition Function Gives Helmholtz's Free Energy

The energy E_n of the microscopic state n may depend on the volume V of the system, and we denote it as $E_n(V)$. The **partition function**, which is the normalization constant for probability, becomes a function of temperature and volume:

$$Z(T, V) \equiv \sum_n \mathrm{e}^{-\frac{E_n(V)}{k_B T}} \tag{3.68}$$

where the sum is over all the possible states of the system. Differentiating this with respect to temperature T, we obtain

$$\begin{aligned} \frac{\partial Z(T, V)}{\partial T} &= \sum_n \frac{E_n(V)}{k_B T^2} \mathrm{e}^{-\frac{E_n(V)}{k_B T}} \\ &= \frac{1}{k_B T^2} Z(T, V) \langle E \rangle \end{aligned} \tag{3.69}$$

where

$$\langle E \rangle \equiv \sum_n p_n E_n = U \tag{3.70}$$

is the average energy. This gives a formula that gives the average energy as the derivative of the partition function with respect to temperature:

$$\boxed{\frac{\partial \ln Z(T, V)}{\partial T} = \frac{1}{k_B T^2} U} \tag{3.71}$$

Similarly, the partial derivative with respect to volume gives

$$\left(\frac{\partial Z(T,V)}{\partial V}\right)_T = -\sum_n \frac{1}{k_BT}\frac{dE_n(V)}{dV}\,\mathrm{e}^{-\frac{E_n}{k_BT}}$$
$$= -\frac{1}{k_BT}Z(T,V)\left\langle\frac{dE}{dV}\right\rangle \tag{3.72}$$

Here, the average of $\frac{dE}{dV}$ can be interpreted as minus the pressure. This is because it is the change in energy with respect to volume while preserving the probability distribution over the states. Assuming a large system we obtain

$$\left\langle\frac{dE}{dV}\right\rangle = \left(\frac{\partial U}{\partial V}\right)_S = -p \tag{3.73}$$

The second equality is the thermodynamic equation derived from the first law $dU = TdS - pdV$. Therefore, we obtain a formula that gives the pressure as the partial derivative of the partition function with respect to volume:

$$\boxed{\frac{\partial \ln Z(T,V)}{\partial V} = \frac{p}{k_BT}} \tag{3.74}$$

From Eqs. (3.71) and (3.74), we obtain

$$d\ln Z(T,V) = \frac{1}{k_BT^2}UdT + \frac{p}{k_BT}dV$$
$$= \frac{1}{k_BT^2}\Big(\underbrace{(U-TS)}_{=F}dT + TSdT + TpdV\Big)$$
$$= \frac{1}{k_BT}\left(F\frac{dT}{T} - dF\right) = d\left(-\frac{F}{k_BT}\right) \tag{3.75}$$

This implies that the logarithm of the partition function is the Helmholtz free energy:[18]

$$\ln Z(T,V) = -\frac{F(T,V)}{k_BT} \Longrightarrow \boxed{Z(T,V) = \exp\left[-\frac{F(T,V)}{k_BT}\right]} \tag{3.76}$$

[18] Strictly speaking, we obtain $\ln Z$ up to an additive constant. This is the additive constant ambiguity of the entropy.

From this, the entropy can be expressed as follows:

$$\begin{aligned} S &= \frac{U - F}{T} = \frac{1}{T} \sum_n p_n \left(E_n + k_B T \ln Z\right) \\ &= \frac{1}{T} \sum_n p_n(-k_B T) \ln p_n \\ &= k_B \sum_n p_n \ln\left(1/p_n\right) = k_B \langle \ln\left(1/p_n\right) \rangle \end{aligned} \tag{3.77}$$

Since $1/p_n$ is proportional to the number of states that the system can take in thermal equilibrium, this equation gives the famous formula found on Boltzmann's tombstone:

$$S = k_B \ln W$$

At absolute zero temperature, the ground state of energy is realized with probability 1, so the entropy is zero.

Specific Heat Gives Energy Fluctuations

The specific heat (heat capacity to be more precise) at constant volume can be calculated from the partition function as follows:

$$\begin{aligned} C_V &\equiv \left(\frac{\partial U}{\partial T}\right)_V \\ &= \frac{\partial}{\partial T}\left(\frac{1}{Z(T, V)} \sum_n E_n \mathrm{e}^{-\frac{E_n}{k_B T}}\right) \\ &= -\frac{\partial \ln Z}{\partial T} U + \frac{1}{Z} \sum_n \frac{E_n^2}{k_B T^2} \mathrm{e}^{-\frac{E_n}{k_B T}} \\ &= -\frac{1}{k_B T^2} U^2 + \frac{1}{k_B T^2} \langle E^2 \rangle \\ &= \frac{1}{k_B T^2} \left(\langle E^2 \rangle - U^2\right) \\ &= \frac{1}{k_B T^2} \langle (E - U)^2 \rangle \end{aligned} \tag{3.78}$$

The last line gives the *average of squared fluctuations of energy*, or the mean squared fluctuation.

We have thus obtained the relation between the specific heat and the fluctuations in the energy of the system. Let us note two important points:

1. **Relative fluctuation of energy**—Suppose that our system is macroscopic with a large degree of freedom $N \gg 1$. Both the average energy and specific heat are of order N:

$$U = \mathrm{O}(N), \quad C_V = \mathrm{O}(N) \tag{3.79}$$

   Then, the mean squared fluctuation of energy is also of order N:

$$\langle (E_n - U)^2 \rangle = k_B T^2 C_V = \mathrm{O}(N) \tag{3.80}$$

   Hence, the relative fluctuation of energy is only of order $\frac{1}{\sqrt{N}}$:

$$\frac{\sqrt{\langle (E_n - U)^2 \rangle}}{U} = \mathrm{O}\left(\frac{1}{\sqrt{N}}\right) \tag{3.81}$$

   Therefore, for macroscopic systems, we can ignore the fluctuations in energy, resulting in the thermodynamic energy that we denote as U.
2. **Specific heat is always positive**—Since the specific heat is proportional to the mean squared fluctuation of energy, it is always positive:

$$\boxed{C_V > 0} \tag{3.82}$$

Example 1: Two-State System

Consider a simple system consisting of two states with energies $E_1 = -\varepsilon$ and $E_2 = \varepsilon > 0$. The partition function is given by

$$Z(T) = \sum_{n=1,2} \mathrm{e}^{-\frac{E_n}{k_B T}} = \mathrm{e}^{\frac{\varepsilon}{k_B T}} + \mathrm{e}^{-\frac{\varepsilon}{k_B T}} = 2\cosh\frac{\varepsilon}{k_B T} \tag{3.83}$$

Hence, the probability of the system having energy $\mp\varepsilon$ is

$$p_{\mp} = \frac{\mathrm{e}^{\pm\frac{\varepsilon}{k_B T}}}{Z(T)} \tag{3.84}$$

The average energy is

$$\langle E \rangle = \frac{1}{Z(T)}\varepsilon\left(-\mathrm{e}^{\frac{\varepsilon}{k_B T}} + \mathrm{e}^{-\frac{\varepsilon}{k_B T}}\right) = -\varepsilon \tanh\frac{\varepsilon}{k_B T} \tag{3.85}$$

At low temperatures, the system prefers the lower energy state, but at high temperatures, the probabilities of both states become almost equal:

$$\langle E\rangle \longrightarrow \begin{cases} -\varepsilon & (k_B T \ll \varepsilon) \\ 0 & (k_B T \gg \varepsilon) \end{cases} \tag{3.86}$$

Example 2: Quantum Harmonic Oscillator

The energy of a quantum harmonic oscillator with the natural angular frequency ω is given by

$$E_n = \left(n + \frac{1}{2}\right)\hbar\omega \quad (n = 0, 1, 2, \ldots) \tag{3.87}$$

Therefore, the partition function of a single harmonic oscillator is

$$Z(T) = \sum_{n=0}^{\infty} \mathrm{e}^{-\left(n+\frac{1}{2}\right)\frac{\hbar\omega}{k_B T}} = \frac{\mathrm{e}^{-\frac{\hbar\omega}{2k_B T}}}{1 - \mathrm{e}^{-\frac{\hbar\omega}{k_B T}}} = \frac{1}{2\sinh\frac{\hbar\omega}{2k_B T}} \tag{3.88}$$

Hence, the average energy is

$$\begin{aligned} \langle E\rangle &= \frac{1}{Z(T)} \sum_{n=0}^{\infty} \left(n + \frac{1}{2}\right)\hbar\omega\, \mathrm{e}^{-\left(n+\frac{1}{2}\right)\frac{\hbar\omega}{k_B T}} \\ &= \hbar\omega\Big(\frac{1}{2} + \underbrace{\langle n\rangle}_{=\frac{1}{\mathrm{e}^{\frac{\hbar\omega}{k_B T}}-1}}\Big) \\ &= \frac{1}{2}\hbar\omega\coth\frac{\hbar\omega}{2k_B T} \end{aligned} \tag{3.89}$$

Note the behavior at low and high temperatures:

$$\langle E\rangle \longrightarrow \begin{cases} \hbar\omega\left(\frac{1}{2} + \mathrm{e}^{-\frac{\hbar\omega}{k_B T}}\right) & \text{for} \quad T \ll \frac{\hbar\omega}{k_B} \\ k_B T & \text{for} \quad T \gg \frac{\hbar\omega}{k_B} \end{cases} \tag{3.90}$$

At high temperatures, the equipartition law of classical statistical mechanics holds ($\frac{1}{2}k_B T$ for each of the potential and kinetic energies).

Example 3: Einstein's Model for Solids

Following Einstein, we regard a crystal lattice made up of N atoms as a collection of $3N$ independent quantum harmonic oscillators, each with the same eigenfrequency

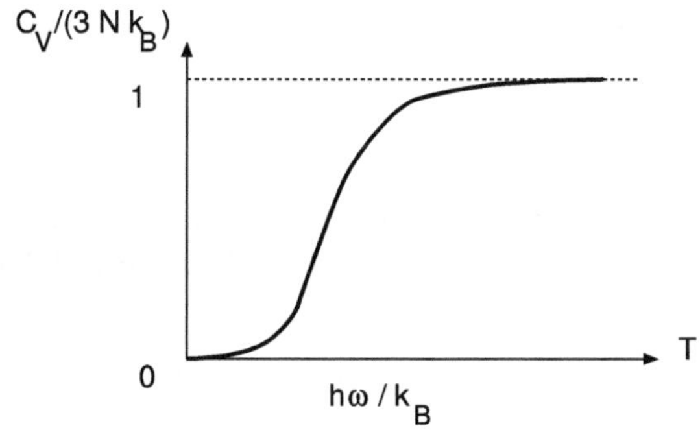

Fig. 3.9 Einstein's solid model explains Dulong-Petit's law

ω. (Since each atom can move in three directions, there are $3N$ of them.) The result for one harmonic oscillator can simply be multiplied by $3N$. Thus, the average energy is

$$U(T) = 3N \cdot \frac{1}{2}\hbar\omega \coth \frac{\hbar\omega}{2k_B T}, \tag{3.91}$$

and the specific heat is

$$C_V = \left(\frac{\partial U}{\partial T}\right)_V = 3Nk_B \left(\frac{\frac{\hbar\omega}{2k_B T}}{\sinh \frac{\hbar\omega}{2k_B T}}\right)^2$$

$$\xrightarrow{k_B T \gg \hbar\omega} 3Nk_B \quad \text{Dulong-Petit's Law} \tag{3.92}$$

$$\xrightarrow{k_B T \ll \hbar\omega} 3Nk_B \left(\frac{\hbar\omega}{k_B T}\right)^2 e^{-\frac{\hbar\omega}{k_B T}} \tag{3.93}$$

See Fig. 3.9. The behavior at low temperatures is more accurately described by Debye's model, which will be studied in Exercise 7-3.[19]

The Helmholtz free energy is

$$F(T, V) = 3Nk_B T \ln\left(2 \sinh \frac{\hbar\omega(V/N)}{2k_B T}\right), \tag{3.94}$$

which gives the equation of state as

$$p(T, V) = -\left(\frac{\partial F}{\partial V}\right)_V = \frac{3}{2}\left(-\omega'(V/N)\right) \frac{1}{\sinh \frac{\hbar\omega(V/N)}{2k_B T}} \tag{3.95}$$

[19] At low temperatures, the specific heat is proportional to T^3.

For the pressure to be positive, the angular frequency must be a decreasing function of volume.

3.2.2 *The Grand Canonical Distribution Is Characterized by Temperature T and Chemical Potential μ*

Let us take time to become familiar with the use of a chemical potential.

When considering a system consisting of one type of particle, instead of specifying the number of particles, we can specify its chemical potential μ.

Let us consider a system in thermal equilibrium with a heat bath characterized by temperature T and chemical potential μ. The relative probability of this system taking a microscopic state n with energy E_n and particle number N is given by

$$\frac{1}{\mathcal{Z}(T,\mu,V)}\exp\left(\frac{-E_n+N\mu}{k_B T}\right) \tag{3.96}$$

The normalization constant is defined as

$$\mathcal{Z}(T,\mu,V) \equiv \sum_N \exp\left(\frac{N\mu}{k_B T}\right) Z_N(T,V) \tag{3.97}$$

where $Z_N(T,V)$ is the partition function for the states with exactly N particles.[20] $\mathcal{Z}$ is called the **grand partition function**.

Let us derive the above result for the grand canonical distribution.

Derivation of the Grand Canonical Distribution

Let $S(U,V,N)$ be the entropy of the heat bath as a function of energy, volume, and particle number. We define the **chemical potential** μ as

$$\mu \equiv -T\left(\frac{\partial S}{\partial N}\right)_{U,V} \tag{3.98}$$

This implies

$$dS = \frac{dU}{T} + \frac{p}{T}dV - \frac{\mu}{T}dN \tag{3.99}$$

[20] In Sect. 3.2.1 we have fixed N tacitly.

We can rewrite this as

$$dU = TdS - pdV + \mu dN \tag{3.100}$$

Hence, we can also define the chemical potential as

$$\mu = \left(\frac{\partial U}{\partial N}\right)_{S,V} \tag{3.101}$$

Now, consider a closed system consisting of the heat bath and the system of interest. The total energy U_0 and the total number of particles N_0 are conserved. Therefore, for a system with energy E_n and particle number N, the entropy of the heat bath is

$$S(U_0 - E_n, N_0 - N) = S(U_0, N_0) - \frac{E_n}{T} + \frac{N\mu}{T} \tag{3.102}$$

Using the principle of equal weight, we find that the relative probability of the system is given by

$$\exp\left(\frac{-E_n + N\mu}{k_B T}\right) \tag{3.103}$$

This distribution is called the **grand canonical distribution**.

Fluctuations in Particle Number

In the case of the grand canonical distribution, the particle number is not constant, and we can consider its fluctuations. The derivation is almost identical to that of energy fluctuations in the canonical distribution. The partial derivatives of the grand partition function are given by

$$\frac{\partial \mathcal{Z}}{\partial \mu} = \sum_N \frac{N}{k_B T} \exp\left(\frac{N\mu}{k_B T}\right) Z_N(T, V) = \frac{1}{k_B T}\langle N\rangle \cdot \mathcal{Z} \tag{3.104}$$

$$\frac{\partial^2 \mathcal{Z}}{(\partial \mu)^2} = \sum_N \left(\frac{N}{k_B T}\right)^2 \exp\left(\frac{N\mu}{k_B T}\right) Z_N(T) = \frac{1}{(k_B T)^2}\langle N^2\rangle \cdot \mathcal{Z} \tag{3.105}$$

The second-order partial derivative can also be calculated as

$$\begin{aligned}\frac{\partial^2 \mathcal{Z}}{(\partial \mu)^2} &= \frac{1}{k_B T}\left[\frac{\partial \langle N\rangle}{\partial \mu}\mathcal{Z} + \langle N\rangle \frac{1}{k_B T}\langle N\rangle \mathcal{Z}\right] \\ &= \frac{1}{(k_B T)^2}\left[k_B T \frac{\partial \langle N\rangle}{\partial \mu} + \langle N\rangle^2\right]\mathcal{Z}\end{aligned} \tag{3.106}$$

Therefore, we obtain the mean squared fluctuation of the particle number as

$$\langle (N - \langle N \rangle)^2 \rangle = k_B T \frac{\partial \langle N \rangle}{\partial \mu} > 0 \tag{3.107}$$

From this, we conclude the following:

1. When $\langle N \rangle \gg 1$, the relative fluctuation in particle number is of the order $\frac{1}{\sqrt{\langle N \rangle}}$:

$$\frac{\sqrt{\langle (N - \langle N \rangle)^2 \rangle}}{\langle N \rangle} = \mathrm{O}\left(\frac{1}{\sqrt{\langle N \rangle}}\right) \tag{3.108}$$

   Therefore, when $\langle N \rangle \gg 1$, the fluctuations in particle number can be ignored.
2. The average number of particles $\langle N \rangle$ is an increasing function of μ:

$$\frac{\partial \langle N \rangle}{\partial \mu} > 0 \tag{3.109}$$

Thermodynamic Interpretation of the Grand Partition Function

In the macroscopic limit, where fluctuations can be ignored, the grand partition function can be interpreted thermodynamically. By differentiating the grand partition function with respect to temperature or volume, we obtain the following expressions:

$$\frac{\partial \ln \mathcal{Z}(T, V, \mu)}{\partial T} = \frac{U - N\mu}{k_B T^2} \tag{3.110}$$

$$\frac{\partial \ln \mathcal{Z}(T, V, \mu)}{\partial V} = \frac{p}{k_B T} \tag{3.111}$$

Furthermore, differentiation with respect to the chemical potential yields the particle number:

$$\frac{\partial \ln \mathcal{Z}(T, V, \mu)}{\partial \mu} = \frac{N}{k_B T} \tag{3.112}$$

Therefore, we obtain

$$\begin{aligned}
d \ln \mathcal{Z} &= \frac{(U - N\mu) dT}{k_B T^2} + \frac{p dV}{k_B T} + \frac{N d\mu}{k_B T} \\
&= -\frac{dT}{k_B T^2}(N\mu - U + TS) + \frac{1}{k_B T}(N d\mu + S dT + p dV) \\
&= -\frac{dT}{k_B T^2}(N\mu - U + TS) + \frac{1}{k_B T}(N d\mu + \mu dN + S dT + p dV - \mu dN)
\end{aligned}$$

$$= -\frac{dT}{k_B T^2}(N\mu - F) + \frac{1}{k_B T}(d(N\mu) - dF)$$
$$= d\left(\frac{1}{k_B T}(N\mu - F)\right) \tag{3.113}$$

As a result,[21] we obtain

$$\ln \mathcal{Z} = \frac{1}{k_B T}(N\mu - F) \tag{3.114}$$

Relation Between the Chemical Potential μ and the Gibbs Free Energy (Mainly for Gases and Liquids)

In a homogeneous system whose state is determined by temperature T, pressure p, and particle number N (e.g., a gas or a liquid), the Gibbs free energy should be a state function that depends only on the **extensive variables**, which are quantities that double when the size of the system doubles, such as volume, energy, and free energy. Therefore, G/N is an **intensive variable** that depends only on T and p:

$$G(T, p, N) = Ng(T, p) \tag{3.115}$$

Hence, we obtain

$$\left(\frac{\partial G}{\partial N}\right)_{T,p} = g(T, p) \tag{3.116}$$

On the other hand, from the thermodynamic identity (3.100), we find

$$d(\underbrace{U - TS + pV}_{=G}) = -SdT + Vdp + \mu dN \tag{3.117}$$

Therefore, we obtain

$$\mu = \left(\frac{\partial G}{\partial N}\right)_{T,p} \tag{3.118}$$

Using Eq. (3.116), we can conclude

$$\mu = g(T, p) \Longrightarrow G(T, p, N) = N\mu \tag{3.119}$$

[21] Here, we have used $dF = -SdT - pdV + \mu dN$, which can be derived from Eq. (3.100).

In other words, the chemical potential is the Gibbs energy per particle. Consequently, from Eq. (3.114), we obtain

$$\ln \mathcal{Z} = \frac{1}{k_B T}(\underbrace{N\mu}_{=G} - F) = \frac{pV}{k_B T} \tag{3.120}$$

Example: Quantum Mechanics of an Ideal Gas

Let us consider a gas of free particles confined within a cubic volume $V \equiv L^3$. When L is sufficiently large, the boundary conditions imposed on the walls of the box are not important. Therefore, it is permissible to impose whatever convenient boundary conditions for calculations. Adopting the periodic boundary conditions, the wave vectors of the particles are quantized as follows:

$$\vec{k} = \frac{2\pi}{L}\vec{n}, \quad \vec{n} = (n_1, n_2, n_3) \tag{3.121}$$

where n_1, n_2, n_3 are arbitrary integers, both positive and negative. Consequently, the number of states within the volume element $\Delta k_x \Delta k_y \Delta k_z$ in the wave vector space is given by

$$\left(\frac{L}{2\pi}\right)^3 \Delta k_x \Delta k_y \Delta k_z$$

Now, focusing on the state with wave vector $\vec{k}$, the energy of a free particle in this state is given by

$$\varepsilon_{\vec{k}} \equiv \frac{\hbar^2 \vec{k}^2}{2m} \tag{3.122}$$

The average number of particles in this state depends on whether they are Fermi particles or Bose particles.

- **Fermi particles**—Only $n = 0, 1$ are allowed, and the average number is given by

$$N_{\vec{k}} = \frac{e^{\frac{\mu - \epsilon_k}{k_B T}}}{1 + e^{\frac{\mu - \epsilon_k}{k_B T}}} = \frac{1}{e^{\frac{\epsilon_k - \mu}{k_B T}} + 1} \tag{3.123}$$

- **Bose particles**—All $n = 0, 1, 2, \cdots$ are allowed, and the average number is given by

$$N_{\vec{k}} = \frac{\sum_{n=1}^{\infty} n \left(e^{\frac{\mu-\epsilon_k}{k_B T}}\right)^n}{\sum_{n=0}^{\infty} \left(e^{\frac{\mu-\epsilon_k}{k_B T}}\right)^n} = \frac{1}{e^{\frac{\epsilon_k-\mu}{k_B T}} - 1} \tag{3.124}$$

We consider a gas dilute if $N_{\vec{k}} \ll 1$. This condition is achieved if

$$e^{\frac{\mu}{k_B T}} \ll 1 \tag{3.125}$$

holds. In this case, for both Fermi and Bose particles, we have

$$N_{\vec{k}} \simeq e^{\frac{\mu-\epsilon_k}{k_B T}} \ll 1 \tag{3.126}$$

Therefore, the total number of particles can be calculated as follows:

$$\begin{aligned} N &= V \int \frac{d^3k}{(2\pi)^3} e^{\frac{\mu-\epsilon_k}{k_B T}} \\ &= V e^{\frac{\mu}{k_B T}} \int \frac{d^3k}{(2\pi)^3} e^{-\frac{\hbar^2 k^2}{2mk_B T}} \\ &= V e^{\frac{\mu}{k_B T}} \left(\frac{\sqrt{2mk_B T}}{2\pi\hbar}\right)^3 \underbrace{\left(\int_{-\infty}^{\infty} dx\, e^{-x^2}\right)^3}_{=\pi^{\frac{3}{2}}} \\ &= V e^{\frac{\mu}{k_B T}} \left(\frac{mk_B T}{2\pi\hbar^2}\right)^{\frac{3}{2}} \end{aligned} \tag{3.127}$$

Hence, the chemical potential is determined as

$$\mu = k_B T \ln\left[\frac{N}{V}\left(\frac{mk_B T}{2\pi\hbar^2}\right)^{-\frac{3}{2}}\right] \tag{3.128}$$

Now, define the **quantum number density**[22] as

[22] Also known as quantum concentration. The energy of a particle with the De Broglie wavelength of $n_Q(T)^{-\frac{1}{3}}$ is roughly $k_B T$.

$$n_Q(T) \equiv \left(\frac{mk_BT}{2\pi\hbar^2}\right)^{\frac{3}{2}} \propto T^{\frac{3}{2}} \tag{3.129}$$

so that

$$\mu = k_BT \ln \frac{\frac{N}{V}}{n_Q(T)} \Longrightarrow e^{\frac{\mu}{k_BT}} = \frac{\frac{N}{V}}{n_Q(T)} \tag{3.130}$$

Thus, to be able to treat the gas as dilute, the number density must be much smaller than the quantum number density, as given by Eq. (3.125).

Now, for a dilute gas, the grand partition function is approximately given by

$$\begin{aligned}\mathcal{Z} &\simeq \prod_{\vec{k}} \left(1 + e^{\frac{\mu-\epsilon_k}{k_BT}}\right) \\ &\simeq \exp\left[V \int \frac{d^3k}{(2\pi)^3} e^{\frac{\mu-\epsilon_k}{k_BT}}\right] \\ &= e^N \end{aligned} \tag{3.131}$$

and, from Eq. (3.120), we obtain the equation of state as

$$pV = Nk_BT \tag{3.132}$$

Thus, the Gibbs free energy, as determined from Eq. (3.128), is

$$G = N\mu = Nk_BT \ln\left[\frac{p}{k_BT}\frac{1}{n_Q(T)}\right] \tag{3.133}$$

The additive constants ε_0, s_0 in energy and entropy that were left undetermined by thermodynamics are now fixed. Substituting (3.129) into the above, we obtain

$$\begin{aligned}G &= Nk_BT \ln\left[\frac{p}{k_BT}\left(\frac{2\pi\hbar^2}{mk_BT}\right)^{\frac{3}{2}}\right] \\ &= Nk_BT\left[\ln p - \ln k_BT + \frac{3}{2}\ln\frac{2\pi\hbar^2}{mk_BT}\right]\end{aligned} \tag{3.134}$$

Comparing this with Eq. (3.43) for $c_V = \frac{3}{2}$, we obtain

$$\varepsilon_0 = 0 \tag{3.135a}$$

$$s_0 = k_B \left(\frac{5}{2} - \frac{3}{2} \ln \frac{2\pi \hbar^2}{mk_B} \right) \tag{3.135b}$$

3.2.3 *To Understand Cooperative Phenomena, the Mean Field Approximation Is the First Step*

The mean field approximation is used often in condensed matter physics and elementary particle physics. All graduate students need to familiarize themselves with it.

In Example 6-5, we have considered Landau's phenomenological model for ferromagnets.[23] We have observed that even if no external magnetic field is applied, if the temperature is sufficiently low, the magnetization is nonvanishing, resulting in a lower Gibbs free energy.

$$G(T)|_{H=0} = \begin{cases} F(T, M = 0) = 0 & \text{if} \quad T > T_c \\ F(T, M \neq 0) < 0 & \text{if} \quad T < T_c \end{cases} \tag{3.136}$$

(Fig. 3.10) Let us see what is going on at $T < T_c$. When the temperature is low, it is more important to reduce energy than to increase entropy. Aligning magnetization vectors reduces entropy, increasing free energy. On the other hand the magnetization within the system acts as an external magnetic field, making the magnetization non-zero, thereby reducing energy and decreasing free energy.

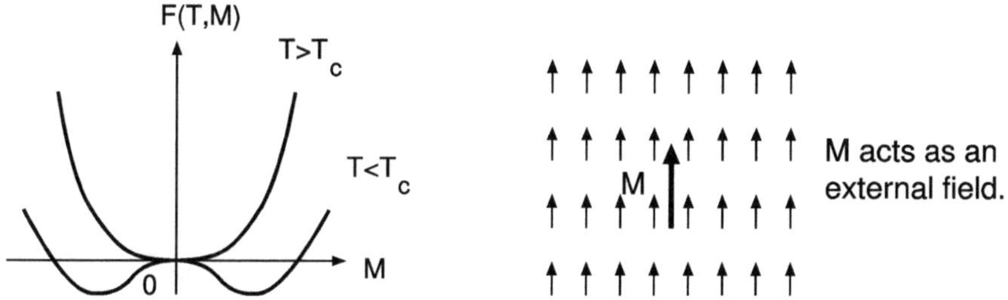

Fig. 3.10 The Helmholtz free energy of a ferromagnet as a function of average magnetization M

[23] In Sect. 3.3.6 we derive this model using the Bragg-Williams approximation equivalent to the mean field approximation.

To qualitatively understand cooperative phenomena, the mean field approximation described here is very handy. Though it may not always provide a good quantitative approximation, it is very useful for seeing what kind of phases are possible and for predicting the temperature dependence of the system. Let us now look at a concrete example.

Example: Ising Model

Consider a D-dimensional cubic lattice. Each lattice site (or point) has a spin with magnetic moment, and its direction is assumed to be either in the direction of a certain coordinate axis or in the opposite direction. The energy of this entire lattice is given by

$$E = -K \sum_{\langle i,j \rangle} s_i s_j - h \sum_i s_i \,, \tag{3.137}$$

where $s_i = \pm 1$ is the normalized magnetic spin and i represents each lattice site. $\langle i, j \rangle$ represents all neighboring pairs of the spins. h is the external magnetic field (Fig. 3.11).

Let us focus on one spin s_i, which has $2D$ neighboring spins. Ignoring the magnetic field, the contribution of this spin to the energy depends only on the relative orientations of the $2D$ neighboring spins. We now replace the sum of the values of the $2D$ neighboring spins with their average value, or **mean field**:

$$\sum_{\text{neighbor } j} s_j \to \alpha \langle s \rangle \tag{3.138}$$

Here,

$$\alpha = 2D \tag{3.139}$$

is the number of neighboring spins. For a non-cubic lattice, α may not be $2D$. Under this replacement, the contribution of the spin s_i to the energy is

$$E(s_i) = -K\alpha s_i \langle s \rangle - h s_i = -s_i (K\alpha \langle s \rangle + h) \tag{3.140}$$

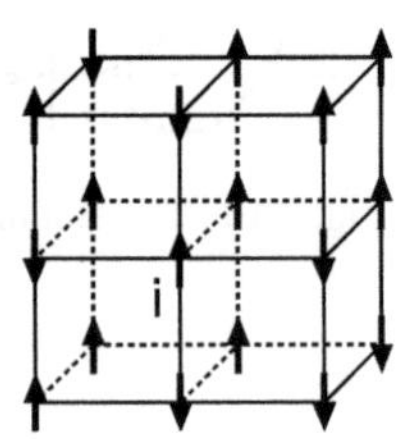

Fig. 3.11 Ising model

In order for the replacement to be consistent, the average s_i calculated using the energy $E(s_i)$ above must be equal to $\langle s\rangle$. Therefore, we must find

$$\langle s\rangle = \frac{e^{\frac{K\alpha\langle s\rangle+h}{k_BT}} - e^{-\frac{K\alpha\langle s\rangle+h}{k_BT}}}{e^{\frac{K\alpha\langle s\rangle+h}{k_BT}} + e^{-\frac{K\alpha\langle s\rangle+h}{k_BT}}} = \tanh\frac{K\alpha\langle s\rangle+h}{k_BT} \tag{3.141}$$

This defines $\langle s\rangle$ implicitly as a function of T and h. In the absence of an external magnetic field, the above equation yields

$$\langle s\rangle = \tanh\left(\frac{K\alpha}{k_BT}\langle s\rangle\right) \tag{3.142}$$

We define

$$T_c \equiv \frac{K\alpha}{k_B} \tag{3.143}$$

For $T > T_c$ Eq. (3.142) has no solution other than $\langle s\rangle = 0$. For $T < T_c$, however, there exists another solution where $\langle s\rangle \neq 0$, and this solution has a lower energy. Therefore, spontaneous magnetization occurs when $T < T_c$. T_c is the **critical temperature** (Fig. 3.12).

Consider a temperature slightly below the critical temperature.

$$\frac{K\alpha}{k_BT} = \frac{K\alpha}{k_B(T_c+T-T_c)} \simeq \underbrace{\frac{K\alpha}{k_BT_c}}_{=1}\left(1-\frac{T-T_c}{T_c}\right) \tag{3.144}$$

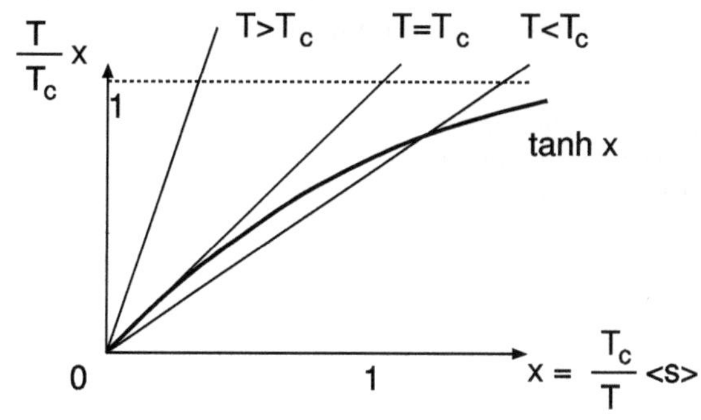

Fig. 3.12 When $T < T_c$, there exists a solution $x > 0$ to $\frac{T}{T_c}x = \tanh x$

Denoting $\Delta T \equiv T - T_c$, and using the approximation $\tanh x \simeq x - \frac{x^3}{3}$ for $|x| \ll 1$, Eq. (3.142) gives

$$\begin{aligned}\langle s\rangle &= \tanh\left(1-\frac{\Delta T}{T_c}\right)\langle s\rangle \\ &\simeq \left(1-\frac{\Delta T}{T_c}\right)\langle s\rangle - \frac{1}{3}\left\{\left(1-\frac{\Delta T}{T_c}\right)\langle s\rangle\right\}^3 \\ &\simeq \left(1-\frac{\Delta T}{T_c}\right)\langle s\rangle - \frac{1}{3}\langle s\rangle^3 \end{aligned} \tag{3.145}$$

Assuming $\langle s\rangle \neq 0$, we get

$$\frac{-\Delta T}{T_c}\langle s\rangle \simeq \frac{1}{3}\langle s\rangle^3 \tag{3.146}$$

For $\Delta T < 0$, this gives the **spontaneous magnetization**

$$\langle s\rangle \simeq \pm\sqrt{\frac{3(-\Delta T)}{T_c}} \propto (-\Delta T)^{\frac{1}{2}} \tag{3.147}$$

(Fig. 3.13) Below the critical temperature, spontaneous magnetization occurs. Near the critical temperature, it is proportional to the square root of ΔT, as has been derived from Landau's model for ferromagnetism in Example 6-5. In actual magnetic materials, spontaneous magnetization is better approximated by $(-\Delta T)^{\frac{1}{3}}$.[24]

Examples 7

1 **Model for a thermal bath**—Consider a system made up of N quantum harmonic oscillators with angular frequency ω. The interaction energy between the oscillators is assumed to be negligible. Let U be the total energy of the system (excluding the zero-point energy).

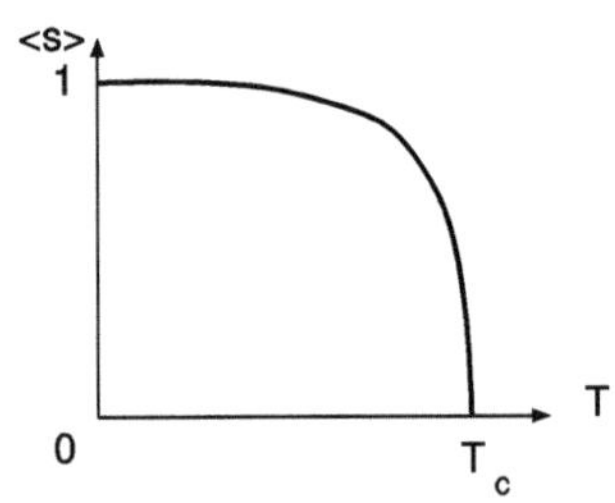

Fig. 3.13 Temperature dependence of spontaneous magnetization

[24] The actual exponent of the power law varies depending on the type of a magnet.

(a) Find the number of states of the system $e^{\frac{S(U)}{k_B}}$. Assume $N \gg 1$ and $\frac{U}{\hbar\omega} \gg 1$.
Hint—Use Stirling's approximation: $\ln N! \approx N(\ln N - 1)$.

(b) Express the absolute temperature T, defined by $\frac{1}{T} \equiv \frac{dS(U)}{dU}$, as a function of U.
Comment—The equipartition theorem is valid.

(c) Determine the heat capacity of the system.

2 **Magnetic moment**—Consider N spins that interact with each other only to exchange energy. Each spin is assumed to have a magnetic moment μ in the direction of the spin. Furthermore, assume that each spin points in the direction of the magnetic field or exactly opposite to it.

(a) Determine the partition function $Z(T, H)$ under the temperature T and external magnetic field H, and compute the Gibbs free energy

$$G(T, H) \equiv -k_B T \ln Z(T, H)$$

(b) Find the entropy $S(T, H)$.

(c) Determine the magnetization $M(T, H)$.

(d) Compute the energy $U(T, H) \equiv G(T, H) + TS + HM$.
Hint: Since we are not considering the interaction among the magnetic moments, the energy vanishes.

(e) Find the enthalpy $E(T, H) \equiv U(T, H) - HM$.[25]

(f) Compute the squared fluctuation of enthalpy $\left\langle (E - \langle E \rangle)^2 \right\rangle$, and derive its relation to the specific heat $C(T, H)$.

(g) Initially, the system's temperature is T, and the magnetic field is H. Find the system's temperature T' after the magnetic field is adiabatically changed from H to H'. If $0 < H' < H$, how do T' and T compare?
Hint: Note that the entropy is a function of $\frac{\mu H}{T}$.

3 T, p **distribution**—Consider a macroscopic system with volume V. If the energy of the system in quantum state n is denoted as $E_n(V)$, then the partition function under temperature T and pressure p is given by

$$Z(T, p) \equiv \int_0^\infty dV \sum_n e^{-\frac{1}{k_B T}(E_n(V) + pV)}$$

(a) Show

$$Z(T, p) = e^{-\frac{G(T,p)}{k_B T}}$$

where $G(T, p)$ is the Gibbs free energy.

[25] This is usually called energy.

Hint: Calculate $\frac{\partial Z(T,p)}{\partial T}$ and $\frac{\partial Z(T,p)}{\partial p}$, and show that $G \equiv -k_B T \ln Z$ satisfies

$$dG = -SdT + Vdp$$

where V is the average volume.

(b) Denoting the average volume as $\langle V \rangle$, derive the relation between the squared fluctuation of the volume $\langle (V - \langle V \rangle)^2 \rangle$ and the compressibility $-\frac{p}{V}\left(\frac{\partial \langle V \rangle}{\partial p}\right)_T$.

Comment: This shows that the compressibility is positive.

(c) If the system's degrees of freedom is N, demonstrate that the relative fluctuation of the volume is of the order $1/\sqrt{N}$:

$$\frac{\sqrt{\langle V^2 \rangle - \langle V \rangle^2}}{\langle V \rangle} = O\left(\frac{1}{\sqrt{N}}\right)$$

(d) For an ideal gas, compute the compressibility and the volume fluctuation.

4 **Langmuir's formula**—Consider an adsorption site in thermal equilibrium with a monatomic ideal gas at temperature T and pressure p.

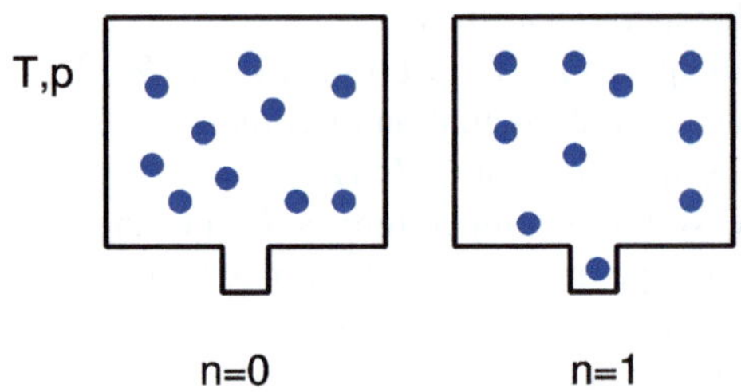

(a) Given an atomic mass m, calculate the chemical potential $\mu(T, p)$ of the ideal gas.

Hint: You may use Eq. (3.128).

Assume that the energy of an atom is $-\varepsilon$ if adsorbed on the site and the energy is zero if not adsorbed. Assume also that no more than one atom is adsorbed on the site.

(b) Calculate the probability that an atom is adsorbed on the site. Plot this probability as a function of pressure at a constant temperature.

5 **Number density in a gravitational field**—Consider a dilute gas in equilibrium at temperature T under the gravitational potential $\phi(\vec{x})$. When the chemical potential is μ, the number density of the gas is proportional to

$$\mathrm{e}^{-\frac{m\phi(\vec{x})}{k_B T}}$$

where m is the mass of the particle. **Hint**—You will see that the chemical potential μ is replaced by $\mu - m\phi(\vec{x})$.

6 **Curie temperature**—Consider the classical magnetic spin $\vec{\mu}$. The magnitude of $\vec{\mu}$ is a constant, but its direction is free. The energy under an external magnetic field H is given by

$$E = -\vec{\mu} \cdot \vec{H} = -\mu_z H$$

(a) Determine the average $\langle \mu_z \rangle$ under the temperature T and external magnetic field H.

Next, consider a cubic lattice made of the magnetic spins. The total energy of the system, with g as a positive constant, is given by

$$E = -g \sum_{\langle i,j \rangle} \vec{\mu}_i \cdot \vec{\mu}_j - H \sum_i \mu_{i,z}$$

where the initial sum is taken over the nearest spin pairs, and the external magnetic field acts in the z direction.

(b) Using the mean field approximation, derive the equation that gives the average $\langle \mu_z \rangle$.
(c) Find the Curie temperature T_c below which spontaneous magnetization occurs.
(d) For $T \simeq T_c$ and $H = 0$, determine $\langle \mu_z \rangle$ as a function of T.
(e) For $T > T_c$, determine the magnetic susceptibility $\frac{\partial \langle \mu_z \rangle}{\partial H} |_{H=0}$.
(f) For $T = T_c$, determine $\langle \mu_z \rangle$ as a function of H.

Solutions to Examples 7

1 (a) The energy of each harmonic oscillator is $n\hbar\omega$ $(n = 0, 1, 2, \cdots)$. Therefore, the number of states of the system is equal to the number of ways to distribute $\frac{U}{\hbar\omega}$ into N groups. Hence,

$$e^{\frac{S(U)}{k_B}} = \frac{\left(\frac{U}{\hbar\omega} + N - 1\right)!}{(N-1)!\left(\frac{U}{\hbar\omega}\right)!}$$

Therefore, by approximating the logarithm of this with the Stirling formula, we obtain

$$\frac{S(U)}{k_B} \simeq \left(\frac{U}{\hbar\omega} + N - 1\right) \ln\left(\frac{U}{\hbar\omega} + N - 1\right) - (N-1)\ln(N-1) - \frac{U}{\hbar\omega} \ln \frac{U}{\hbar\omega}$$

$$\simeq \left(\frac{U}{\hbar\omega}+N\right)\ln\frac{U}{\hbar\omega}-N\ln N-\frac{U}{\hbar\omega}\ln\frac{U}{\hbar\omega}=N\ln\frac{U}{N\hbar\omega}$$

(b) $\frac{1}{T}=k_B\frac{N}{U}$, so $T=\frac{U}{Nk_B}$

(c) From $U=Nk_BT$, we have $C=\frac{dU}{dT}=Nk_B$.

2 (a) Using $Z(T,H)=\left(e^{\frac{\mu H}{k_BT}}+e^{-\frac{\mu H}{k_BT}}\right)^N$, we get

$$G(T,H)=-Nk_BT\ln\left(e^{\frac{\mu H}{k_BT}}+e^{-\frac{\mu H}{k_BT}}\right)$$

(b)

$$S(T,H)=-\frac{\partial G(T,H)}{\partial T}=Nk_B\left[\ln\left(e^{\frac{\mu H}{k_BT}}+e^{-\frac{\mu H}{k_BT}}\right)-\frac{\mu H}{k_BT}\tanh\frac{\mu H}{k_BT}\right]$$

(c)

$$M(T,H)=N\mu\tanh\frac{\mu H}{k_BT}$$

(d) 0

(e) $E=-HM=-N\mu H\tanh\frac{\mu H}{k_BT}$

(f) The mean squared fluctuation is

$$\left\langle (E-\langle E\rangle)^2\right\rangle=N(\mu H)^2\left(1-\tanh^2\frac{\mu H}{k_BT}\right)=N(\mu H)^2\frac{1}{\cosh^2\frac{\mu H}{k_BT}}$$

On the other hand, the heat capacity is

$$C=T\frac{\partial S}{\partial T}=N\frac{(\mu H)^2}{k_BT^2}\frac{1}{\cosh^2\frac{\mu H}{k_BT}}=\frac{1}{k_BT^2}\left\langle (E-\langle E\rangle)^2\right\rangle$$

(g) Under the adiabatic condition, $\frac{\mu H}{k_BT}$ is constant. So, T changes in proportion to H:

$$T'=T\frac{H'}{H}$$

3 (a)

$$\begin{cases}\frac{\partial\ln Z}{\partial T} = \frac{1}{k_BT^2}\langle E_n+pV\rangle=\frac{1}{k_BT^2}(U+pV)\\ \frac{\partial\ln Z}{\partial p} = -\frac{V}{k_BT}\end{cases}$$

Therefore,

$$d\left(-\frac{G}{k_BT}\right)=\frac{1}{k_BT^2}\underbrace{(U+pV)}_{G+TS}dT-\frac{Vdp}{k_BT}\Longrightarrow dG=-SdT+Vdp$$

(b)

$$\langle V\rangle=\frac{1}{Z}\int_0^{\infty}dV\,V\sum_n Ve^{-\frac{1}{k_BT}(E_n(V)+pV)}$$

Hence,

$$-\left.\frac{\partial\langle V\rangle}{\partial p}\right|_T=k_BT\left(\left\langle V^2\right\rangle-\langle V\rangle^2\right)=\frac{1}{k_BT}\left\langle(V-\langle V\rangle)^2\right\rangle$$

(c) Since $\frac{\partial\langle V\rangle}{\partial p}=\mathrm{O}(N)$, we have

$$\frac{\left\langle(V-\langle V\rangle)^2\right\rangle}{\langle V\rangle^2}=\mathrm{O}\left(\frac{1}{N}\right)$$

(d) Using $V=\frac{Nk_BT}{p}$, we obtain

$$\left\langle(V-\langle V\rangle)^2\right\rangle=-k_BT\frac{\partial V}{\partial p}=\frac{N(k_BT)^2}{p^2}=\frac{1}{N}\langle V\rangle^2$$

4 (a) From (3.128),

$$\mu(T,p)=k_BT\ln\left[\frac{p}{k_BT}\frac{1}{n_Q(T)}\right]$$

where

$$n_Q(T)\equiv\left(\frac{mk_BT}{2\pi\hbar^2}\right)^{\frac{3}{2}}$$

(b) The probability of an atom adsorbing to a site is given by

$$P(T,p)=\frac{e^{\frac{\epsilon+\mu}{k_BT}}}{1+e^{\frac{\epsilon+\mu}{k_BT}}}=\frac{1}{1+e^{-\frac{\epsilon}{k_BT}}\frac{k_BTn_Q(T)}{p}}=\frac{p}{p+p_0(T)}$$

where $p_0(T)\equiv e^{-\frac{\epsilon}{k_BT}}k_BTn_Q(T)$ is an increasing function of T. Therefore, as the temperature increases, adsorption becomes less likely.

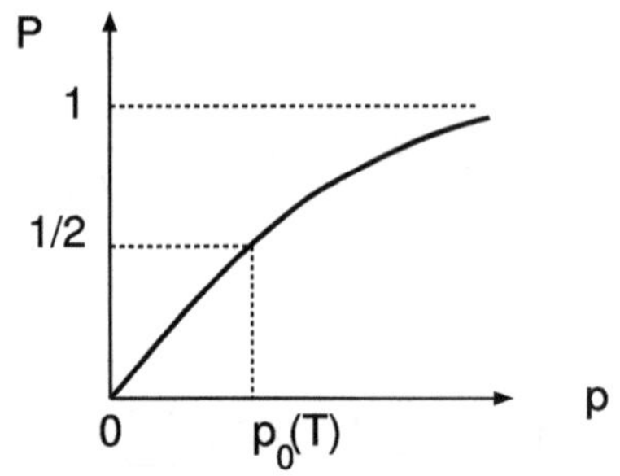

5 The Boltzmann weight for kinetic energy ε is given as $\mathrm{e}^{-\frac{\varepsilon+m\phi-\mu}{k_BT}}$, where $\varepsilon + m\phi$ is the total energy. This weight is obtained from $\mathrm{e}^{-\frac{\varepsilon-\mu}{k_BT}}$ by replacing μ by $\mu - m\phi(\vec{x})$.

6 (a)

$$\langle \mu_z \rangle = \frac{\int_0^\pi \sin\theta d\theta \int_0^{2\pi} d\phi\, \mu \cos\theta \cdot \mathrm{e}^{\frac{\mu H}{k_BT}\cos\theta}}{\int_0^\pi \sin\theta d\theta \int_0^{2\pi} d\phi\, \mathrm{e}^{\frac{\mu H}{k_BT}\cos\theta}}$$

$$= \mu \frac{\int_{-1}^1 dx\, x\mathrm{e}^{\frac{\mu H}{k_BT}x}}{\int_{-1}^1 dx\, \mathrm{e}^{\frac{\mu H}{k_BT}x}} = \mu\left(\coth\frac{\mu H}{k_BT} - \frac{k_BT}{\mu H}\right)$$

(b) Since there are 6 nearest neighboring spins, we can replace H by $6g\langle \mu_z \rangle + H$. Therefore,

$$\langle \mu_z \rangle = \mu\left(\coth\frac{\mu(6g\langle \mu_z \rangle + H)}{k_BT} - \frac{k_BT}{\mu(6g\langle \mu_z \rangle + H)}\right)$$

(c) Define $T_c \equiv \frac{2g\mu^2}{k_B}$ and $x \equiv \frac{\langle \mu_z \rangle}{\mu}$. Then, for $H = 0$, we obtain

$$x = f\left(\frac{3T_c}{T}x\right)$$

where $f(x) \equiv \coth x - 1/x$. For $x \ll 1$, we find $f(x) \simeq \frac{x}{3} - \frac{x^3}{45}$. Hence, T_c is the critical temperature.

(d) For $T < T_c$ and $T \simeq T_c$, we find

$$x = f\left(3\left(1 + \frac{T_c - T}{T_c}\right)x\right) \simeq \left(1 + \frac{T_c - T}{T_c}\right)x - \frac{3}{5}x^3$$

This gives $x^2 \simeq \frac{5}{3}\frac{T_c-T}{T_c}$, and consequently

$$\langle \mu_z \rangle \simeq \mu\sqrt{\frac{5}{3}\frac{T_c-T}{T_c}} \propto \sqrt{T_c-T}$$

(e) For $T > T_c$ in a weak magnetic field, we obtain

$$\langle \mu_z \rangle \simeq \frac{\mu}{3}\frac{\mu(6g\langle \mu_z \rangle + H)}{k_B T} \Longrightarrow \left(1-\frac{T_c}{T}\right)\langle \mu_z \rangle \simeq \frac{\mu^2 H}{3k_B T}$$

Therefore,

$$\langle \mu_z \rangle \simeq \frac{\mu^2}{3k_B(T-T_c)}H \propto \frac{H}{T-T_c}$$

(f) At $T = T_c$, we find

$$x = f\left(3x + \frac{\mu H}{k_B T_c}\right) \simeq x + \frac{\mu H}{3k_B T_c} - \frac{3}{5}x^3$$

which yields

$$\langle \mu_z \rangle = \mu\left(\frac{5\mu}{9k_B T_c}H\right)^{\frac{1}{3}} \propto H^{\frac{1}{3}}$$

Exercises 7

1 **Spin** J—Consider a quantum spin $\vec{s}$ with magnitude $J = 0, \hbar/2, \hbar, 3\hbar/2, \ldots$. Its magnetic moment is given by

$$\vec{\mu} = \mu\frac{\vec{s}}{J\hbar}$$

Find the average magnetization in an external magnetic field H. Additionally, determine the limit as $J \to \infty$.
Hint: The limit should coincide with the result of Example 7-6(a).

2 **Blackbody radiation**—Consider a photon gas at absolute temperature T. Electromagnetic waves with wave vector $\vec{k}$ and angular frequency $\omega \equiv |\vec{k}|c$ correspond to photons with momentum $\vec{p} = \hbar\vec{k}$ and energy $\hbar\omega$. The photons follow Bose statistics, and the chemical potential is zero.

(a) Consider a sufficiently large volume V. Count the number of single-photon states contained in a small volume element $\Delta k_x \Delta k_y \Delta k_z$ in the wave vector space.

Hint—Use periodic boundary conditions; do not forget the two degrees of freedom for polarization.

(b) Determine the average energy carried by the photons between angular frequencies ω and $\omega + \Delta\omega$.

(c) Calculate the energy per unit volume of the photon gas, and check the dimensionality of the obtained result.

(d) Predict the entropy per unit volume using dimensional analysis and compare it with the calculated result.

3 **Debye model**—To address the shortcomings at low temperatures of Einstein's model for solids, Debye proposed the following model.

Like electromagnetic waves, lattice vibrations must also be quantized, and the quantized lattice vibrations are called phonons. Phonons also follow Bose statistics, but unlike photons, they have one longitudinal mode in addition to two transverse modes for a given wave vector. Denoting the transverse phase velocity as c_t and the longitudinal phase velocity as c_l, the energy of a phonon with wave vector $\vec{k}$ is given by[26]

$$\varepsilon_t(\vec{k}) = \hbar k c_t, \quad \varepsilon_l(\vec{k}) = \hbar k c_l$$

(a) Show that the number of phonon states with energies between ε and $\varepsilon + d\varepsilon$ is given by

$$V\frac{3}{\bar{c}^3}\frac{\varepsilon^2 d\varepsilon}{2\pi^2\hbar^3}$$

where V is the total volume of the lattice and $\bar{c}$ is defined as

$$\frac{1}{\bar{c}^3} \equiv \frac{1}{3}\left(\frac{2}{c_t^3} + \frac{1}{c_l^3}\right)$$

representing the average velocity.

(b) The total number of modes of lattice vibrations must be equal to the total degrees of freedom of lattice vibrations, which is $3N$. Calculate the maximum energy of phonons, $k_B\Theta$. (Here, Θ is called the Debye temperature.)

(c) Calculate the Debye temperature for copper (Cu) with the mass number 64, mass density 8.9 g/cm^3, and $\bar{c} = 2.4 \times 10^5$ cm/s.

(d) Determine the free energy of lattice vibrations as a function of T, V, and N.
Hint—Express it using the Debye function defined as follows:

[26] Generally, $c_l > c_t$.

$$D(x) \equiv \frac{3}{x^3} \int_0^x \frac{z^3 dz}{e^z - 1}$$

(e) Show that the heat capacity at constant volume is given by

$$C_V = 3Nk_B \left(4D\left(\frac{\Theta}{T}\right) - \frac{3\frac{\Theta}{T}}{e^{\frac{\Theta}{T}} - 1} \right)$$

Therefore, C_V/N becomes a universal function of Θ/T. The Einstein model also exhibits the same property if ω is set equal to Θ.

(f) Use $D(x) \simeq \frac{\pi^4}{5x^3}$ for $x \gg 1$ to approximate the specific heat C_V for low temperatures $T \ll \Theta$.

(g) Approximate the specific heat C_V for high temperatures $T \gg \Theta$.

4 **Electrons in solids**—Consider a solid with the density of one-electron state per unit volume and per unit energy, given by

$$D(\varepsilon) = \begin{cases} A & \text{for} \quad -\varepsilon_F < \varepsilon < 0 \\ 0 & \text{for} \quad 0 < \varepsilon < \varepsilon_g \\ A & \text{for} \quad \varepsilon > \varepsilon_g \end{cases}$$

(See figure) Here, $\varepsilon_F \gg \varepsilon_g$, and at $T = 0$, all states with $\varepsilon < 0$ are completely filled, while the states above ε_g are completely empty. Consider temperatures T lower than $\frac{\varepsilon_g}{k_B}$, and neglect terms of the order $e^{-\frac{\varepsilon_F}{k_B T}}$.

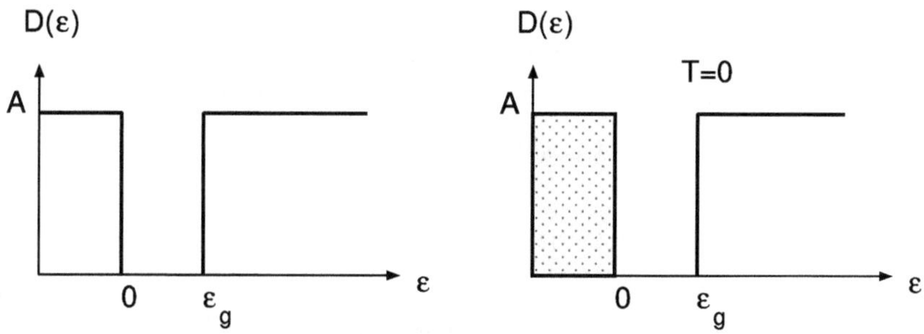

(a) Determine the temperature dependence of the chemical potential μ.
Hint—Ensure that the number density does not depend on temperature.

(b) Find the number density of electrons with energy above ε_g.

(c) Calculate the specific heat per unit volume.

5 **Thermal electrons**—Consider a dilute gas of electrons in thermal equilibrium with a solid at temperature T and chemical potential μ. The electron mass is

denoted as m.

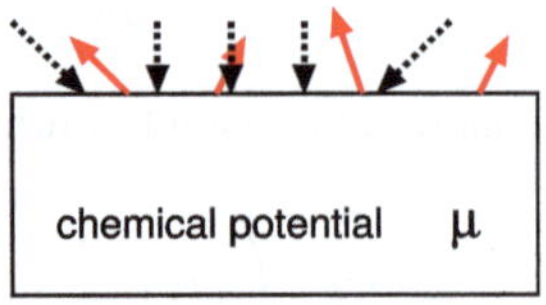

(a) For a dilute gas of electrons per unit volume, find the number of electrons with momentum

$$\vec{p} \in [p_x, p_x + \Delta p_x] \times [p_y, p_y + \Delta p_y] \times [p_z, p_z + \Delta p_z]$$

Hint—See Example 7-4(a).

(b) Determine the number of gas electrons hitting the surface per unit area per unit time.

Hint—Denoting the result of (a) as

$$n(\vec{p})\, \Delta p_x \Delta p_y \Delta p_z$$

show that the answer is given by

$$\int_0^\infty dp_x \int_{-\infty}^\infty dp_y \int_{-\infty}^\infty dp_z\, \frac{p_x}{m} n(\vec{p})$$

(c) Assuming that all the electrons hitting the solid are absorbed, how many thermal electrons are emitted from a unit area of the surface per unit time?

(d) Given $\mu = -1\,\text{eV}$ and $T = 2000\,\text{K}$, calculate the number of thermal electrons emitted per unit time from a unit surface of the solid.

6 **Pauli paramagnetism**—Consider N free electrons at absolute zero temperature in a volume V.

(a) Find the chemical potential ϵ_F. **Hint**: The electron has spin $\frac{1}{2}$ with two spin states.

The magnetic moment of an electron with spin $\vec{s}$ is given by

$$\vec{\mu} = -\mu_B \frac{2\vec{s}}{\hbar} \qquad \left(\mu_B \equiv \frac{e\hbar}{2mc}\right)$$

When a magnetic field $H > 0$ is applied in the z direction, the magnetic energy is given by

$$E_m = -\mu_z H$$

Here, assume $\mu_B H \ll \varepsilon_F$.

(b) Determine the magnetic field dependence of the chemical potential μ. (Distinguish the chemical potential μ from the magnetic moment $\vec{\mu}$ or μ_z.)
(c) Find the magnetic field dependence of the total energy.
(d) Calculate the total magnetization.
(e) Explain the sign of the total magnetization.

7 **Lattice gas**—As a model to explain ferromagnetism, you have already learned the Ising model. Here, we show that the same model can be used to understand the first-order phase transition between liquid and gas.

Consider a three-dimensional cubic lattice consisting of cubes of volume a^3. Put a spin $s_i = \pm 1$ in each cube, where the up state (+) of the spin is interpreted as the presence of an atom and the down state (−) as the absence. Therefore, the number of spins in the up state represents the number of atoms, denoted as N:

$$N = \frac{1}{2}\sum_i (s_i + 1)$$

where the sum extends over all cubes. Now, let s denote the average value of s_i. We then *interpret* $s > 0$ *as a liquid and* $s < 0$ *as a gas.*

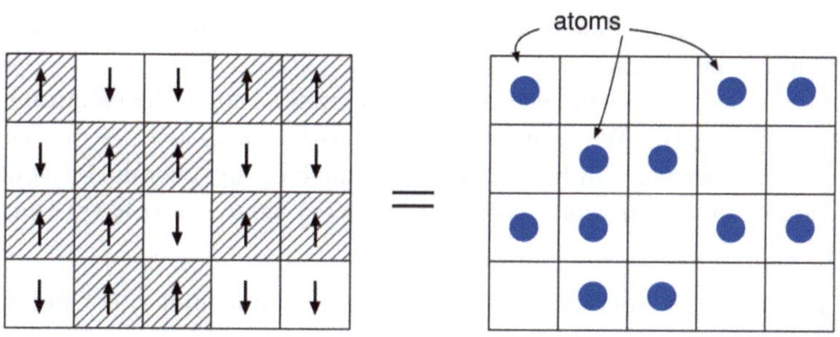

The energy of the system is given by

$$E = -K \sum_{\langle i,j \rangle} (s_i + 1)(s_j + 1)$$

where $K > 0$, and the sum extends over all nearest neighbor pairs.

(a) Explain the physical meaning of the energy defined above. **Hint**: Is it attraction or repulsion between two atoms?

Now, we introduce a chemical potential μ:

$$E - N\mu = -K \sum_{\langle i,j \rangle} (s_i + 1)(s_j + 1) - \frac{\mu}{2}\sum_i (s_i + 1)$$

(b) Derive the following expression:

$$E - N\mu = -K \sum_{\langle i,j \rangle} s_i s_j - \left(6K + \frac{\mu}{2}\right) \sum_i s_i - \sum_i \left(3K + \frac{\mu}{2}\right)$$

Comment: Thus, $6K + \frac{\mu}{2}$ corresponds to an external magnetic field.

(c) Using the mean field approximation (or the Bragg-Williams approximation), find the temperature T_c and chemical potential μ_c at the critical point.

(d) Calculate the temperature dependence of the vapor pressure $p_{tr}(T)$ for T slightly lower than T_c.
Hint: Use $pV = N\mu - E + TS$. The Ising model corresponds to the case of zero external magnetic field.

(e) Find the discontinuity in the number density on the coexistence curve $p = p_{tr}(T)$ when T is slightly lower than T_c.
Hint: For $p = p_{tr}(T) + 0$, it is in the liquid phase, and for $p = p_{tr}(T) - 0$, it is in the gas phase. The discontinuity in the number density corresponds to the spontaneous magnetization of the Ising model.

8 **Thomas-Fermi screening**—The aim of this problem is to consider charge screening by a degenerate electron gas with number density n_0 at absolute zero temperature. We assume that the electron charges are neutralized by a fixed uniform ion distribution.

Now, if we denote the electrostatic potential as $\phi(\vec{r})$, the electron's potential energy is $-e\phi(\vec{r})$. We assume that $\phi(\vec{r})$ goes to zero as $r \to \infty$. The number density of electrons can be obtained by replacing the chemical potential in the absence of the electrostatic potential by $\epsilon_F(\vec{r}) \equiv \mu - (-e)\phi(\vec{r})$ (see Example 7-5).

(a) Determine the relation between the chemical potential $\epsilon_F(\vec{r})$ and the number density $n(\vec{r})$.

Let us assume that there is an impurity charge Ze at $\vec{r} = 0$. Then, the electrostatic potential $\phi(\vec{r})$ satisfies the Poisson equation:

$$-\nabla^2\phi(\vec{r}) = 4\pi e\Big(-n(\vec{r}) + \underbrace{n_0}_{\text{ion number density}} + Z\delta^{(3)}(\vec{r})\Big) \qquad (1)$$

(b) Assuming $|e\phi| \ll \epsilon_F$, expand $n(\vec{r})$ to first order in ϕ and approximate the Poisson equation.

(c) Solve the equation obtained above, and find the distance, called the screening distance, over which the impurity charge affects the charge distribution.

(e) Calculate the screening distance for $n_0 = 10^{20}\,\text{cm}^{-3}$.

9 **Debye screening**—We next consider the charge screening by a dilute electron gas. We consider a dilute electron gas with number density n_0 and a uniform ion

distribution that neutralizes the electrons' charge. Let T denote the temperature of the electron gas.

Let $\phi(\vec{r})$ be the electrostatic potential due to a charge Ze at $\vec{r} = 0$. As derived in Example 7-5, the number density of electrons $n(\vec{r})$ under the electrostatic potential $\phi(\vec{r})$ is proportional to the Boltzmann factor

$$e^{\frac{e\phi(\vec{r})}{k_B T}}$$

Here, the electrostatic potential also satisfies the Poisson equation (1) given in the previous problem.

(a) Assuming $|e\phi(\vec{r})| \ll k_B T$, linearize the Poisson equation satisfied by $\phi(\vec{r})$.
(b) Solve the equation obtained above under the following boundary conditions:

$$\phi(\vec{r}) \longrightarrow \begin{cases} \frac{Ze}{r} & (r \to 0) \\ 0 & (r \to \infty) \end{cases}$$

(c) Determine the screening distance and explain its temperature dependence.

10 **Bose-Einstein condensation**—Consider a system of spinless Bose particles of mass m in three-dimensional space under a harmonic potential

$$V(\vec{r}) = \frac{m\omega^2}{2}\vec{r}^{\,2}$$

(a) Assuming $k_B T \gg \hbar\omega$, calculate the maximum number of particles $N_0(T)$ that the system can accommodate.
Hint—Use quantum mechanics. The maximum chemical potential equals the zero-point energy $\mu = \frac{3}{2}\hbar\omega$. Approximate the sum representing $N_0(T)$ by an integral. Use $\int_0^\infty dn\, \frac{n^2}{e^n - 1} = 2\zeta(3) \simeq 2.404$.
(b) Assuming $N \gg 1$, find the temperature T_{tr} defined by $N_0(T_{tr}) = N$.
(c) Calculate the ratio of particles occupying the ground state $\frac{N - N_0(T)}{N}$.
Comment: Such Bose-Einstein condensation can be realized in the laboratory using alkali atomic gases.
(d) For Na atoms, take $N = 10^{10}$ and $\hbar\omega = 10^{-9}$ eV. Estimate T_{tr}.

11 **Free energy of the Ising model** —We have already learned how to understand the Ising model using the mean field approximation in Sect. 3.2.3. Recall that in the Ising model the total energy of the system is given by

$$E = -K \sum_{\langle i,j \rangle} s_i s_j \qquad (s_i = \pm 1)$$

where the sum is over all the pairs of nearest neighbors. In the mean field approximation the magnetization M under an external magnetic field H is given by

$$M \equiv \mu \sum_i \langle s_i \rangle = N\mu \tanh \frac{K\alpha \frac{M}{N\mu} + \mu H}{k_B T} \tag{M}$$

where α is the number of nearest neighbors.

(a) Show that the probabilities of a spin in the up or down states are, respectively, given by

$$p_\uparrow = \frac{1 + \frac{M}{N\mu}}{2}, \quad p_\downarrow = \frac{1 - \frac{M}{N\mu}}{2}$$

(b) Neglecting the spin-spin correlations, we assume that each spin is independently in the up state with probability $p_\uparrow$ and in the down state with probability $p_\downarrow$.[27] Show that the energy is given by

$$U = \frac{\alpha N K}{2}\left(-\left(p_\uparrow^2 + p_\downarrow^2\right) + 2p_\uparrow p_\downarrow\right) = -\frac{\alpha K M^2}{2N\mu^2} \tag{U}$$

(c) Use Eqs. (M) and (U) to find the Helmholtz free energy $F(T, M)$ by integrating

$$\begin{cases} \frac{\partial}{\partial T}\frac{F(T,M)}{T} &= -\frac{U}{T^2} \\ \frac{\partial}{\partial M}\frac{F(T,M)}{T} &= \frac{H}{T} \end{cases}$$

Comment: The resulting free energy is the same as the one obtained through the Bragg-Williams approximation explained in Sect. 3.3.6.

3.3 Supplementary Notes

3.3.1 Partial Derivatives and Exact Differentials

Partial derivatives and exact differentials play a crucial role in thermodynamics; they are the essential tools for deriving and understanding the relations among various thermodynamic quantities.

[27] In fact there is an important correlation so that the probability of a spin up is greater if the neighboring spin is up and is smaller if the neighboring spin is down.

Let $z(x, y)$ be a differentiable function. Consider an infinitesimal change of $z(x, y)$ given by

$$dz(x, y) = u(x, y)dx + v(x, y)dy \tag{3.148}$$

If the **integrability condition**

$$\frac{\partial u(x, y)}{\partial y} = \frac{\partial v(x, y)}{\partial x} \tag{3.149}$$

is satisfied, then dz is an exact differential, i.e.,

$$u(x, y) = \frac{\partial z(x, y)}{\partial x}, \quad v(x, y) = \frac{\partial z(x, y)}{\partial y}. \tag{3.150}$$

The integrability condition amounts to

$$\frac{\partial}{\partial y}\frac{\partial}{\partial x}z(x, y) = \frac{\partial}{\partial x}\frac{\partial}{\partial y}z(x, y) \tag{3.151}$$

which is a property we usually take for granted.

In thermodynamics we often use the notation

$$\left(\frac{\partial z}{\partial x}\right)_y \equiv \frac{\partial z(x, y)}{\partial x}$$

to emphasize that z is considered as a function of x, y, and the derivative is taken with respect to x while y is fixed. The integrability of dz implies that the integral of dz from a point A to a point B is path independent, i.e., the integral along a curve connecting A to B does not depend on the choice of the path:[28]

$$\int_A^B dz = z(B) - z(A) \tag{3.152}$$

Hence, the integral along any closed curve C vanishes:

$$\oint_C dz = 0 \tag{3.153}$$

Now, suppose that we can regard x as a function of y and z, i.e., $x(y, z)$. Then we can write

$$dx = \left(\frac{\partial x}{\partial y}\right)_z dy + \left(\frac{\partial x}{\partial z}\right)_y dz \tag{3.154}$$

[28] A proof follows from defining the gradient vector $\nabla z = (\frac{\partial z}{\partial x}, \frac{\partial z}{\partial y})$ and applying Stokes's theorem. See Sect. 2.4.3.

Substituting this into dz we obtain

$$dz = \left(\frac{\partial z}{\partial x}\right)_y \left(\left(\frac{\partial x}{\partial y}\right)_z dy + \left(\frac{\partial x}{\partial z}\right)_y dz\right) + \left(\frac{\partial z}{\partial y}\right)_x dy$$
$$= \left(\left(\frac{\partial z}{\partial x}\right)_y \left(\frac{\partial x}{\partial y}\right)_z + \left(\frac{\partial z}{\partial y}\right)_x\right) dy + \left(\frac{\partial z}{\partial x}\right)_y \left(\frac{\partial x}{\partial z}\right)_y dz \tag{3.155}$$

Comparing the coefficients of dz and dy on both sides, we obtain two important relations for partial derivatives:

$$\left(\frac{\partial z}{\partial x}\right)_y \left(\frac{\partial x}{\partial z}\right)_y = 1\,, \tag{3.156}$$

$$\left(\frac{\partial z}{\partial x}\right)_y \left(\frac{\partial x}{\partial y}\right)_z \left(\frac{\partial y}{\partial z}\right)_x = -1 \tag{3.157}$$

Maxwell Relations

Fundamental thermodynamic relations can be derived using the properties of partial differentials. Let us consider the first law again:

$$dU = TdS - pdV$$

Since U is a function of state, dU is an exact differential. From Eq. (3.150), we obtain

$$\left(\frac{\partial U}{\partial S}\right)_V = T \quad , \quad \left(\frac{\partial U}{\partial V}\right)_S = -p \tag{3.158}$$

Maxwell's relation follows from the symmetry of second-order derivatives (3.151):

$$\frac{\partial U(S,V)}{\partial S \partial V} = \frac{\partial U(S,V)}{\partial V \partial S} \quad \Rightarrow \quad \left(\frac{\partial T}{\partial V}\right)_S = -\left(\frac{\partial p}{\partial S}\right)_V \tag{3.159}$$

Other Maxwell relations can be derived using the same technique from the thermodynamic potentials (state functions) such as S, F, and G.

3.3.2 *Choosing Independent Variables*

In thermodynamics, a proper choice of independent variables is crucial. Here, we explain the relation among three choices, and introduce the Legendre transformations that connect them.

Three Choices for Two Independent Variables

$\boxed{T, V}$

The thermodynamic state of a uniform and isotropic system can be uniquely determined by specifying the absolute temperature T and volume V. In this case, the function that determines the thermodynamic state is the Helmholtz free energy $F(T, V)$, as explained in Sect. 3.1 for thermodynamics. From this function, entropy S and pressure p can be obtained as partial derivatives:

$$dF(T, V) = -S(T, V)dT - p(T, V)dV \Longleftrightarrow \begin{cases} S(T, V) = -\frac{\partial F(T,V)}{\partial T} \\ p(T, V) = -\frac{\partial F(T,V)}{\partial V} \end{cases} \tag{3.160}$$

Furthermore, by differentiating the entropy with respect to temperature, we obtain the specific heat at constant volume:

$$C_V(T, V) = T\frac{\partial S(T, V)}{\partial T} \tag{3.161}$$

We obtain the isothermal compressibility as

$$\beta_T(T, V) = -\frac{1}{V\frac{\partial p(T,V)}{\partial V}} \tag{3.162}$$

$\boxed{T, p}$

It is also possible to choose pressure instead of volume as an independent variable. In this case, the function that determines the thermodynamic state of the system is the Gibbs free energy $G(T, p)$. From this function, entropy S and volume V can be obtained as

$$dG(T, p) = -S(T, p)dT - V(T, p)dp \Longleftrightarrow \begin{cases} S(T, p) = -\frac{\partial G(T,p)}{\partial T} \\ V(T, p) = \frac{\partial G(T,p)}{\partial p} \end{cases} \tag{3.163}$$

Furthermore, by differentiating the entropy with respect to temperature, we obtain the specific heat at constant pressure:

$$C_p(T, p) = T\left(\frac{\partial S(T, p)}{\partial T}\right)_p \tag{3.164}$$

The isothermal compressibility is

$$\beta_T(T, p) = -\frac{1}{V}\frac{\partial V(T, p)}{\partial p} \tag{3.165}$$

$\boxed{U, V}$

In considering thermodynamics from the perspective of statistical mechanics, the most important function is the entropy $S(U, V)$ expressed as a function of energy U and volume V. From this, we obtain

$$dS = \frac{1}{T}dU + \frac{p}{T}dV \Longleftrightarrow \begin{cases} \frac{\partial S(U,V)}{\partial U} = \frac{1}{T} \\ \frac{\partial S(U,V)}{\partial V} = \frac{p}{T} \end{cases} \tag{3.166}$$

This can be rewritten as

$$dU = TdS - pdV \tag{3.167}$$

which gives the first law of thermodynamics (energy conservation). At constant entropy (i.e., under an adiabatic change), we find

$$dU = -pdV \quad (S\text{constant}) \tag{3.168}$$

and we obtain an alternative expression for pressure:

$$p = -\frac{\partial U(S, V)}{\partial V} = -\left(\frac{\partial U}{\partial V}\right)_S \tag{3.169}$$

Relation Among $F(T, V), G(T, p), S(U, V)$

We have summarized three different ways of choosing two independent variables. In the following we summarize the fundamental thermodynamic functions corresponding to each choice:

Independent variables	Fundamental thermodynamic function
T, V	$F(T, V)$
T, p	$G(T, p)$
U, V	$S(U, V)$

This section will explain how any one of these three thermodynamic functions can be used to obtain the other two.

$\boxed{F(T, V) \longrightarrow G(T, p), S(U, V)}$

Starting with the Helmholtz free energy $F(T, V)$ as a function of T and V, we define p, S, and U as follows:

$$p \equiv -\frac{\partial F(T, V)}{\partial V}, \quad S \equiv -\frac{\partial F(T, V)}{\partial T}$$
$$U \equiv F + TS = F(T, V) - T\frac{\partial F(T, V)}{\partial T} \tag{3.170}$$

Defining the Gibbs free energy first as a function of T and V:

$$G \equiv F + pV = F(T, V) - V\frac{\partial F(T, V)}{\partial V} \tag{3.171}$$

we find its infinitesimal variation as

$$\begin{aligned} dG &= dF + Vdp + pdV = -SdT - pdV + pdV + Vdp \\ &= -SdT + Vdp \end{aligned} \tag{3.172}$$

This shows that it is more natural to use T and p as independent variables for G.

Furthermore, from the definition of U, we obtain

$$S = \frac{1}{T}(U - F) \tag{3.173}$$

Varying this, we get

$$\begin{aligned} dS &= -\frac{dT}{T}S + \frac{1}{T}(dU - dF) \\ &= \frac{1}{T}(-SdT + dU + SdT + pdV) \\ &= \frac{1}{T}(dU + pdV) \end{aligned} \tag{3.174}$$

This implies that it is more natural to use U and V as independent variables for S.

To summarize, we have shown that starting with $F(T, V)$ as a function of T and V, we obtain $G(T, p)$ with independent variables T and p using Eq. (3.171). Similarly, using Eq. (3.173), we obtain $S(U, V)$ with independent variables U and V.

$$\boxed{G(T, p) \longrightarrow F(T, V), S(U, V)}$$

Starting with the Gibbs free energy $G(T, p)$ as a function of T and p, we define V, S, and U as functions of T and p:

$$V \equiv \frac{\partial G(T, p)}{\partial p}, \quad S \equiv -\frac{\partial G(T, p)}{\partial T}$$
$$U \equiv G + TS - pV = G(T, p) - T\frac{\partial G(T, p)}{\partial T} - p\frac{\partial G(T, p)}{\partial p} \tag{3.175}$$

Defining the Helmholtz free energy first as a function of T and p:

$$F \equiv G(T, p) - pV = G(T, p) - p\frac{\partial G(T, p)}{\partial p} \tag{3.176}$$

we find its infinitesimal variation as

$$dF = -SdT + Vdp - Vdp - pdV = -SdT - pdV \tag{3.177}$$

This shows that it is more natural to use T and V as independent variables for F.

Furthermore, from the definition of U, we obtain

$$S = \frac{1}{T}(U - G + pV) \tag{3.178}$$

Varying this, we get

$$\begin{aligned} dS &= -\frac{dT}{T}S + \frac{1}{T}(dU + SdT - Vdp + Vdp + pdV) \\ &= \frac{dU}{T} + \frac{p}{T}dV \end{aligned} \tag{3.179}$$

This shows that it is more natural to use U and V as independent variables for S.

In summary, we have shown that starting with $G(T, p)$ as a function of T and p, we obtain $F(T, V)$ with independent variables T and V using Eq. (3.176). Similarly, using Eq. (3.178), we obtain $S(U, V)$ with independent variables U and V.

$$\boxed{S(U, V) \longrightarrow F(T, V), G(T, p)}$$

Starting with the entropy $S(U, V)$ as a function of U and V, we define T and p as follows:

$$\frac{1}{T} \equiv \frac{\partial S(U, V)}{\partial U}, \quad \frac{p}{T} \equiv \frac{\partial S(U, V)}{\partial V} \tag{3.180}$$

We then define the free energy functions

$$F \equiv U - TS(U,V) = U - \frac{S(U,V)}{\frac{\partial S(U,V)}{\partial U}} \tag{3.181}$$

$$G \equiv U - TS(U,V) + pV = U - \frac{S(U,V)}{\frac{\partial S(U,V)}{\partial U}} + V\frac{\frac{\partial S(U,V)}{\partial V}}{\frac{\partial S(U,V)}{\partial U}} \tag{3.182}$$

first as functions of U and V.

Let us consider the variation of F first. From Eq. (3.180), we find

$$dS = \frac{1}{T}dU + \frac{p}{T}dV \tag{3.183}$$

which implies

$$dU = TdS - pdV \tag{3.184}$$

Therefore, we obtain

$$dF = dU - SdT - TdS = TdS - pdV - SdT - TdS = -SdT - pdV \tag{3.185}$$

This implies that it is more natural to use T and V as independent variables for F. Similarly, varying G we obtain

$$dG = -SdT + Vdp \tag{3.186}$$

This implies that it is more natural to use T and p as independent variables for G.

Legendre Transformation

The transformations among F, G, S are examples of what is known as the Legendre transformation. Its construction is straightforward. Given a function $f(x)$ of x, we define $L_f(y)$ by

$$L_f(y) = f(x) - xy\,, \tag{3.187}$$

where x is chosen as a function of y so that

$$y = f'(x) \tag{3.188}$$

is satisfied. Then, $L_f(y)$ is defined as a function of y, and it satisfies

$$\frac{dL_f(y)}{dy} = \underbrace{(f'(x) - y)}_{=0}\frac{dx}{dy} - x = -x \tag{3.189}$$

We can derive this also as follows:

$$dL_f = df - ydx - xdy = ydx - ydx - xdy = -xdy \tag{3.190}$$

Let us verify that our examples are indeed Legendre transformations. To obtain the Gibbs free energy $G(T, p)$ from the Helmholtz free energy $F(T, V)$, we treat T as a constant. Then, by setting $x = V$, $f(x) = F(x, T)$, and $y = -p$, we perform a Legendre transformation:

$$L_f(y) = f(x) - xy = G(T, -y) \tag{3.191}$$

To obtain entropy $S(U, V)$, we treat V as a constant. We set $x = \frac{1}{T}$, $f(x) = -xF\left(\frac{1}{x}, V\right)$, and $y = -U$. This leads to

$$L_f(y) = f(x) - xy = S(-y, V) \tag{3.192}$$

You can try to perform Legendre transformations from $G(T, p)$ and $S(U, V)$ on your own.

To gain a deeper understanding of Legendre transformations, it is essential to grasp their geometric meaning, which is related to **minimization problems**. Let us reconsider Legendre transformations from this perspective. Given a function $f(x)$ with the variable x, we introduce another variable y and create a two-variable function:

$$g(x, y) \equiv f(x) - xy \tag{3.193}$$

The value of x that minimizes (or maximizes) $g(x, y)$ for any given y is determined as a function of y, denoted as $X(y)$. The Legendre transformation of $f(x)$ is then defined as the minimum value of $g(x, y)$:

$$L_f(y) \equiv g(X(y), y) = f(X(y), y) - X(y)y \tag{3.194}$$

From (3.193), the minimization gives

$$\frac{\partial g(x, y)}{\partial x} = 0 \Longleftrightarrow f'(x) = y \tag{3.195}$$

This leads to the inverse relation:

$$y = f'(X(y)) \tag{3.196}$$

Hence, $X(y)$ is the inverse function of $f'(x)$. Furthermore, the variation of $L_f(y)$ is given by

$$dL_f(y) = \frac{dL_f(y)}{dy}dy = f'(X(y))dX(y) - ydX(y) - X(y)dy = -X(y)dy \tag{3.197}$$

This shows that $-X(y)$ is thc derivative of $L_f(y)$:

$$-X(y) = \frac{dL_f(y)}{dy} \tag{3.198}$$

Not any function $f(x)$ can be Legendre transformed without restrictions. From Eq. (3.196), $X(y)$ is the inverse function of $f(x)$, so $f'(x)$ within the range of interest must be a monotonic function:

$$f''(x) > 0 \text{ or } f''(x) < 0 \tag{3.199}$$

In the case of $f''(x) > 0$, $f(x)$ is said to be **convex**, and we have tried to explain the geometrical meaning in Fig. 3.14. $f(x)$ is **concave** if $f''(x) < 0$. In essence, the Legendre transformation $L_f(y_0)$ involves finding the value of x, denoted by x_0, where $f(x)$ gives the given slope y_0. We obtain the Legendre transform $f(x_0) - y_0x_0$ by shifting the straight line y_0x up and down until it touches the graph of $f(x)$. Hence, the value of $L_f(y_0)$ corresponds to the displacement between the straight line y_0x and the straight line with the same slope y_0 that touches the graph of $f(x)$.

It is worth noting that performing the Legendre transformation twice brings us back the original function. By definition, we find

$$\begin{aligned} L_f(y) &= f(x) - yx \text{ where } y = f'(x) \\ L_{L_f}(z) &= L_f(y) - zy \text{ where } z = L'_f(y) \end{aligned} \tag{3.200}$$

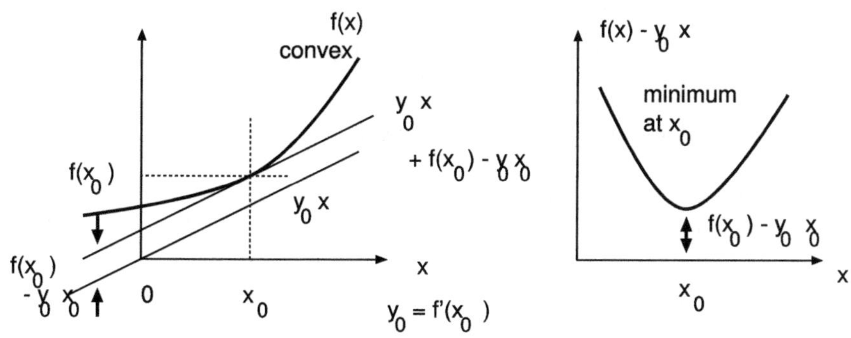

Fig. 3.14 Geometric understanding of the Legendre transform

Since $L'_f(y) = -x$, we obtain

$$L_{L_f}(z) = f(x) - yx + xy = f(x) = f(-z) \tag{3.201}$$

Thus, ignoring the change in the sign of the variable, performing the Legendre transformation twice on a function brings us back the original function.

Let us apply the Legendre transformation and its relation with minimization or maximization to the thermodynamic functions. Starting from $F(T, V)$, we construct $G(T, p)$ via the Legendre transformation:

$$G(T, p) = \mathrm{Min}_V\,(F(T, V) + pV) \tag{3.202}$$

where p is an arbitrary variable and the right-hand side means minimizing $F(T, V) + pV$ with respect to V. The existence of this Legendre transformation is ensured by the convexity (positive compressibility):

$$\frac{\partial^2 F(T, V)}{\partial V^2} = -\frac{\partial p(T, V)}{\partial V} > 0 \tag{3.203}$$

Similarly, starting from $F(T, V)$, we construct $S(U, V)$ via the Legendre transformation:

$$S(U, V) = \mathrm{Max}_T \left(-\frac{1}{T}F(T, V) + \frac{1}{T}U\right) \tag{3.204}$$

where V is an arbitrary variable. This requires the concavity (positive specific heat):

$$\frac{\partial^2}{\partial \frac{1}{T}^2}\frac{F(T, V)}{T} = -T^2\frac{\partial U(T, V)}{\partial T} = -T^2 C_V(T, V) < 0 \tag{3.205}$$

Conversely, starting from $S(U, V)$, we construct $F(T, V)$ via the inverse Legendre transformation

$$F(T, V) = -T\mathrm{Max}_U \left(S(U, V) - \frac{U}{T}\right) = \mathrm{Min}_U\,(U - TS(U, V)) \tag{3.206}$$

This requires the convexity (positive specific heat):

$$-\frac{\partial^2 S(U, V)}{\partial U^2} = -\frac{\partial}{\partial U}\frac{1}{T} = \frac{1}{T^2}\frac{\partial T}{\partial U} = \frac{1}{T^2 \frac{\partial U(T,V)}{\partial T}} > 0 \tag{3.207}$$

Using Eqs. (3.202) and (3.206), we obtain

$$G(T, p) = \mathrm{Min}_{U,V}\,(U - TS(U, V) + pV) \tag{3.208}$$

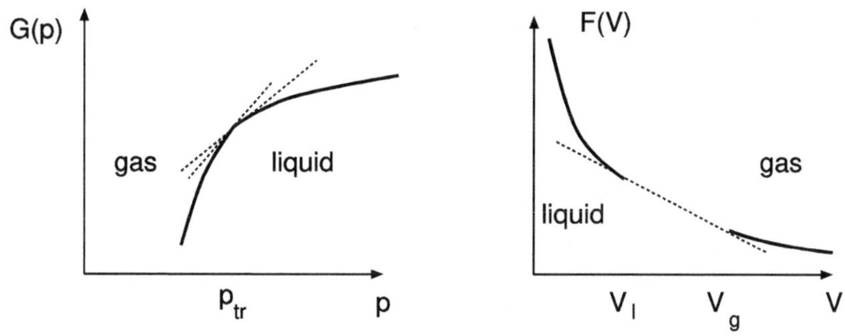

Fig. 3.15 Gibbs energy of the liquid-gas system below the critical temperature

The statistical mechanical interpretation of the minimization (3.206) has been explained in Sect. 3.1. Minimization (3.208) implies that under given temperature and pressure, the probability density for the energy U and volume V of the system is given by

$$\exp\left[-\frac{1}{k_B T}\left(U - TS(U, V) + pV\right)\right]$$

as has been explained in Sect. 3.2.

Finally, let us introduce a concrete example. We consider a liquid-gas system below the critical temperature. The pressure dependence of the Gibbs energy $G(p)$ is as shown on the left-hand side of Fig. 3.15. We note three things about this figure:

1. $G(p)$ is a continuous function.
2. At the transition pressure p_{tr}, the volume $V = G'(p)$ is discontinuous.
3. $G''(p) = \left(\frac{\partial V}{\partial p}\right)_T < 0$, i.e., $G(p)$ is a concave function of p.

Performing a Legendre transformation on the Gibbs energy gives us the Helmholtz energy:

$$F(V) = \mathrm{Min}_p \left(G(p) - pV\right) \tag{3.209}$$

The corresponding plot is as shown on the right-hand side of Fig. 3.15. About this figure, we also note three things:

1. $F(V)$ is not defined for $V_l < V < V_g$. This is because in this range, $G(p) - pV$ does not have a minimum.
2. The tangents of $F(V)$ at $V = V_l$ and $V = V_g$ coincide, meaning that the pressure $p = -F'(V)$ is the same.
3. $F''(V) = -\left(\frac{\partial p}{\partial V}\right)_T > 0$, i.e., the curve is convex.

3.3.3 Summary of Partition Functions

In the previous subsection, we have explained how the choice of independent variables determines the selection of fundamental thermodynamic functions. Consequently, the choice of independent variables determines the fundamental partition function, since the partition function is the exponential of the fundamental thermodynamic function.

We assume that the heat bath satisfies the first law of thermodynamics:

$$dU = TdS - pdV$$

1. **T, V Distribution**—Let T be the temperature of the heat bath, and we hold the volume of the system fixed at V. The total energy of the heat bath and the system is kept at U_0.

   When the system is in a microscopic state n with energy $E_n(V)$, the energy of the heat bath is $U_0 - E_n(V)$. Therefore, the entropy of the heat bath is given by

$$S_{\text{res}}(U_0 - E_n) = S_{\text{res}}(U_0) - \frac{E_n}{T}$$

   From this, we obtain the partition function:

$$Z(T, V) \equiv \sum_n \mathrm{e}^{-\frac{E_n(V)}{k_B T}}$$

   In the macroscopic limit, the partition function gives the Helmholtz free energy of the system:

$$Z(T, V) = \mathrm{e}^{-\frac{F(T,V)}{k_B T}}$$

   The Helmholtz free energy $F(T, V)$ satisfies

$$dF = -SdT - pdV$$

2. **T, p Distribution**—Let T be the temperature of the heat bath and p be its pressure. The total energy U_0 and total volume V_0 of the heat bath and the system are kept fixed.

   When the system is in a microscopic state n with volume V and energy $E_n(V)$, the volume of the heat bath is $V_0 - V$, and the energy is $U_0 - E_n(V)$. Therefore, the entropy of the heat bath is given by

$$S_{\text{res}}(U_0 - E_n, V_0 - V) = S_{\text{res}}(U_0, V_0) - \frac{E_n + pV}{T}$$

From this, we obtain the partition function:

$$Z'(T, p) \equiv \int_0^\infty dV \sum_n e^{-\frac{E_n(V)+pV}{k_BT}} = \int_0^\infty dV\, Z(T, V)e^{-\frac{pV}{k_BT}}$$

In the macroscopic limit, the partition function gives the Gibbs free energy of the system:

$$Z'(T, p) = e^{-\frac{G(T,p)}{k_BT}}$$

The Gibbs free energy $G(T, p)$ satisfies

$$dG = -SdT + Vdp$$

3. **T, M Distribution**—Let T be the temperature of the heat bath. The total energy of the heat bath and the system is kept at U_0.

   We only consider the microscopic states with a fixed magnetization M. When the system is in a state n with energy E_n, the energy of the heat bath is $U_0 - E_n$. Therefore, the entropy of the heat bath is given by

$$S_{\text{res}}(U_0 - E_n) = S_{\text{res}}(U_0) - \frac{E_n}{T}$$

   From this, we obtain the partition function:

$$Z(T, M) \equiv \sum_{n;M_n=M} e^{-\frac{E_n}{k_BT}},$$

   where the sum is restricted to the states with magnetization $M_n = M$. In the macroscopic limit, the partition function gives the Helmholtz free energy of the system:

$$Z'(T, M) = e^{-\frac{F(T,M)}{k_BT}}$$

   The Helmholtz free energy $F(T, M)$ satisfies

$$dF = -SdT + HdM$$

   where H is the external magnetic field.
4. **T, H Distribution**—Let T be the temperature of the heat bath. We introduce an electric circuit to create an external magnetic field H. It is assumed that there is no heat loss from the electric circuit.

   Suppose the system is in a microscopic state with energy E_n and magnetization M_n. When the magnetization of the system changes by ΔM, the work done by the electric circuit on the system is given by

$$W = H\Delta M$$

(See Sect. 3.3.4.) Therefore, among the energy E_n possessed by the system, HM_n is obtained from the electric circuit and $E_n - HM_n$ from the heat bath. Usually, this $E_n - HM_n$ is referred to as the system's energy (more precisely, enthalpy). Hence, the entropy of the heat bath is given by

$$S_{\text{res}}(U_0 - (E_n - HM_n)) = S_{\text{res}}(U_0) - \frac{E_n - HM_n}{T}$$

From this, we obtain the partition function:

$$Z'(T, H) \equiv \sum_n \mathrm{e}^{-\frac{E_n - HM_n}{k_B T}} = \int_{-\infty}^{\infty} dM\, Z(T, M)\mathrm{e}^{\frac{HM}{k_B T}}$$

In the macroscopic limit, the partition function gives the Gibbs free energy of the system:

$$Z'(T, H) = \mathrm{e}^{-\frac{G(T,H)}{k_B T}}$$

The Gibbs free energy $G(T, H)$ satisfies

$$dG = -SdT - MdH$$

3.3.4 Magnetic Energy of a Magnetic Material

Let us consider an infinitely long solenoid with a radius of a. When a current I per unit length flows through this solenoid, a uniform magnetic field

$$H_e = \frac{4\pi}{c} I$$

is created inside the solenoid (left-hand side of Fig. 3.16).

Now, let us consider filling the interior of the solenoid with a uniform magnetic material. We assume that a uniform magnetization M is induced in the magnetic material in the same direction as the external magnetic field. Therefore, a uniform loop current is generated inside the magnetic material, and a current per unit length

$$I_{\text{ind}} = cM$$

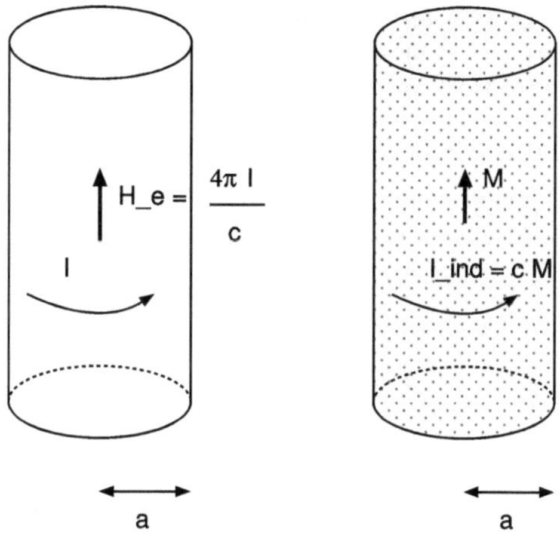

Fig. 3.16 The external magnetic field H_e induces current I_{ind} on the surface of the magnetic material

is induced on the surface (Example 5-3). Hence, the total magnetic field inside the magnetic material is given by

$$B = \frac{4\pi(I + I_{\text{ind}})}{c} = H_e + 4\pi M$$

(Right-hand side of Fig. 3.16)

Now, let us consider changing the current in the solenoid from I to $I + \Delta I$ over a small time interval Δt. During this time, the external magnetic field changes as

$$H_e \longrightarrow H_e + \frac{4\pi \Delta I}{c}$$

Consequently, the total magnetic field inside the magnetic material changes as

$$B \longrightarrow B + \Delta B$$

where

$$\Delta B = \frac{4\pi \Delta I}{c} + \Delta M$$

Therefore, the time variation of the magnetic flux through the solenoid induces an electric field on the surface of the solenoid:

$$E = \frac{1}{2\pi a} \cdot \frac{\Delta B}{c\Delta t}\pi a^2 = \frac{a}{2c}\frac{\Delta B}{\Delta t}$$

which acts in the opposite direction to the current (Fig. 3.17).

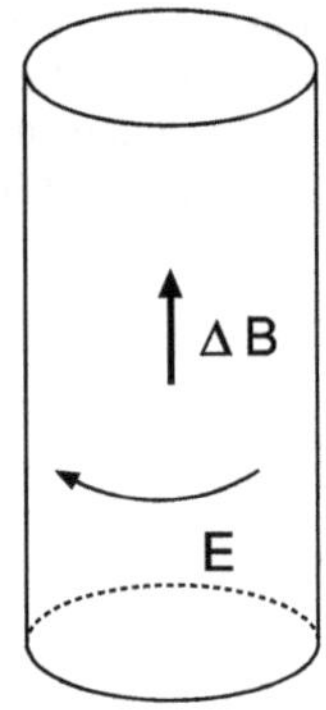

Fig. 3.17 Electric field E induced by the change of the magnetic flux

Hence, the work done by the current through the solenoid must be

$$\Delta W = \Delta t \cdot EI \cdot 2\pi a = \Delta t \cdot \frac{a}{2c}\frac{\Delta B}{\Delta t} \cdot \underbrace{I}_{=\frac{cH_e}{4\pi}} \cdot 2\pi a = \pi a^2 \cdot \frac{H_e dB}{4\pi}$$

where πa^2 is the cross section of the cylindrical magnetic material. Therefore, the infinitesimal change ΔU in the unit volume of the magnetic material is

$$\Delta U = \frac{\Delta W}{\pi a^2} = \boxed{\frac{H_e dB}{4\pi}}$$

Now, in the absence of the magnetic material, the energy density U_0 of the magnetic field H_e satisfies

$$\Delta U_0 = \frac{H_e dH_e}{4\pi}$$

Integrating this we obtain

$$\boxed{U_0 = \frac{H_e^2}{8\pi}}$$

Defining $U' \equiv U - U_0$, we obtain

$$\boxed{\Delta U' = H_e dM}$$

Considering the thermodynamics of magnetic materials, it is U' that is usually considered instead of U. Taking into account thermal energy, the first law of thermodynamics, with entropy S and temperature T, becomes

$$\boxed{dU' = TdS + H_e dM}$$

We consider this further in Exercise 5-4.

3.3.5 Landau-Ginzburg Model

In Sect. 2.3, we have demonstrated that the permeability of type II superconductors is given by

$$\frac{1}{\mu(k)} = 1 + \frac{1}{k^2 \lambda_L^2} \tag{3.210}$$

where λ_L is London's penetration depth. Though we have understood that this implies superconductivity, it may remain mysterious where this permeability comes from. In this note, without going into full details, we use the semi-phenomenological Landau-Ginzburg model to derive it.[29]

In addition to the scalar and vector potentials $(\phi, \vec{A})$ that represent the electromagnetic field, we introduce a classical complex field

$$\Psi(\vec{x}, t) = \sqrt{n}\, e^{i\varphi(\vec{x},t)}$$

to represent the conduction electrons in the medium. Here, $n > 0$ is a constant. The Lagrangian L that describes the electromagnetic field and the complex field is given by the spatial integral of the Lagrangian density $\mathcal{L}$:

$$\begin{aligned}
\mathcal{L} &= \frac{1}{8\pi}\left(\vec{E}^2 - \vec{B}^2\right) + \Psi^* i(\hbar\partial_t - iq\phi)\Psi \\
&\quad - \frac{1}{2m}\left(\hbar\nabla - i\frac{q}{c}\vec{A}\right)\Psi^* \cdot \left(\hbar\nabla + i\frac{q}{c}\vec{A}\right)\Psi - \rho_{ext}\phi + \frac{1}{c}\vec{j}_{ext}\cdot\vec{A} \\
&= \frac{1}{8\pi}\left(\vec{E}^2 - \vec{B}^2\right) + n(-\hbar\partial_t\varphi + q\phi) - \frac{n}{2m}\left(\hbar\nabla\varphi + \frac{q}{c}\vec{A}\right)^2 \\
&\quad - \rho_{ext}\phi + \frac{1}{c}\vec{j}_{ext}\cdot\vec{A}
\end{aligned} \tag{3.211}$$

where ρ_{ext} and $\vec{j}_{ext}$ represent the charge and current densities of sources other than the electrons. This Lagrangian is valid as long as the wave numbers and

[29] For the BCS theory, you may refer to graduate textbooks like R. Schrieffer's "Introduction to Superconductivity" (Advanced Books).

frequencies of the fields are not too large, and it can be derived from the fundamental Lagrangian that describes electrons in a metal. The Lagrangian (3.211) defines the Landau-Ginzburg model named after the physicists who first proposed it to explain superconductivity.

The Lagrangian has two peculiar aspects. One is the presence of the Planck constant $\hbar \equiv \frac{h}{2\pi}$, which is necessary to make the dimensions right, and suggests that superconductivity is a quantum mechanical phenomenon. The other is the electric charge

$$q = 2e \tag{3.212}$$

being twice the elementary charge e. This is because Ψ represents not individual electrons but pairs of electrons known as **Cooper pairs**. Therefore, the mass m represents the effective mass of a Cooper pair. $|\Psi|^2 = n$ gives the number density of Cooper pairs and $2n$ the electron number density.

Deriving the equations of motion from the Lagrangian (3.211) is straightforward. Here, we only give the results. First, the electric and magnetic fields $\vec{E}$ and $\vec{B}$, determined by the potentials

$$\vec{E} = -\nabla\phi - \frac{1}{c}\partial_t \vec{A}, \quad \vec{B} = \nabla \times \vec{A}, \tag{3.213}$$

satisfy automatically the Maxwell equations

$$\nabla \cdot \vec{B} = 0 \tag{3.214}$$

$$\nabla \times \vec{E} + \frac{1}{c}\partial_t \vec{B} = 0 \tag{3.215}$$

From the variation of the scalar potential, we obtain

$$\nabla \cdot \vec{E} = 4\pi \left(-qn + \rho_{ext}\right), \tag{3.216}$$

which implies that the charge density due to Cooper pairs is given by

$$\rho_{ind} \equiv -qn = -2en \tag{3.217}$$

From the variation of the vector potential, we get

$$\nabla \times \vec{B} - \frac{1}{c}\partial_t \vec{E} = \frac{4\pi}{c}\left(-\frac{qn}{m}(\hbar\nabla\varphi + \frac{q}{c}\vec{A}) + \vec{j}_{ext}\right), \tag{3.218}$$

which implies that the current density due to Cooper pairs is given by

$$\vec{j}_{ind} \equiv -\frac{qn}{m}\left(\hbar\nabla\varphi + \frac{q}{c}\vec{A}\right) \tag{3.219}$$

Finally, by varying the field of Cooper pairs φ, we obtain **London's equation**:

$$\nabla^2\varphi + \frac{q}{\hbar c}\nabla \cdot \vec{A} = 0, \tag{3.220}$$

which amounts to the conservation of the induced current: $\nabla \cdot \vec{j}_{ind} = 0$.

In the superconductor, in addition to the electrons, there are ions that neutralize the electrons' charges. To account for the charge density of the ions, we replace ρ_{ext} by $qn + \rho_{ext}$ in Eq. (3.216), which is rewritten as

$$\nabla \cdot \vec{E} = 4\pi\rho_{ext} \tag{3.221}$$

Taking the curl of Eq. (3.218) and using Eq. (3.215), we obtain

$$\left(\frac{1}{c^2}\partial_t^2 - \nabla^2 + \frac{4\pi q^2 n}{mc^2}\right)\vec{B} = \frac{4\pi}{c}\nabla \times \vec{j}_{ext} \tag{3.222}$$

The time-independent version of this equation has been derived in Example 5-5. Comparing Eq. (3.222) with

$$\left(\frac{1}{c^2}\partial_t^2 - \nabla^2 + \frac{1}{\lambda_L^2}\right)\vec{B} = \frac{4\pi}{c}\nabla \times \vec{j}_{ext} \tag{3.223}$$

obtained from the permeability (3.210), we conclude

$$\frac{1}{\lambda_L^2} = \frac{4\pi q^2 n}{mc^2} \tag{3.224}$$

For instance, if we take the mass m to be twice the electron mass and $n \approx 10^{20}/\mathrm{cm}^3$, then $\lambda_L \approx 10^{-5}$ cm. This represents the depth of the region into which the magnetic field penetrates inside a superconductor. (See Example 5-5.)

Finally, to conclude this note, we derive **magnetic flux quantization**. Consider a superconductor with a hole. Let Φ be the magnetic flux passing through this hole (Fig. 3.18).

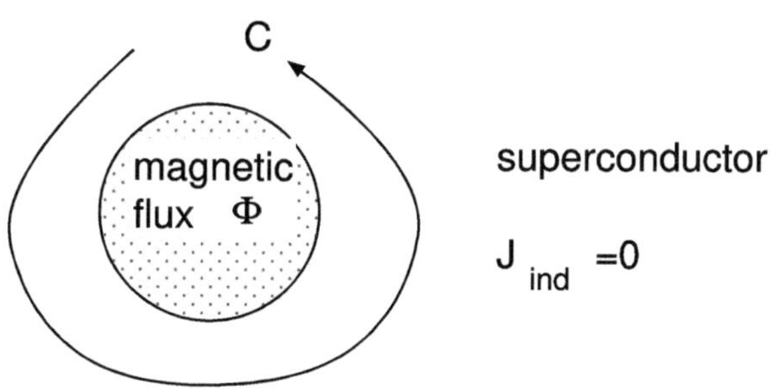

Fig. 3.18 Magnetic flux quantization

Assuming that the complex field $\Psi(\vec{x}, t)$ is single valued, its phase φ must be unique except for a multiple of 2π. Therefore, the line integral along a closed path C going around the hole is given by

$$\oint_C \nabla\varphi \cdot d\vec{x} = 2\pi l \quad (l \in \mathbf{Z}) \tag{3.225}$$

Since the magnetic field penetrates into the superconductor only over the length of λ_L, we can take the current to vanish along C:

$$\hbar\nabla\varphi + \frac{q}{c}\vec{A} = 0 \tag{3.226}$$

Hence, the magnetic flux Φ passing through the hole is given by

$$\Phi \equiv \oint_C \vec{A} \cdot d\vec{x} = -\frac{\hbar c}{q}\oint_C \nabla\varphi = -\frac{2\pi\hbar c}{q} l \tag{3.227}$$

The magnetic flux is thus quantized as an integral multiple of the superconducting **magnetic flux quantum** defined by

$$\Phi_0 \equiv \frac{2\pi\hbar c}{q} = \frac{\pi\hbar c}{e} \simeq 2.07 \times 10^{-7}\ \text{esu} \tag{3.228}$$

in the cgs-Gauss units. In the SI units, we obtain[30]

$$\Phi_0(\text{SI}) = \frac{\pi\hbar}{e_{\text{SI}}} = 2.067\,833\,848\cdots \times 10^{-15}\ \text{Wb} \tag{3.229}$$

where $e_{\text{SI}} = \sqrt{4\pi\epsilon_0}\, e \equiv 1.602\,176\,634 \times 10^{-19}\ \text{C}$ is the elementary charge in the SI. Note that Φ_0 is a defined quantity in the SI units.

3.3.6 *Derivation of Landau's Free Energy from the Bragg-Williams Approximation*

In Example 6-5, we have learned Landau's phenomenological theory for ferromagnets. Denoting the magnetization by M, the Helmholtz free energy density for a ferromagnet is assumed to be

$$F(T, M) = a(T)M^2 + bM^4 \tag{3.230}$$

30 $\Phi_0 = \sqrt{\frac{4\pi}{\mu_0}}\,\Phi_0(\text{SI})$, and $e = \frac{1}{\sqrt{4\pi\epsilon_0}} e_{\text{SI}} = c\sqrt{\frac{\mu_0}{4\pi}}\, e_{\text{SI}}$.

Using this, we have calculated the Gibbs free energy density for a ferromagnet in an external magnetic field H as

$$G(T, H) = \min_M (F(T, M) - MH) \tag{3.231}$$

where we choose the magnetization M that minimizes the right-hand side. We have demonstrated that at low temperatures where $a(T) < 0$, spontaneous magnetization occurs. Furthermore, in Example 7-6, we have shown that the spontaneous magnetization obtained from Landau's theory is reproduced by the mean-field approximation applied to the Ising model.

In this note, we accomplish two things:

1. Introduce the Bragg-Williams approximation to compute $F(T, M)$, and demonstrate that this approximation is equivalent to the mean-field approximation.
2. Derive the Helmholtz free energy (3.230) for small magnetization.

We consider the Ising model consisting of N_0 spins. Each spin s can take values ± 1. Let α be the number of nearest neighbors. $\alpha = 2D$ for the D-dimensional cubic lattice. The total energy is given as

$$E = -K \sum_{\langle i,j \rangle} s_i s_j \tag{3.232}$$

where the sum is over the pairs of nearest neighbors.

In the Bragg-Williams approximation, similar to the mean-field approximation, we neglect spin correlations and describe the system in terms of its average behavior. Let N be the number of up spins so that the number of down spins is $N_0 - N$. Focusing on a single spin, the probability that it is an up spin is N/N_0, and that for a down spin is $1 - N/N_0$. In the Bragg-Williams approximation, we assume that for a nearest neighbor pair, the probabilities are given as follows: the probability that both spins are up is $(N/N_0)^2$, the probability that both spins are down is $(1 - N/N_0)^2$, and the probability that one spin is up and the other is down is $2\,(1 - N/N_0)\,N/N_0$. The three add up to 1. We emphasize that we are ignoring correlations among the spins.

Therefore, the total energy of the system, considering that there are $N_0 \frac{\alpha}{2}$ nearest neighbor pairs, is obtained as

$$\begin{aligned} N_0 E(N) &= N_0 \frac{\alpha}{2} K \left(\left(\frac{N}{N_0} \right)^2 + \left(1 - \frac{N}{N_0} \right)^2 - 2 \left(1 - \frac{N}{N_0} \right) \frac{N}{N_0} \right) \\ &= N_0 \frac{\alpha K}{2} \left(\frac{2N}{N_0} - 1 \right)^2 \end{aligned} \tag{3.233}$$

where $E(N)$ is the energy per spin. On the other hand, the entropy is given as

$$\frac{1}{k_B}N_0 S(N) = \ln\binom{N_0}{N} \simeq N_0 \ln N_0 - N\ln N - (N_0 - N)\ln(N_0 - N)$$
$$= -N_0\left[\frac{N}{N_0}\ln\frac{N}{N_0} + \left(1-\frac{N}{N_0}\right)\ln\left(1-\frac{N}{N_0}\right)\right] \tag{3.234}$$

where we have used Stirling's approximation $\ln N! \simeq N(\ln N - 1)$, valid for $N \gg 1$.

Therefore, if we define the average spin by

$$s \equiv \frac{N}{N_0} - \left(1-\frac{N}{N_0}\right) = \frac{2N}{N_0} - 1 \tag{3.235}$$

then

$$\frac{N}{N_0} = \frac{1}{2}(1+s), \quad 1-\frac{N}{N_0} = \frac{1}{2}(1-s) \tag{3.236}$$

Furthermore, if we denote the magnetic moment of the spin as μ, the magnetization per spin is given by

$$M = \mu s \tag{3.237}$$

Hence, using M, we can rewrite the energy and entropy per spin as follows:

$$U(T, M) = -K\alpha\frac{s^2}{2} = -\frac{K\alpha}{2\mu^2}M^2, \tag{3.238}$$

$$\frac{1}{k_B}S(T, M) = -\left(\frac{1}{2}(1+s)\ln\frac{1}{2}(1+s) + \frac{1}{2}(1-s)\ln\frac{1}{2}(1-s)\right)$$
$$= -\left[\frac{1}{2}\left(1+\frac{M}{\mu}\right)\ln\frac{1}{2}\left(1+\frac{M}{\mu}\right) + \frac{1}{2}\left(1-\frac{M}{\mu}\right)\ln\frac{1}{2}\left(1-\frac{M}{\mu}\right)\right] \tag{3.239}$$

Therefore, the Helmholtz free energy per spin is given by

$$F(T, M) \equiv U - TS$$
$$= -\frac{K\alpha}{2\mu^2}M^2$$
$$+ k_BT\left[\frac{1}{2}\left(1+\frac{M}{\mu}\right)\ln\frac{1}{2}\left(1+\frac{M}{\mu}\right) + \frac{1}{2}\left(1-\frac{M}{\mu}\right)\ln\frac{1}{2}\left(1-\frac{M}{\mu}\right)\right] \tag{3.240}$$

Note that $F(T, M)$ does not depend on the sign of M:

$$\boxed{F(T, -M) = F(T, M)} \tag{3.241}$$

Then, the Gibbs free energy in the presence of an external magnetic field H is given by

$$G(T, H) = \mathrm{Min}_M \left(F(T, M) - HM\right) \tag{3.242}$$

The value of M that minimizes the right-hand side can be determined from Eq. (3.240):

$$\frac{\partial F(T, M)}{\partial M} = H \Longleftrightarrow -K\alpha\frac{M}{\mu} - \mu H + \frac{k_B T}{2}\ln\frac{1+\frac{M}{\mu}}{1-\frac{M}{\mu}} = 0$$

$$\Longleftrightarrow \boxed{\frac{M}{\mu} = \tanh\frac{K\alpha\frac{M}{\mu} + \mu H}{k_B T}} \tag{3.243}$$

The final expression is the same as Eq. (3.141) in Sect. 3.2.3 (where μH is denoted as h). This shows that the Bragg-Williams approximation explained above and the mean-field approximation explained in Sect. 3.2.3 are equivalent.

Furthermore, to understand the relation with Landau's phenomenological theory, let us approximate Eq. (3.240) for M much smaller than μ (i.e., $|s| \ll 1$). This gives

$$F(T, M) \simeq \frac{1}{2\mu^2}\left(k_B T - K\alpha\right) M^2 + \frac{k_B T}{12\mu^4} M^4 - k_B T \ln 2 \tag{3.244}$$

Hence, the values of $a(T)$ and b in Eq. (3.230) are determined as

$$a(T) = \frac{k_B}{2\mu^2}\left(T - T_c\right), \quad b = \frac{k_B T_c}{12\mu^4}, \tag{3.245}$$

where $T_c \equiv \frac{K\alpha}{k_B}$ is the critical temperature. We have thus derived Landau's phenomenological theory from the Bragg-Williams approximation of the Ising model.

Chapter 4
Quantum Mechanics

Without quantum mechanics, it is impossible to understand modern physics. Quantum mechanics was developed to explain the vast amount of spectroscopic data accumulated from the nineteenth to the early twentieth centuries. In order to aim for breakthroughs in the twenty-first century, we need to master the basics of quantum mechanics.

4.1 Formulation of Quantum Mechanics by Operators

Three Points to Remember

1. Formulation by operators—Quantum states are vectors, and physical quantities are hermitian operators.
2. Quantum harmonic oscillator—The energy eigenvalue of a quantum harmonic oscillator with the angular frequency ω is given by $E_n = \left(n + \frac{1}{2}\right) \hbar\omega$.
3. One-dimensional potential problem—Besides numerical methods, the variational method and WKB approximation are available.

4.1.1 Formulation by Operators—Quantum States Are Vectors, and Physical Quantities Are Hermitian Operators

In a typical undergraduate course, quantum mechanics is formulated mainly via wave functions. Graduate students, however, should understand the more abstract formulation based on vector spaces and linear operators. We summarize the mathematics necessary for the formulation of quantum mechanics in Sect. 4.6.1.

H. Sonoda, K. Ülker, *Basic Physics for Graduate Students*,
https://doi.org/10.1007/978-981-95-6607-5_4

State Vectors

The quantum state of a physical system is represented by a state vector, and the entire set of these vectors forms a complex vector space equipped with inner products. Whether its dimension is finite or infinite depends on the physical system. We introduce an orthonormal basis, i.e., a set of orthonormal vectors $\{|n\rangle\}$ for the vector space. Here, we consider basis vectors indexed by integers $n = 0, 1, 2, \cdots$. Therefore, the inner product is given by[1]

$$\langle m \mid n \rangle = \delta_{m,n} \tag{4.1}$$

Any state vector (also called a **ket vector**) can be expanded in terms of the basis vectors as[2]

$$|\psi\rangle = \sum_{n=0}^{\infty} c_n |n\rangle \qquad (c_n \in \mathbf{C}) \tag{4.2}$$

Due to the orthonormality of the basis vectors, the complex coefficient c_n is given by the inner product

$$c_n = \langle n \mid \psi \rangle \tag{4.3}$$

The vector **conjugate** to the state vector (4.2) (called a **bra vector**) is given by

$$\langle \psi | = \sum_{n=0}^{\infty} c_n^* \langle n | \tag{4.4}$$

Assume that $|\phi\rangle$ and $|\psi\rangle$ are two normalized state vectors:

$$\langle \phi \mid \phi \rangle = \langle \psi \mid \psi \rangle = 1 \tag{4.5}$$

We interpret the complex number

$$\langle \psi \mid \phi \rangle$$

as the **probability amplitude** that the quantum state ϕ is also a state ψ. The probability is given by the square of the absolute value of the probability amplitude:

$$|\langle \psi \mid \phi \rangle|^2$$

[1] If n is a continuous variable, the Kronecker delta is replaced by the delta function. For example, in the case of a coordinate x, the inner product is given by $\langle x \mid x' \rangle = \delta(x - x')$.

[2] For the continuous coordinate, this becomes $|\psi\rangle = \int dx\, |x\rangle \langle x \mid \psi \rangle = \int dx\, |x\rangle\, \psi(x)$.

Since $\langle\psi \mid \phi\rangle = \langle\phi \mid \psi\rangle^*$, we find

1. The probability that a system prepared in the state ϕ is measured to be in the state ψ
2. The probability that a system prepared in the state ψ is measured to be in the state ϕ

arc cqual.

Let us look at this more concretely using a particle on the x-axis under a harmonic potential. (We will discuss the quantum harmonic oscillator more fully in the next subsection.) Denote the n-th energy eigenstate of the harmonic oscillator by $|n\rangle$ $(n = 0, 1, 2, \cdots)$, and the eigenstate of the x-coordinate as $|x\rangle$ $(x \in \mathbf{R})$. The former is normalized as $\langle n \mid m\rangle = \delta_{nm}$ and the latter as $\langle x \mid x'\rangle = \delta(x - x')$. We find

1. $|\langle x \mid n\rangle|^2 dx$ represents the probability for the particle in the n-th eigenstate is found in the interval $(x, x + dx)$.
2. $|\langle n \mid x\rangle|^2$ represents the relative probability that a particle at coordinate x is in the n-th eigenstate.

Physical Quantities

Physical quantities such as position, momentum, and energy are represented by hermitian linear operators. A **hermitian operator** $\hat{O}$ satisfies, for any vectors $|\phi\rangle$ and $|\psi\rangle$,

$$\langle\phi|\,\hat{O}\,|\psi\rangle = \langle\psi|\,\hat{O}\,|\phi\rangle^* \tag{4.6}$$

The **hermitian conjugate** operator $\hat{O}^\dagger$ of $\hat{O}$ is defined as

$$\langle\phi|\,\hat{O}^\dagger\,|\psi\rangle = \langle\psi|\,\hat{O}\,|\phi\rangle^* \tag{4.7}$$

Hence, for $\hat{O}$ to be hermitian, we must find[3]

$$\hat{O}^\dagger = \hat{O} \tag{4.8}$$

It is an easy exercise to show that all the eigenvalues of a hermitian operator $\hat{O}$ are real. Given an eigenstate $|\psi\rangle$ of $\hat{O}$ with an eigenvalue λ

$$\hat{O}\,|\psi\rangle = \lambda\,|\psi\rangle\ , \tag{4.9}$$

[3] Hermitian operators are also called self-adjoint operators. In physics we often do not distinguish hermitian operators from self-adjoint operators. In functional analysis, however, the two are distinguished.

we find

$$\lambda \langle \phi \mid \psi \rangle = \langle \phi | \hat{O} | \psi \rangle \tag{4.10}$$

for any state $|\phi\rangle$. Taking the complex conjugate, we find

$$\lambda^* \langle \psi \mid \phi \rangle = \langle \psi | \hat{O}^\dagger | \phi \rangle = \langle \psi | \hat{O} | \phi \rangle \tag{4.11}$$

using the hermiticity of $\hat{O}$. Taking $\phi = \psi$, we obtain

$$\lambda^* \langle \psi \mid \psi \rangle = \lambda \langle \psi \mid \psi \rangle \tag{4.12}$$

Since $\langle \psi \mid \psi \rangle > 0$ for the nonzero state, it follows that $\lambda = \lambda^*$, i.e., real. We can also show that the eigenvector $|\phi\rangle$ with eigenvalue λ of a hermitian operator $\hat{O}$ is orthogonal to $|\psi\rangle$ with a different eigenvalue λ':

$$\lambda \langle \psi \mid \phi \rangle = \langle \psi | \hat{O} | \phi \rangle = \lambda' \langle \psi \mid \phi \rangle \tag{4.13}$$

where we have used

$$\langle \psi | \hat{O} = \lambda' \langle \psi | \tag{4.14}$$

If $\lambda \neq \lambda'$, this implies

$$\langle \psi \mid \phi \rangle = 0 \tag{4.15}$$

If two different vectors $|\phi\rangle$ and $|\psi\rangle$ share the same eigenvalue λ, we can modify $|\psi\rangle$ into

$$|\psi'\rangle \equiv |\psi\rangle - |\phi\rangle \langle \phi \mid \psi \rangle \tag{4.16}$$

so that $\hat{O} |\psi'\rangle = \lambda |\psi'\rangle$ but $\langle \phi \mid \psi' \rangle = 0$.

Generally, given a hermitian operator $\hat{O}$, it is possible to create a complete set of orthonormal basis vectors for the vector space, using all of its eigenvectors. Writing it as $\{|n\rangle\}_{n=0,1,2,\cdots}$, it satisfies

$$\hat{O} |n\rangle = \lambda_n |n\rangle \quad (\lambda_n \in \mathbf{R}) \tag{4.17}$$

and

$$\langle n \mid m \rangle = \delta_{n,m} \tag{4.18}$$

As has been shown above, two eigenvectors with distinct eigenvalues are orthogonal to each other, and even if they have the same eigenvalue (called degeneracy), appropriate linear combinations can be made orthogonal to each other.

A physical quantity is represented by a hermitian operator $\hat{O}$. In an eigenstate $|n\rangle$ of $\hat{O}$ with eigenvalue λ_n, the result of measurement is always λ_n. Measuring the same physical quantity in state $|\psi\rangle = \sum_n c_n |n\rangle$, the probability of obtaining λ_n is given by $|c_n|^2$. **Values other than the eigenvalues are never obtained.**

Temporal Evolution of States

The temporal evolution of states is governed by the Schrödinger equation:

$$\boxed{i\hbar \frac{d}{dt} |\psi(t)\rangle = \hat{H} |\psi(t)\rangle} \tag{4.19}$$

Here, the Hamiltonian $\hat{H}$ is a hermitian operator. In general, $\hat{H}$ depends on time, but for now, we consider the time independent case.

Let $|n\rangle$ be the eigenstate of the Hamiltonian $\hat{H}$ with energy eigenvalue E_n:

$$\hat{H} |n\rangle = E_n |n\rangle \tag{4.20}$$

Since the eigenstates $\{|n\rangle\}$ form a complete orthonormal basis, we can expand the state $|\psi(t)\rangle$ in terms of the eigenstates as

$$|\psi(t)\rangle = \sum_n \psi_n(t) |n\rangle \tag{4.21}$$

where $\psi_n(t) \equiv \langle n \mid \psi(t)\rangle$ are the expansion coefficients. The Schrödinger equation then becomes

$$i\hbar \sum_n \frac{d\psi_n(t)}{dt} |n\rangle = \sum_n E_n \psi_n(t) |n\rangle \tag{4.22}$$

This gives

$$\frac{d\psi_n(t)}{dt} = -i\frac{E_n}{\hbar}\psi_n(t) \tag{4.23}$$

Therefore, the time dependence of each coefficient is

$$\psi_n(t) = \mathrm{e}^{-i\frac{E_n}{\hbar}t}\psi_n(0) \tag{4.24}$$

Example: Two-Dimensional Vector Space

Let $|\uparrow\rangle$ be a spin-up state and $|\downarrow\rangle$ be a spin-down state.[4]

$$\langle\uparrow \mid \uparrow\rangle = \langle\downarrow \mid \downarrow\rangle = 1$$
$$\langle\uparrow \mid \downarrow\rangle = \langle\downarrow \mid \uparrow\rangle = 0$$

An arbitrary state is given as a superposition (linear combination) of $|\uparrow\rangle$ and $|\downarrow\rangle$:

$$|\psi\rangle = a\,|\uparrow\rangle + b\,|\downarrow\rangle \tag{4.25}$$

where a and b are complex coefficients. If $\langle\, \psi \mid \psi\,\rangle = 1$, we obtain $|a|^2 + |b|^2 = 1$.

Physical quantities are given by hermitian operators. Any hermitian operator $\hat{O}$ corresponds one-to-one with a 2-by-2 hermitian matrix. If we define real numbers α, β and a complex number γ by

$$\alpha = \langle\uparrow|\,\hat{O}\,|\uparrow\rangle\,, \qquad \beta = \langle\downarrow|\,\hat{O}\,|\downarrow\rangle$$
$$\gamma = \langle\uparrow|\,\hat{O}\,|\downarrow\rangle\,, \qquad \gamma^* = \langle\downarrow|\,\hat{O}\,|\uparrow\rangle = \langle\uparrow|\,\hat{O}\,|\downarrow\rangle^*$$

the corresponding hermitian matrix is given by

$$M_O \equiv \begin{pmatrix} \alpha & \gamma \\ \gamma^* & \beta \end{pmatrix} \quad \alpha, \beta \in \mathbf{R} \tag{4.26}$$

Using the matrix M_O, the hermitian operator $\hat{O}$ is expressed as

$$\begin{aligned} \hat{O} &= (|\uparrow\rangle\,, |\downarrow\rangle) M_O \begin{pmatrix} \langle\uparrow| \\ \langle\downarrow| \end{pmatrix} \\ &= \alpha\,|\uparrow\rangle\,\langle\uparrow| + \beta\,|\downarrow\rangle\,\langle\uparrow| + \gamma\,|\uparrow\rangle\,\langle\downarrow| + \gamma^*\,|\downarrow\rangle\,\langle\uparrow| \end{aligned} \tag{4.27}$$

The hermitian operator $\hat{O}$ has two eigenvectors that form an orthonormal basis:

$$\hat{O}\,|\lambda_i\rangle = \lambda_i\,|\lambda_i\rangle \quad (i = 1, 2) \tag{4.28}$$

where $\lambda_i \in \mathbf{R}$. The two eigenstates are orthogonal to each other:

$$\langle\lambda_i \mid \lambda_j\rangle = \delta_{ij} \tag{4.29}$$

[4] This two-dimensional state vector space is mathematically quite simple but has very broad applications in physics. Not only does it represent a particle with spin $\frac{1}{2}$, but it can also represent an oscillation between two types of neutrinos and describe covalent bonds and lasers. For more details, refer to the Feynman Lectures Vol. III (Quantum Mechanics).

We introduce three hermitian matrices, known as the **Pauli matrices**, which are useful in applications:

$$\boxed{\sigma_x = \begin{pmatrix} 0 & 1 \\ 1 & 0 \end{pmatrix}, \quad \sigma_y = \begin{pmatrix} 0 & -i \\ i & 0 \end{pmatrix}, \quad \sigma_z = \begin{pmatrix} 1 & 0 \\ 0 & -1 \end{pmatrix}} \tag{4.30}$$

All of their eigenvalues are ± 1. Let us summarize the normalized eigenvectors in a table.

Pauli matrix	Eigenvector with eigenvalue 1	Eigenvector with eigenvalue -1
σ_x	$\frac{1}{\sqrt{2}}\begin{pmatrix} 1 \\ 1 \end{pmatrix}$	$\frac{1}{\sqrt{2}}\begin{pmatrix} 1 \\ -1 \end{pmatrix}$
σ_y	$\frac{1}{\sqrt{2}}\begin{pmatrix} 1 \\ i \end{pmatrix}$	$\frac{1}{\sqrt{2}}\begin{pmatrix} 1 \\ -i \end{pmatrix}$
σ_z	$\begin{pmatrix} 1 \\ 0 \end{pmatrix}$	$\begin{pmatrix} 0 \\ 1 \end{pmatrix}$

The hermitian operators corresponding to the Pauli matrices are

$$\hat{\sigma}_x = (|\uparrow\rangle, |\downarrow\rangle)\sigma_x \begin{pmatrix} \langle\uparrow| \\ \langle\downarrow| \end{pmatrix} = |\uparrow\rangle\langle\downarrow| + |\downarrow\rangle\langle\uparrow| \tag{4.31}$$

$$\hat{\sigma}_y = (|\uparrow\rangle, |\downarrow\rangle)\sigma_y \begin{pmatrix} \langle\uparrow| \\ \langle\downarrow| \end{pmatrix} = -i\,|\uparrow\rangle\langle\downarrow| + i\,|\downarrow\rangle\langle\uparrow| \tag{4.32}$$

$$\hat{\sigma}_z = (|\uparrow\rangle, |\downarrow\rangle)\sigma_z \begin{pmatrix} \langle\uparrow| \\ \langle\downarrow| \end{pmatrix} = |\uparrow\rangle\langle\uparrow| - |\downarrow\rangle\langle\downarrow| \tag{4.33}$$

Any 2-by-2 hermitian matrix M can be expressed as a linear combination

$$M = a\mathbf{1}_2 + \vec{a}\cdot\vec{\sigma} \quad (a, a_x, a_y, a_z \in \mathbf{R})$$

where $\mathbf{1}_2$ is the identity matrix. The eigenvalues of M are $a \pm |\vec{a}|$.

Example: An Infinite Dimensional Vector Space

Let $\hat{x}$ be the hermitian operator denoting the x-coordinate of a particle. Denoting the eigenstate with eigenvalue x by $|x\rangle$, we obtain

$$\hat{x}\,|x\rangle = x\,|x\rangle, \quad \langle x \mid y\rangle = \delta(x-y), \quad \int_{-\infty}^{\infty} dx\ |x\rangle\langle x| = 1 \tag{4.34}$$

Any state $|\psi\rangle$ can be represented as a linear combination of $|x\rangle$:

$$|\psi\rangle = \int_{-\infty}^{\infty} dx\ |x\rangle \underbrace{\langle x \mid \psi\rangle}_{=\psi(x)} = \int_{-\infty}^{\infty} dx\ |x\rangle\, \psi(x) \tag{4.35}$$

where the complex coefficient $\psi(x)$ is called a **wave function**.[5] The inner product is given by

$$\langle \phi \mid \psi\rangle = \int dx\, \langle \phi \mid x\rangle \langle x \mid \psi\rangle = \int dx\, \phi^*(x)\psi(x) \tag{4.36}$$

Let $\hat{T}(\alpha)$ be a unitary operator that translates the particle along the x-axis by α. It is then given by

$$\hat{T}(\alpha)\,|x\rangle = |x+\alpha\rangle\,, \quad \langle x|\,\hat{T}(\alpha) = \langle x-\alpha| \tag{4.37}$$

Note that $\hat{T}(\alpha)^\dagger = \hat{T}(-\alpha) = \hat{T}(\alpha)^{-1}$. The momentum operator $\hat{p}$ is a hermitian operator defined by

$$\hat{T}(\epsilon) = 1 - \frac{i}{\hbar}\epsilon\,\hat{p} \tag{4.38}$$

where ϵ is an infinitesimal distance. Thus, from Eq. (4.37), we get

$$\hat{p}\,|x\rangle = i\hbar\frac{d}{dx}\,|x\rangle \tag{4.39}$$

For any ket state $|\psi\rangle$, we have

$$\langle x|\,\hat{T}(\alpha)\,|\psi\rangle = \langle x-\alpha \mid \psi\rangle = \psi(x-\alpha) \tag{4.40}$$

This leads to

$$\langle x|\,\hat{p}\,|\psi\rangle = -i\hbar\frac{d}{dx}\psi(x) \tag{4.41}$$

Thus, the momentum eigenstate $|p\rangle$ is given by

$$\langle x \mid p\rangle = e^{\frac{i}{\hbar}px} \tag{4.42}$$

Here, $|p\rangle$ is given a particular normalization for the sake of convenience.

---

[5] The state vector $|\psi\rangle$ refers to the state whose wave function is given by ψ. Note the difference between the function ψ and its value $\psi(x)$ at some x. For instance, when we mention the exponential function e^x, we are referring to the "map" that assigns the value e^x to the variable x, not the value of the function at some specific x.

Now, according to the definition (4.42), the momentum eigenstates satisfy

$$\hat{p}\,|p\rangle = p\,|p\rangle\,, \quad \langle p \mid q\rangle = 2\pi\hbar\;\delta(p-q), \quad \int_{-\infty}^{\infty}\frac{dp}{2\pi\hbar}\,|p\rangle\,\langle p| = 1 \tag{4.43}$$

Under a translation, the momentum eigenstates undergo only a phase change:

$$\hat{T}(\alpha)\,|p\rangle = \mathrm{e}^{-\frac{i}{\hbar}p\alpha}\,|p\rangle \tag{4.44}$$

To see how the position operator acts on the eigenstates of momentum, we only need to follow definitions:

$$\begin{aligned}\hat{x}\,|p\rangle &= \int dx\;\;|x\rangle\,\langle x|\,\hat{x}\,|p\rangle = \int dx\;\;|x\rangle\,x\,\langle x \mid p\rangle = \int dx\;\;|x\rangle\,x\mathrm{e}^{i\frac{px}{\hbar}} \\ &= -i\hbar\frac{d}{dp}\int dx\;\;|x\rangle\,\mathrm{e}^{\frac{i}{\hbar}px} = -i\hbar\frac{d}{dp}\,|p\rangle\end{aligned} \tag{4.45}$$

This is to be compared with Eq. (4.39).

Conserved Quantities

We wish to consider the conditions for a physical quantity to be conserved. Let $\hat{O}$ be a hermitian operator representing a physical quantity. Consider its eigenstate $|\psi\rangle$ with eigenvalue λ:

$$\hat{O}\,|\psi\rangle = \lambda\,|\psi\rangle \tag{4.46}$$

After a small time interval Δt, the state vector changes to

$$|\psi(\Delta t)\rangle = |\psi\rangle - i\frac{\Delta t}{\hbar}\hat{H}\,|\psi\rangle \tag{4.47}$$

For the physical quantity $\hat{O}$ to be conserved, this state must still be an eigenstate of $\hat{O}$ with the same eigenvalue λ:

$$\hat{O}\,|\psi(\Delta t)\rangle = \lambda\,|\psi(\Delta t)\rangle \tag{4.48}$$

Let us determine the conditions for this to hold. The left-hand side of the above equation is

$$\begin{aligned}\text{(LHS)} &= \hat{O}\left(|\psi\rangle - i\frac{\Delta t}{\hbar}\hat{H}\,|\psi\rangle\right) = \left(1 - i\frac{\Delta t}{\hbar}\hat{H}\right)\hat{O}\,|\psi\rangle + i\frac{\Delta t}{\hbar}[\hat{H},\hat{O}]\,|\psi\rangle \\ &= \lambda\,|\psi(\Delta t)\rangle + i\frac{\Delta t}{\hbar}[\hat{H},\hat{O}]\,|\psi\rangle\end{aligned} \tag{4.49}$$

For the conservation law (4.48) to hold, it is sufficient to have

$$\boxed{\left[\hat{H}, \hat{O}\right] = 0} \tag{4.50}$$

In other words, a hermitian operator that commutes with the Hamiltonian represents a conserved quantity. Hence, the Hamiltonian itself, i.e., energy, is a conserved quantity.

Equation (4.50) is indeed a reasonable condition. If this is met, as a basis of the state vector space, we can choose vectors that diagonalize both the Hamiltonian and $\hat{O}$ simultaneously. For instance, in the case of a free particle, eigenstates of momentum, which is conserved, are also eigenstates of energy. Thus, as time progresses, only the phase of the state changes.

4.1.2 *Quantum Harmonic Oscillator—The Energy Eigenvalues of a Quantum Harmonic Oscillator with the Angular Frequency ω Are Given by* $E_n = \left(n + \frac{1}{2}\right)\hbar\omega$

The quantization of a harmonic oscillator becomes especially important when quantizing continuous systems. For instance, the quantization of electromagnetic fields or lattice vibrations is a simple application of the quantization of a harmonic oscillator.

The Hamiltonian of a quantum harmonic oscillator is given by

$$\hat{H} = \frac{1}{2m}\hat{p}^2 + \frac{1}{2}m\omega^2\hat{x}^2 \tag{4.51}$$

We introduce an **annihilation operator** $\hat{a}$ and its hermitian conjugate, a **creation operator**, $\hat{a}^\dagger$ by

$$\hat{a} = \mathrm{e}^{i\theta}\left(\frac{1}{\sqrt{2\hbar m\omega}}\hat{p} - i\sqrt{\frac{m\omega}{2\hbar}}\hat{x}\right) \tag{4.52}$$

$$\hat{a}^\dagger = \mathrm{e}^{-i\theta}\left(\frac{1}{\sqrt{2\hbar m\omega}}\hat{p} + i\sqrt{\frac{m\omega}{2\hbar}}\hat{x}\right) \tag{4.53}$$

where θ is an arbitrary real number. These are constructed to satisfy the commutation relation:

$$[\hat{a}, \hat{a}^\dagger] = \frac{1}{2\hbar}\left([\hat{p}, i\hat{x}] - [i\hat{x}, \hat{p}]\right) = 1 \tag{4.54}$$

Using $\hat{a}$, $\hat{a}^\dagger$, we can express $\hat{x}$, $\hat{p}$ as

$$\sqrt{\frac{2}{\hbar m\omega}}\,\hat{p} = \hat{a}\mathrm{e}^{-i\theta} + \hat{a}^{\dagger}\mathrm{e}^{i\theta} \tag{4.55}$$

$$-i\sqrt{\frac{2m\omega}{\hbar}}\,\hat{x} = \hat{a}\mathrm{e}^{-i\theta} - \hat{a}^{\dagger}\mathrm{e}^{i\theta} \tag{4.56}$$

Thus, the Hamiltonian becomes

$$\begin{aligned}\hat{H} &= \frac{1}{4}\hbar\omega\left(\hat{a}\mathrm{e}^{-i\theta} + \hat{a}^{\dagger}\mathrm{e}^{i\theta}\right)^2 - \frac{1}{4}\hbar\omega\left(\hat{a}\mathrm{e}^{-i\theta} - \hat{a}^{\dagger}\mathrm{e}^{i\theta}\right)^2 \\ &= \frac{\hbar\omega}{4}\left(\hat{a}\hat{a}^{\dagger} + \hat{a}^{\dagger}\hat{a} + \hat{a}\hat{a}^{\dagger} + \hat{a}^{\dagger}\hat{a}\right) \\ &= \hbar\omega\left(\hat{a}^{\dagger}\hat{a} + \frac{1}{2}\right)\end{aligned} \tag{4.57}$$

The operator $\hat{a}^{\dagger}\hat{a}$ is positive (to be precise it should be said non-negative, but that is a bit wordy). Therefore, the ground state $|\Omega\rangle$ can be defined as

$$\hat{a}\,|\Omega\rangle = 0 \tag{4.58}$$

Note that as we will demonstrate shortly, the ground state must necessarily be annihilated by $\hat{a}$ since $\hat{a}$ has the effect of reducing energy by $\hbar\omega$. If not, there would be states with energy lower than the ground state, leading to a contradiction. Thus, we obtain

$$\hat{H}\,|\Omega\rangle = \frac{\hbar\omega}{2}\,|\Omega\rangle \tag{4.59}$$

This implies that the energy of the ground state is not zero but $\frac{\hbar\omega}{2}$. This is referred to as the **zero-point energy**. It arises due to the uncertainty in momentum and position, resulting in this energy.

The n-th energy eigenstate can be obtained by applying the creation operator n times on the ground state:

$$\boxed{|n\rangle \equiv \frac{1}{\sqrt{n!}}\left(\hat{a}^{\dagger}\right)^n|\Omega\rangle} \tag{4.60}$$

Let us verify these results.

- $|n\rangle$ **is an energy eigenstate**—When $n = 0$, the above formula gives the ground state. Now, assume that $|\psi\rangle$ is an energy eigenstate with energy $E = \hbar\omega(n + 1/2)$:

$$\hat{a}^{\dagger}\hat{a}\,|\psi\rangle = n\,|\psi\rangle \tag{4.61}$$

From the commutation relation

$$[\hat{a}^\dagger\hat{a}, \hat{a}^\dagger] = \hat{a}^\dagger[\hat{a}, \hat{a}^\dagger] + [\hat{a}^\dagger, \hat{a}^\dagger]\hat{a} = \hat{a}^\dagger \tag{4.62}$$

it follows that

$$\hat{a}^\dagger\hat{a}\hat{a}^\dagger |\psi\rangle = \left([\hat{a}^\dagger\hat{a}, \hat{a}^\dagger] + \hat{a}^\dagger\hat{a}^\dagger\hat{a}\right)|\psi\rangle = (n+1)\hat{a}^\dagger |\psi\rangle \tag{4.63}$$

This shows that $\hat{a}^\dagger$ increases the eigenvalue of $\hat{a}^\dagger\hat{a}$ by 1. Therefore, the state given by (4.60) is higher in energy by $n\hbar\omega$ than the ground state and becomes an eigenstate of energy $\hbar\omega\left(n + \frac{1}{2}\right)$.

- $|n\rangle$ **is normalized**—Assume that the states up to some n in (4.60) are normalized states. Then, from the commutation relation (4.54), we obtain

$$\left\langle n \mid \hat{a}\hat{a}^\dagger \mid n\right\rangle = \left\langle n \mid [\hat{a}, \hat{a}^\dagger] + \hat{a}^\dagger\hat{a} \mid n\right\rangle = (n+1)\langle n \mid n\rangle = n+1 \tag{4.64}$$

Therefore,

$$|n+1\rangle \equiv \frac{1}{\sqrt{n+1}}\hat{a}^\dagger |n\rangle = \frac{1}{\sqrt{(n+1)!}}(\hat{a}^\dagger)^{n+1}|\Omega\rangle \tag{4.65}$$

is normalized.

Furthermore, as a result of Eq. (4.65), we obtain an important relation

$$\boxed{\left\langle n+1 \mid \hat{a}^\dagger \mid n\right\rangle = \sqrt{n+1}, \quad \langle n \mid \hat{a} \mid n+1\rangle = \sqrt{n+1}} \tag{4.66}$$

Before concluding this section, let us determine the wave function of the ground state. From here on, we will assume $\theta = 0$. The wave function is defined by

$$\varphi_0(x) \equiv \langle x \mid \Omega\rangle \tag{4.67}$$

From $\hat{a}|\Omega\rangle = 0$, we obtain

$$\begin{aligned}
&\langle x|\left(\frac{1}{\sqrt{2\hbar m\omega}}\hat{p} - i\sqrt{\frac{m\omega}{2\hbar}}\hat{x}\right)|\Omega\rangle = 0 \\
\Longleftrightarrow\ &\left(\frac{1}{\sqrt{m\omega}}(-i\hbar)\frac{d}{dx} - i\sqrt{m\omega}x\right)\varphi_0(x) = 0 \\
\Longleftrightarrow\ &\left(\frac{d}{dx} + \frac{m\omega}{\hbar}x\right)\varphi_0(x) = 0 \\
\Longleftrightarrow\ &\varphi_0(x) = \mathcal{N}\,\mathrm{e}^{-\frac{m\omega}{2\hbar}x^2}
\end{aligned} \tag{4.68}$$

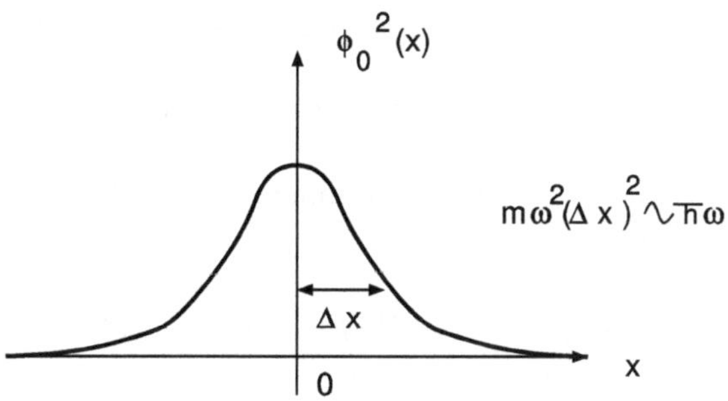

Fig. 4.1 Ground state of the harmonic oscillator. Note that the spread of the wave function, Δx, is the range of classical motion with the zero point energy

The normalization constant, $\mathcal{N}$, can be obtained by integration:

$$1 = \int_{-\infty}^{\infty} dx\ \varphi_0(x)^2 = \mathcal{N}^2 \int_{-\infty}^{\infty} dx\ \mathrm{e}^{-\frac{m\omega}{\hbar}x^2}$$

$$= \mathcal{N}^2 \sqrt{\frac{\hbar}{m\omega}} \int_{-\infty}^{\infty} \mathrm{e}^{-x^2} dx = \mathcal{N}^2 \sqrt{\frac{\pi\hbar}{m\omega}} \tag{4.69}$$

Thus, we obtain (Fig. 4.1)

$$\boxed{\varphi_0(x) = \left(\frac{m\omega}{\pi\hbar}\right)^{\frac{1}{4}} \mathrm{e}^{-\frac{m\omega}{2\hbar}x^2}} \tag{4.70}$$

4.1.3 One-Dimensional Potential Problem—Besides Numerical Methods, the Variational Method and WKB Approximation Are Available

In the one-dimensional case, besides solving the Schrödinger equation numerically, one can also use the variational method and the WKB approximation. They are particularly useful for a semi-quantitative understanding of the system.

Variational Method

This is very effective in finding the energy of the ground state. One needs to minimize the expectation value of the Hamiltonian using a trial wave function with parameters. Even if the wave function itself is not a very good approximation of the real wave function, the approximate ground state energy obtained is often good. There is a reason for this. We consider a state approximating the ground state $|\Omega\rangle$ of energy E_0 given as

$$|\psi\rangle = |\Omega\rangle + \epsilon\, |\Phi\rangle \tag{4.71}$$

where $|\Phi\rangle$ is a normalized superposition of higher energy states, orthogonal to $|\Omega\rangle$.

$$\langle\Omega \mid \Omega\rangle = \langle\Phi \mid \Phi\rangle = 1, \quad \langle\Omega \mid \Phi\rangle = 0 \tag{4.72}$$

Assume ϵ is a small real number. Then, the expectation of the energy in the state $|\psi\rangle$ is

$$\left\langle \hat{H} \right\rangle_\psi \equiv \frac{\langle\psi|\,\hat{H}\,|\psi\rangle}{\langle\psi \mid \psi\rangle} = \frac{E_0 + \epsilon^2 \langle\Phi|\,\hat{H}\,|\Phi\rangle}{1+\epsilon^2} = E_0 + \mathrm{O}(\epsilon^2) \tag{4.73}$$

We see that the energy gives the true ground energy to the accuracy of ϵ^2. Of course, the energy obtained in this way is greater than the true value. Let us consider a specific example.

Example: $V(x) = \kappa|x|$

Let $\psi(x,\alpha) \equiv \mathrm{e}^{-\frac{\alpha x^2}{2}}$ be the trial wave function with parameter α. We minimize the expectation value of the energy with respect to α . First, we compute the expectation value:

$$\begin{aligned}
\langle\psi|\,\hat{H}\,|\psi\rangle &= \int_{-\infty}^{\infty} dx\, \mathrm{e}^{-\frac{\alpha x^2}{2}} \left(-\frac{\hbar^2}{2m}\frac{d^2}{dx^2} + \kappa|x|\right) \mathrm{e}^{-\frac{\alpha x^2}{2}} \\
&= \frac{\hbar^2}{2m}\int_{-\infty}^{\infty} dx \left(\frac{d}{dx}\mathrm{e}^{-\frac{\alpha x^2}{2}}\right)^2 + 2\kappa \int_0^\infty dx\, x\mathrm{e}^{-\alpha x^2} \\
&= \frac{\hbar^2}{2m} \underbrace{\int_{-\infty}^{\infty} dx\, \alpha^2 x^2 \mathrm{e}^{-\alpha x^2}}_{-\alpha^2 \frac{d}{d\alpha} \underbrace{\int_{-\infty}^{\infty} dx\, \mathrm{e}^{-\alpha x^2}}_{\sqrt{\frac{\pi}{\alpha}}}} + \underbrace{2\kappa \int_0^\infty dx\, x\mathrm{e}^{-\alpha x^2}}_{\kappa \int_0^\infty dy\, \mathrm{e}^{-\alpha y} = \frac{\kappa}{\alpha}} \\
&= \underbrace{\frac{\hbar^2}{4m}\sqrt{\pi}\sqrt{\alpha}}_{K.E.} + \underbrace{\frac{\kappa}{\alpha}}_{P.E.}
\end{aligned} \tag{4.74}$$

On the other hand,

$$\langle\psi \mid \psi\rangle = \int_{-\infty}^{\infty} dx\, \mathrm{e}^{-\alpha x^2} = \sqrt{\frac{\pi}{\alpha}} \tag{4.75}$$

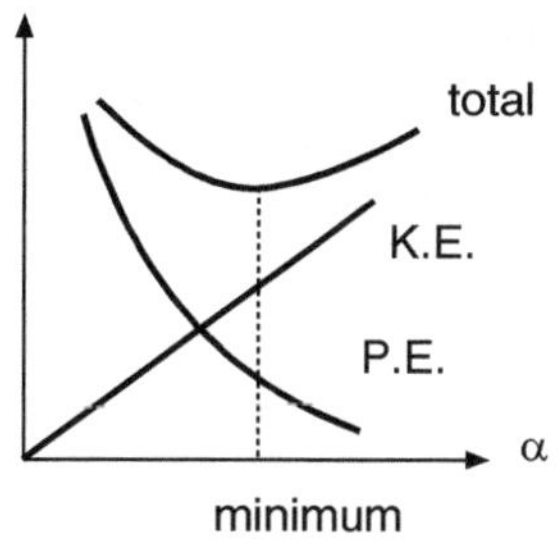

Fig. 4.2 $\left\langle \hat{H} \right\rangle_\psi$ as a function of α

Therefore, the expectation value is

$$\left\langle \hat{H} \right\rangle_\psi = \frac{\langle \psi | \hat{H} | \psi \rangle}{\langle \psi | \psi \rangle} = \underbrace{\frac{\hbar^2}{4m}\alpha}_{K.E.} + \underbrace{\frac{\kappa}{\sqrt{\pi}} \frac{1}{\sqrt{\alpha}}}_{P.E.} \tag{4.76}$$

The spread of the wave function is $\delta x \approx \frac{1}{\sqrt{\alpha}}$. Therefore, from Heisenberg's uncertainty principle, we can expect that the kinetic energy is of the order of $\frac{\hbar^2}{m}\alpha$ and the potential energy is of the order of $\frac{\kappa}{\sqrt{\alpha}}$. The above result matches our expectations (Fig. 4.2).

To find the minimum of the expectation value, we need to find where its derivative with respect to α vanishes:

$$\frac{d}{d\alpha} \left\langle \hat{H} \right\rangle_\psi = \frac{\hbar^2}{4m} - \frac{1}{2} \frac{\kappa}{\sqrt{\pi}} \frac{1}{\alpha^{\frac{3}{2}}} = 0 \tag{4.77}$$

We obtain

$$\alpha^{\frac{3}{2}} = \frac{2m\kappa}{\sqrt{\pi}\hbar} \Longrightarrow \alpha = \left(\frac{2m\kappa}{\sqrt{\pi}\hbar^2} \right)^{\frac{2}{3}} \tag{4.78}$$

From this, we obtain

$$\begin{aligned} \text{Min}_\alpha \left\langle \hat{H} \right\rangle_\psi &= \frac{\hbar^2}{4m} \left(\frac{2m\kappa}{\sqrt{\pi}\hbar^2} \right)^{\frac{2}{3}} + \frac{\kappa}{\sqrt{\pi}} \left(\frac{\sqrt{\pi}\hbar^2}{2m\kappa} \right)^{\frac{1}{3}} \\ &= \underbrace{\frac{3 \cdot 2^{\frac{2}{3}}}{4\pi^{\frac{1}{3}}}}_{\simeq 0.813} \left(\frac{\hbar\kappa}{\sqrt{m}} \right)^{\frac{2}{3}} \end{aligned} \tag{4.79}$$

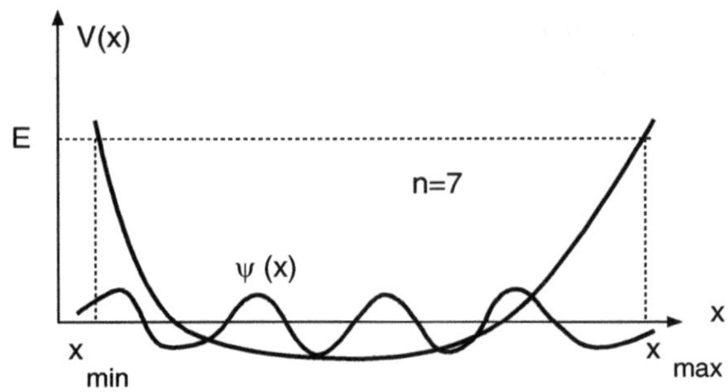

Fig. 4.3 The n in the quantization condition represents the number of zeros in the wave function

WKB Approximation

The WKB (Wentzel-Kramers-Brillouin) approximation is particularly useful for high-energy eigenstates under a bound state potential.

Consider a particle of mass m with energy E under the potential $V(x)$. If the equation

$$E = V(x) \tag{4.80}$$

has two solutions $x_{\min}(E), x_{\max}(E)$, the classically allowed region of motion is $[x_{\min}(E), x_{\max}(E)]$ (Fig. 4.3). In this region, the coordinate-dependent wave number $k(x)$ is defined by

$$k(x) \equiv \frac{\sqrt{2m(E - V(x))}}{\hbar} \tag{4.81}$$

If the potential does not change much over the range of wavelengths, i.e.,

$$\left| \frac{k'(x)}{k(x)} \right| \ll k(x) \tag{4.82}$$

is satisfied, then the wave function can be approximately given by

$$\sin\left[\int^x dx\, k(x) + \text{const}\right]$$

This approximation starts failing around $x \simeq x_{\min}(E), x_{\max}(E)$. However, if E is sufficiently high, this approximation can be considered valid throughout $[x_{\min}(E), x_{\max}(E)]$. If the number of nodes in the wave function is denoted by $n = 0, 1, 2, \cdots$, the phase changes by $\pi(n + \frac{1}{2})$ between $x_{\min}(E)$ and $x_{\max}(E)$ (Figure 4.3). From this, we obtain the Bohr-Sommerfeld quantization condition:

$$\boxed{\frac{1}{\hbar}\int_{x_{\min}(E)}^{x_{\max}(E)} dx\, \sqrt{2m(E - V(x))} = \pi\left(n + \frac{1}{2}\right)} \qquad (n = 0, 1, 2, \cdots) \tag{4.83}$$

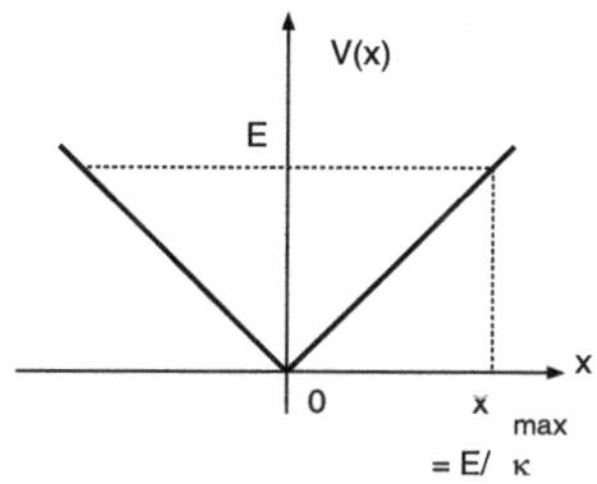

Fig. 4.4 Classical motion is restricted to the region $|x| < x_{\max}(E)$

The correction of $\frac{1}{2}$ to the n on the right side of the quantization condition represents the contribution from the region where $x \simeq x_{\min}(E)$ or $x \simeq x_{\max}(E)$. This can be derived by considering the connection of solutions in the regions where $E > V(x)$ and $E < V(x)$.[6]

Example: $V(x) = \kappa|x|$

Let us apply the WKB method to the potential to which we have applied the variational method (Fig. 4.4). The phase change over the region of classical motion is given by

$$\frac{2}{\hbar}\int_0^{x_{\max}(E)} dx\sqrt{2m(E - V(x))} = \frac{2}{\hbar}\int_0^{\frac{E}{\kappa}} dx\sqrt{2m(E - \kappa x)}$$
$$= \frac{2E}{\hbar\kappa}\int_0^1 dy\sqrt{2mE}\sqrt{1-y} = \frac{2E\sqrt{2mE}}{\hbar\kappa}\frac{2}{3} = \frac{4}{3}\frac{\sqrt{2m}}{\hbar\kappa}E^{\frac{3}{2}} \tag{4.84}$$

This gives

$$\frac{4}{3}\frac{\sqrt{2m}}{\hbar\kappa}E^{\frac{3}{2}} = \pi\left(n + \frac{1}{2}\right) \Longrightarrow E_n = \left(\frac{3\pi}{4\sqrt{2}}\right)^{\frac{2}{3}}\left(\frac{\hbar\kappa}{\sqrt{m}}\right)^{\frac{2}{3}}\left(n + \frac{1}{2}\right)^{\frac{2}{3}} \tag{4.85}$$

Even in the case of $n = 0$, where the approximation is not expected to be very good, we obtain

$$\left(\frac{3\pi}{4\cdot 2\sqrt{2}}\right)^{\frac{2}{3}} \simeq 0.885 > \frac{3\cdot 2^{\frac{2}{3}}}{4\pi^{\frac{1}{3}}} \simeq 0.813 \tag{4.86}$$

The approximation is not too bad though it is not as good as what we get from the variational method.

[6] For example, see D. Griffiths, *Quantum Mechanics* 8.3, E. Abers, *Quantum Mechanics* 7.7.2, L. Schiff, *Quantum Mechanics* sect. 34.

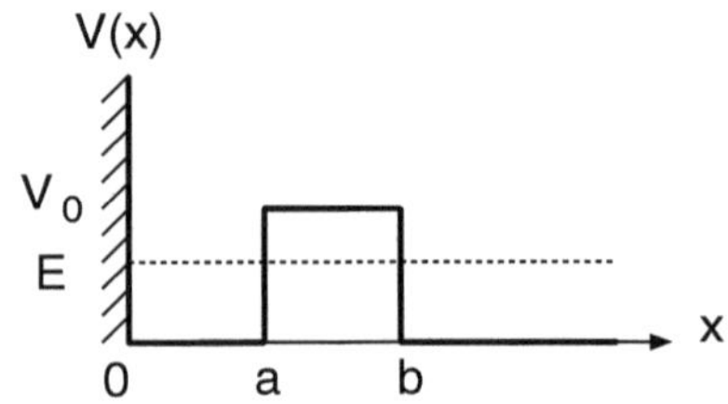

Fig. 4.5 Tunneling effect

Appendix: WKB Approximation in Classically Forbidden Regions

In the regions forbidden classically, i.e., $E < V(x)$, the wave function can be roughly described as a linear combination of exponential functions

$$\exp\left[\pm\frac{1}{\hbar}\int^x dx\,\sqrt{2m(V(x)-E)}\right] \tag{4.87}$$

Consider, for instance, the potential shown in Fig. 4.5.

The wave function must vanish at $x = 0$, and in the region $0 < x < a$ it is given by

$$\psi(x) = A\sin\frac{\sqrt{2mE}}{\hbar}x \tag{4.88}$$

From the normalization condition, we obtain

$$A^2 \simeq \frac{2}{a} \tag{4.89}$$

Applying the Bohr-Sommerfeld quantization condition in the region $[0, a]$, we obtain

$$\frac{\sqrt{2mE}}{\hbar}a = \pi\left(n+\frac{1}{2}\right) \quad (n = 0, 1, 2, \cdots)$$

Actually, these states are unstable; the particle will eventually tunnel through the potential barrier V_0 and escape to the region $x > b$. Let us calculate this transition probability using the WKB approximation.

Consider the forbidden region $a < x < b$ of classical motion. The wave function is given by the exponentially decaying function

$$\psi(x) \simeq B\exp\left[-\frac{\sqrt{2m(V_0-E)}}{\hbar}(x-a)\right] \tag{4.90}$$

where

$$B^2 \approx A^2 \simeq \frac{2}{a} \tag{4.91}$$

For $x > b$ the wave function is given by the outgoing wave

$$C\,\mathrm{e}^{i\frac{\sqrt{2mE}}{\hbar}(x-b)} \tag{4.92}$$

where

$$|C| \approx |B| \exp\left[-\frac{\sqrt{2m(V_0-E)}}{\hbar}(b-a)\right] \tag{4.93}$$

Thus,

$$|C|^2 \approx \frac{2}{a} \exp\left[-2\frac{\sqrt{2m(V_0-E)}}{\hbar}(b-a)\right] \tag{4.94}$$

is obtained. Therefore, the probability flow at $x = b$ (see Exercise 8-2) per unit time is

$$\frac{1}{\tau} = |C|^2\sqrt{\frac{2E}{m}} \approx \underbrace{\frac{2}{a}\sqrt{\frac{2E}{m}}}_{=\frac{2}{T}} \exp\left[-2\frac{\sqrt{2m(V_0-E)}}{\hbar}(b-a)\right] \tag{4.95}$$

Here, T is the time a classical particle takes to move from $x = 0$ to $x = a$. We have been sloppy about numerical factors. Fermi, in *Nuclear Physics* (University of Chicago Press) IIIA–C, gives a result half as big:

$$\frac{1}{\tau} = \frac{1}{T} \exp\left[-2\frac{\sqrt{2m(V_0-E)}}{\hbar}(b-a)\right] \tag{4.96}$$

See Example 8-8. See also Sect. 4.6.7 for a more general treatment.

Examples 8

1. **Coordinate and momentum acting on the wave functions**—We have shown

$$\hat{p}\,|x\rangle = i\hbar\frac{d}{dx}\,|x\rangle$$

$$\hat{x}\,|p\rangle = -i\hbar\frac{d}{dp}\,|p\rangle$$

Show

$$\langle x|\,\hat{p}\,|\psi\rangle = -i\hbar\frac{d}{dx}\,\langle\, x \mid \psi\,\rangle = -i\hbar\frac{d}{dx}\psi(x)$$

$$\langle p|\hat{x}\,|\psi\rangle = i\hbar\frac{d}{dp}\,\langle\, p \mid \psi\,\rangle = i\hbar\frac{d}{dp}\tilde{\psi}(p)$$

where

$$\tilde{\psi}(p) = \langle\, p \mid \psi\,\rangle = \int dx\,\langle\, p \mid x\,\rangle\langle\, x \mid \psi\,\rangle = \int dx\, e^{-i\frac{px}{\hbar}}\,\psi(x)$$

is the Fourier transform of the wave function $\psi(x) \equiv \langle\, x \mid \psi\,\rangle$.

2. **Propagator of a free particle**—Consider a particle moving freely on the x-axis. The Hamiltonian is given by

$$\hat{H} = \frac{\hat{p}^2}{2m}$$

   (a) Calculate the probability amplitude

$$K(x,t;y) \equiv \langle x|\,\mathrm{e}^{-\frac{i}{\hbar}\hat{H}t}\,|y\rangle$$

   Hint: Use the momentum representation.

   (b) Given the wave function $\phi(x)$ at time $t = 0$, give the wave function at a later time $t > 0$ using ϕ and K.

   (c) Given $\phi(x) = \mathrm{e}^{-\frac{x^2}{4\sigma^2}}$, find the wave function at time $t > 0$.

3. **Harmonic Oscillator under a Uniform Force (Part 1)**—Consider a harmonic oscillator under a uniform force f. The Hamiltonian is given by

$$\hat{H} = \frac{\hat{p}^2}{2m} + \frac{m\omega^2}{2}\hat{x}^2 - f\hat{x}$$

   (a) Solve the classical equations of motion.

   (b) Let $\psi_n(x)$ be the wave function of the nth excited state for the case $f = 0$. Give the wave function of the nth excited state for the case $f \neq 0$.

   (c) Using the annihilation operator

$$\hat{a} \equiv \frac{i}{\sqrt{2m\hbar\omega}}\,\hat{p} + \sqrt{\frac{m\omega}{2\hbar}}\,\hat{x}$$

   and its hermitian conjugate, express the Hamiltonian.

   (d) The ground state $|\Omega\rangle_0$ for the case $f = 0$ is defined by

$$\hat{a}\,|\Omega\rangle_0 = 0$$

How can you define the ground state $|\Omega\rangle_f$ for the case $f \neq 0$ using the annihilation operator $\hat{a}$?

4. **The WKB approximation and variational method**—Find the energy eigenvalues of the Hamiltonian

$$\hat{H} = \frac{\hat{p}^2}{2m} + \alpha\,\hat{x}^4 \qquad (\alpha > 0)$$

approximately using the WKB method. Further, compare the energy of the ground state obtained by the WKB method with the value obtained by the variation method using the test function

$$\psi_\sigma(x) \equiv \mathrm{e}^{-\frac{x^2}{2\sigma}} \qquad (\sigma > 0)$$

5. **Time dependence of a two-state system**—Consider a system where the Hamiltonian is given by the following 2-by-2 hermitian matrix:

$$H = \epsilon_0 \mathbf{1}_2 + \epsilon_1 \sigma_x = \begin{pmatrix} \epsilon_0 & \epsilon_1 \\ \epsilon_1 & \epsilon_0 \end{pmatrix}$$

(a) Find the energy eigenstates.

(b) If the initial state at time $t = 0$ is $\begin{pmatrix} 1 \\ 0 \end{pmatrix}$, calculate the probability that the state is $\begin{pmatrix} 0 \\ 1 \end{pmatrix}$ at a later time $t > 0$.

6. **Delta function potential (Part 1)**—Consider a particle of mass m on the x-axis. For the potential energy

$$V(x) = -V_0 \delta(x) \qquad (V_0 > 0)$$

find the energy and wave function of the bound state.
Hint: There always exists only one bound state.

7. **Ground State**—Consider a particle of mass m under the potential below:

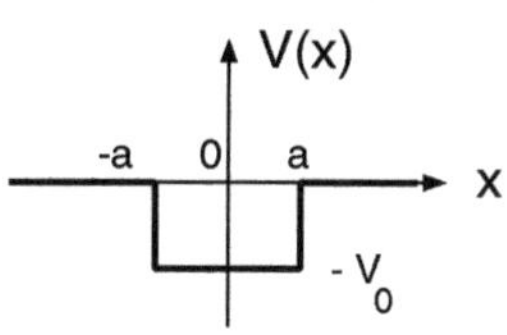

For the existence of a bound state, what is the condition on V_0 (or Ka)? Can we interpret the result using the uncertainty principle?

8. **Penetration of a rectangular barrier**—We consider the transmission of a particle of mass m on the x-axis through a rectangular potential barrier whose potential is given by

$$V(x) = \begin{cases} 0 & x < 0 \\ V_0 = \frac{\hbar^2\kappa^2}{2m} & (0 < x < a) \\ 0 & x > a \end{cases}$$

Let $\hbar k$, where $k < \kappa$, be the momentum of the particle coming from the left toward the potential barrier. The wave function is given in the following form:

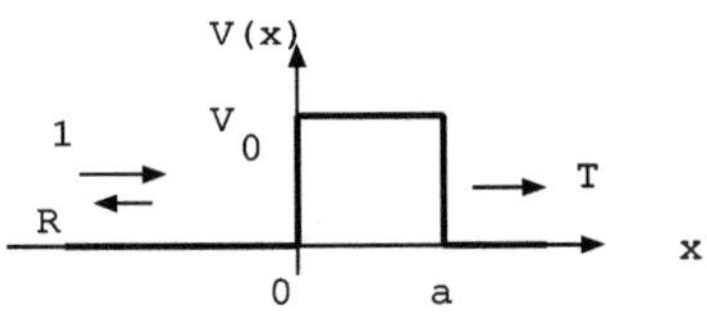

A rectangular potential barrier

$$\psi(x) = \begin{cases} e^{ikx} + Re^{-ikx} & (x < 0) \\ Ae^{-\sqrt{\kappa^2-k^2}x} + Be^{\sqrt{\kappa^2-k^2}x} & (0 < x < a) \\ Te^{ik(x-a)} & (x > a) \end{cases}$$

Compute T and the probability of the transmission $|T|^2$.
Hints: Use the continuity of $\psi(x)$ and $\psi'(x)$ to compute A, B in terms of T. You can determine T by expressing the amplitude 1 of the incoming wave in terms of T.

Solutions to Examples 8

1. Taking the hermitian conjugate of the two equations given, we obtain

$$\langle x|\,\hat{p} = -i\hbar\frac{d}{dx}\,\langle x|\,, \quad \langle p|\,\hat{x} = i\hbar\frac{d}{dp}\,\langle p|$$

Hence, we obtain

$$\langle x|\,\hat{p}\,|\psi\rangle = -i\hbar\frac{d}{dx}\langle\, x \mid \psi\,\rangle = -i\hbar\frac{d}{dx}\psi(x)$$

$$\langle p|\,\hat{x}\,|\psi\rangle = i\hbar\frac{d}{dp}\langle\, p \mid \psi\,\rangle = i\hbar\frac{d}{dp}\tilde{\psi}(p)$$

2. (a) Using the momentum representation and $\int \frac{dp}{2\pi\hbar} \left|p\right\rangle \left\langle p\right| = 1$, we get

$$K(x,t;y) = \int \frac{dp}{2\pi\hbar} \langle\, x \mid p \,\rangle \mathrm{e}^{-\frac{it}{\hbar}\frac{p^2}{2m}} \langle\, p \mid y \,\rangle$$
$$= \int \frac{dp}{2\pi\hbar} \exp\left[-\frac{it}{\hbar}\frac{p^2}{2m} + \frac{i}{\hbar}p(x-y)\right]$$

Completing the square in the exponent, we obtain

$$K(x,t;y) = \mathrm{e}^{\frac{i}{\hbar}\frac{m(\vec{x}-\vec{y})^2}{2t}} \int \frac{dp}{2\pi\hbar} \exp\left[-\frac{it}{2m\hbar}\left(p - m\frac{x-y}{t}\right)^2\right]$$
$$= \sqrt{\frac{m}{2\pi i\hbar t}}\, \mathrm{e}^{\frac{i}{\hbar}\frac{m(x-y)^2}{2t}}$$

(b) Inserting $\int dy \left|y\right\rangle \left\langle y\right| = 1$, we obtain

$$\langle x|\, \mathrm{e}^{-\frac{i}{\hbar}\hat{H}t}\, |\phi\rangle = \int dy\, \langle x|\, \mathrm{e}^{-\frac{i}{\hbar}\hat{H}t}\, |y\rangle \langle y \mid \phi\rangle = \int dy\, K(x,t;y)\phi(y)$$

(c) As the integral is Gaussian, we can complete the square to compute

$$\phi(x,t) = \int dy \sqrt{\frac{m}{2\pi i\hbar t}}\, \mathrm{e}^{\frac{i}{\hbar}\frac{m(x-y)^2}{2t} - \frac{y^2}{4\sigma^2}}$$
$$= \frac{1}{\sqrt{1 + i\frac{\hbar t}{2m\sigma^2}}} \exp\left[-\frac{x^2}{4\sigma^2}\frac{1}{1 + i\frac{\hbar t}{2m\sigma^2}}\right]$$

From this, the probability density is

$$|\phi(x,t)|^2 \propto \exp\left[-\frac{x^2}{2\sigma^2}\frac{1}{1 + \frac{\hbar^2 t^2}{4m^2\sigma^4}}\right]$$

Thus, the spread of the wave function grows as $\sigma\sqrt{1 + \frac{\hbar^2 t^2}{4m^2\sigma^4}}$ as time passes. This is due to the spread of momentum $\frac{\hbar}{2\sigma}$ given at $t = 0$.

3. (a) The potential energy can be rewritten as

$$V(x) = \frac{m\omega^2}{2}\left(x - \frac{f}{m\omega^2}\right)^2 - \frac{f^2}{2m\omega^2}$$

Thus,

$$x(t) = \frac{f}{m\omega^2} + A\cos(\omega t + \phi)$$

(b) $\psi_n\left(x - \frac{f}{m\omega^2}\right)$

(c) Define

$$\hat{a}' \equiv \frac{i}{\sqrt{2m\hbar\omega}}\hat{p} + \sqrt{\frac{m\omega}{2\hbar}}\left(\hat{x} - \frac{f}{m\omega^2}\right) = \hat{a} - f\frac{1}{\sqrt{2\hbar m\omega^3}}$$

With this, we obtain

$$\hat{H} = \hbar\omega\left(\hat{a}'^{\dagger}\hat{a}' + \frac{1}{2}\right) - \frac{f^2}{2m\omega^2}$$

Therefore,

$$\hat{H} = \hbar\omega\left(\hat{a}^{\dagger}\hat{a} - \frac{f}{\sqrt{2\hbar m\omega^3}}\left(\hat{a} + \hat{a}^{\dagger}\right) + \frac{1}{2}\right)$$

(d) Since $\hat{a}'\,|\Omega\rangle_f = 0$, we have

$$\hat{a}\,|\Omega\rangle_f = f\frac{1}{\sqrt{2\hbar m\omega^3}}\,|\Omega\rangle_f$$

Therefore, the ground state is an eigenstate of $\hat{a}$ (see Exercise 8-5).

4. Let E_n be the n-th energy eigenvalue. The range of classical motion is $|x| \le x_n \equiv \left(\frac{E_n}{\alpha}\right)^{\frac{1}{4}}$. Thus, in the WKB approximation, we find

$$\int_0^{x_n} p\,dx = \frac{\pi}{2}\hbar\left(n + \frac{1}{2}\right)$$

where

$$\int_0^{x_n} p\,dx = \int_0^{x_n} dx\,\sqrt{2m(E_n - \alpha x^4)} = \sqrt{2m}\alpha^{-\frac{1}{4}}E_n^{\frac{3}{4}}\underbrace{\int_0^1 ds\,\sqrt{1-s^4}}_{\simeq 0.874}$$

From this, we obtain

$$E_n = \left(\frac{1}{0.874}\frac{\pi}{2\sqrt{2}}\left(n + \frac{1}{2}\right)\right)^{\frac{4}{3}}\frac{\hbar^{\frac{4}{3}}\alpha^{\frac{1}{3}}}{m^{\frac{2}{3}}} \propto \left(n + \frac{1}{2}\right)^{\frac{4}{3}}$$

Especially, the ground state energy ($n = 0$) is

$$E_0 \simeq 0.546 \frac{\hbar^{\frac{4}{3}} \alpha^{\frac{1}{3}}}{m^{\frac{2}{3}}}$$

On the other hand, taking $\psi_\sigma(x) \equiv \mathrm{e}^{-\frac{x^2}{2\sigma}}$ as the wave function, the expected values for the kinetic and potential energies are given as

$$\left\langle \frac{p^2}{2m} \right\rangle_\sigma = \frac{\hbar^2}{4m\sigma}$$

$$\left\langle \alpha x^4 \right\rangle_\sigma = \frac{3}{4}\alpha\sigma^2$$

Thus, the total energy

$$\langle E \rangle_\sigma = \frac{\hbar^2}{4m\sigma} + \frac{3}{4}\alpha\sigma^2$$

is minimized when $\sigma = \left(\frac{\hbar^2}{6m\alpha}\right)^{\frac{1}{3}}$, yielding the minimum value:

$$\mathrm{Min}_\sigma \langle E \rangle_\sigma = \underbrace{\frac{9}{4 \cdot 6^{\frac{2}{3}}}}_{\simeq 0.681} \cdot \frac{\hbar^{\frac{4}{3}} \alpha^{\frac{1}{3}}}{m^{\frac{2}{3}}}$$

This is slightly larger than the value from the WKB.

5. (a) The state with energy $\epsilon_0 - \epsilon_1$ is given by $|1\rangle \equiv \frac{1}{\sqrt{2}}\begin{pmatrix} 1 \\ -1 \end{pmatrix}$. The state with energy $\epsilon_0 + \epsilon_1$ is $|2\rangle \equiv \frac{1}{\sqrt{2}}\begin{pmatrix} 1 \\ 1 \end{pmatrix}$.
   (b) The initial state is $|\psi(0)\rangle = \frac{1}{\sqrt{2}}(|1\rangle + |2\rangle)$. Hence, the state at time t is

$$|\psi(t)\rangle = \frac{1}{\sqrt{2}}\left(\mathrm{e}^{-\frac{i}{\hbar}(\epsilon_0 - \epsilon_1)t}\,|1\rangle + \mathrm{e}^{-\frac{i}{\hbar}(\epsilon_0 + \epsilon_1)t}\,|2\rangle\right) = \mathrm{e}^{-\frac{i}{\hbar}\epsilon_0 t}\begin{pmatrix} \cos\epsilon_1 t \\ -i\sin\epsilon_1 t \end{pmatrix}$$

   The probability that this state is $\begin{pmatrix} 0 \\ 1 \end{pmatrix}$ is given by $\sin^2\epsilon_1 t$.

6. Writing the energy eigenvalue as $E = -\frac{\hbar^2\kappa^2}{2m}$, and $V_0 = \frac{\hbar^2}{2ma}$ (where a has the dimension of a length), the Schrödinger equation becomes

$$\psi''(x) = -\kappa^2\psi(x) - \frac{1}{a}\psi(0)\delta(x)$$

The delta function implies the boundary conditions:

$$\begin{cases} \psi(0+) &= \psi(0-) \\ \psi'(0+) &= \psi'(0-) - \frac{1}{a}\psi(0) \end{cases}$$

Since the wave function vanishes as $|x| \to \infty$, we must find

$$\psi(x) = \begin{cases} e^{\kappa x} & (x < 0) \\ A\,e^{-\kappa x} & (x > 0) \end{cases}$$

From the boundary conditions, we get

$$A = 1, \quad 2\kappa = \frac{1}{a}$$

Thus, the energy eigenvalue for the bound state is

$$E = -\frac{\hbar^2}{8ma^2}$$

7. The wave function of the ground state must be even. Defining the energy as $-\frac{\hbar^2\kappa^2}{2m} < 0$, the wave function in the interval $-a < x < a$ is

$$\psi(x) = \cos\sqrt{K^2 - \kappa^2}\,x\,,$$

and that in $|x| > a$ is given by $Ae^{-\kappa|x|}$. For the wave function and its derivative to be continuous at $x = a$, we must find

$$\begin{cases} \cos\sqrt{K^2 - \kappa^2}\,a &= Ae^{-\kappa a} \\ \sqrt{K^2 - \kappa^2}\sin\sqrt{K^2 - \kappa^2}\,a &= \kappa Ae^{-\kappa a} \end{cases}$$

Taking the ratio of the two equations, we get

$$\sqrt{(Ka)^2 - (\kappa a)^2}\tan\sqrt{(Ka)^2 - (\kappa a)^2} = \kappa a$$

Defining $s \equiv \sqrt{(Ka)^2 - (\kappa a)^2}$, we can rewrite this as

$$s\,\tan s = \sqrt{(Ka)^2 - s^2}$$

Regardless of how small Ka is, this equation **always** has a solution, indicating the presence of a bound state. (A symmetric excited state exists when $Ka > \pi/2$.)

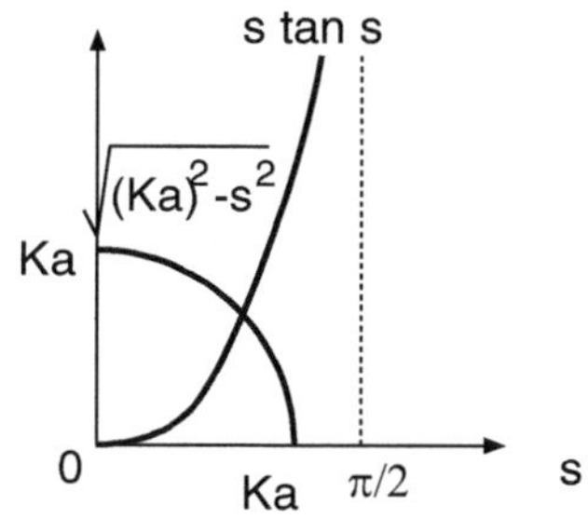

Can we understand this result using the uncertainty principle? Assuming that the wave function has a size of order a, we estimate the kinetic energy as $\frac{\hbar^2}{2ma^2}$ and the potential energy as $-\frac{\hbar^2 K^2}{2m}$. For a bound state to exist, we must find

$$\frac{\hbar^2}{2ma^2} < \frac{\hbar^2 K^2}{2m} \Longleftrightarrow 1 < Ka$$

Hence, Ka must be greater than 1. But we have shown above that a bound state exists for any Ka. The mistake lies in evaluating the typical wave number as $\frac{1}{a}$. Since the wave function is even, the wave number can be arbitrarily small. (In a three-dimensional space, the evaluation of the wave number as $\frac{1}{a}$ is correct, and a bound state exists only if Ka is greater than about 1.)

8. The continuity of $\psi(x)$ and $\psi'(x)$ at $x = a$ gives

$$Ae^{-\sqrt{\kappa^2-k^2}a} + Be^{\sqrt{\kappa^2-k^2}a} = T$$

$$\sqrt{\kappa^2-k^2}\left(-Ae^{-\sqrt{\kappa^2-k^2}a} + Be^{\sqrt{\kappa^2-k^2}a}\right) = ikT$$

This gives

$$Ae^{-\sqrt{\kappa^2-k^2}a} = \frac{1}{2}\left(1 - i\frac{k}{\sqrt{\kappa^2-k^2}}\right)T$$

$$Be^{\sqrt{\kappa^2-k^2}a} = \frac{1}{2}\left(1 + i\frac{k}{\sqrt{\kappa^2-k^2}}\right)T$$

Similarly, from the continuity at $x = 0$, we obtain

$$1 + R = A + B, \quad ik(1-R) = \sqrt{\kappa^2-k^2}\,(-A+B)$$

This gives

$$1 = \frac{1}{2}\left(\left(1 + i\frac{\sqrt{\kappa^2 - k^2}}{k}\right)A + \left(1 - i\frac{\sqrt{\kappa^2 - k^2}}{k}\right)B\right)$$

Substituting the expressions of A, B in terms of T into the above, we obtain

$$i\frac{\sqrt{\kappa^2 - k^2}}{k}\left\{\left(1 - i\frac{k}{\sqrt{\kappa^2 - k^2}}\right)^2 e^{\sqrt{\kappa^2 - k^2}a} - \left(1 + i\frac{k}{\sqrt{\kappa^2 - k^2}}\right)^2 e^{-\sqrt{\kappa^2 - k^2}a}\right\}T = 4$$

For $e^{-\sqrt{\kappa^2 - k^2}a} \ll 1$, this gives approximately

$$T \simeq -4i\frac{k}{\sqrt{\kappa^2 - k^2}}\frac{e^{-\sqrt{\kappa^2 - k^2}a}}{\left(1 - i\frac{k}{\sqrt{\kappa^2 - k^2}}\right)^2}$$

Hence, we obtain the probability of penetration, called **transparency**, as

$$|T|^2 \simeq 16\frac{k^2(\kappa^2 - k^2)}{\kappa^4}e^{-2\sqrt{\kappa^2 - k^2}a}$$

This is proportional to $e^{-2\sqrt{\kappa^2 - k^2}a}$ as expected from the WKB approximation. Note that the mode that grows in $0 < x < a$ has a nonvanishing amplitude B of the order $e^{-2\sqrt{\kappa^2 - k^2}a}$.

Exercises 8

1. **Uncertainty principle**—Using the uncertainty principle, determine the dependence of the Bohr radius and the Rydberg energy on physical constants (e, m_e, $\hbar$). **Hint**: Consider the spatial spread of the state wave function as Δx, and evaluate the kinetic and potential energy. Find the Δx that minimizes the total energy. You may use the variational method, but you do not have to go into a great detail. (The kinetic energy is $\sim\frac{\hbar^2}{2m_e(\Delta x)^2}$, and the potential energy is $\sim -\frac{e^2}{4\pi\epsilon_0\Delta x}$. Minimize the total.)
2. **Conservation of probability**—Consider a particle of mass m moving along the x-axis under the influence of a potential energy $V(x)$. If we write the wave function as $\psi(x, t)$, its time dependence is determined by the Schrödinger equation:

$$i\hbar\partial_t\psi(x, t) = \left(-\frac{\hbar^2}{2m}\partial_x^2 + V(x)\right)\psi(x, t)$$

(a) The probability density ρ and the probability current j are defined by

$$\rho(x,t) \equiv \psi^*\psi(x,t), \quad j(x,t) \equiv \frac{-i\hbar}{2m}\left(\psi^*\partial_x\psi - \partial_x\psi^* \cdot \psi\right)$$

Show that the equation of continuity

$$\partial_t\rho + \partial_x j = 0$$

is satisfied.

(b) For the potential energy given by

$$V(x) = K\delta(x) \quad (K > 0)$$

calculate the transmission and reflection probabilities for a particle with energy $E(> 0)$.

Hint: Solve the Schrödinger equation using a wave function of the form:

$$\psi(x) = \begin{cases} \mathrm{e}^{ikx} + r(E)\mathrm{e}^{-ikx} & (x < 0) \\ t(E)\mathrm{e}^{ikx} & (x > 0) \end{cases}$$

Note that $\psi(x)$ is continuous at $x = 0$, but $\frac{d\psi(x)}{dx}$ is discontinuous.

(c) For the above example, confirm that the current $j(x)$ is continuous at $x = 0$.

3. **Hermite polynomials**—The creation operator $\hat{a}^\dagger$ for a quantum harmonic oscillator with angular frequency ω is given in terms of position $\hat{x}$ and momentum $\hat{p}$ as

$$\hat{a}^\dagger \equiv \frac{-i}{\sqrt{2m\hbar\omega}}\,\hat{p} + \sqrt{\frac{m\omega}{2\hbar}}\,\hat{x}$$

where m is the mass. Show that the normalized wave function for the n-th excited state is given by

$$\langle\, x \mid n \,\rangle = \frac{1}{\sqrt{n!}}\langle x \left|(\hat{a}^\dagger)^n\right| \Omega\rangle = \frac{\pi^{-\frac{1}{4}}}{\sqrt{n!2^n x_0}} \exp\left(-\frac{x^2}{2x_0^2}\right) H_n\left(\frac{x}{x_0}\right)$$

where $m\omega^2 x_0^2 \equiv \hbar\omega$,

$$H_n(x) \equiv (-)^n \exp(x^2)\frac{d^n}{dx^n}\exp(-x^2)$$

is the n-th Hermite polynomial and $|\Omega\rangle$ is the normalized ground state.

Hint: First derive

$$\langle x|\hat{a}^{\dagger}|\psi\rangle = \frac{x_0}{\sqrt{2}}\left(-\frac{d}{dx}+\frac{x}{x_0}\right)\psi(x)$$
$$= \frac{x_0}{\sqrt{2}}\exp\left(\frac{x^2}{2x_0^2}\right)(-)\frac{d}{dx}\left(\exp\left(-\frac{x^2}{2x_0^2}\right)\psi(x)\right)$$

4. **Landau levels**—As has been derived in Example 3-4 of Electromagnetism, the Hamiltonian (energy) of a charged particle with mass m in a vector potential $\vec{A}(\vec{x})$ is given by

$$H = \frac{1}{2m}\left(\vec{p} - \frac{e}{c}\vec{A}(\vec{x})\right)^2$$

To quantize this we only need to replace the momentum $\vec{p}$ by the differential operator $-i\hbar\nabla$.

(a) Under a given vector potential $\vec{A}(\vec{x})$, let $\psi(x)$ be the wave function of the energy eigenstate with energy E. Show how the wave function transforms under the gauge transformation of the vector potential:

$$\vec{A}(\vec{x}) \longrightarrow \vec{A}'(\vec{x}) = \vec{A}(\vec{x}) + \nabla\chi(\vec{x})$$

Consider a vector potential $\vec{A}(\vec{x})$ that gives a uniform magnetic field $\vec{B} = (0, 0, B)$ in the z-direction:

$$\vec{A}(\vec{x}) = (0, Bx, 0)$$

(b) Compute the commutator $\left[\hat{p}_x, \hat{p}_y - \frac{e}{c}A_y(\hat{x})\right]$.

(c) Show

$$\hat{H} = \hat{H}_\omega + \frac{\hat{p}_z^2}{2m}$$

where $\hat{H}_\omega$ is the Hamiltonian of a harmonic oscillator with angular frequency $\omega \equiv \frac{eB}{mc}$.

Hint: Using the result from (b), construct an operator $\hat{X}$ that satisfies the commutation relation:

$$\left[\hat{p}_x, \hat{X}\right] = -i\hbar$$

Then, using $\hat{X}$ and $\hat{p}_x$, construct $\hat{H}_\omega$.

(d) Construct operators $\hat{Y}$ and $\hat{P}_Y$ using linear combinations of $\hat{x}$, $\hat{p}_x$, $\hat{y}$, $\hat{p}_y$ such that they commute with $\hat{X}$ and $\hat{p}_x$ and satisfy

$$[\hat{Y}, \hat{P}_Y] = i\hbar$$

(e) Determine all the eigenstates and their corresponding eigenvalues.
Hint: Do not forget to diagonalize $\hat{p}_z$ and $\hat{P}_Y$ (or $\hat{Y}$). The Hamiltonian is independent of $\hat{P}_Y$.

5. **Harmonic oscillator under uniform force (Part 2)**—Define the **coherent state** $|a\rangle$ (where a belongs to the complex numbers) by

$$|a\rangle \equiv \mathrm{e}^{a\hat{a}^\dagger} |\Omega\rangle$$

(a) Show $\hat{a}\,|a\rangle = a\,|a\rangle$.
(b) Derive the first-order differential equation in x satisfied by $\langle\, x \mid a\,\rangle$. Use the equality $a\,\langle\, x \mid a\,\rangle = \langle x|\,\hat{a}\,|a\rangle$.
(c) Using the equality $\frac{d}{da}\,\langle\, x \mid a\,\rangle = \langle x|\,\hat{a}^\dagger\,|a\rangle$, derive the first-order differential equation in a satisfied by $\langle\, x \mid a\,\rangle$.
(d) Solve the equations obtained in (5, 5) to determine the dependence of $\langle\, x \mid a\,\rangle$ on x and a.
Hint: $|\Omega\rangle$ is normalized.
(e) Determine the constant a such that the ground state $|\Omega\rangle_f$ obtained in Exercise 8-3 is a coherent state $|a\rangle$.

Now, let us consider the time evolution of a coherent state whose Hamiltonian is

$$\hat{H} = \hbar\omega\left(\hat{a}^\dagger\hat{a} + \frac{1}{2}\right)$$

If the initial state at time $t = 0$ is $|a\rangle$, the state at a time $t > 0$ is given by

$$\mathrm{e}^{-\frac{it}{\hbar}\hat{H}}\,|a\rangle = \mathrm{e}^{-i\frac{\omega t}{2}}\mathrm{e}^{-i\omega t\,\hat{a}^\dagger\hat{a}}\,\mathrm{e}^{a\hat{a}^\dagger}\,\mathrm{e}^{i\omega t\,\hat{a}^\dagger\hat{a}}\,|\Omega\rangle$$

(f) Derive the following equation:

$$\mathrm{e}^{-i\omega t\,\hat{a}^\dagger\hat{a}}\hat{a}^\dagger\,\mathrm{e}^{i\omega t\,\hat{a}^\dagger\hat{a}} = \hat{a}^\dagger\mathrm{e}^{-i\omega t}$$

(g) Using the result above, derive

$$\mathrm{e}^{-i\omega t\,\hat{a}^\dagger\hat{a}}\,\mathrm{e}^{a\hat{a}^\dagger}\,\mathrm{e}^{i\omega t\,\hat{a}^\dagger\hat{a}} = \mathrm{e}^{a\mathrm{e}^{-i\omega t}\hat{a}^\dagger}$$

Hint: Check by differentiation with respect to a.

From the above results, the time evolution of the coherent state is given by

$$e^{-\frac{it}{\hbar}\hat{H}}\,|a\rangle = e^{-i\frac{\omega t}{2}}\left|ae^{-i\omega t}\right\rangle$$

(h) Compute $e^{-i\frac{t}{\hbar}\hat{H}}\,|\Omega\rangle_f$ and interpret the result physically.
Hint: Compare the result obtained from the classical theory.

6. **Parity conservation**—Consider a particle of mass m moving along the x-axis under the potential energy $V(x)$. Defining the ket vector $|x\rangle$ to represent a state localized at x, the **parity** transformation is defined as

$$\hat{P}\,|x\rangle \equiv |-x\rangle$$

From the definition, it is evident that

$$\hat{P}^2 = 1, \quad \hat{P}^{-1} = \hat{P}$$

(a) Show that $\hat{P}$ is a unitary transformation.
Hint: For any state $|\psi\rangle$, demonstrate

$$\left\langle \hat{P}\psi \middle| \hat{P}\psi \right\rangle = \langle\, \psi \mid \psi\,\rangle$$

(b) Determine the transformation rules for the coordinate $\hat{x}$ and momentum $\hat{p}$ operators:

$$\begin{aligned} \hat{x} &\longrightarrow \hat{P}\hat{x}\hat{P}^{-1} = ? \\ \hat{p} &\longrightarrow \hat{P}\hat{p}\hat{P}^{-1} = ? \end{aligned}$$

Hint: Remember that $\hat{P}^{-1} = \hat{P}$.
(c) For the Hamiltonian

$$\hat{H} = \frac{\hat{p}^2}{2m} + V(\hat{x})$$

determine the condition for $\hat{P}$ to be a conserved quantity.

Now, consider two types of particles. We represent the state in which particle 1 is at x_1 and particle 2 at x_2 by the product[7] $|x_1, x_2\rangle \equiv |x_1\rangle\,|x_2\rangle$. The parity

[7] More precisely, the direct product.

transformation for particle 1 is defined as before:

$$\hat{P}\,|x_1\rangle = |-x_1\rangle$$

However, the parity transformation for particle 2 is defined with a reversed sign:[8]

$$\hat{P}\,|x_2\rangle = -\,|-x_2\rangle$$

Thus, the parity transformation for the two-particle state is given by

$$\hat{P}\,|x_1, x_2\rangle = -\,|-x_1, -x_2\rangle$$

(d) Consider a two-particle state given by

$$|\psi\rangle \equiv \int_{-\infty}^{\infty} dx_1 \int_{-\infty}^{\infty} dx_2\, \psi(x_1, x_2)\, |x_1, x_2\rangle$$

Determine the conditions that the wave function $\psi(x_1, x_2)$ must satisfy for the cases $\hat{P}\,|\psi\rangle = \pm\,|\psi\rangle$.

(e) Provide an example of $\psi(x_1, x_2)$ satisfying $\hat{P}\,|\psi\rangle = \pm\,|\psi\rangle$.

7. **Delta function potential (Part 2)**—Consider a particle of mass m on the x-axis.

(a) For the potential energy

$$V(x) = -V_0\,(\delta(x - x_0) + \delta(x + x_0)) \qquad (V_0 > 0, x_0 > 0)$$

show that a bound state with positive parity always exists. Determine its energy and wave function.

(b) Determine the minimum value x_{min} of x_0 for which a bound state with negative parity exists under a given V_0. In addition, explain qualitatively why no bound state exists if $x_0 < x_{\text{min}}$.

8. **Bloch's theorem**—Consider a particle of mass m under a periodic potential energy $V(x)$ with a period a, i.e., $V(x + a) = V(x)$.

(a) Define the unitary operator $\hat{T}$ by

$$\hat{T}\,|x\rangle = |x - a\rangle$$

Show that the Hamiltonian $\hat{H} = \frac{\hat{p}^2}{2m} + V(x)$ commutes with $\hat{T}$.

[8] Particle 1 is called a parity-plus particle and particle 2 a parity-minus particle.

Comment: Hence, $\hat{H}$ and $\hat{T}$ can be diagonalized simultaneously.

(b) Show that the eigenstate of $\hat{T}$ with eigenvalue e^{ika}, where k is unique up to an integral multiple of $\frac{2\pi}{a}$, has the wave function:

$$\psi(x) = e^{ikx}\phi_k(x)$$

where $\phi_k(x)$ is a periodic function with period a.

9. **Periodic delta function potential**—Consider a particle of mass m under the periodic potential

$$V(x) = V_0 \sum_{n=-\infty}^{\infty} \delta(x - na)$$

We wish to find the energy eigenstate with eigenvalue e^{ika} for $\hat{T}$ defined in Exercise 8-8.

(a) Determine the energy eigenvalue as a function of k for the case $V_0 > 0$.
Hint: Let $\psi_k(x) = e^{ikx}\phi_k(x) = e^{i\kappa(k)x} + C_k e^{-i\kappa(k)x}$ $(0 < x < a)$ be the wave function for the energy eigenvalue $E_k = \frac{\hbar^2}{2m}\kappa(k)^2$. $\phi_k(x)$ can be extended outside $0 < x < a$ by periodicity. For example, in $-a < x < 0$, we find $\phi_k(x) = e^{i(\kappa(k)-k)(x+a)} + C_k e^{-i(\kappa(k)+k)(x+a)}$. You will find $\kappa(k)$ determined implicitly by

$$\cos ka = \cos \kappa a + \frac{mV_0 a}{\hbar^2}\frac{\sin \kappa a}{\kappa a}$$

For $ka = \pi n$ $(n = 1, 2, \cdots)$ there are two κ. See Example 11-5.

(b) Determine the energy eigenvalue for the case $V_0 < 0$. **Hint**: Note that negative energy eigenvalues $-\frac{\hbar^2}{2m}\kappa(k)^2$ are also possible. The lowest energy state corresponds to $k = 0$.

10. **Supersymmetry**—We consider the Hamiltonian

$$\hat{H} = \hat{Q}^2$$

where

$$\hat{Q} \equiv \frac{1}{\sqrt{2m}}\left(\sigma_x \hat{p} - \sigma_y W(x)\right)$$

with the Pauli matrices $\vec{\sigma}$. The asymptotic form of the real-valued function $W(x)$ is given by

$$W(x) \overset{|x|\to\infty}{\longrightarrow} W_0 x \quad (W_0 > 0)$$

(a) Show that the ground state wave function is given by

$$\psi(x) = \begin{pmatrix} \phi(x) \\ 0 \end{pmatrix}, \quad \phi(x) = \exp\left[-\frac{1}{\hbar}\int^x dx'\, W(x')\right]$$

What is the energy eigenvalue?
Hint: Show $\hat{Q}\psi = 0$

(b) Show that the excited states always appear in pairs with $\sigma_z = \pm 1$.
Hint: Given an energy eigenstate $\begin{pmatrix} \phi(x) \\ \psi(x) \end{pmatrix}$ with $E > 0$, $\frac{1}{\sqrt{E}} Q \begin{pmatrix} \phi(x) \\ \psi(x) \end{pmatrix}$ is an eigenstate with the same energy.

4.2 All You Want to Know About Angular Momentum

Three Points to Remember

1. The angular momentum operator $\hat{J}_i\,(i = x, y, z)$ generates rotations about the i-th coordinate axis.
2. The irreducible representation of the commutation relation $[\hat{J}_i, \hat{J}_j] = i\hbar\epsilon_{ijk}\hat{J}_k$ is characterized by a spin $s = 0, \frac{1}{2}, 1, \frac{3}{2}, \cdots$. The dimension of each representation is $2s + 1$.
3. The sum of angular momenta j_1 and j_2 gives the total angular momentum

$$j = j_1 + j_2, j_1 + j_2 - 1, \cdots, |j_1 - j_2|$$

4.2.1 *The Angular Momentum Operators $\hat{J}_{x,y,z}$ Generate Rotations About the x, y, z-Axes*

In both classical and quantum mechanics the angular momentum generates rotations in space. The geometrical relations of rotations appear in quantum mechanics as the commutation relations of the angular momentum operators.

Consider a quantum system in three-dimensional space. Let $\hat{R}_z(\theta)$ be the operator that rotates the system about the z-axis by an angle θ. Under this rotation, the normalization of states remains unchanged, and this operator is unitary:

$$\| \hat{R}_z(\theta)\,|\psi\rangle \|^2 = \langle\psi \mid \psi\rangle \Longrightarrow \hat{R}_z^\dagger(\theta)\hat{R}_z(\theta) = 1 \tag{4.97}$$

From the composition rule of rotations

$$\hat{R}_z(\theta)\hat{R}_z(\phi) = \hat{R}_z(\theta + \phi) \tag{4.98}$$

we obtain

$$\hat{R}_z^\dagger(\theta) = \hat{R}_z(-\theta) \tag{4.99}$$

For infinitesimally small θ, we obtain

$$\hat{R}_z(\theta) = 1 - i\frac{\theta}{\hbar}\hat{J}_z \tag{4.100}$$

where $\hat{J}_z$ is a hermitian operator because we obtain, by substituting (4.100) into the unitarity (4.97),

$$\left(1 + i\frac{\theta}{\hbar}\hat{J}_z^\dagger\right)\left(1 - i\frac{\theta}{\hbar}\hat{J}_z\right) = 1 + i\frac{\theta}{\hbar}\left(\hat{J}_z^\dagger - \hat{J}_z\right) = 1 \Longrightarrow \hat{J}_z^\dagger = \hat{J}_z \tag{4.101}$$

Integrating (4.100) over θ, we obtain

$$\hat{R}_z(\theta) = \mathrm{e}^{-i\frac{\theta}{\hbar}\hat{J}_z} \tag{4.102}$$

for a finite θ.

Similarly, we denote the operators generating infinitesimal rotations around the x and y axes by $\hat{J}_x$ and $\hat{J}_y$. The vector $\hat{\vec{J}}$ is the operator for angular momentum in quantum mechanics. The unitary operators representing finite rotations around the x and y axes are given by

$$\hat{R}_x(\theta) = \mathrm{e}^{-i\frac{\theta}{\hbar}\hat{J}_x}, \quad \hat{R}_y(\theta) = \mathrm{e}^{-i\frac{\theta}{\hbar}\hat{J}_y} \tag{4.103}$$

Example: Point Particle in Three-Dimensional Space

The orbital angular momentum is given by

$$\hat{\vec{L}} \equiv \hat{\vec{r}} \times \hat{\vec{p}} \tag{4.104}$$

as in classical mechanics. Let us verify this.

$$\begin{aligned}\frac{1}{\hbar}\hat{L}_z\,|\vec{x}\rangle &= \frac{1}{\hbar}\left(\hat{x}\hat{p}_y - \hat{y}\hat{p}_x\right)|x, y, z\rangle \\ &= \frac{1}{\hbar}\left(xi\hbar\frac{\partial}{\partial y} - yi\hbar\frac{\partial}{\partial x}\right)|x, y, z\rangle\end{aligned} \tag{4.105}$$

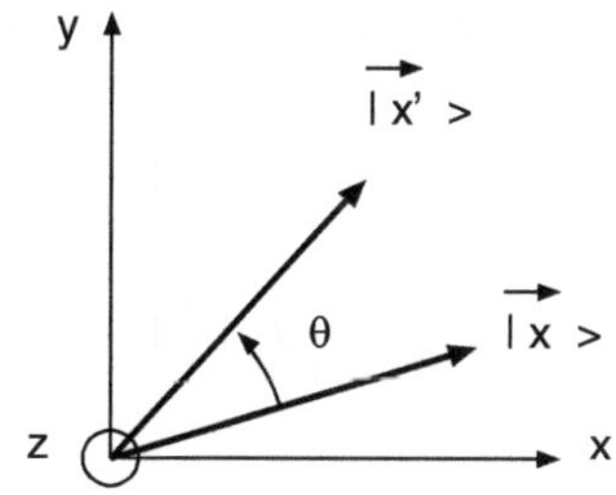

Fig. 4.6 Rotation around the z axis

Therefore,

$$\left(1 - i\frac{\theta}{\hbar}\hat{L}_z\right)|\vec{x}\rangle = |\vec{x}\rangle + \theta(x\partial_y - y\partial_x)\,|x, y, z\rangle$$
$$= |x - \theta y, y + \theta x, z\rangle \tag{4.106}$$

This implies that $\hat{J}_z$ indeed gives a rotation around the z axis (Fig. 4.6).

Commutation Relations

The three components $\hat{J}_x$, $\hat{J}_y$, and $\hat{J}_z$ of the angular momentum vector $\hat{\vec{J}}$ do not commute with each other. They do not commute for a geometric reason. For example, a rotation around the x-axis followed by a rotation around the y-axis does not produce the same result as the compound rotation in the reverse order.

To see this, consider two products, where ϵ and η are infinitesimal:

$$\left(1 - i\frac{\epsilon}{\hbar}\hat{J}_x\right)\left(1 - i\frac{\eta}{\hbar}\hat{J}_y\right) \quad \text{and} \quad \left(1 - i\frac{\eta}{\hbar}\hat{J}_y\right)\left(1 - i\frac{\epsilon}{\hbar}\hat{J}_x\right)$$

In the following calculations, terms with ϵ^2 and η^2 are ignored, but terms with $\epsilon\eta$ are retained. First, by rotating the $|\vec{x}\rangle$ around the y-axis by η, we get

$$\left(1 - i\frac{\eta}{\hbar}\hat{J}_y\right)|\vec{x}\rangle = |x + \eta z, y, z - \eta x\rangle \tag{4.107}$$

Rotating it further around the x-axis by ϵ, we obtain

$$\left(1 - i\frac{\epsilon}{\hbar}\hat{J}_x\right)\left(1 - i\frac{\eta}{\hbar}\hat{J}_y\right)|\vec{x}\rangle = |x + \eta z, y - \epsilon(z - \eta x), z - \eta x + \epsilon y\rangle \tag{4.108}$$

If we reverse the order of rotation, we get

$$\left(1 - i\frac{\epsilon}{\hbar}\hat{J}_x\right)|\vec{x}\rangle = |x, y - \epsilon z, z + \epsilon y\rangle \tag{4.109}$$

$$\left(1 - i\frac{\eta}{\hbar}\hat{J}_y\right)\left(1 - i\frac{\epsilon}{\hbar}\hat{J}_x\right)|\vec{x}\rangle = |x + \eta(z + \epsilon y), y - \epsilon z, z + \epsilon y - \eta x\rangle \tag{4.110}$$

Rotating this further by $\epsilon\eta$ around the z-axis, we obtain

$$\begin{aligned}
&\left(1 - i\frac{\epsilon\eta}{\hbar}\hat{J}_z\right)\left(1 - i\frac{\eta}{\hbar}\hat{J}_y\right)\left(1 - i\frac{\epsilon}{\hbar}\hat{J}_x\right)|\vec{x}\rangle \\
&= |x + \eta(z + \epsilon y) - \epsilon\eta y, y - \epsilon z + \epsilon\eta x, z + \epsilon y - \eta x\rangle \\
&= |x + \eta z, y - \epsilon(z - \eta x), z - \eta x + \epsilon y\rangle \\
&= \left(1 - i\frac{\epsilon}{\hbar}\hat{J}_x\right)\left(1 - i\frac{\eta}{\hbar}\hat{J}_y\right)|\vec{x}\rangle
\end{aligned} \tag{4.111}$$

Thus, a rotation by ϵ around the x-axis followed by a rotation by η around the y-axis is equivalent to a rotation by η around the y-axis followed by a rotation by ϵ around the x-axis and followed further by a rotation by $\epsilon\eta$ around the z-axis.

Since $\vec{x}$ in Eq. (4.111) is arbitrary, we get a general operator relation:

$$\left(1 - i\frac{\epsilon}{\hbar}\hat{J}_x\right)\left(1 - i\frac{\eta}{\hbar}\hat{J}_y\right) = \left(1 - i\frac{\epsilon\eta}{\hbar}\hat{J}_z\right)\left(1 - i\frac{\eta}{\hbar}\hat{J}_y\right)\left(1 - i\frac{\epsilon}{\hbar}\hat{J}_x\right) \tag{4.112}$$

By expanding both sides and comparing the terms of order $\epsilon\eta$, we derive the commutation relation

$$[\hat{J}_x, \hat{J}_y] = i\hbar\hat{J}_z \tag{4.113}$$

Similarly, we can derive the commutation relations

$$[\hat{J}_y, \hat{J}_z] = i\hbar\hat{J}_x, \quad [\hat{J}_z, \hat{J}_x] = i\hbar\hat{J}_y \tag{4.114}$$

Thus, we have derived the commutation relation[9]

$$[\hat{J}_i, \hat{J}_j] = i\hbar \sum_{k=x,y,z} \epsilon_{ijk}\hat{J}_k \tag{4.115}$$

[9] ϵ_{ijk} is antisymmetric with respect to i, j, k, and $\epsilon_{xyz} = 1$. The summation symbol $\sum_k$ is often omitted. So, we may write $[\hat{J}_i, \hat{J}_j] = i\epsilon_{ijk}\hat{J}_k$.

As is evident from the derivation, the commutation relation has a geometric meaning.

The operator for the square of angular momentum, $\hat{\vec{J}}^2 \equiv \hat{J}_x^2 + \hat{J}_y^2 + \hat{J}_z^2$, is a scalar, invariant under rotations, and commutes with each component of the angular momentum:

$$\left[\hat{\vec{J}}^2, \hat{J}_i\right] = 0 \quad (i = x, y, z) \tag{4.116}$$

Therefore, for any state vector $|\psi\rangle$, we find

$$\hat{R}_i(\theta)\hat{\vec{J}}^2 |\psi\rangle = \hat{\vec{J}}^2 \hat{R}_i(\theta) |\psi\rangle \quad (i = x, y, z) \tag{4.117}$$

4.2.2 *The Irreducible Representation of the Commutation Relations of the Angular Momentum Operators Is Characterized by a Half-Integral Spin*

We recall that the annihilation and creation operators, $\hat{a}$ and $\hat{a}^\dagger$, of the harmonic oscillator satisfy the following commutation relations:

$$\left[\hat{a}^\dagger, \hat{a}^\dagger\hat{a}\right] = -\hat{a}^\dagger, \quad \left[\hat{a}, \hat{a}^\dagger\hat{a}\right] = \hat{a} \tag{4.118}$$

To make an analogy, we introduce two operators:

$$\hat{J}_\pm \equiv \hat{J}_x \pm i\hat{J}_y \tag{4.119}$$

These satisfy the commutation relation:

$$\begin{aligned}[\hat{J}_\pm, \hat{J}_z] &= [\hat{J}_x \pm i\hat{J}_y, \hat{J}_z] \\ &= -i\hbar\hat{J}_y \mp \hbar\hat{J}_x = \mp\hbar\hat{J}_\pm\end{aligned} \tag{4.120}$$

This implies that $\hat{J}_\pm$ raises (or lowers) the eigenvalue of $\hat{J}_z$ by $\hbar$. To see why, let $|m\rangle$ be the eigenstate of $\hat{J}_z$ with eigenvalue $\hbar m$:

$$\hat{J}_z |m\rangle = \hbar m |m\rangle \tag{4.121}$$

Then, the state $\hat{J}_\pm |m\rangle$ has the eigenvalue $\hbar(m \pm 1)$:

$$\hat{J}_z\hat{J}_\pm |m\rangle = \Big(\underbrace{[\hat{J}_z, \hat{J}_\pm]}_{=\pm\hbar\hat{J}_\pm} + \hat{J}_\pm \underbrace{\hat{J}_z}_{=\hbar m}\Big) |m\rangle = \hbar(m \pm 1)\hat{J}_\pm |m\rangle \tag{4.122}$$

The commutator of $\hat{J}_+$ and $\hat{J}_-$ is not a simple number, unlike the annihilation and creation operators of a harmonic oscillator:

$$\left[\hat{J}_+, \hat{J}_-\right] = \left[\hat{J}_x + i\hat{J}_y, \hat{J}_x - i\hat{J}_y\right] = -i\left[\hat{J}_x, \hat{J}_y\right] + i\left[\hat{J}_y, \hat{J}_x\right] = 2\hbar\hat{J}_z \tag{4.123}$$

Compare this with the commutation relation:

$$\left[\hat{a}, \hat{a}^\dagger\right] = 1 \tag{4.124}$$

In the case of a harmonic oscillator, the ground state is defined by $\hat{a}\,|\Omega\rangle = 0$, and the excited state $|n\rangle$ satisfying $\hat{a}^\dagger\hat{a} = n$ is obtained by applying the creation operator $\hat{a}^\dagger$ n times on the ground state $|\Omega\rangle$. n has no upper bound. In the case of angular momentum, as we will show, similar results can be obtained. A big difference is, however, that n has an upper bound in the case of angular momentum.

Let us consider $\mathcal{S}$ to be a finite-dimensional subspace of ket vectors. We assume that under the action of angular momentum operators, $\mathcal{S}$ is *closed*. This means for any $|\psi\rangle \in \mathcal{S}$

$$\hat{R}_i(\theta)\,|\psi\rangle \in \mathcal{S} \quad (i = x, y, z) \tag{4.125}$$

where θ is arbitrary. Since the operator $\hat{J}_z$ is hermitian, it can be diagonalized. Given that $\mathcal{S}$ is finite, there should be at least one state $|J\rangle$ with the maximum eigenvalue $\hbar J$ for $\hat{J}_z$. (If there are multiple, the following considerations can be applied to each.) Therefore,

$$\hat{J}_z\,|J\rangle = \hbar J\,|J\rangle \tag{4.126}$$

From Eq. (4.122), the operator $\hat{J}_+$ raises the eigenvalue of $\hat{J}_z$ by $\hbar$. Since we have just assumed that $|J\rangle$ has the maximum eigenvalue for $\hat{J}_z$, we must find

$$\hat{J}_+\,|J\rangle = 0 \tag{4.127}$$

Starting from the eigenstate $|J\rangle$, we can create states with lower eigenvalues of $\hat{J}_z$ by applying $\hat{J}_-$ successively:

$$\hat{J}_z\,\underbrace{\hat{J}_-\cdots\hat{J}_-}_{k}\,|J\rangle = \hbar(J-k)\,\underbrace{\hat{J}_-\cdots\hat{J}_-}_{k}\,|J\rangle \tag{4.128}$$

We now introduce positive constants $c_k > 0$ so that

$$|J-k\rangle \equiv \frac{c_k}{\hbar^k}\,\underbrace{\hat{J}_-\cdots\hat{J}_-}_{k}\,|J\rangle \quad (k = 0, 1, \cdots) \tag{4.129}$$

are normalized. $\hat{J}_+$ raises the eigenvalue of $\hat{J}_z$ by $\hbar$, and we find

$$\hat{J}_+^l |J-k\rangle \propto |J-k+l\rangle \tag{4.130}$$

Let us define $\mathcal{S}' \subset \mathcal{S}$ as the subspace consisting of all states that can be obtained from $|J\rangle$ by rotations. This subspace is spanned by $|J-k\rangle$ $(k = 0, 1, 2, \cdots)$. Since $\mathcal{S}$ is assumed to be finite, $\mathcal{S}'$ is also finite. Hence, there must be an upper limit for k. In other words, there exists a maximum positive integer k_{max} such that $|J-k_{\text{max}}\rangle$ is non-zero, but

$$\hat{J}_- |J-k_{\text{max}}\rangle = 0 \tag{4.131}$$

In the following we wish to show

$$k_{\text{max}} = 2J \tag{4.132}$$

Let us show first that

$$k_{\text{max}} \geq 2J \tag{4.133}$$

Consider the state obtained by rotating $|J\rangle$ by π around the x-axis:

$$\hat{R}_x(\pi)|J\rangle = \mathrm{e}^{-i\frac{\pi}{\hbar}\hat{J}_x}|J\rangle \tag{4.134}$$

From the commutation relation

$$\hat{R}_x(\pi)^\dagger \hat{J}_z \hat{R}_x(\pi) = -\hat{J}_z \tag{4.135}$$

(which can geometrically be understood as $\hat{R}_x(\pi)^\dagger \hat{R}_z(\theta)\hat{R}_x(\pi) = \hat{R}_z(-\theta)$), we obtain

$$\hat{J}_z \hat{R}_x(\pi)|J\rangle = -\hbar J \hat{R}_x(\pi)|J\rangle \tag{4.136}$$

Therefore, $\mathcal{S}'$ must contain $|-J\rangle$. Hence, from the inequality

$$J - k_{\text{max}} \leq -J\,, \tag{4.137}$$

we obtain the inequality (4.133). Furthermore, for $|-J\rangle$ to be represented as $|J-k\rangle$ for some integer k, $2J$ must be a non-negative integer:

$$J = 0, \frac{1}{2}, 1, \frac{3}{2}, 2, \cdots \tag{4.138}$$

Using the same reasoning as above, it can also be said that $|-J-1\rangle = 0$. This is because if it were not zero, then $\hat{R}_x(\pi)\,|-J-1\rangle$ would satisfy

$$\hat{J}_z\hat{R}_x(\pi)\,|-J-1\rangle = \hbar(J+1)\hat{R}_x(\pi)\,|-J-1\rangle \tag{4.139}$$

which contradicts our assumption that $|J\rangle$ has the maximum eigenvalue of $\hat{J}_z$. Thus, we have shown Eq. (4.132).

Therefore, we have drawn the following conclusions:

1. The vector space $\mathcal{S}$, which is closed under rotation, is characterized by a half-integer $J = 0, \frac{1}{2}, 1, \frac{3}{2}, 2, \cdots$.
2. $\mathcal{S}$ is spanned by $2J+1$ basis vectors $|m\rangle$ $(m = J, J-1, \cdots, -(J-1), -J)$, satisfying

$$\hat{J}_z\,|m\rangle = \hbar m\,|m\rangle \tag{4.140}$$

3. The eigenvalues of $\hat{J}_z$ can be raised and lowered by $\hat{J}_\pm$:

$$\hat{J}_+\,|m\rangle \propto |m+1\rangle\,, \quad \hat{J}_-\,|m\rangle \propto |m-1\rangle \tag{4.141}$$

The matrix elements of $\hat{J}_\pm$, if we choose the phases of the $2J+1$ states appropriately, are given by

$$\boxed{\langle m+1|\,\hat{J}_+\,|m\rangle = \langle m|\,\hat{J}_-\,|m+1\rangle = \hbar\sqrt{(J-m)(J+m+1)}} \tag{4.142}$$

as we will derive in Exercise 9-2.

Eigenvalues of the Scalar Operator $\hat{\vec{J}}^2$

The scalar operator $\hat{\vec{J}}^2$ commutes with $\hat{J}_i (i = x, y, z)$, and it can be diagonalized simultaneously with $\hat{J}_z$ for example. Let us show that $\hat{\vec{J}}^2$ is a constant in the subspace $\mathcal{S}$. First, using $[\hat{J}_+, \hat{J}_-] = 2\hbar\hat{J}_z$, we obtain

$$\hat{\vec{J}}^2 = \frac{1}{2}\left(\hat{J}_+\hat{J}_- + \hat{J}_-\hat{J}_+\right) + \hat{J}_z^2 = \hat{J}_-\hat{J}_+ + \hbar\hat{J}_z + \hat{J}_z^2 \tag{4.143}$$

Thus, from $\hat{J}_+\,|J\rangle = 0$, we obtain

$$\hat{\vec{J}}^2\,|J\rangle = \hbar^2 J(J+1)\,|J\rangle \tag{4.144}$$

Since $\hat{J}_-$ and $\hat{\vec{J}}^2$ commute, $|m\rangle \propto \hat{J}_-^{J-m}|J\rangle$ $(-J \leq m \leq J)$ is also an eigenstate of $\hat{\vec{J}}^2$ with the same eigenvalue $\hbar^2 J(J+1)$.

Example: Spin $\frac{1}{2}$ $(J = \frac{1}{2})$

$$\frac{1}{\hbar}\hat{J}_z|\uparrow\rangle = \frac{1}{2}|\uparrow\rangle \tag{4.145a}$$

$$\frac{1}{\hbar}\hat{J}_-|\uparrow\rangle = |\downarrow\rangle \tag{4.145b}$$

$$\frac{1}{\hbar}\hat{J}_z|\downarrow\rangle = -\frac{1}{2}|\downarrow\rangle \tag{4.145c}$$

Spherical Harmonics

Consider the state vectors defined on the unit sphere. Using spherical coordinates θ, ϕ, we denote the basis vectors as $|\theta, \phi\rangle$. With the normalization

$$\langle\theta, \phi \mid \theta', \phi'\rangle = \frac{1}{\sin\theta}\delta(\theta - \theta')\delta(\phi - \phi') \tag{4.146}$$

we get

$$\int_0^{\pi} \sin\theta d\theta \int_{-\pi}^{\pi} d\phi \langle\theta, \phi \mid \theta', \phi'\rangle = 1 \tag{4.147}$$

We introduce the eigenstates $|l, m\rangle$ of the orbital angular momentum $\hat{L}_z$ and magnitude $\hat{\vec{L}}^2$ by

$$\hat{\vec{L}}^2|l, m\rangle = \hbar^2 l(l+1)|l, m\rangle \tag{4.148}$$

$$\hat{L}_z|l, m\rangle = \hbar m|l, m\rangle \tag{4.149}$$

Spherical harmonics are defined as the wave functions of these states:

$$\boxed{Y_{l,m}(\theta, \phi) \equiv \langle\theta, \phi \mid l, m\rangle} \tag{4.150}$$

Any $|\theta, \phi\rangle$ can be obtained from the state localized in the $+z$ direction, $|+z\rangle \equiv |\theta = 0\rangle$, in terms of a rotation (Fig. 4.7):

$$|\theta, \phi\rangle = \hat{R}_z(\phi)\hat{R}_y(\theta)|+z\rangle \tag{4.151}$$

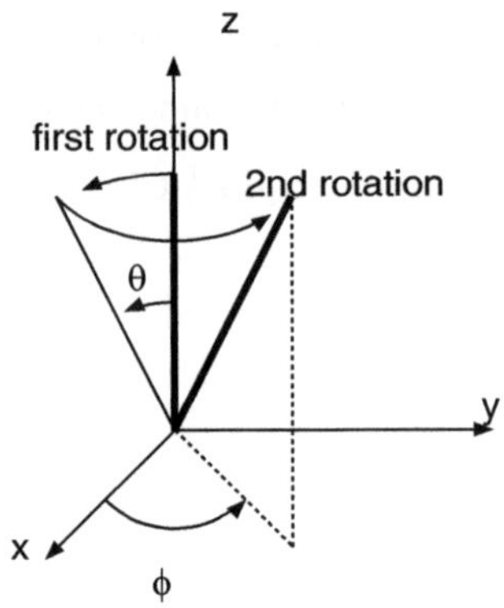

Fig. 4.7 First rotate around the y axis and then around the z axis

Thus, we obtain

$$Y_{l,m}(\theta, \phi) = \langle +z|\, \hat{R}_y(-\theta)\hat{R}_z(-\phi)\, |l, m\rangle = \mathrm{e}^{im\phi}\, \langle +z|\, \hat{R}_y(-\theta)\, |l, m\rangle \tag{4.152}$$

We learn more about the spherical harmonics in Exercises 9-3 and 9-4.

4.2.3 Remember the Rule for Summing Two Angular Momenta

The rule for combining two angular momenta is one of the most important results in elementary quantum mechanics. The rule for composition is an application of the representation of angular momentum given in the previous section.

Introduce two physical systems with fixed magnitudes of angular momenta. One system (system 1) is a state vector space spanned by $|j_1, m_1\rangle$ $(m_1 = -j_1, \cdots, j_1)$ with spin j_1, and the other system (system 2) is a state vector space spanned by $|j_2, m_2\rangle$ $(m_2 = -j_2, \cdots, j_2)$ with spin j_2. The system obtained by "combining" these two is a vector space spanned by the **direct products**:

$$|j_1, m_1\rangle\, |j_2, m_2\rangle = |j_1, m_1; j_2, m_2\rangle$$

Thus, the dimension of the state vector space is $(2j_1 + 1)(2j_2 + 1)$.

Example: Angular Momentum of an Electron

Let us consider a single electron. An electron has two angular momenta: orbital angular momentum $\hat{\vec{L}}$ and spin angular momentum $\hat{\vec{s}}$. Therefore, for an electron with orbital angular momentum l, the state of total angular momentum is given as a linear combination of the products

$$|l, m\rangle \left|\frac{1}{2}, \pm\frac{1}{2}\right\rangle \tag{4.153}$$

Now, let us return to the general case to continue our discussion. Let $\hat{\vec{J}}_1$ and $\hat{\vec{J}}_2$ be the angular momentum operators for systems 1 and 2, respectively. Then, the total angular momentum for the combined system is given by their sum:

$$\hat{\vec{J}} = \hat{\vec{J}}_1 + \hat{\vec{J}}_2 \tag{4.154}$$

This seems obvious, but let us verify this according to the original definition of angular momentum. If system 1 is in state $|\psi\rangle_1$ and system 2 is in state $|\phi\rangle_2$, then the state for the entire system is given by the direct product $|\psi\rangle_1 |\phi\rangle_2$. Rotating the whole system around the i-axis by θ is equivalent to rotating each subsystem, so

$$\begin{aligned} \hat{R}_i(\theta)\left\{|\psi\rangle_1 |\phi\rangle_2\right\} &= \left(\hat{R}_{1,i}(\theta)\,|\psi\rangle_1\right)\left(\hat{R}_{2,i}(\theta)\,|\phi\rangle_2\right) \\ &= \left(\mathrm{e}^{-i\frac{\theta}{\hbar}\hat{J}_{1,i}}\,|\psi\rangle_1\right)\left(\mathrm{e}^{-i\frac{\theta}{\hbar}\hat{J}_{2,i}}\,|\phi\rangle_2\right) \end{aligned} \tag{4.155}$$

For an infinitesimally small angle θ, we obtain

$$\left(1 - i\frac{\theta}{\hbar}\hat{J}_i\right)\left\{|\psi\rangle_1 |\phi\rangle_2\right\} = |\psi\rangle_1 |\phi\rangle_2 - i\frac{\theta}{\hbar}\left[\left(\hat{J}_{1,i}\,|\psi\rangle_1\right)|\phi\rangle_2 + |\psi\rangle_1\left(\hat{J}_{2,i}\,|\phi\rangle_2\right)\right] \tag{4.156}$$

This gives Eq. (4.154).

Now, the important thing is

$$\hat{J}_z\,|j_1, m_1;\, j_2, m_2\rangle = \hbar(m_1 + m_2)\,|j_1, m_1;\, j_2, m_2\rangle \tag{4.157}$$

Therefore, there are $(2j_1 + 1)(2j_2 + 1)$ eigenstates of $\hat{J}_z$. We would like to classify these states and understand what magnitudes of the total angular momentum they contain. This can be easily understood using the method introduced in the previous section.

For the following discussion, assume that $j_1 \geq j_2$ without loss of generality. The first thing we should do is to enumerate how many linearly independent states there are for each eigenvalue of $\hat{J}_z$. We should count the number of cases where $m_1 + m_2$

are equal, and the results are as follows:

Eigenvalue of $\hat{J}_z$	Number of states
(Maximum) $j_1 + j_2$	1
$j_1 + j_2 - 1$	2
$j_1 + j_2 - 2$	3
$\vdots$	$\vdots$
$j_1 - j_2$	$2j_2 + 1$
$(j_1 - j_2) - 1$	$2j_2 + 1$
$\vdots$	$\vdots$
$-(j_1 - j_2)$	$2j_2 + 1$
$-(j_1 + j_2) + 2j_2 - 1$	$2j_2$
$\vdots$	$\vdots$
$-(j_1 + j_2) + 2$	3
$-(j_1 + j_2) + 1$	2
(Minimum) $-(j_1 + j_2)$	1
Total	$(2j_1 + 1)(2j_2 + 1)$

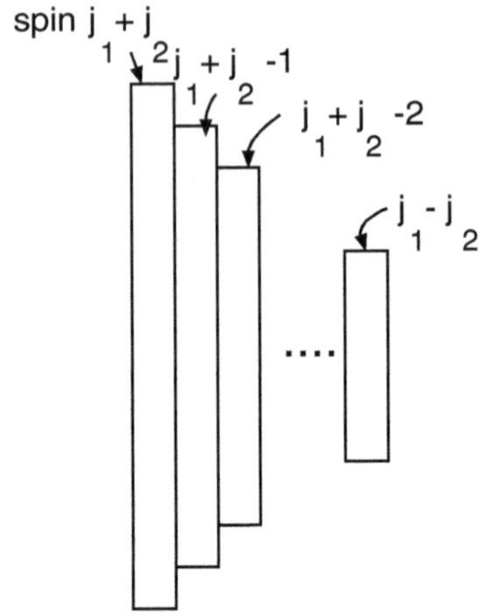

To classify these states by the magnitude of the total angular momentum, we first focus on the maximum eigenvalue of $\hat{J}_z$. The maximum eigenvalue is $j_1 + j_2$ and the number of states is 1. From this, we understand that there are $2(j_1 + j_2) + 1$ states with the total angular momentum $j_1 + j_2$. Looking next at the states with the second largest eigenvalue of $\hat{J}_z$, we find there are two. One of these belongs to the previously considered states with total angular momentum $j_1 + j_2$. The other, which is expected to become zero when acted upon by the operator $\hat{J}_+$, corresponds to the total angular momentum $j_1 + j_2 - 1$. From this, it is clear that there are $2(j_1 + j_2 - 1) + 1$ states with this angular momentum. Continuing in a similar fashion, we conclude that the possible magnitudes for the total angular momentum are

$$\boxed{j_1 + j_2, j_1 + j_2 - 1, \cdots, j_1 - j_2} \tag{4.158}$$

Adding up the number of states for each magnitude, we get

$$\sum_{j=j_1-j_2}^{j_1+j_2} (2j + 1) = (2j_1 + 1)(2j_2 + 1) \tag{4.159}$$

as expected.

Table 4.1 $\frac{1}{2} \times \frac{1}{2}$

$\frac{1}{\hbar}\hat{J}_z$	State
1	$\lvert\uparrow,\uparrow\rangle$
0	$\lvert\uparrow,\downarrow\rangle, \lvert\downarrow,\uparrow\rangle$
-1	$\lvert\downarrow,\downarrow\rangle$

Example: spin $\frac{1}{2}$ $\times$ spin $\frac{1}{2}$ $\rightarrow$ spin $1 +$ spin 0 (**Also Written as** $\frac{1}{2} \otimes \frac{1}{2} = 1 \oplus 0$)

Let

$$\hat{J}_- \equiv \hat{J}_{1-} + \hat{J}_{2-} \tag{4.160}$$

We find

$$\frac{1}{\hbar}\hat{J}_- \lvert\uparrow,\uparrow\rangle = \lvert\uparrow,\downarrow\rangle + \lvert\downarrow,\uparrow\rangle = \sqrt{2}\frac{\lvert\uparrow,\downarrow\rangle + \lvert\downarrow,\uparrow\rangle}{\sqrt{2}} \tag{4.161}$$

Therefore, the states with a total angular momentum of 1 are

$$\lvert\uparrow,\uparrow\rangle, \frac{1}{\sqrt{2}}\left(\lvert\uparrow,\downarrow\rangle + \lvert\downarrow,\uparrow\rangle\right), \lvert\downarrow,\downarrow\rangle , \tag{4.162}$$

and the remaining state

$$\frac{1}{\sqrt{2}}\left(\lvert\uparrow,\downarrow\rangle - \lvert\downarrow,\uparrow\rangle\right) \tag{4.163}$$

has a total angular momentum of zero.

As an application, let us consider a hydrogen molecule with two electrons. The total spin of the two electrons can be either 1 or 0. In the case of 1, the spin wave function is symmetric with respect to the exchange of electrons, so the orbital wave function must be antisymmetric due to the Fermi statistics. This state is called **ortho**. In the case of 0, the spin wave function is antisymmetric, and the orbital wave function must be symmetric; this is called **para**.

Clebsch-Gordan Coefficients

When composing angular momenta j_1 and j_2 to create a total angular momentum J, the eigenstates of $\hat{J}_z \equiv \hat{J}_{1,z} + \hat{J}_{2,z}$ can be expanded as

$$\lvert(j_1, j_2)J, M\rangle = \sum_{m_1+m_2=M} \lvert j_1, m_1; j_2, m_2\rangle \cdot \langle j_1, m_1; j_2, m_2 \mid (j_1, j_2)J, M\rangle \tag{4.164}$$

Here, the real coefficients

$$\langle j_1, m_1; j_2, m_2 \mid (j_1, j_2)J, M\rangle$$

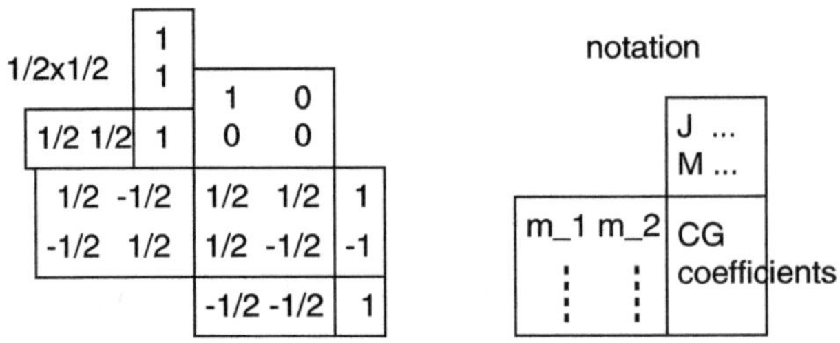

Fig. 4.8 Table of Clebsch-Gordan coefficients

are called **Clebsch-Gordan coefficients.**[10] We refer you to Sect. 4.6.3 for how to determine the Clebsch-Gordan coefficients and for the related Wigner-Eckard theorem.

Example: $\frac{1}{2} \otimes \frac{1}{2} \to 1 \oplus 0$

$$|1, 1\rangle = |\uparrow, \uparrow\rangle \tag{4.165a}$$

$$|1, 0\rangle = \frac{1}{\sqrt{2}} (|\uparrow, \downarrow\rangle + |\downarrow, \uparrow\rangle) \tag{4.165b}$$

$$|1, -1\rangle = |\downarrow, \downarrow\rangle \tag{4.165c}$$

$$|0, 0\rangle = \frac{1}{\sqrt{2}} (|\uparrow, \downarrow\rangle - |\downarrow, \uparrow\rangle) \tag{4.165d}$$

We can give this result in a table in Fig. 4.8 (square roots are omitted. $-1/2$ means $-\sqrt{1/2}$):

A table of the Clebsch-Gordan coefficients up to $2 \otimes 2$ is given after Exercises 9 (See Fig. 4.9).

Examples 9

1. **Precession of a spin**—Consider a spin-$\frac{1}{2}$ particle for which the Hamiltonian is given by

$$\hat{H} = -\omega \hat{J}_z$$

[10] P. Gordan (1837–1912) was a French mathematician. Do not confuse him with the physicist W. Gordon (1893–1939) well known for the Klein-Gordon equation.

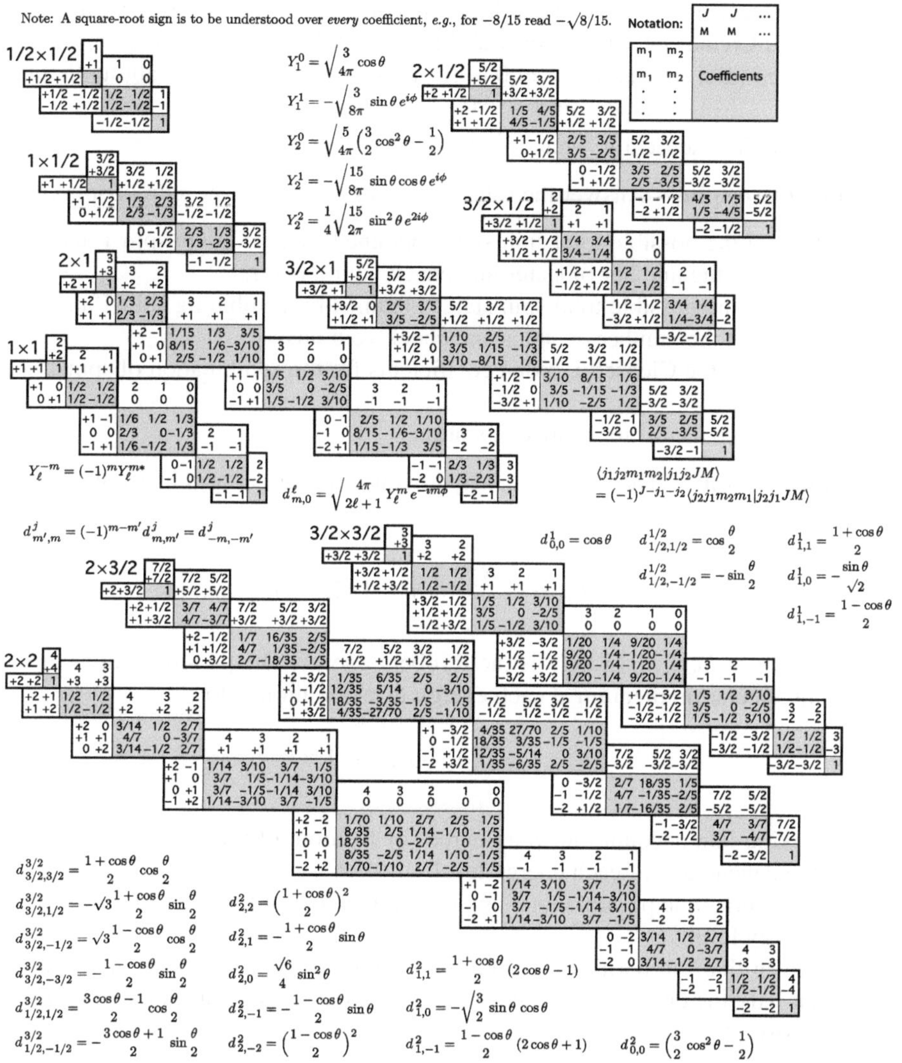

Fig. 4.9 Clebsch-Gordan Coefficients, Spherical Harmonics, and d Functions. The sign convention is that of Wigner (*Group Theory*, Academic Press, New York, 1959), also used by Condon and Shortley (*The Theory of Atomic Spectra*, Cambridge University Press, New York, 1953), Rose (*Elementary Theory of Angular Momentum*, Wiley, New York, 1957), and Cohen (*Tables of the Clebsch-Gordan Coefficients*, North American Rockwell Science Center, Thousand Oaks, Calif., 1974). Table of Clebsch-Gordan coefficients. (Reproduced from http://pdg.lbl.gov)

(a) Construct states $|\pm x\rangle$ where the spin is directed in the $\pm x$ direction.
Hint: Create states that satisfy $\hat{J}_x |\pm x\rangle = \pm\frac{\hbar}{2} |\pm x\rangle$ in terms of $|\pm z\rangle$.
(b) If the initial state at time $t = 0$ is given by $|+x\rangle$, find the probability that the state is $|-x\rangle$ at a later time $t > 0$.
Hint: Geometrically, what does the operator $e^{-i\frac{t}{\hbar}\hat{H}}$ do?

2. **Sum of two spins**—Answer the following questions about the sum of spins.

(a) List all the possible total spin states obtained by combining spin 1 and spin 2. Count the number of states and check your answer.
(b) List all the possible total spin states obtained by combining spin 1 and spin $\frac{1}{2}$. Count the number of states and check your answer.
(c) Compute the Clebsch-Gordan coefficients for the combination of spin 1 and spin $\frac{1}{2}$.
Hint: Use the orthogonality of states.

3. **Hydrogen atom**—The electron in a hydrogen atom has energy eigenvalues given by

$$E_n = -\frac{Ry}{n^2} \quad (n = 1, 2, \cdots)$$

where $Ry \equiv hcR \simeq 13.6\,\text{eV}$ is the Rydberg energy. The possible values of the orbital angular momentum at each energy level are limited to

$$l = 0, 1, \cdots, n-1$$

(a) Taking into account the spin degrees of freedom of the electron, determine the degeneracy of each energy level.
(b) Enumerate the possible total angular momentum of the electron for the n-th energy level.
(c) Given that the parity of a state with orbital angular momentum l is $(-)^l$, classify the energy eigenstates of level n by total angular momentum j and parity.

4. **Decay of the ρ meson**—The ρ^+ meson is a spin-1 particle that mainly decays into a spinless π^+ meson and a spinless π^0 meson. In the rest frame of the ρ meson, where the total momentum is zero, the final state of the two mesons is given by

$$\left|\pi^+\pi^0\right\rangle = \int \frac{d^3p}{(2\pi\hbar)^3} f(\vec{p}) \left|\pi^+(\vec{p})\right\rangle \left|\pi^0(-\vec{p})\right\rangle$$

where a π meson with momentum $\vec{p}$ is given as $|\pi(\vec{p})\rangle$.

(a) Given that the state of the π meson transforms as

$$|\pi(\vec{p})\rangle \longrightarrow |\pi(\vec{p} + \vec{\epsilon} \times \vec{p})\rangle$$

under an infinitesimal rotation, show that the total angular momentum operator $\hat{\vec{J}}$ acts on the state as

$$\hat{\vec{J}}\left|\pi^+\pi^0\right\rangle = \int \frac{d^3p}{(2\pi\hbar)^3}\left(-i\hbar\vec{p} \times \frac{\partial}{\partial\vec{p}} f(\vec{p})\right)|\pi^+(\vec{p})\rangle\left|\pi^0(-\vec{p})\right\rangle$$

If the decaying ρ meson has its spin pointed in the $+z$ direction, we find

$$\hat{J}_z\left|\pi^+\pi^0\right\rangle = \hbar\left|\pi^+\pi^0\right\rangle, \quad \hat{J}_+\left|\pi^+\pi^0\right\rangle = 0$$

from the conservation of angular momentum.

(b) Show that $f(\vec{p})$ is given by

$$f(\vec{p}) = (p_x + ip_y)g(p)$$

where $g(p)$ is a function depending only on the magnitude of $\vec{p}$.
Hint: The orbital angular momentum is given by $-i\hbar\vec{x} \times \frac{\partial}{\partial\vec{x}}$ in the coordinate representation. In the momentum representation, it is given by $-i\hbar\vec{p} \times \frac{\partial}{\partial\vec{p}}$. Thus, by replacing $\vec{x}$ by $\vec{p}$ in the l-th order polynomial $r^l Y_{l,m}(\theta, \phi)$, you obtain the wave function with angular momentum of magnitude l and z component m in the momentum representation.

(c) Plot the angular distribution of the two π mesons in the final state.

5. **Landé's g factor and the anomalous Zeeman effect**—Consider electrons (possibly multiple) in an atom. Let $\hat{\vec{L}}$ represent the orbital angular momentum and $\hat{\vec{S}}$ the spin angular momentum. Let $\hat{\vec{J}} = \hat{\vec{L}} + \hat{\vec{S}}$ be the total angular momentum of the electrons. Let L and S be the magnitudes of the orbital and spin angular momenta. The magnitude of the total angular momentum is given by

$$J = L + S, \cdots, |L - S|$$

Denoting the normalized eigenstates of the total angular momentum (i.e., of $\hat{\vec{J}}^2$ and $\hat{J}_z$) as $|J, M\rangle$, we aim to compute the matrix element

$$\langle J, M| \hat{S}_z |J, M\rangle = (g - 1)\hbar M$$

(where g is called Landé's g factor). Here, we assume the Wigner-Eckard theorem derived in Sect. 4.6.3:

$$\langle J, M'| \hat{\vec{S}} |J, M\rangle = c(J, L, S)\langle J, M'| \hat{\vec{J}} |J, M\rangle$$

The coefficient $c(J, L, S)$ is independent of M, M'.

(a) Derive the relation:

$$\langle J, M| \hat{\vec{J}} \cdot \hat{\vec{S}} |J, M\rangle = c(J, L, S)\hbar^2 J(J+1)$$

(b) Derive the relation:

$$c(J, L, S) = \frac{1}{2J(J+1)} (-L(L+1) + J(J+1) + S(S+1))$$

Hint: $\vec{L}^2 = (\vec{J} - \vec{S})^2$.

(c) Show

$$g = \frac{3}{2} + \frac{(S-L)(S+L+1)}{2J(J+1)}$$

(d) Given the hamiltonian

$$\hat{H} = \mu_B B \frac{1}{\hbar} \left(\hat{L}_z + 2\hat{S}_z\right)$$

under an external magnetic field B, find the expectation value of $\hat{H}$ in the state $|J, M\rangle$. (μ_B is a magnetic moment called the Bohr magneton.)

Solutions to Examples 9

1. (a) From $(\hat{J}_x \pm i\hat{J}_y)|\pm z\rangle = 0$ and $(\hat{J}_x \mp i\hat{J}_y)|\pm z\rangle = \hbar\,|\mp z\rangle$, we obtain

$$\hat{J}_x |\pm z\rangle = \frac{\hbar}{2} |\mp z\rangle$$

Thus,

$$|\pm x\rangle = \frac{1}{\sqrt{2}} (|+z\rangle \pm |-z\rangle)$$

(b) The state at time t is

$$|\psi(t)\rangle = \mathrm{e}^{-i\frac{t}{\hbar}\hat{H}} |+x\rangle$$

From the relation $\hat{H}|\pm z\rangle = \mp\frac{1}{2}\hbar\omega |\pm z\rangle$, we get

$$\begin{aligned}|\psi(t)\rangle &= \frac{1}{\sqrt{2}} \left(\mathrm{e}^{i\frac{\omega t}{2}} |+z\rangle + \mathrm{e}^{-i\frac{\omega t}{2}} |-z\rangle\right) \\ &= \cos\frac{\omega t}{2} |+x\rangle + i\sin\frac{\omega t}{2} |-x\rangle\end{aligned}$$

Thus, the probability is $|\langle -x \mid \psi(t)\rangle|^2 = \sin^2\frac{\omega t}{2} = \frac{1}{2}(1-\cos\omega t)$. This result is consistent with the operator $e^{-\frac{i}{\hbar}\hat{H}t} = e^{i\omega t\frac{\hat{J}_z}{\hbar}}$ rotating the system about the z-axis by angle $-\omega t$.

2. (a) $1 \otimes 2 = 1 \oplus 2 \oplus 3$. When combining spin-1 (with 3 degrees of freedom) and spin-2 (with 5 degrees of freedom), we get a total of $3 \times 5 = 15$ degrees of freedom. On the other hand, when summing the degrees of freedom for spins 1, 2, and 3, we get $3 + 5 + 7 = 15$. Two ways of counting agree.

(b)

$$\begin{aligned} 1 \otimes \tfrac{1}{2} &= \tfrac{1}{2} \oplus \tfrac{3}{2} \\ 3 \times 2 &= 2 + 4 \end{aligned}$$

(c) We note

$$\text{Spin } 1 \begin{cases} \frac{1}{\hbar}\hat{J}_-^{(1)}\,|1\rangle = \sqrt{2}\,|0\rangle \\ \frac{1}{\hbar}\hat{J}_-^{(1)}\,|0\rangle = \sqrt{2}\,|-1\rangle \end{cases} \qquad \text{Spin } \frac{1}{2} \left\{ \frac{1}{\hbar}\hat{J}_-^{(2)}\left|\frac{1}{2}\right\rangle = \left|-\frac{1}{2}\right\rangle \right.$$

Starting with $\left|\left(1, \frac{1}{2}\right); \frac{3}{2}, \frac{3}{2}\right\rangle = |1\rangle\left|\frac{1}{2}\right\rangle$, we get

$$\begin{aligned} \frac{1}{\hbar}\hat{J}_-\left|\left(1, \frac{1}{2}\right)\frac{3}{2}, \frac{3}{2}\right\rangle &= \frac{1}{\hbar}\hat{J}_-^{(1)}\,|1\rangle\left|\frac{1}{2}\right\rangle + |1\rangle\,\frac{1}{\hbar}\hat{J}_-^{(2)}\left|\frac{1}{2}\right\rangle \\ &= \sqrt{2}\,|0\rangle\left|\frac{1}{2}\right\rangle + |1\rangle\left|-\frac{1}{2}\right\rangle \end{aligned}$$

Normalizing, we get

$$\left|\left(1, \frac{1}{2}\right)\frac{3}{2}, \frac{1}{2}\right\rangle = \sqrt{\frac{2}{3}}\,|0\rangle\left|\frac{1}{2}\right\rangle + \sqrt{\frac{1}{3}}\,|1\rangle\left|-\frac{1}{2}\right\rangle$$

Next, we compute

$$\begin{aligned} \frac{1}{\hbar}\hat{J}_-\left|\left(1, \frac{1}{2}\right)\frac{3}{2}, \frac{1}{2}\right\rangle &= \sqrt{\frac{2}{3}}\left(\sqrt{2}\,|-1\rangle\left|\frac{1}{2}\right\rangle + |0\rangle\left|-\frac{1}{2}\right\rangle\right) + \sqrt{\frac{2}{3}}\,|0\rangle\left|-\frac{1}{2}\right\rangle \\ &= \frac{2}{\sqrt{3}}\left(|-1\rangle\left|\frac{1}{2}\right\rangle + \sqrt{2}\,|0\rangle\left|-\frac{1}{2}\right\rangle\right) \end{aligned}$$

Normalizing, we get

$$\left|\left(1, \frac{1}{2}\right)\frac{3}{2}, -\frac{1}{2}\right\rangle = \sqrt{\frac{1}{3}}\,|-1\rangle\left|\frac{1}{2}\right\rangle + \sqrt{\frac{2}{3}}\,|0\rangle\left|-\frac{1}{2}\right\rangle$$

Lastly

$$\left|\left(1,\frac{1}{2}\right)\frac{3}{2},-\frac{3}{2}\right\rangle = |-1\rangle\left|-\frac{1}{2}\right\rangle$$

3. (a) The degeneracy for the n-th level due to the orbital angular momentum is

$$\sum_{l=0}^{n-1}(2l+1) = n^2$$

Including the degree of freedom for spin, the degeneracy is twice this, i.e., $2n^2$.

(b) Combining the orbital angular momentum l and the spin $1/2$, we obtain total angular momentum $l \pm 1/2$ (only $1/2$ for $l = 0$). Denoting the n-th level with the total angular momentum j and orbital angular momentum l as nl_j, we get the following states:

$$\begin{aligned}
&1s_{\frac{1}{2}}\\
&2s_{\frac{1}{2}}, 2p_{\frac{1}{2}}, 2p_{\frac{3}{2}}\\
&3s_{\frac{1}{2}}, 3p_{\frac{1}{2}}, 3p_{\frac{3}{2}}, 3d_{\frac{3}{2}}, 3d_{\frac{5}{2}}\\
&\ldots\\
&ns_{\frac{1}{2}}, np_{\frac{1}{2}}, \cdots, nl_{l-\frac{1}{2}}, nl_{l+\frac{1}{2}}, \cdots, n(n-1)_{n-\frac{1}{2}}
\end{aligned}$$

(c) The states $n(l-1)_{l-\frac{1}{2}}$ (with parity $(-)^{l-1}$) and $nl_{l-\frac{1}{2}}$ (with parity $(-)^{l}$) both have total angular momentum $j = l - \frac{1}{2}$ and are called **parity doublets** (the degeneracy is broken by the so-called radiative corrections).

4. (a) Writing the momentum eigenstate of one particle as $|\vec{p}\rangle$, the transformation of $|\vec{p}\rangle$ under rotation is the same as that of the coordinate eigenstate $|\vec{x}\rangle$. Hence,

$$\hat{\vec{x}} \times \hat{\vec{p}}\,|\vec{p}\rangle = i\hbar\vec{p} \times \frac{d}{d\vec{p}}\,|\vec{p}\rangle$$

Therefore,

$$\begin{aligned}
\hat{\vec{x}} \times \hat{\vec{p}} \int \frac{d^3p}{(2\pi)^3}\phi(\vec{p})\,|\vec{p}\rangle &= \int \frac{d^3p}{(2\pi)^3}\phi(\vec{p}) i\hbar\vec{p} \times \frac{d}{d\vec{p}}\,|\vec{p}\rangle\\
&= \int \frac{d^3p}{(2\pi)^3}(-i)\hbar\vec{p} \times \left(\frac{d}{d\vec{p}}\phi(\vec{p})\right)|\vec{p}\rangle
\end{aligned}$$

Since the transformation of $|\vec{p}\rangle\,|-\vec{p}\rangle$ is the same as that of $|\vec{p}\rangle$, we obtain

$$\hat{\vec{J}}\int\frac{d^3p}{(2\pi\hbar)^3}f(\vec{p})\left|\pi^+(\vec{p})\right\rangle\left|\pi^0(-\vec{p})\right\rangle$$
$$=\int\frac{d^3p}{(2\pi)^3}\left(-i\hbar\vec{p}\times\frac{\partial}{\partial\vec{p}}f(\vec{p})\right)\left|\pi^+(\vec{p})\right\rangle\left|\pi^0(-\vec{p})\right\rangle$$

(b) The wave function of a particle with angular momentum 1 and the z-component 1 has the form $(x+iy)\phi(r)$ (since $rY_{1,1}=x+iy$). Therefore, replacing $\vec{x}$ by $\vec{p}$, we obtain

$$f(\vec{p})=(p_x+ip_y)g(p)$$

(c) The angular distribution is proportional to $|f(\vec{p})|^2d\Omega$. Therefore, from $p_x^2+p_y^2=p^2\sin^2\theta$, the angular distribution is proportional to

$$\sin^2\theta d\cos\theta d\phi=(1-\cos^2\theta)d\cos\theta d\phi$$

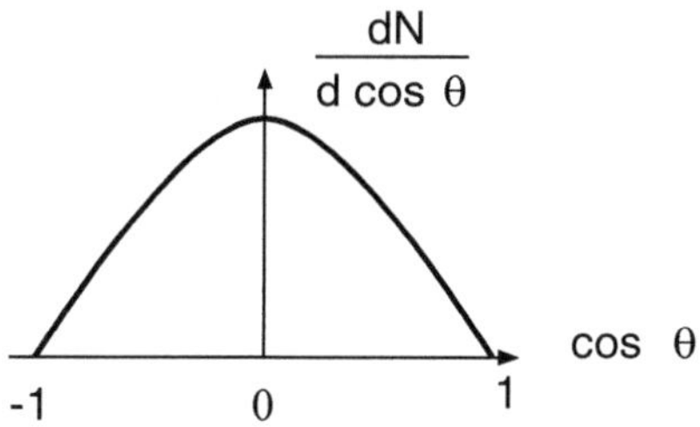

In other words, the probability of π mesons emerging perpendicular to the z-axis is high.

5. (a)

$$\langle J,M|\,\hat{\vec{J}}\cdot\hat{\vec{S}}\,|J,M\rangle=c(J,L,M)\,\langle J,M|\,\hat{\vec{J}}^{\,2}\,|J,M\rangle$$
$$=c(J,L,M)\hbar^2J(J+1)$$

(b) From $\vec{L}^{\,2}=\vec{J}^{\,2}+\vec{S}^{\,2}-2\vec{J}\cdot\vec{S}$, we obtain

$$\langle J,M|\,\hat{\vec{J}}\cdot\hat{\vec{S}}\,|J,M\rangle=\frac{\hbar^2}{2}\left(J(J+1)+S(S+1)-L(L+1)\right)$$

Thus

$$c(J,L,M)=\frac{J(J+1)+S(S+1)-L(L+1)}{2J(J+1)}$$

(c)

$$g = 1 + c(J, L, M) = \frac{3}{2} + \frac{(S-L)(S+L+1)}{2J(J+1)}$$

(d) Since

$$\hat{L}_z + 2\hat{S}_z = \hat{J}_z + \hat{S}_z$$

we obtain

$$\mu_B B \langle J, M| \frac{1}{\hbar} \left(\hat{L}_z + 2\hat{S}_z\right) |J, M\rangle$$
$$= \mu_B B \langle J, M| \frac{1}{\hbar} \left(\hat{J}_z + \hat{S}_z\right) |J, M\rangle = g\mu_B BM$$

Exercises 9

1. **Spin $\frac{1}{2}$**—The angular momentum operator $\hat{J}_i$ $(i = x, y, z)$ is introduced as an operator generating rotations. A rotation of a small angle ϵ around the i-axis is given by

$$\hat{R}_i(\epsilon) = 1 - i\frac{\epsilon}{\hbar}\hat{J}_i$$

(a) Show that a rotation by a finite angle ϕ is given by

$$\hat{R}_i(\phi) = \mathrm{e}^{-i\frac{\phi}{\hbar}\hat{J}_i}$$

Hint: Differentiate the above equation with respect to ϕ and check its validity.

(b) For the states $|\psi\rangle$ with spin $j = 0, \frac{1}{2}, 1, \cdots$, find $\hat{R}_i(2\pi)|\psi\rangle$.
Comment: This explains why the magnitude of the orbital angular momentum must be an integral multiple of $\hbar$.

Let us consider the case of spin $\frac{1}{2}$, where the normalized up-spin state is $|\uparrow\rangle$, and the normalized down-spin state is $|\downarrow\rangle \equiv \frac{1}{\hbar}(\hat{J}_x - i\hat{J}_y)|\uparrow\rangle$.

(c) Explain geometrically that the spin state pointing in the direction of an arbitrary three-dimensional unit vector $\hat{e} \equiv (\sin\theta\cos\phi, \sin\theta\sin\phi, \cos\theta)$ is given by

$$|\hat{e}\rangle = \hat{R}_z(\phi)\hat{R}_y(\theta)|\uparrow\rangle$$

and give it as a linear combination of $|\uparrow\rangle$ and $|\downarrow\rangle$.

(d) Consider the up-spin state $|\uparrow\rangle$. Calculate the probability of the state being $|\pm\hat{e}\rangle$.

2. **Matrix elements of** $\hat{J}_{\pm}$—Given the angular momentum operator $\hat{\vec{J}}$, we define the raising and lowering operators by

$$\hat{J}_{\pm} \equiv \hat{J}_x \pm i\,\hat{J}_y$$

We introduce a normalized state $|J\rangle$ (where $J = 0, \frac{1}{2}, 1, \cdots$) that satisfies the two conditions

$$\hat{J}_z\,|J\rangle = \hbar J\,|J\rangle\,, \quad \hat{J}_+\,|J\rangle = 0$$

We then define the normalized state $|m\rangle$ by

$$|m\rangle \equiv N_m\,\hat{J}_-^{J-m}\,|J\rangle \quad (m = J, J-1, \cdots, -J+1, -J)$$

where $N_m > 0$. These satisfy

$$\hat{J}_z\,|m\rangle = \hbar m\,|m\rangle$$

(a) Derive the following equation:

$$\hat{\vec{J}}^{\,2} = \hat{J}_+\hat{J}_- - \hbar\hat{J}_z + \hat{J}_z^2 = \hat{J}_-\hat{J}_+ + \hbar\hat{J}_z + \hat{J}_z^2$$

(b) Compute the matrix element

$$\langle m|\,\hat{J}_+\hat{J}_-\,|m\rangle$$

using $\langle\, m \mid m\,\rangle = 1$. **Hint**: Rewrite using $\hat{J}_z$ and $\hat{\vec{J}}^{\,2} = \hbar^2 J(J+1)$.

(c) From the result above, compute the matrix elements:

$$\langle m-1|\,\hat{J}_-\,|m\rangle\,, \quad \langle m+1|\,\hat{J}_+\,|m\rangle$$

(d) From the above results, determine the normalization factor N_m.

3. **Spherical harmonics (Part 1)**—Consider the state $|l,l\rangle$ given by the wave function

$$\langle\,\vec{x}|l,l\,\rangle = c_l \times (x+iy)^l f(r) \quad (l = 0, 1, 2, \cdots)$$

where $f(r)$ is a function dependent only on the radius, and it satisfies the normalization condition

$$\int_0^\infty dr\, r^{2(l+1)} |f(r)|^2 = 1$$

(a) Derive the following:

$$\hat{L}_z\,|l,l\rangle = \hbar l\,|l,l\rangle\,, \quad \hat{L}_+\,|l,l\rangle = 0$$

(b) Determine $|c_l|^2$ to normalize $|l,l\rangle$.

The state $|l,m\rangle$ $(-l \le m \le l)$ is defined as

$$|l,m\rangle = N_m \hat{L}_-^{l-m}\,|l,l\rangle$$

where N_m is the positive constant obtained in the previous problem for $J = l$. The spherical harmonics are then defined by

$$Y_{l,m}(\theta,\phi) \equiv \frac{1}{r^l f(r)} \langle\, \vec{x} | l, m\,\rangle$$

The phase of c_l is determined such that

$$Y_{l,0}(0,0) > 0$$

(c) Show that $r^l Y_{l,m}(\theta,\phi)$ is a polynomial of degree l in x, y, z.
Hint: This is obvious for $m = l$ by definition. Show that $\hat{L}_-$ does not change the degree of the polynomial. Express the angular momentum operators $\hat{L}_z, \hat{L}_\pm \equiv \hat{L}_x \pm i\hat{L}_y$ as differential operators in coordinates $z, x_\pm \equiv x \pm iy$, acting on the wave functions.

(d) Determine $Y_{l,m}(\theta,\phi)$ for the cases $l = 1, 2, 3$.
Hint: It is simpler to express $r^l Y_{l,m}$ using x, y, z, not the spherical coordinates.

Finally, consider the parity transformation. Under the parity transformation $\hat{x} \to -\hat{x}$, note that the angles transform as

$$\theta \longrightarrow \pi - \theta, \quad \phi \longrightarrow \phi + \pi$$

(e) Express $Y_{l,m}(\pi - \theta, \phi + \pi)$ in terms of $Y_{l,m}(\theta,\phi)$.
Hint: $r^l Y_{lm}(\theta,\phi)$ is an l-th degree polynomial in x, y, z. How does it change under $\vec{x} \to -\vec{x}$?

4. **Spherical harmonics (Part 2)**—We consider how the states $|l,m\rangle$, introduced in the previous problem, transform under an arbitrary rotation.

(a) For $m \neq 0$, show that $\left\langle\, r\hat{z} \mid l, m \,\right\rangle = 0$.

Hint: That is, prove $Y_{l,m}(0, \phi) = 0$. Use the invariance of $r\hat{z}$ under the rotations about the z-axis.

(b) When the spherical coordinates of $\vec{r}$ are (r, θ, ϕ), derive

$$Y_{l,m}(\theta, \phi) = Y_{l,0}(0, 0) \langle l, 0| \hat{R}_y(-\theta)\hat{R}_z(-\phi) |l, m\rangle$$

Hint: Calculate $\langle \vec{r} \mid l, m \rangle$ as follows:

$$\langle \vec{r} \mid l, m \rangle = \langle r\hat{z}| \hat{R}_y(-\theta)\hat{R}_z(-\phi) |l, m\rangle$$

(c) Using $\int d\Omega \sum_{m=-l}^{l} |Y_{l,m}|^2 = 2l + 1$ and $Y_{l,0}(0, 0) > 0$, determine $Y_{l,0}(0, 0)$.

From the above, we obtain

$$\hat{R}_z(\phi)\hat{R}_y(\theta) |l, 0\rangle = \sqrt{\frac{4\pi}{2l+1}} \sum_{m=-l}^{l} Y_{l,m}(\theta, \phi)^* |l, m\rangle$$

5. **Central force**—Consider a particle of mass μ under the influence of a central force potential $V(r)$. Here, r denotes the distance from the center of the potential.

(a) Show that the Hamiltonian $\hat{H}$ commutes with the angular momentum $\hat{\vec{L}}$.
Comment: Therefore, angular momentum is conserved, and it is possible to diagonalize energy and $\hat{\vec{L}}^2$, $\hat{L}_z$ simultaneously.

(b) Show that the wave function of a state with angular momentum l can be written in the form:

$$\psi(\vec{r}) = \frac{\chi(r)}{r} Y_{l,m}(\theta, \phi)$$

where $Y_{l,m}(\theta, \phi)$ is a spherical harmonic.

(c) For the energy eigenstate with energy E, show that $\chi(r)$ satisfies the one-dimensional Schrödinger equation:

$$-\frac{\hbar^2}{2\mu}\frac{d^2}{dr^2}\chi(r) + V_{\text{eff}}(r)\chi(r) = E\chi(r)$$

where the effective potential is defined as

$$V_{\text{eff}}(r) \equiv V(r) + \frac{\hbar^2 l(l+1)}{2\mu r^2}$$

Hint: In spherical coordinates, the Laplacian is expressed as

$$\nabla^2 = \frac{1}{r}\frac{d^2}{dr^2}r + \frac{\hat{\vec{L}}^2}{r^2}\frac{1}{\hbar^2}$$

6. **Scattering of carbon nuclei**—Carbon nuclei are directed at a thin carbon film, inducing the following nuclear reaction:

$$\mathrm{C}^{12} + \mathrm{C}^{12} \longrightarrow \mathrm{Ne}^{20*} + \alpha_1$$

Here, Ne^{20*} is an excited state of neon with spin j. Both the carbon nucleus and the α particle have zero spin. The motion of the neon nucleus can be neglected as it is minimal, and hence its orbital angular momentum can be considered zero.

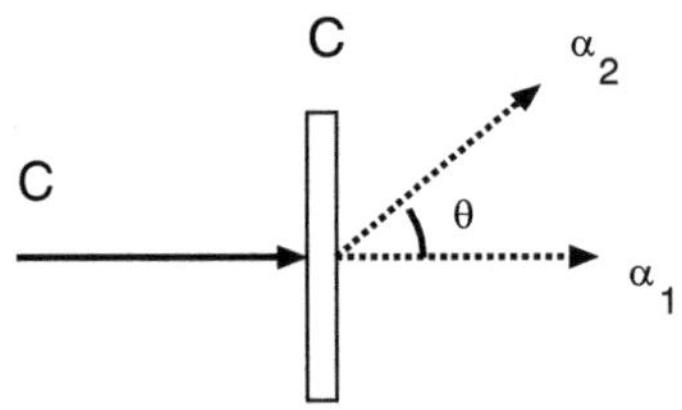

(a) As depicted in the above figure, explain why the angular momentum of the neon nucleus is in the state $|j, 0\rangle$ when an α-particle is detected in the direction of the carbon nucleus.
Hint: Does the state change under the rotations around the z-axis in the direction of α_1?

The produced Ne^{20*} is unstable and decays as

$$\mathrm{Ne}^{20*} \longrightarrow \mathrm{O}^{16} + \alpha_2$$

We assume that the spin of the oxygen nucleus is zero and that its orbital angular momentum is also zero.

(b) Show that the angular distribution of the α particle produced by decay is given by the square of the j-th Legendre polynomial, $P_j(\cos\theta)^2$.
(c) For $j = 0, 1, 2$, plot $P_j(\cos\theta)^2$.
Comment: By measuring this angular distribution, the spin of the neon excited state can be determined.

7. **Spin of the photon**—We represent the state of a single photon with momentum $\vec{p}$ and polarization $\vec{\epsilon}$ (where $|\vec{\epsilon}| = 1$ and $\vec{\epsilon} \cdot \vec{p} = 0$) as

$$|\vec{p}, \vec{\epsilon}\rangle$$

When two unit vectors $\vec{\epsilon}_1$ and $\vec{\epsilon}_2$ are orthogonal to $\vec{p}$ and mutually orthogonal, the polarization vector $\vec{\epsilon}$ can be expressed as a linear combination of $\vec{\epsilon}_1$ and $\vec{\epsilon}_2$:

$$\vec{\epsilon} = \cos\theta \cdot \vec{\epsilon}_1 + \sin\theta \cdot \vec{\epsilon}_2$$

Correspondingly, the photon state can be expressed as a linear combination of $|\vec{p}, \vec{\epsilon}_1\rangle$ and $|\vec{p}, \vec{\epsilon}_2\rangle$:

$$|\vec{p}, \vec{\epsilon}\rangle = \cos\theta\, |\vec{p}, \vec{\epsilon}_1\rangle + \sin\theta\, |\vec{p}, \vec{\epsilon}_2\rangle$$

Thus,

$$|\vec{p}, -\vec{\epsilon}\rangle = -\,|\vec{p}, \vec{\epsilon}\rangle$$

implying that the polarization $\vec{\epsilon}$ and $-\vec{\epsilon}$ are physically indistinguishable.
Under a spatial rotation

$$\vec{r} \longrightarrow R\vec{r}$$

the state of the photon transforms as

$$|\vec{p}, \vec{\epsilon}\rangle \longrightarrow |R\vec{p}, R\vec{\epsilon}\rangle$$

(a) Take the z-axis in the direction of momentum, and construct the eigenstate of $\hat{J}_z$ as a superposition of $|p\hat{z}, \hat{x}\rangle$ and $|p\hat{z}, \hat{y}\rangle$. List the possible eigenvalues. **Hint**: Complex coefficients are required.

(b) Show that the following two-photon state is invariant under the rotations about the z-axis:

$$|\psi\rangle \equiv |p\hat{z}, \hat{x}\rangle\, |-p\hat{z}, \hat{x}\rangle + |p\hat{z}, \hat{y}\rangle\, |-p\hat{z}, \hat{y}\rangle$$

(c) Let $\vec{\epsilon}$ be a unit vector orthogonal to $\hat{z}$. When one of the two photons in state $|\psi\rangle$ is polarized in the direction of $\vec{\epsilon}$, show that the other photon is also polarized in the direction of $\vec{\epsilon}$.
Hint: Show that $|\psi\rangle$ can be rewritten as

$$|\psi\rangle = |p\hat{z}, \vec{\epsilon}\rangle\, |-p\hat{z}, \vec{\epsilon}\rangle + |p\hat{z}, \hat{z} \times \vec{\epsilon}\rangle\, |-p\hat{z}, \hat{z} \times \vec{\epsilon}\rangle$$

(d) Show that the following two-photon state is also invariant under the rotations about the z-axis.

$$|\phi\rangle \equiv |p\hat{z}, \hat{x}\rangle\, |-p\hat{z}, \hat{y}\rangle + |p\hat{z}, \hat{y}\rangle\, |-p\hat{z}, \hat{x}\rangle$$

(e) Suppose that one of the two photons in $|\phi\rangle$ is polarized in the direction of $\vec{\epsilon}$. Show that the other photon is polarized in the direction of $\hat{z} \times \vec{\epsilon}$.

Now, under the reflection with the zx plane as the mirror, the photon states transform as

$$|\pm p\hat{z}, \hat{x}\rangle \longrightarrow |\pm p\hat{z}, \hat{x}\rangle$$
$$|\pm p\hat{z}, \hat{y}\rangle \longrightarrow |\pm p\hat{z}, -\hat{y}\rangle$$

(f) Find the mirror reflection of $|\psi\rangle$ and $|\phi\rangle$.

(g) The π^0 meson is a spin-zero particle that decays into two photons. Given that the polarization vectors of the two photons are always orthogonal to each other in the rest frame of the π^0 meson, determine the parity of the π^0. (Assume parity is conserved.)
Hint: The π^0 is invariant under rotations. If its parity is $\pm$, its state changes by $\pm$ under the mirror reflection. Note that the mirror reflection combined with the 180° rotation around the y-axis gives parity (Fig. 4.9).

4.3 Hydrogen Atom

Three Points to Remember

1. The energy eigenstate of the electron in a hydrogen atom is characterized by three quantum numbers: the principal quantum number n, the orbital angular momentum l, and the magnetic quantum number m. The energy depends only on the principal quantum number as

$$E_n = -\frac{Ry}{n^2} \quad (n = 1, 2, \cdots)$$

where $Ry = hcR \equiv \frac{1}{2}\alpha^2 mc^2 \simeq 13.6\,\text{eV}$ is the Rydberg energy.
2. The periodic table of elements results from the quantization of angular momentum and Pauli's exclusion principle.
3. Particles with an integer spin follow Bose-Einstein statistics, and those with a half-integer spin follow Fermi-Dirac statistics.

4.3.1 The Energy Eigenstate of the Electron in a Hydrogen Atom Is Characterized by Three Quantum Numbers: n, l, m

The Hamiltonian of the electron in a hydrogen atom is given by

$$\hat{H} = \frac{\hat{\vec{p}}^{\,2}}{2m} - \frac{e^2}{\hat{r}} \tag{4.166}$$

where m is the electron's mass and r is the distance between the electron and the atomic nucleus (proton). The Hamiltonian can also be expressed as a differential operator acting on the wave function:

$$\hat{H} = -\frac{\hbar^2}{2m}\nabla^2 - \frac{e^2}{r} \tag{4.167}$$

Strictly speaking, m is not the electron mass m_e, but is the reduced mass of the electron and the proton. Since $\frac{m_e}{m_p} \simeq \frac{1}{2000}$,

$$m \equiv \frac{m_e m_p}{m_e + m_p} = \frac{m_e}{1 + \frac{m_e}{m_p}} \simeq m_e \tag{4.168}$$

The Hamiltonian is invariant under rotations, and it commutes with the orbital angular momentum operator $\hat{\vec{L}}$:

$$[\hat{H}, \hat{\vec{L}}] = 0 \tag{4.169}$$

Hence, the orbital angular momentum is conserved. As a result, as has been shown in Exercise 9-5, the energy eigenstates are obtained in the form:

$$\psi(\vec{r}) = \frac{\chi(r)}{r} Y_{l,m'}(\theta, \phi) \quad (l = 0, 1, 2, \cdots; -l \le m' \le l) \tag{4.170}$$

where $Y_{l,m'}$ is a spherical harmonic function.[11] For the energy eigenvalue E, the radial function $\chi(r)$ satisfies the one-dimensional Schrödinger equation:

$$-\frac{\hbar^2}{2m}\frac{d^2}{dr^2}\chi(r) + \left(\frac{\hbar^2 l(l+1)}{2mr^2} - \frac{e^2}{r}\right)\chi(r) = E\chi(r) \tag{4.171}$$

[11] We use m' for the magnetic quantum number to distinguish it from the electron mass m.

Here,

$$U_{\mathrm{eff}}(r) \equiv \frac{\hbar^2 l(l+1)}{2mr^2} - \frac{e^2}{r} \quad (l = 0, 1, 2, \cdots) \tag{4.172}$$

is the effective potential, and unlike in classical mechanics the square of the orbital angular momentum is quantized as $\hbar^2 l(l+1)$. Note that the effective potential is independent of the magnetic quantum number m' (the z-component of angular momentum). Therefore, the wave function $\chi(r)$ depends only on the principal quantum number n and the angular momentum l.

Equation (4.171) is too complex as it stands; let us rewrite it in the dimensionless form. First, dividing it by e^2, we obtain

$$\begin{aligned}
&-\frac{1}{2}\frac{\hbar^2}{me^2}\frac{d^2}{dr^2}\chi + \left(\frac{1}{2}\frac{\hbar^2}{me^2}\frac{l(l+1)}{r^2} - \frac{1}{r}\right)\chi = \frac{E}{e^2}\chi \\
&\overset{\times\frac{\hbar^2}{me^2}}{\Longrightarrow} -\frac{1}{2}\left(\frac{\hbar^2}{me^2}\right)^2 \frac{d^2}{dr^2}\chi + \left(\frac{1}{2}\left(\frac{\hbar^2}{me^2}\right)^2 \frac{l(l+1)}{r^2} - \frac{\hbar^2}{me^2}\frac{1}{r}\right)\chi \\
&= \frac{\hbar^2 E}{me^4}\chi
\end{aligned} \tag{4.173}$$

Define the **Bohr radius** as[12]

$$a \equiv \frac{\hbar^2}{me^2} \simeq 5.29 \times 10^{-9}\,\mathrm{cm} \tag{4.174}$$

and define the **Rydberg energy** as[13]

$$Ry = hcR \equiv \frac{1}{2}m(\alpha c)^2 = \frac{1}{2}mc^2\left(\frac{e^2}{\hbar c}\right)^2 \simeq 13.6\,\mathrm{eV} \tag{4.175}$$

Measuring the radius r in units of a and the energy E in units of Ry:

$$\frac{r}{a} \to r, \quad \frac{E}{Ry} \to E \tag{4.176}$$

we can rewrite (4.171) as

$$\boxed{-\frac{d^2}{dr^2}\chi + \left(\frac{l(l+1)}{r^2} - \frac{2}{r}\right)\chi = E\chi} \tag{4.177}$$

---

[12] For $m = m_e$, a should be denoted as a_∞, corresponding to $m_p \to \infty$.

[13] R is the **Rydberg constant**. For $m = m_e$ it should be written as R_∞.

The dimensionless quantity in Eq. (4.175)

$$\boxed{\alpha \equiv \frac{e^2}{\hbar c} = \frac{e_{\text{SI}}^2}{4\pi\epsilon_0 \hbar c} \simeq \frac{1}{137}} \tag{4.178}$$

is called the **fine structure constant**, an important physical quantity worth memorizing. The Rydberg energy is about α^2 times the electron's mass energy because the electron's velocity within the atom is of the order of α times the speed of light c.

Now, Eq. (4.177) is a second-order differential equation with $r = 0$ as a regular singular point. (This is a differential equation of Fuchs type.) Therefore, near the origin, the solution can be obtained in the form of a power series:

$$\chi(r) = r^p g(r), \quad \text{where } g(r) = 1 + \sum_{n=1}^{\infty} g_n r^n \tag{4.179}$$

The indicial equation for the power p is

$$p(p-1) = l(l+1) \Longrightarrow p = l+1, -l \tag{4.180}$$

The probability density $|\psi|^2 = \frac{|\chi|^2}{r^2} Y_{l,m'}^2$ must be integrable near the origin, so we must have

$$p = l + 1 \tag{4.181}$$

Therefore,

$$\boxed{\chi(r) = r^{l+1} g(r)} \Longrightarrow \frac{1}{r}\chi(r) \propto r^l \tag{4.182}$$

From Eq. (4.179), we obtain

$$\chi'' = r^{l+1}\left(g'' + 2(l+1)\frac{1}{r}g' + l(l+1)\frac{1}{r^2}g\right) \tag{4.183}$$

Thus, Eq. (4.177) gives

$$-g'' - 2(l+1)\frac{1}{r}g' - l(l+1)\frac{1}{r^2}g + \left(l(l+1)\frac{1}{r^2} - \frac{2}{r}\right)g = Eg, \tag{4.184}$$

which can be rewritten as

$$g'' + \frac{2(l+1)}{r}g' + \left(\frac{2}{r} + E\right)g = 0 \tag{4.185}$$

This differential equation can be solved for any E. Under the condition $g(0) = 1$, the solution is uniquely determined. (Since it is a second-order differential equation, there are two linearly independent solutions, but the one diverging at the origin is discarded.) Since we are interested in the bound states of the electron, we take $E < 0$. For the wave function to be normalizable, $g(r)$ must satisfy

$$\int_0^\infty dr\, \chi(r)^2 = \int_0^\infty dr\, r^{2(l+1)} g(r)^2 < \infty \tag{4.186}$$

This condition restricts the allowed values of E, as we will show shortly.

Before starting detailed calculations, let us examine the asymptotic behavior of $g(r)$ as $r \to \infty$. Equation (4.185) becomes asymptotically

$$g'' + Eg = 0 \tag{4.187}$$

and therefore

$$g(r) \sim \mathrm{e}^{\pm\sqrt{-E}\,r} \tag{4.188}$$

For bound states, $g(r) \sim e^{-\sqrt{-E}\,r}$, i.e., $g(r)$ approaches zero exponentially as $r \to \infty$.

Now, let us define $f(r)$ by

$$g(r) = \mathrm{e}^{-\sqrt{-E}\,r} f(r) \tag{4.189}$$

Then, we obtain

$$g' = \mathrm{e}^{-\sqrt{-E}\,r}\left(f' - \sqrt{-E}\,f\right) \tag{4.190}$$

$$g'' = \mathrm{e}^{-\sqrt{-E}\,r}\left(f'' - 2\sqrt{-E}\,f' - Ef\right) \tag{4.191}$$

Thus, the differential equation for f becomes[14]

$$rf'' + 2(l + 1 - r\sqrt{-E})f' + 2(1 - (l+1)\sqrt{-E})f = 0 \tag{4.192}$$

Substituting the power series

$$f(r) = 1 + \sum_{n=1} f_n r^n \tag{4.193}$$

into the differential equation, we obtain

[14] This is Kummer's differential equation, and the solutions are confluent hypergeometric functions.

$$\sum_n f_n n(n-1)r^{n-1} + 2(l+1)\sum_n n f_n r^{n-1} - 2\sqrt{-E}\sum_n n f_n r^n$$
$$+ 2(1-(l+1)\sqrt{-E})\sum_n f_n r^n = 0 \tag{4.194}$$

from which we get the recursion relation

$$f_{n+1}(n+1)(n+2(l+1)) = 2f_n(\sqrt{-E}(n+l+1)-1) \tag{4.195}$$

The ratio of the successive coefficients satisfies

$$\frac{f_{n+1}}{f_n} = \frac{2(\sqrt{-E}(n+l+1)-1)}{(n+1)(n+2(l+1))} \overset{n\to\infty}{\longrightarrow} \frac{2\sqrt{-E}}{n} \tag{4.196}$$

Therefore, if $f(r)$ is an infinite series, it must behave asymptotically as

$$f(r) \overset{r\gg 1}{\sim} \mathrm{e}^{2\sqrt{-E}\,r} \Longrightarrow g \sim \mathrm{e}^{\sqrt{-E}\,r} \tag{4.197}$$

Hence, to obtain a normalizable bound state, $f(r)$ must be a finite-degree polynomial. For this to occur, the right-hand side of Eq. (4.195) must vanish at some $n' = 0, 1, 2, \cdots$:

$$\sqrt{-E}\,\underbrace{(n'+l+1)}_{\equiv n} = 1 \tag{4.198}$$

Then, $f(r)$ is an n'-degree polynomial. We define the **principal quantum number** n by

$$n \equiv n' + l + 1 \tag{4.199}$$

Let us summarize the results obtained so far. Given an orbital angular momentum $l = 0, 1, 2, \cdots$, the energy eigenvalues are determined by the principal quantum number $n \geq l+1$ as

$$\boxed{E_n = -\frac{Ry}{n^2}} \tag{4.200}$$

The wave function of the energy eigenstates is given in spherical coordinates as

$$\psi_{n,l,m'}(\vec{r}) = \frac{1}{r}\chi_{n,l}(r)Y_{l,m'}(\theta,\phi) \tag{4.201}$$

where

$$R_{n,l}(r) \equiv \frac{1}{r}\chi_{n,l}(r) = r^l F_{n,l}(r) \mathrm{e}^{-\frac{r}{n}} \tag{4.202}$$

with $F_{n,l}(r)$ being an $(n-(l+1))$-degree polynomial $F_{n,l}(r) = \sum_{k=0}^{n-(l+1)} f_k r^k$. The ratio of the successive coefficients is given by

$$\frac{f_{k+1}}{f_k} = \frac{2\left(\frac{k+l+1}{n} - 1\right)}{(k+1)(k+2(l+1))} = \frac{2}{n}\frac{k+l+1-n}{(k+1)(k+2(l+1))} \tag{4.203}$$

The normalization condition for the wave function is given by

$$\int_0^\infty dr\, \chi_{n,l}(r)\chi_{n',l}(r) = \delta_{n,n'} \tag{4.204}$$

Let us examine the cases of orbital angular momentum $l = 0, 1, 2$.

S-Wave ($l = 0$)

$$\chi_{10} = 2r\mathrm{e}^{-r}$$
$$\chi_{20} = \frac{1}{\sqrt{2}} r \left(1 - \frac{r}{2}\right) \mathrm{e}^{-\frac{r}{2}}$$
$$\chi_{30} = \frac{2}{3\sqrt{3}} r \left(1 - \frac{2}{3}r + \frac{2}{27}r^2\right) \mathrm{e}^{-\frac{r}{3}}$$

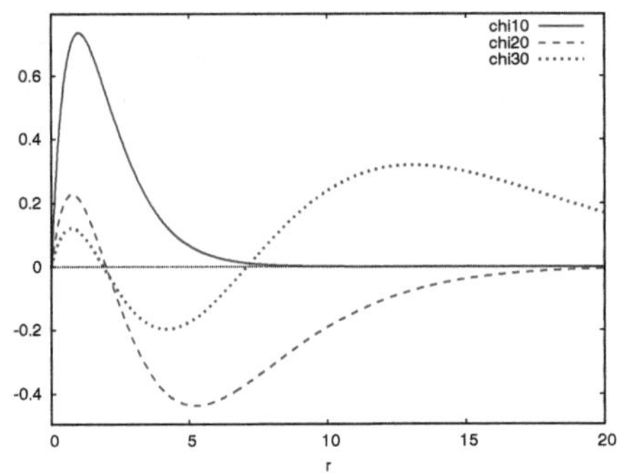

P-Wave ($l = 1$)

$$\chi_{21} = \frac{1}{2\sqrt{6}} r^2 \mathrm{e}^{-\frac{r}{2}}$$
$$\chi_{31} = \frac{8}{27\sqrt{6}} r^2 \left(1 - \frac{r}{6}\right) \mathrm{e}^{-\frac{r}{3}}$$
$$\chi_{41} = \frac{\sqrt{5}}{16\sqrt{3}} r^2 \left(1 - \frac{r}{4} + \frac{r^2}{80}\right) \mathrm{e}^{-\frac{r}{4}}$$

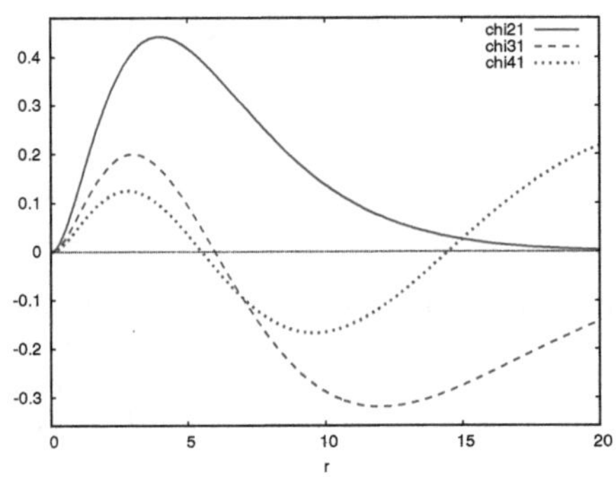

D-Wave ($l = 2$)

$$\chi_{32} = \frac{4}{81\sqrt{30}} r^3 e^{-\frac{r}{3}}$$
$$\chi_{42} = \frac{1}{64\sqrt{5}} r^3 \left(1 - \frac{r}{12}\right) e^{-\frac{r}{4}}$$

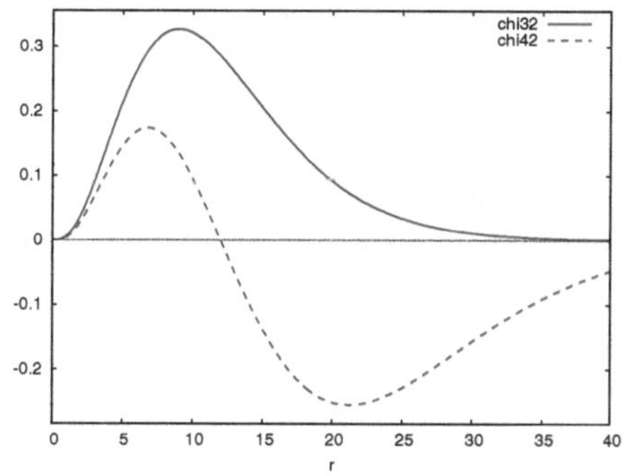

Examining the dependence of the wave function on the principal quantum number n for each angular momentum l, we find that **the number of nodes in the wave function increases by one for each increase in the principal quantum number** n. This is a consequence of the orthogonality of the wave functions.

Furthermore, the allowed angular momenta for each principal quantum number n are given by

$$\boxed{l = 0, 1, \cdots, n-1} \tag{4.205}$$

Thus, the degeneracy (total number of energy eigenstates) of the energy level E_n is given by

$$\sum_{l=0}^{n-1} (2l + 1) = n^2 \tag{4.206}$$

4.3.2 The Periodic Table of Elements Results from the Quantization of Angular Momentum and Pauli's Exclusion Principle

Consider an atom with atomic number Z. There are Z electrons, and the potential felt by each electron is the sum of the potentials created by the nucleus and the other $Z - 1$ electrons. Roughly speaking, this potential can be approximated by

Table 4.2 Energy eigenstates of the electron in a hydrogen atom

Principal quantum number	Angular momentum states	Degeneracy
$n = 1$	1s	1
$n = 2$	2s, 2p	$1 + 3 = 4$
$n = 3$	3s, 3p, 3d	$1 + 3 + 5 = 9$
$n = 4$	4s, 4p, 4d, 4f	$1 + 3 + 5 + 7 = 16$
n	$l = 0, 1, \cdots, n-1$	n^2

a spherically symmetric potential.[15] This average potential is common to all the electrons.

Now consider the energy eigenstates of electrons under this average potential. Since it is a central force potential, the orbital angular momentum is conserved, and the eigenstates are described by the magnitude of the angular momentum l and the magnetic quantum number m. The effective potential depends on l. Here, the principal quantum number $n = 0, 1, 2, \cdots$ is introduced such that $n - (l + 1)$ represents the number of nodes of the wave function excluding the origin. In the case of the Coulomb potential discussed in the previous section, the energy eigenvalue happens to depend only on n, but this is exceptional; generally, the eigenvalues depend on both the principal quantum number n and the orbital angular momentum l. For a given l, higher n corresponds to higher energy, and for a given n, higher l corresponds to higher energy.

According to Pauli's exclusion principle, at most two electrons with opposite spins can occupy the same orbital state.[16] Therefore, by considering the quantum numbers of the highest energy electron, we can classify the elements and obtain the following periodic table:

Table 4.3 The periodic table classified by the quantum number (nl) of the electron with the highest energy

n											
1	s										s
2	s	s				p	p	p	p	p	p
3	s	s	← transition elements		→	p	p	p	p	p	p
4	s	s	3d	···	3d	p	p	p	p	p	p
5	s	s	4d	···	4d	p	p	p	p	p	p

As you all know, this table was originally introduced by Mendeleev in the nineteenth century. Its understanding awaited quantum mechanics. This table is called the **periodic table** because the chemical properties of elements are determined by the properties of the outermost electrons in the orbitals. Hence, we find:

1. Elements belonging to the same column have similar chemical properties.
2. Transition elements belonging to the same row have similar chemical properties.

For the transition elements, the wave functions of the s state electrons extend in space more than those of the d state electrons. This results in similar chemical properties among them.

---

[15] A more rigorous treatment of this concept is given by the Hartree-Fock approximation. Detailed explanations are found in H. Bethe, R. Jackiw, *Intermediate Quantum Mechanics* (Westview Press).

[16] In reality, the energy depends slightly on the relative orientation of the orbital and spin angular momenta.

4.3.3 *Particles with an Integer Spin Follow Bose-Einstein Statistics, and Those with a Half-Integer Spin Follow Fermi-Dirac Statistics*

All particles have spin (or intrinsic) angular momentum in addition to orbital angular momentum. Classically, only extended particles such as molecules can rotate, but in quantum mechanics, even point-like particles such as electrons can have spin. There are two types of particles, those following Bose-Einstein statistics and those following Fermi-Dirac statistics. When there are more than two identical particles, the entire wave function must satisfy the following symmetry requirements:

- The wave function of a system consisting of $N \geq 2$ identical Bose particles must be **symmetric** with respect to the interchange of two particles.
- The wave function of a system consisting of $N \geq 2$ identical Fermi particles must be **antisymmetric** with respect to the interchange of two particles.

Let us elaborate a little on this. Consider the wave function of two particles in three dimensions $\psi_n(\vec{r}_1, s_1; \vec{r}_2, s_2)$, where $s_{1,2} = \uparrow, \downarrow$ is for spin. Since the particles are indistinguishable, the probability distribution must be the same when two particles are interchanged:

$$|\psi_n(\vec{r}_1, s_1; \vec{r}_2, s_2)|^2 = |\psi_n(\vec{r}_2, s_2; \vec{r}_1, s_1)|^2$$

This implies

$$\psi_n(\vec{r}_2, s_2; \vec{r}_1, s_1) = e^{i\alpha}\psi_n(\vec{r}_1, s_1; \vec{r}_2, s_2)$$

where α is a real phase. By interchanging the particles twice, we must have the same wave function; hence we must find $e^{2i\alpha} = 1$ or $e^{i\alpha} = \pm 1$. Therefore, the wave function must be either symmetric $e^{i\alpha} = 1$ or antisymmetric $e^{i\alpha} = -1$.

The relation between the spin and statistics was first explained by W. Pauli using relativistic quantum field theory[17] (1940).

Example 1: Electrons in Atoms

$$|\text{He}\rangle = \underbrace{|1s\uparrow\rangle}_{\text{first electron}}\ \underbrace{|1s\downarrow\rangle}_{\text{second electron}} - |1s\downarrow\rangle\,|1s\uparrow\rangle \tag{4.207}$$

$$\begin{aligned}|\text{Li}\rangle = {} & |1s\uparrow\rangle\,|1s\downarrow\rangle\,|2s\uparrow\rangle - |1s\downarrow\rangle\,|1s\uparrow\rangle\,|2s\uparrow\rangle \\ & - |1s\uparrow\rangle\,|2s\uparrow\rangle\,|1s\downarrow\rangle + |1s\downarrow\rangle\,|2s\uparrow\rangle\,|1s\uparrow\rangle \\ & + |2s\uparrow\rangle\,|1s\uparrow\rangle\,|1s\downarrow\rangle - |2s\uparrow\rangle\,|1s\downarrow\rangle\,|1s\uparrow\rangle\end{aligned} \tag{4.208}$$

[17] No non-relativistic proof exists for the spin-statistics theorem.

In the Hartree-Fock approximation, the wave function of N electrons can be expressed in the form of a Slater determinant:

$$|\psi\rangle = \det \begin{bmatrix} |1\rangle & |1\rangle & \cdots & |1\rangle \\ |2\rangle & |2\rangle & \cdots & |2\rangle \\ \vdots & \vdots & \cdots & \vdots \\ |N\rangle & |N\rangle & \cdots & |N\rangle \end{bmatrix} \tag{4.209}$$

where the k-th column represents the state of the k-th electron.

Example 2: Hydrogen Molecule

The proton has spin $\frac{1}{2}$. Therefore, the total spin of two protons is either 1 or 0. **Ortho** and **para** denote the states of the orbital part of the wave function.

- **Ortho (for spin** 1)—Spin states are symmetric.

$$|\uparrow\rangle\,|\uparrow\rangle\,, \quad \frac{1}{\sqrt{2}}\left(|\uparrow\rangle\,|\downarrow\rangle + |\downarrow\rangle\,|\uparrow\rangle\right), \quad |\downarrow\rangle\,|\downarrow\rangle \tag{4.210}$$

  Therefore, the orbital part of the wave function must be antisymmetric, thus the name "ortho." The orbital angular momentum l is restricted to odd integers $1, 3, 5, \cdots$. The probability amplitude for two protons to be at the same point is zero.
- **Para (for spin** 0)—Spin states are antisymmetric.

$$\frac{1}{\sqrt{2}}\left(|\uparrow\rangle\,|\downarrow\rangle - |\downarrow\rangle\,|\uparrow\rangle\right) \tag{4.211}$$

  Therefore, the orbital part of the wave function must be symmetric, thus the name "para." The orbital angular momentum l is restricted to even integers $l = 0, 2, 4, \cdots$.

See Fig. 4.10.

The difference in orbital angular momentum affects the specific heat of the hydrogen molecular gas. Historically, the proton's spin $\frac{1}{2}$ was discovered from the specific heat at low temperatures of hydrogen molecular gas. (See Example 10-4.)

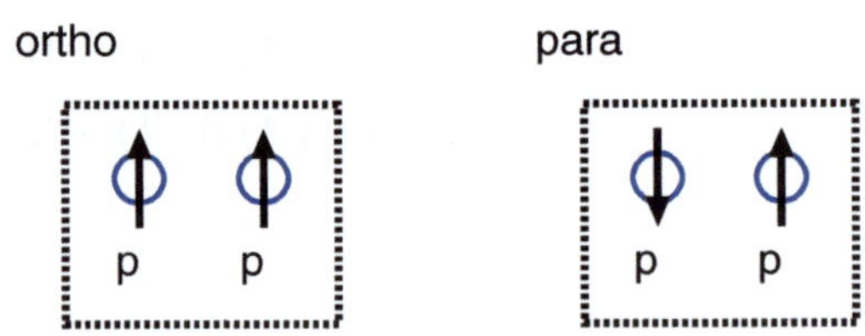

Fig. 4.10 Hydrogen molecule has two states: ortho and para

Example 3: Oxygen Molecule O_2

The oxygen nucleus has zero spin, so the total orbital angular momentum of two nuclei must be even. Therefore, only $l = 0, 2, 4, \cdots$ are allowed.

Examples 10

1. **Virial theorem**—Let us prove the following virial theorem for the energy eigenstates of the electron in a hydrogen atom:

$$\left\langle \frac{e^2}{\hat{r}} \right\rangle = 2 \left\langle \frac{\hat{\vec{p}}^{\,2}}{2m} \right\rangle$$

   (a) Compute the commutator $\left[\hat{H}, \hat{\vec{p}} \cdot \hat{\vec{r}}\right]$ where $\hat{H} = \frac{\hat{\vec{p}}^{\,2}}{2m} - \frac{e^2}{\hat{r}}$.
   (b) Show that the expectation value of the above commutator in any energy eigenstate vanishes, and use it to prove the virial theorem.
   (c) Use the virial theorem to compute the expectation value $\left\langle \frac{1}{\hat{r}} \right\rangle$ in the energy eigenstate with energy $-\frac{Ry}{n^2}$.

   Comment: This virial theorem is the same as the virial theorem in classical mechanics. Refer to Sect. 4.6.2.

2. **Rydberg atom**—Show that the wave function of a state with principal quantum number n and orbital angular momentum $l = n - 1$ can be given by

$$R(r) = \frac{\chi(r)}{r}$$

   where

$$\chi(r) \equiv (\text{Normalization constant}) \cdot r^n \mathrm{e}^{-\frac{r}{n}} \quad (n = 1, 2, \cdots)$$

   in terms of the radius in units of the Bohr radius. Use this result to find the average orbital radius $\langle r \rangle$. **Comment**: States with $n \gg 1$ and $l = n - 1$ have relatively long lifetimes, as roughly estimated in Exercise 10-1; they are often used in atomic physics experiments.

3. **Carbon atom**—In the ground state of carbon, there are two electrons in the $2p$ state. Classify the possible states of these two electrons by total orbital angular momentum L and total spin S.
   Hint: The symmetry of the orbital part of the wave function is determined by L.
   Comment: The energy of the two electrons depends slightly on the values of L and S. According to Hund's rule, the state with "first the largest S, and then the largest L" has the lowest energy.

4. **Specific heat of the hydrogen molecular gas**—The Hamiltonian representing the rotational energy of a hydrogen molecule is given by the moment of inertia I

and the relative orbital angular momentum $\hat{\vec{L}}$ of two hydrogen atoms:

$$\hat{H}_{rot} = \frac{\hat{\vec{L}}^2}{2I}$$

(a) Estimate $\frac{I}{\hbar^2}$ in units of eV^{-1} assuming the distance between two hydrogen atoms as $d \sim 0.74 \times 10^{-8}$cm.
(b) Estimate the minimum temperature required to excite a rotational motion.
(c) Express the partition function for para and ortho hydrogen molecules as an infinite series.
(d) At high temperatures $k_B T \gg \frac{\hbar^2}{I}$, determine the ratio of para and ortho hydrogen molecules in a molecular gas.
Hint: Consider replacing the infinite series from the previous question with an integral. (No need to compute the integral.)
(e) When a high-temperature hydrogen molecular gas is cooled rapidly, transitions between ortho and para molecules take place slowly, and the ratio calculated in the previous question remains. Find the specific heat of the molecular gas at low temperature.
Hint: At low temperatures the partition function can be approximated by keeping the two lowest energy states: $Z(T) \simeq N_1 e^{-\frac{E_1}{k_B T}} + N_2 e^{-\frac{E_2}{k_B T}}$, where $E_2 > E_1$ and $N_{1,2}$ is the number of states. We then obtain the approximate specific heat as $C(T) \simeq \frac{N_2}{N_1} \frac{(E_2 - E_1)^2}{k_B T^2} e^{-\frac{E_2 - E_1}{k_B T}}$.
Comment: In 1927 D. Dennison used this reasoning to determine the proton spin from the specific heat of a hydrogen molecular gas at low temperatures.

Solutions to Examples 10

1. (a) The Hamiltonian is given by $\hat{H} = \hat{\vec{p}}^{\,2}/(2m) - e^2/\hat{r}$. We find

$$\left[\hat{H}, \hat{\vec{p}} \cdot \hat{\vec{r}}\right] = \left[\hat{H}, \hat{\vec{p}}\right] \cdot \hat{\vec{r}} + \hat{\vec{p}} \cdot \left[\hat{H}, \hat{\vec{r}}\right]$$

$$= -e^2 \left[\frac{1}{\hat{r}}, \hat{\vec{p}}\right] \cdot \hat{\vec{r}} + \hat{\vec{p}} \cdot \left[\frac{\hat{\vec{p}}^{\,2}}{m}, \hat{\vec{r}}\right]$$

$$= i\hbar \left(e^2 \frac{1}{\hat{r}} - \frac{\hat{\vec{p}}^{\,2}}{m}\right)$$

(b) For an energy eigenstate $|E\rangle$ with energy E,

$$\langle E| [\hat{H}, \hat{\vec{p}} \cdot \hat{\vec{r}}] |E\rangle = \langle E| \Big(\underbrace{\hat{H}}_{=E} \hat{\vec{p}} \cdot \hat{\vec{r}} - \hat{\vec{p}} \cdot \hat{\vec{r}} \underbrace{\hat{H}}_{=E} \Big) |E\rangle = 0$$

Therefore,

$$\langle E| \left(\frac{e^2}{\hat{r}} - \frac{\hat{\vec{p}}^{\,2}}{m} \right) |E\rangle = 0$$

which implies

$$\left\langle \frac{e^2}{\hat{r}} \right\rangle = 2 \left\langle \frac{\hat{\vec{p}}^{\,2}}{2m} \right\rangle$$

(c) For the energy eigenstate with energy $E_n = -\frac{Ry}{n^2}$, we find

$$E_n = \left\langle \hat{H} \right\rangle = \left\langle \frac{\vec{p}^{\,2}}{2m} - \frac{e^2}{\hat{r}} \right\rangle = -\frac{1}{2} \left\langle \frac{e^2}{\hat{r}} \right\rangle$$

Hence,

$$\left\langle \frac{1}{\hat{r}} \right\rangle = \frac{2Ry}{e^2 n^2} = \frac{1}{n^2 a}$$

where $a = \frac{e^2}{2Ry}$ is the Bohr radius.

2. The equation satisfied by $\chi_{n,l}$ is

$$-\frac{d^2}{dr^2}\chi_{n,l} + \left(\frac{l(l+1)}{r^2} - \frac{2}{r} \right) \chi_{n,l} = -\frac{1}{n^2}\chi_{n,l}$$

It is easy to check $\chi_{n,n-1} \propto r^n \mathrm{e}^{-\frac{r}{n}}$. Then,

$$\int_0^\infty dr \, \left(r^n \mathrm{e}^{-\frac{r}{n}} \right)^2 = \int_0^\infty dr \, r^{2n} \mathrm{e}^{-\frac{2r}{n}} = \left(\frac{n}{2} \right)^{2n+1} (2n)!$$

Therefore, the normalized wave function is

$$\chi_{n,n-1}(r) = \frac{2^n}{n^{n+1}\sqrt{(2n-1)!}} r^n \mathrm{e}^{-\frac{r}{n}}$$

Hence,

$$\langle \, r \, \rangle = \frac{1}{\left(\frac{n}{2}\right)^{2n+1} (2n)!} \int_0^\infty dr \, r^{2n+1} \mathrm{e}^{-\frac{2r}{n}} = \frac{n}{2}(2n+1) = n\left(n + \frac{1}{2} \right)$$

3. The total angular momentum of the two p electrons must be $S = 1$ if symmetric and $S = 0$ if antisymmetric. The total orbital angular momentum must be $L =$

2, 0 if symmetric and $L = 1$ if antisymmetric. Therefore, the states that are antisymmetric as a whole are ${}^3P$, ${}^1D$, or ${}^1S$ (where the notation ${}^{2S+1}L$ is used). According to Hund's rule, the ground state is the ${}^3P$ state with the largest spin.

4. (a) The moment of inertia is given by $I = \frac{1}{2}m_p d^2$. Therefore,

$$\frac{I}{\hbar^2} = \frac{1}{2}m_p c^2 \frac{d^2}{(\hbar c)^2} \simeq \frac{1}{2} \times 9.4 \times 10^8\,\text{eV}\left(\frac{0.74 \times 10^{-8}\,\text{cm}}{2.0 \times 10^{-5}\,\text{eV}\cdot\text{cm}}\right)^2 \simeq 64\,\text{eV}^{-1}$$

(b) Since the magnitude of angular momentum is of the order of $\hbar$, the magnitude of rotational energy is approximately $\frac{\hbar^2}{I} \simeq \frac{1}{64}$ eV. Using $\frac{1\,\text{eV}}{k_B} \simeq 10^4$ K, the absolute temperature required to excite rotational motions is approximately 150 K.

(c) The orbital angular momentum of para hydrogen is even, and that of ortho hydrogen is odd. Therefore, the partition function for the spin 0 para hydrogen molecule is

$$Z_{\text{para}}(T) = \sum_{n=0}^{\infty}(4n+1)e^{-\frac{n(2n+1)\hbar^2}{Ik_BT}},$$

and that for the spin 1 ortho hydrogen molecule is

$$Z_{\text{ortho}}(T) = 3\sum_{n=0}^{\infty}(4n+3)e^{-\frac{(2n+1)(n+1)\hbar^2}{Ik_BT}}$$

Hence, the specific heat per molecule is given by

$$C_{\text{para}}(T) = T\frac{d^2}{dT^2}\left(k_B T \ln Z_{\text{para}}(T)\right)$$

$$C_{\text{ortho}}(T) = T\frac{d^2}{dT^2}\left(k_B T \ln Z_{\text{ortho}}(T)\right)$$

(d) At high temperatures where $k_B T \gg \frac{\hbar^2}{I}$, the two sums are approximately the same, and we find

$$Z_{\text{para}}(T) \simeq \frac{1}{3}Z_{\text{ortho}}(T)$$

Therefore, the ratio of para to ortho is 1:3. This ratio is kept at low temperatures if the gas is cooled too rapidly.

If you want, we can compute the sum approximately by replacing it by an integral as

$$\sum_{n=0}^{\infty}(4n+1)\mathrm{e}^{-\frac{n(2n+1)\hbar^2}{Ik_BT}} \simeq \sum_{n=0}^{\infty}(4n+3)\mathrm{e}^{-\frac{(2n+1)(n+1)\hbar^2}{Ik_BT}}$$
$$\simeq 4\frac{Ik_BT}{\hbar^2}\underbrace{\int_0^{\infty}dx\,x\mathrm{e}^{-2x^2}}_{=\frac{1}{4}} = \frac{Ik_BT}{\hbar^2}$$

(e) At low temperatures where $k_BT \ll \frac{\hbar^2}{I}$, including only the first excited state, we obtain

$$Z_{\text{para}}(T) \simeq 1 + 5\mathrm{e}^{-\frac{3\hbar^2}{Ik_BT}}$$
$$Z_{\text{ortho}}(T) \simeq 3\left(3\mathrm{e}^{-\frac{\hbar^2}{Ik_BT}} + 7\mathrm{e}^{-\frac{6\hbar^2}{Ik_BT}}\right)$$

Thus, using the hint, we obtain

$$C_{\text{para}}(T) \simeq 5\frac{\left(3\frac{\hbar^2}{I}\right)^2}{k_BT^2}e^{-\frac{3\hbar^2}{Ik_BT}}, \quad C_{\text{ortho}}(T) \simeq \frac{7}{3}\frac{\left(5\frac{\hbar^2}{I}\right)^2}{k_BT^2}e^{-\frac{5\hbar^2}{Ik_BT}}$$

Hence, the specific heat per molecule is

$$C(T) \equiv \frac{1}{4}C_{\text{para}}(T) + \frac{3}{4}C_{\text{ortho}}(T)$$
$$\simeq \frac{1}{4}k_B\left(\frac{\hbar^2}{Ik_BT}\right)^2\left(45\,\mathrm{e}^{-\frac{3\hbar^2}{Ik_BT}} + 175\,\mathrm{e}^{-\frac{5\hbar^2}{Ik_BT}}\right)$$

Plotting $\frac{1}{k_B}C(T)$ against $x \equiv \frac{Ik_BT}{\hbar^2}$ we obtain

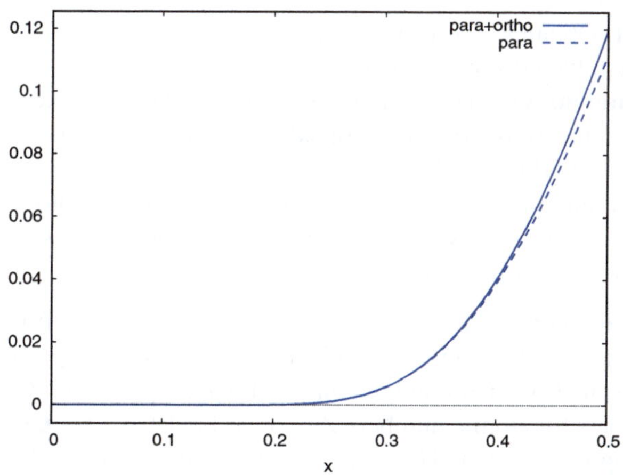

Para hydrogen molecules, which occupy a quarter of the total, contribute dominantly to the specific heat.

Exercises 10

1. **Bohr model**—We consider the classical motion of an electron with mass m in the Coulomb potential of the hydrogen nucleus given by

$$V(r) = -\frac{e^2}{r}$$

The orbit of the electron is assumed to be an ellipse with major axis a and minor axis $b \leq a$.

(a) Show that for the total energy $E < 0$ and orbital angular momentum $M > 0$, the major and minor axes are given by

$$a = \frac{e^2}{-2E}, \quad b = \frac{M}{\sqrt{-2mE}}$$

Hint: It is sufficient to assume an elliptical orbit and express E and M in terms of a and b using the conservation laws of energy and angular momentum. See Example 1-7.

(b) From Bohr's quantization conditions

$$\oint \vec{p} \cdot d\vec{x} = 2\pi n\hbar \quad (n = 1, 2, \cdots)$$

$$\oint M d\theta = 2\pi\hbar k \quad (k = 1, 2, \cdots, n)$$

express E and M in terms of n and k. Here, the integrals are taken over one period.
Hint: Obtain $k \leq n$ from $b \leq a$.

(c) Sketch the allowed orbits for $n = 1, 2, 3$.
Comment: Observe what the S orbit ($k = 1$) looks like.

(d) Show the Rydberg atom considered in Example 10-2 corresponds to the classical orbit with $k = n$.

(e) Assuming that the state $(n, k = n)$ mainly decays into $(n - 1, k = n - 1)$, estimate classically the lifetime of the state with $n = k = 50$.
Hint: Use Larmor's formula.

2. **Alkali elements**—Alkali elements are found in the leftmost column of the periodic table. The outermost S electron, due to shielding effects from the other electrons, feels the Coulomb potential of a hydrogen atom. Calculate the sizes of alkali atoms for atomic numbers $Z = 1, 3, 11, 19, 37, 55$. **Comment**: This is **not** a good approximation. The screening is less complete, and the alkali atoms are smaller than our estimate obtained above.

3. **Nitrogen atom**—In the ground state of a nitrogen atom, three electrons occupy the $2p$ state. Classify the states of the three electrons according to the total angular momentum L and total spin S. Also, using Hund's rule, determine the lowest energy state.
   Hint: First list all the combinations of three different 1-electron states satisfying Pauli's exclusion principle ($m(-l \leq m \leq l)$, $\uparrow, \downarrow$). Infer L, S from the maximum values of the z components L_z, S_z.
   Reference: H. Bethe, R. Jackiw, *Intermediate Quantum Mechanics* (Westview Press)
4. **Wave function of an electron in a hydrogen molecule**—Consider a hydrogen molecule consisting of two hydrogen atoms A, B. If we denote the 1-electron state in the ground state of atom A with an up-spin as $|A\rangle\,|\uparrow\rangle$, for instance, then a state, where atom A has an electron with up-spin and atom B an electron with down-spin, is represented as

$$|A\rangle\,|B\rangle\,|\uparrow\rangle\,|\downarrow\rangle$$

   (a) Enumerate all 2-electron states allowed by Pauli's exclusion principle, and classify them by the total spin.

   The total Hamiltonian, denoting the coordinates of the two hydrogen nuclei as $\vec{r}_A, \vec{r}_B$, is given by

$$\hat{H} = \frac{\hat{\vec{p}}_1{}^2}{2m} + \frac{\hat{\vec{p}}_2{}^2}{2m} - \frac{e^2}{|\hat{\vec{r}}_1 - \vec{r}_A|} - \frac{e^2}{|\hat{\vec{r}}_1 - \vec{r}_B|} - \frac{e^2}{|\hat{\vec{r}}_2 - \vec{r}_A|} - \frac{e^2}{|\hat{\vec{r}}_2 - \vec{r}_B|} + \frac{e^2}{|\vec{r}_A - \vec{r}_B|} + \frac{e^2}{|\hat{\vec{r}}_1 - \hat{\vec{r}}_2|}$$

   where $\vec{r}_{A,B}$ are treated as parameters rather than dynamical variables.

   (b) Taking the nuclear distance as $|\vec{r}_A - \vec{r}_B| \simeq 0.74 \times 10^{-8}$cm, compute the expectation value $\left\langle \hat{H} \right\rangle$ for each state obtained in (a).
   Hint: Evaluate integrals numerically if necessary.
   (c) Which state has the smallest energy expectation value? For this minimum value, determine the binding energy. (The binding energy is defined as the total energy $= -2Ry-$ binding energy.)
   Comment: Minimizing the total energy further with respect to $|\vec{r}_A - \vec{r}_B|$ yields the most probable value.
5. **Coulomb wave functions**—In addition to bound states, the Hamiltonian of the electron in a hydrogen atom

$$\hat{H} = \frac{1}{2m}\hat{\vec{p}}^{\,2} - \frac{e^2}{\hat{r}}$$

   allows for scattering states with positive total energy. Let us find them.

Measuring distances in units of the Bohr radius, the Schrödinger equation for the electron is given by

$$\left(\frac{d^2}{dr^2} - \frac{l(l+1)}{r^2} + \frac{2}{r}\right)\chi(r) = -\epsilon\chi(r)$$

where l is the orbital angular momentum and $\epsilon > 0$ is the energy measured in units of the Rydberg energy.

(a) Under the boundary condition $\chi(0) = 0$, find the solution of the differential equation in the form

$$\chi(r) = r^{l+1}\mathrm{e}^{-i\sqrt{\epsilon}\,r} f(r)$$

Hint: Kummer's differential equation

$$xy'' + (\gamma - x)y' - \alpha y = 0$$

has a solution regular at the origin that is given by the confluent hypergeometric function ${}_1F_1(\alpha, \gamma; x)$:

$${}_1F_1(\alpha, \gamma; x) \equiv \sum_{n=0}^{\infty} \frac{\alpha(\alpha+1)\cdots(\alpha+n-1)}{\gamma(\gamma+1)\cdots(\gamma+n-1)} \frac{x^n}{n!}$$

Express $f(r)$ using the confluent hypergeometric function.

(b) Find the asymptotic form of the wave function $\chi(r)$ as $r \to \infty$.
Comment—Note that the influence of the Coulomb potential persists even at infinity.

6. **Method of obtaining wave functions** $\chi_{n,l}$—Let us develop a method for constructing the radial part of the wave function $\chi_{n,l}(r)$ (principal quantum number n and orbital angular momentum l) from $\chi_{n,n-1}(r)$ obtained in Example 10-2. This method is reminiscent of the creation and annihilation operators of the harmonic oscillator $\hat{a}, \hat{a}^\dagger$ or angular momentum ladder operators $\hat{J}_\pm$. First, recall the result of Example 10-2:

$$\chi_{n,n-1}(r) = \frac{2^n}{n^{n+1}\sqrt{(2n-1)!}} r^n \mathrm{e}^{-\frac{r}{n}}$$

Next, introduce two differential operators:

$$\begin{cases} Q_{n-1} & \equiv \frac{d}{dr} - \frac{\chi'_{n,n-1}(r)}{\chi_{n,n-1}(r)} = \frac{d}{dr} - \frac{n}{r} + \frac{1}{n} \\ Q^\dagger_{n-1} & \equiv -\frac{d}{dr} - \frac{\chi'_{n,n-1}(r)}{\chi_{n,n-1}(r)} = -\frac{d}{dr} - \frac{n}{r} + \frac{1}{n} \end{cases}$$

(a) Show

$$\begin{cases} Q_{n-1}^{\dagger} Q_{n-1} &= H_{n-1} - E_n \\ Q_{n-1} Q_{n-1}^{\dagger} &= H_n - E_n \end{cases}$$

where

$$H_l \equiv -\frac{d^2}{dr^2} + \frac{l(l+1)}{r^2} - \frac{2}{r}, \quad E_n \equiv -\frac{1}{n^2}$$

(b) Show that if $H_l \chi_{n,l} = E_n \chi_{n,l}$, then $H_{l-1} Q_{l-1}^{\dagger} \chi_{n,l} = E_n Q_{l-1}^{\dagger} \chi_{n,l}$.
(c) Assuming $\chi_{n,l}$ is normalized as

$$\int_0^{\infty} dr\, \chi_{n,l}(r)^2 = 1$$

show that $\frac{1}{\sqrt{E_n - E_l}} Q_{l-1}^{\dagger} \chi_{n,l}$ is also normalized.

We have thus shown

$$\chi_{n,l}(r) = \frac{Q_l^{\dagger}}{\sqrt{E_n - E_{l+1}}} \frac{Q_{l+1}^{\dagger}}{\sqrt{E_n - E_l}} \cdots \frac{Q_{n-2}^{\dagger}}{\sqrt{E_n - E_{n-1}}} \chi_{n,n-1}(r)$$

(d) Calculate $\chi_{n,l}$ for $n = 1, 2, 3$.

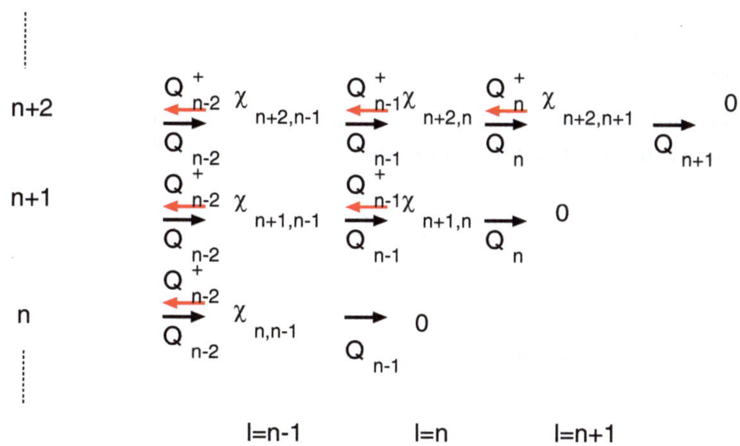

Construction of $\chi_{n,l}$ from the Rydberg state $\chi_{n,n-1}$

4.4 Time-Independent Perturbation Theory

Two Points to Remember

1. When a small time-independent perturbation $\epsilon \hat{V}$ is added to the original Hamiltonian $\hat{H}_0$, the resulting small change in energy eigenvalue is given by the expectation value $\langle \epsilon \hat{V} \rangle$.
2. If there is d-fold degeneracy in $\hat{H}_0$,the small change in the degenerate energy eigenvalue is given by the eigenvalues of a d-dimensional submatrix of $\epsilon \hat{V}$.

4.4.1 *When a Small Time-Independent Perturbation $\epsilon \hat{V}$ Is Added to the Original Hamiltonian, the Resulting Small Change in Energy Eigenvalue Is Given by the Expectation Value $\langle \epsilon \hat{V} \rangle$*

Consider the Hamiltonian given by

$$\hat{H} = \hat{H}_0 + \epsilon \hat{V} \tag{4.212}$$

where $\hat{H}_0$ is a well-understood Hamiltonian and $\epsilon \hat{V}$ is a small perturbation (where $\epsilon \ll 1$ is a small parameter).

Two Examples

- **Anharmonic oscillator**—This is an example discussed in Heisenberg's paper introducing matrix quantum mechanics.

$$\hat{H} = \underbrace{\frac{\hat{p}^2}{2m} + \frac{m\omega^2}{2}\hat{x}^2}_{\hat{H}_0} + \underbrace{g\hat{x}^4}_{\epsilon \hat{V}} \tag{4.213}$$

Estimating the orders of magnitude, we find

$$x^2 \sim \frac{\hbar}{m\omega}, \quad gx^4 \sim g\frac{\hbar^2}{m^2\omega^2} \sim \hbar\omega \cdot \frac{\hbar g}{m^2\omega^3} \tag{4.214}$$

If

$$\epsilon \equiv \frac{\hbar g}{m^2\omega^3} \ll 1 \tag{4.215}$$

we can regard $\epsilon\hat{V} = g\hat{x}^4$ as a small perturbation.

- **Zeeman effect**—The Hamiltonian of the electron in a hydrogen atom under a magnetic field $\vec{B}$ is given by

$$\hat{H} = \frac{\hat{\vec{p}}^2}{2m} - \frac{e^2}{\hat{r}} - \hat{\vec{\mu}} \cdot \vec{B} \tag{4.216}$$

where $\hat{\vec{\mu}}$ is the electron's magnetic moment. $\mu \equiv |\vec{\mu}|$ is approximately the Bohr magneton

$$\mu_B \equiv \frac{e\hbar}{2mc} \simeq 6 \times 10^{-5}\text{eV/Tesla} \tag{4.217}$$

Hence,

$$\epsilon \equiv \frac{\mu_B B}{Ry} \ll 1 \tag{4.218}$$

requires

$$B \ll \frac{Ry}{\mu_B} \simeq \frac{1 \times 14\,\text{eV}}{6 \times 10^{-5}\text{eV/cm}} \simeq 2 \times 10^5\ \text{Tesla} \tag{4.219}$$

Typical experimental magnetic fields range from 1 to 10 Tesla. Hence, the Zeeman effect is usually treated as a perturbation. Let us explore this condition a bit further. Using the Bohr radius $a = \frac{\hbar^2}{me^2}$, we rewrite

$$\epsilon = \frac{Be\hbar}{2mc}\frac{2a}{e^2} = \frac{\pi a^2 B}{\Phi_0} \tag{4.220}$$

where $\Phi_0 \equiv \frac{\pi\hbar c}{e} \simeq 2 \times 10^{-7}$emu is the magnetic flux quantum. Therefore, if the magnetic flux through a circle of Bohr radius is sufficiently small compared with the magnetic flux quantum, the energy due to the magnetic field can be treated as a perturbation.

Now, assuming the eigenstates and eigenvalues of $\hat{H}_0$ are known, let us express the eigenvalues of the total Hamiltonian $\hat{H}$ as a power series in the small parameter ϵ.

We denote the orthonormal basis consisting of eigenstates of $\hat{H}_0$ as $\{|n\rangle\}$:

$$\hat{H}_0\,|n\rangle = E_n^{(0)}\,|n\rangle\,, \quad \langle n \mid m\rangle = \delta_{nm} \tag{4.221}$$

Express the n-th eigenstate of the total Hamiltonian $\hat{H}$ in the form of a power series

$$|\Psi_n\rangle = |n\rangle + \sum_{k=1}^{\infty} \epsilon^k \left|\Psi_n^{(k)}\right\rangle \tag{4.222}$$

Express also its eigenvalues as a power series

$$E_n = E_n^{(0)} + \sum_{k=1}^{\infty} \epsilon^k E_n^{(k)} \tag{4.223}$$

Substituting the power series (4.222) and (4.223) into the eigenvalue equation

$$(\hat{H}_0 + \epsilon \hat{V})\, |\Psi_n\rangle = E_n\, |\Psi_n\rangle \tag{4.224}$$

we obtain

$$\begin{aligned} &\hat{H}_0\, |n\rangle + \epsilon \left\{\hat{V}\, |n\rangle + \hat{H}_0 \left|\Psi_n^{(1)}\right\rangle\right\} + \epsilon^2 \left\{\hat{V} \left|\Psi_n^{(1)}\right\rangle + \hat{H}_0 \left|\Psi_n^{(2)}\right\rangle\right\} + \cdots \\ &= E_n^{(0)}\, |n\rangle + \epsilon \left\{E_n^{(1)}\, |n\rangle + E_n^{(0)} \left|\Psi_n^{(1)}\right\rangle\right\} \\ &\quad + \epsilon^2 \left\{E_n^{(2)}\, |n\rangle + E_n^{(1)} \left|\Psi_n^{(1)}\right\rangle + E_n^{(0)} \left|\Psi_n^{(2)}\right\rangle\right\} + \cdots \end{aligned} \tag{4.225}$$

Comparing the coefficients of each power of ϵ, we obtain

$$\begin{aligned} \hat{V}\, |n\rangle + \hat{H}_0 \left|\Psi_n^{(1)}\right\rangle &= E_n^{(1)}\, |n\rangle + E_n^{(0)} \left|\Psi_n^{(1)}\right\rangle \\ \hat{V} \left|\Psi_n^{(1)}\right\rangle + \hat{H}_0 \left|\Psi_n^{(2)}\right\rangle &= E_n^{(2)}\, |n\rangle + E_n^{(1)} \left|\Psi_n^{(1)}\right\rangle + E_n^{(0)} \left|\Psi_n^{(2)}\right\rangle \\ &\cdots\cdots \\ \hat{V} \left|\Psi_n^{(k-1)}\right\rangle + \hat{H}_0 \left|\Psi_n^{(k)}\right\rangle &= E_n^{(k)}\, |n\rangle + E_n^{(k-1)} \left|\Psi_n^{(1)}\right\rangle + \cdots \\ &\quad + E_n^{(1)} \left|\Psi_n^{(k-1)}\right\rangle + E_n^{(0)} \left|\Psi_n^{(k)}\right\rangle \end{aligned} \tag{4.226}$$

Before proceeding further, we must decide on the normalization of the state $|\Psi_n\rangle$. Since the eigenvalue Eq. (4.224) holds regardless of the normalization of $|\Psi_n\rangle$, we can impose the condition

$$\langle n \mid \Psi_n\rangle = 1 \Longleftrightarrow \left\langle n \mid \Psi_n^{(k)}\right\rangle = 0 \quad (k = 1, 2, \cdots) \tag{4.227}$$

without loss of generality.

Now, let us solve explicitly the equations obtained above.

Order ϵ

The equation to be solved is

$$\hat{V}|n\rangle + \hat{H}_0 \left|\Psi_n^{(1)}\right\rangle = E_n^{(1)}|n\rangle + E_n^{(0)}\left|\Psi_n^{(1)}\right\rangle \tag{4.228}$$

for the unknowns $E_n^{(1)}$ and $\left|\Psi_n^{(1)}\right\rangle$. First, we find the first-order change in eigenvalue $E_n^{(1)}$. To do this, take the inner product of Eq. (4.228) with $|n\rangle$. Using the normalization condition (4.227), we obtain

$$\boxed{\epsilon E_n^{(1)} = \langle n|\,\epsilon\hat{V}\,|n\rangle} \tag{4.229}$$

This equation is the most important result in time-independent perturbation theory, i.e., the first-order change in energy due to the perturbation is the expectation value of the perturbation in the original energy eigenstate.

Example: Anharmonic Oscillator

$$\hat{H} = \underbrace{\frac{\hat{p}^2}{2m} + \frac{m\omega^2}{2}\hat{x}^2}_{\hat{H}_0} + \underbrace{g\hat{x}^4}_{\epsilon\hat{V}} \tag{4.230}$$

It is convenient to introduce the annihilation and creation operators

$$\hat{a} = \frac{i}{\sqrt{2\hbar m\omega}}\,\hat{p} + \sqrt{\frac{m\omega}{2\hbar}}\,\hat{x} \tag{4.231}$$

$$\hat{a}^\dagger = \frac{-i}{\sqrt{2\hbar m\omega}}\,\hat{p} + \sqrt{\frac{m\omega}{2\hbar}}\,\hat{x} \tag{4.232}$$

The unperturbed Hamiltonian is written as

$$\hat{H}_0 = \hbar\omega\left(\hat{a}^\dagger\hat{a} + \frac{1}{2}\right) \tag{4.233}$$

Using

$$\hat{x} = \sqrt{\frac{\hbar}{2m\omega}}\left(\hat{a} + \hat{a}^\dagger\right) \tag{4.234}$$

we obtain

$$\epsilon \hat{V} = g\hat{x}^4 = g\left(\frac{\hbar}{2m\omega}\right)^2 \left(\hat{a} + \hat{a}^\dagger\right)^4 \tag{4.235}$$

In computing the expectation value of $\epsilon \hat{V}$ in the n-th excited state $|n\rangle$, it is important to note that only the terms with the same number of $\hat{a}$ and $\hat{a}^\dagger$ contribute to the expectation value. Therefore, picking up only such terms, we get

$$\begin{aligned} \epsilon \hat{V} \longrightarrow g\left(\frac{\hbar}{2m\omega}\right)^2 \\ \times \left(\hat{a}^{\dagger 2}\hat{a}^2 + \hat{a}^\dagger\hat{a}\hat{a}^\dagger\hat{a} + \hat{a}^\dagger\hat{a}^2\hat{a}^\dagger + \hat{a}\hat{a}^{\dagger 2}\hat{a} + \hat{a}\hat{a}^\dagger\hat{a}\hat{a}^\dagger + \hat{a}^2\hat{a}^{\dagger 2}\right) \end{aligned} \tag{4.236}$$

The matrix elements of $\hat{a}$ and $\hat{a}^\dagger$ have been obtained in the previous section as

$$\langle n-1|\,\hat{a}\,|n\rangle = \langle n|\,\hat{a}^\dagger\,|n-1\rangle = \sqrt{n} \tag{4.237}$$

Thus,

$$\begin{aligned} \langle n|\,\epsilon \hat{V}\,|n\rangle &= g\left(\frac{\hbar}{2m\omega}\right)^2 \big(n(n-1) + n^2 \\ &\quad + n(n+1) + (n+1)n + (n+1)^2 + (n+1)(n+2)\big) \\ &= \hbar\omega\frac{3}{4}\frac{g\hbar}{m^2\omega^3}\left(2n^2 + 2n + 1\right) \ll \hbar\omega \end{aligned} \tag{4.238}$$

In the final expression, the terms dependent on n^2 are due to the nonlinearity of the additional force. The remaining part can be written as

$$\hbar\Delta\omega\left(n + \frac{1}{2}\right)$$

where

$$\Delta\omega \equiv \frac{3}{2}\omega\frac{g\hbar}{m^2\omega^3} \ll \omega$$

This can be interpreted as a small change in the eigenfrequency from ω to $\omega + \Delta\omega$.

We return to a general discussion. We consider the perturbation to the state vector $\left|\Psi_n^{(1)}\right\rangle$. Taking the inner product of Eq. (4.228) with $|m\rangle$ $(m \neq n)$, we get

$$\langle m|\,\hat{V}\,|n\rangle + E_m^{(0)}\left\langle m \mid \Psi_n^{(1)}\right\rangle = E_n^{(0)}\left\langle m \mid \Psi_n^{(1)}\right\rangle \tag{4.239}$$

Hence, we obtain

$$\left\langle m \mid \Psi_n^{(1)} \right\rangle = \frac{1}{E_n^{(0)} - E_m^{(0)}} \langle m| \hat{V} |n\rangle \tag{4.240}$$

Using the normalization condition (4.227), we obtain

$$\boxed{\epsilon \left|\Psi_n^{(1)}\right\rangle = \sum_{m \neq n} |m\rangle \frac{\langle m| \epsilon \hat{V} |n\rangle}{E_n^{(0)} - E_m^{(0)}}} \tag{4.241}$$

For example, in the case of the anharmonic oscillator discussed above, the ground state will mix with $|2\rangle$ and $|4\rangle$, but not with $|1\rangle$ or $|3\rangle$.

Order ϵ^2

The equation to solve is

$$\hat{V} \left|\Psi_n^{(1)}\right\rangle + \hat{H}_0 \left|\Psi_n^{(2)}\right\rangle = E_n^{(2)} |n\rangle + E_n^{(1)} \left|\Psi_n^{(1)}\right\rangle + E_n^{(0)} \left|\Psi_n^{(2)}\right\rangle \tag{4.242}$$

The unknowns are $E_n^{(2)}$ and $\left|\Psi_n^{(2)}\right\rangle$. Similar to the first-order perturbation, we can determine the second-order change in energy by taking the inner product of the above equation with $|n\rangle$:

$$\langle n| \hat{V} \left|\Psi_n^{(1)}\right\rangle = E_n^{(2)} \tag{4.243}$$

Here, we have used $\left\langle n \mid \Psi_n^{(1)} \right\rangle = \left\langle n \mid \Psi_n^{(2)} \right\rangle = 0$ from the normalization condition (4.227). Using Eq. (4.241), we obtain

$$\boxed{\epsilon^2 E_n^{(2)} = \sum_{m \neq n} \frac{\left|\langle m| \epsilon \hat{V} |n\rangle\right|^2}{E_n^{(0)} - E_m^{(0)}}} \tag{4.244}$$

Especially for the ground state $n = 0$, we find

$$\epsilon^2 E_0^{(2)} \leq 0 \tag{4.245}$$

If $\epsilon E_0^{(1)}$ happens to vanish, the ground state energy decreases by perturbation.

Example: Quadratic Stark Effect

Let us consider the ground state energy of the electron in a hydrogen atom under a uniform electric field E. The perturbation Hamiltonian is given by

$$\epsilon \hat{V} = eE\hat{z} \tag{4.246}$$

In the first-order perturbation, the energy remains unchanged:

$$\left\langle 1s \mid \hat{z} \mid 1s \right\rangle = 0 \tag{4.247}$$

This is due to the spherical symmetry of the wave function.

The second-order change is given by

$$\Delta E = \sum_{n \neq 1s} \frac{\left| \langle n | \epsilon \hat{V} \, | 1s \rangle \right|^2}{-Ry - E_n^{(0)}} = e^2 E^2 \sum_{n \neq 1s} \frac{\left| \langle n | \, \hat{z} \, | 1s \rangle \right|^2}{-Ry - E_n^{(0)}} \tag{4.248}$$

Here, the sum over n must include not only bound states but also positive energy states. Since the operator $\hat{z}$ has angular momentum 1, only the p-wave states contribute. Furthermore, since $\hat{z}$ is invariant under rotations about the z-axis, only the states with magnetic quantum number $m = 0$ contribute. Therefore, we obtain

$$\begin{aligned}
\Delta E &= e^2 E^2 \sum_{n=2}^{\infty} \frac{|\left\langle n\text{p}, 0 \mid \hat{z} \mid 1s \right\rangle|^2}{-Ry + \frac{1}{n^2} Ry} + (E_n > 0 \text{ contribution}) \\
&\geq \frac{e^2 E^2}{-\frac{3}{4} Ry} \left(\sum_{n=2}^{\infty} |\left\langle n\text{p}, 0 \mid \hat{z} \mid 1s \right\rangle|^2 + (E_n > 0 \text{ contribution}) \right) \\
&= -\frac{4 e^2 E^2}{3 Ry} \left\langle 1s \mid \hat{z}^2 \mid 1s \right\rangle
\end{aligned} \tag{4.249}$$

This approximation is reasonable because the main contribution comes from $n = 2$.

Using the wave function

$$\phi_{1s}(r) = \frac{2}{a^{\frac{3}{2}}} \mathrm{e}^{-\frac{r}{a}} \frac{1}{\sqrt{4\pi}} \tag{4.250}$$

(where a is the Bohr radius), the average value of $\hat{z}^2$ is calculated as

$$\left\langle 1s \mid \hat{z}^2 \mid 1s \right\rangle = \frac{1}{3} \left\langle 1s \mid \hat{r}^2 \mid 1s \right\rangle \tag{4.251}$$

$$= \frac{4}{3} \frac{1}{a^3} \int_0^{\infty} r^2 dr \, r^2 e^{-2\frac{r}{a}} \tag{4.252}$$

$$= \frac{4}{3}a^2 \underbrace{\int_0^\infty dr\, r^4 e^{-2r}}_{=\frac{3}{4}} \tag{4.253}$$

$$= a^2 \tag{4.254}$$

Thus,

$$\Delta E \geq -\frac{4}{3}\frac{e^2 E^2 a^2}{Ry} = -\frac{4}{3}\left(\frac{eEa}{Ry}\right)^2 Ry = -\frac{8}{3}a^3 E^2 \tag{4.255}$$

(Here, $a^3 \frac{E^2}{8\pi}$ is the energy per volume a^3 of the electric field.)

The right-hand side of Eq. (4.248) can actually be calculated exactly:[18]

$$\Delta E = -\frac{9}{4}a^3 E^2 = -\frac{9}{4}\underbrace{\frac{e^2}{a}}_{2\text{Ry}}\left(\frac{E}{\frac{e}{a^2}}\right)^2$$

The above approximation ($\frac{8}{3}$ vs $\frac{9}{4}$) is found to deviate only by 20%.

4.4.2 If There Is d-Fold Degeneracy in $\hat{H}_0$, the Small Change in the Degenerate Energy Eigenvalue Is Given by the Eigenvalues of a d-Dimensional Submatrix of $\epsilon \hat{V}$

If the unperturbed Hamiltonian $\hat{H}_0$ has degeneracy, the results from the previous section need a slight modification because the degeneracy is often resolved by perturbation. Let $E_n^{(0)}$ be the eigenvalue of $\hat{H}_0$ with d-fold degeneracy. We denote the states by

$$|n, 1\rangle, \cdots, |n, d\rangle .$$

The first-order change in state vectors has been obtained in the previous section as

$$\epsilon \left|\Psi_n^{(1)}\right\rangle = \sum_{m \neq n} |m\rangle \frac{\langle m| \epsilon \hat{V} |n\rangle}{E_n^{(0)} - E_m^{(0)}}, \tag{4.256}$$

[18] An explanation of the method by Dalgarno and Lewis to obtain this result can be found, for example, in L. Schiff, *Quantum Mechanics*, section 33.

Unless

$$\langle n, i|\, \epsilon \hat{V}\, |n, j\rangle = 0 \quad \text{for } i \neq j, \tag{4.257}$$

the vanishing denominator for m degenerate with n results in divergence. This suggests that when applying perturbation theory, the degenerate states must be rearranged such that the d-by-d submatrix of $\epsilon \hat{V}$ is diagonal. Then, the diagonal elements of $\epsilon \hat{V}$ give the first-order change in energy eigenvalues:

$$\boxed{\langle n, i|\, \epsilon \hat{V}\, |n, j\rangle = \epsilon E^{(1)}_{n,i}\, \delta_{i,j}} \tag{4.258}$$

Example: First-Order Stark Effect

We consider again the electron in a hydrogen atom under a uniform electric field E. This time we calculate the change in energy of the $n = 2$ state.

Let us first make an estimate of what we are going to calculate. Since the spread of the wave function is about the Bohr radius a, the electrostatic energy is approximately

$$eEz \sim eEa \tag{4.259}$$

Using Ry $\sim \frac{e^2}{a}$, we obtain

$$\epsilon \sim \frac{eEa}{\frac{e^2}{a}} \sim \frac{E}{\frac{e}{a^2}} \tag{4.260}$$

Here, $\frac{e}{a^2}$ represents the magnitude of the electric field acting on the electron within the atom. If the externally applied electric field is sufficiently small compared with this, it can be considered as a perturbation. Since

$$\frac{e}{a} \simeq 2 \times 13.6\,\text{volt}, \quad a \simeq 5 \times 10^{-9}\,\text{cm} \tag{4.261}$$

we obtain

$$\frac{e}{a^2} \simeq 6 \times 10^9\,\text{Volt/cm} \tag{4.262}$$

This is huge. As long as the electric field E is much smaller than this, the energy of the perturbation is small compared with Ry, and perturbation theory can be used.

Now let us use the formula (4.258) for degenerate perturbation theory. Due to the conservation of angular momentum, the non-zero matrix elements are

$$eE\langle 2p, 0 \mid \hat{z} \mid 2\mathrm{s}\rangle = eE\langle 2\mathrm{s} \mid \hat{z} \mid 2p, 0\rangle \tag{4.263}$$

The wave functions are given by

$$\phi_{2\mathrm{s}}(r) = \frac{1}{a^{\frac{3}{2}}}\frac{1}{\sqrt{2}}\,\mathrm{e}^{-\frac{r}{2a}}\left(1-\frac{r}{2a}\right)\frac{1}{\sqrt{4\pi}} \tag{4.264}$$

$$\phi_{2p,0}(r,\theta) = \frac{1}{a^{\frac{5}{2}}}\frac{r}{2\sqrt{6}}\,\mathrm{e}^{-\frac{r}{2a}}\sqrt{\frac{3}{4\pi}}\cos\theta \tag{4.265}$$

Hence,

$$\begin{aligned}\langle 2p, 0 \mid \hat{z} \mid 2\mathrm{s}\rangle &= \frac{a}{4\sqrt{3}}\int_0^\infty r^2 dr\, r\mathrm{e}^{-r}\left(1-\frac{r}{2}\right)\frac{\sqrt{3}}{4\pi}2\pi\cdot r\cdot\underbrace{\int_{-1}^{1} d\cos\theta\,\cos^2\theta}_{=\frac{2}{3}} \\ &= \frac{a}{12}\underbrace{\int_0^\infty dr\, r^4\left(1-\frac{r}{2}\right)\mathrm{e}^{-r}}_{=4!-5!/2=-36} = -3a\end{aligned} \tag{4.266}$$

Therefore, the eigenvalues of $\epsilon\hat{V}$ are zero for the states $|2p, m = \pm 1\rangle$, and for the states $\frac{|2\mathrm{s}\rangle\pm|2p,0\rangle}{\sqrt{2}}$, they are

$$\boxed{\mp 3\, eEa = \mp 6\,\mathrm{Ry}\,\frac{E}{\frac{e}{a^2}}}$$

See Fig. 4.11.

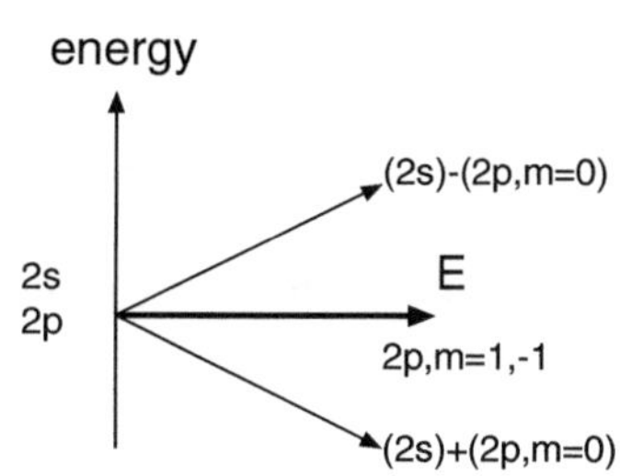

Fig. 4.11 Stark effect

Examples 11

1. **Two-state system**—Consider a system whose Hamiltonian is given by a 2-by-2 hermitian matrix

$$\hat{H} = \hat{H}_0 + \epsilon \hat{V}$$

where

$$\hat{H}_0 \equiv \begin{pmatrix} \epsilon_1 & 0 \\ 0 & \epsilon_2 \end{pmatrix}, \quad \epsilon \hat{V} \equiv \epsilon \begin{pmatrix} 0 & 1 \\ 1 & 0 \end{pmatrix}$$

   (a) In the case of $\epsilon_1 \neq \epsilon_2$, use perturbation theory to find the eigenvalues of $\hat{H}$ up to ϵ^2.
   (b) In the case of $\epsilon_1 = \epsilon_2$, use perturbation theory to find the eigenvalues of $\hat{H}$ up to ϵ.
   (c) Compare the results obtained above with the exact results.

2. **Harmonic oscillator**—For the unperturbed Hamiltonian $\hat{H}_0$ and perturbation energy $\epsilon \hat{V}$ given by

$$\hat{H}_0 \equiv \frac{\hat{p}^2}{2m} + \frac{m\omega_0^2}{2}\hat{x}^2, \quad \epsilon \hat{V} \equiv \frac{m}{2}(\omega^2 - \omega_0^2)\,\hat{x}^2$$

   find the eigenvalues of the Hamiltonian $\hat{H}$ up to the order ϵ^2, and compare them with the exact values. **Hint**—Use $\hat{a}, \hat{a}^\dagger$.

3. **Anharmonic oscillator**—Find the eigenvalues of the Hamiltonian

$$\hat{H} = \frac{\hat{p}^2}{2m} + \frac{m\omega^2}{2}\hat{x}^2 + u\hat{x}^3$$

   up to the order u^2. What are the conditions for using perturbation theory? **Hint**—Use $\hat{a}, \hat{a}^\dagger$.

4. **Zeeman effect**—Consider a hydrogen atom under a uniform magnetic field B in the z direction. The Hamiltonian for the electron, with the Bohr magneton $\mu_B \equiv \frac{e\hbar}{2mc}$ and the spin $\hat{\vec{s}}$, is given by

$$\hat{H} = \frac{1}{2m}\left(\hat{\vec{p}} + \frac{e}{c}\vec{A}(\hat{\vec{r}})\right)^2 - \frac{e^2}{\hat{r}} + \mu_B \frac{2\hat{\vec{s}}}{\hbar} \cdot \vec{B}$$

   We choose the vector potential of the magnetic field as

$$\vec{A}(\vec{r}) = \frac{1}{2}\vec{B} \times \vec{r} = \frac{1}{2}B(-y, x, 0)$$

   (a) How big can B be for perturbation theory to be valid?

(b) Assuming that the condition obtained above is met, show that $\hat{H}$ can be approximated as follows:

$$\hat{H} = \hat{H}_0 + \epsilon \hat{V}$$

where

$$\hat{H}_0 = \frac{1}{2m}\hat{\vec{p}}^{\,2} - \frac{e^2}{\hat{r}}, \quad \epsilon \hat{V} = \mu_B B \frac{1}{\hbar}(\hat{L}_z + 2\hat{s}_z)$$

($\hat{L}_z$ is the z component of orbital angular momentum).

(c) Find the energy shift to the first order of B for states with orbital angular momentum l.

5. **Band gap due to a periodic potential**—Consider a free particle of mass m along the x-axis. The energy of the particle is $\frac{\hbar^2 k^2}{2m}$, and the wave function is

$$\psi_k(x) = \frac{1}{\sqrt{L}}\, e^{ikx}$$

where $L = \int_{-\frac{L}{2}}^{\frac{L}{2}} dx$ is the length of the x-axis, to be taken to infinity. We introduce a weak potential $V(x)$ with period a:

$$V(x + a) = V(x)$$

Find the first-order perturbation energy due to the potential. **Hint**: Be careful with degeneracy when ka is an integer multiple of π.

6. **Effective mass**—We continue on the particle under a periodic potential. If the particle's energy is expressed as a function of wave number k as $E(k)$, we define the effective mass m_{eff} as

$$\frac{\hbar^2}{m_{\text{eff}}} \equiv \frac{d^2}{dk^2} E(k)\bigg|_{k=0}$$

Express m_{eff} up to the second order of V as an infinite series. Which is larger, m or m_{eff}?

7. **More on the band gap**—We continue on Example 11-5. For a small positive ϵ, consider two wave numbers near $ka = n\pi$ ($n = 1, 2, \cdots$):

$$ka = n\pi - \epsilon$$
$$k'a = -n\pi - \epsilon = ka - 2n\pi$$

These two states are almost degenerate and are coupled by the periodic potential $V(x)$.

To construct the Hamiltonian of the system consisting only of these two states, introduce

$$\hat{P} \equiv |k\rangle \langle k| + |k'\rangle\langle k'|$$

(which is called a projection operator), and define the Hamiltonian of this subsystem using the original Hamiltonian $\hat{H} = \frac{\hat{p}^2}{2m} + V(\hat{x})$ as

$$\hat{H}_{\text{eff}} \equiv \hat{P}\hat{H}\hat{P}$$

Diagonalizing this, interpret the results of Example 11-5.

Solutions to Examples 11

1. (a) Let $\begin{pmatrix} 1 \\ 0 \end{pmatrix}$ and $\begin{pmatrix} 0 \\ 1 \end{pmatrix}$ be states $|1\rangle$ and $|2\rangle$, respectively. Then

$$E_1 - \epsilon_1 \simeq \underbrace{\langle 1|\epsilon\hat{V}|1\rangle}_{=0} + \frac{\langle 1|\epsilon\hat{V}|2\rangle^2}{\epsilon_1 - \epsilon_2} = \frac{\epsilon^2}{\epsilon_1 - \epsilon_2}$$

$$E_2 - \epsilon_2 \simeq -\frac{\epsilon^2}{\epsilon_1 - \epsilon_2}$$

(b) To diagonalize $\hat{V}$, states $|1\rangle$ and $|2\rangle$ must be redefined as $1/\sqrt{2}\begin{pmatrix} 1 \\ 1 \end{pmatrix}$ and $1/\sqrt{2}\begin{pmatrix} 1 \\ -1 \end{pmatrix}$, respectively. These are precisely the eigenstates of $\hat{H}$ with eigenvalues $\epsilon_1 \pm \epsilon$.

(c) Using Pauli matrices, the Hamiltonian can be expressed as

$$\hat{H} = \frac{\epsilon_1 + \epsilon_2}{2} + \frac{\epsilon_1 - \epsilon_2}{2}\sigma_z + \epsilon\sigma_x$$

The eigenvalues are

$$\epsilon_\pm \equiv \frac{\epsilon_1 + \epsilon_2}{2} \pm \sqrt{\frac{(\epsilon_1 - \epsilon_2)^2}{4} + \epsilon^2}$$

Therefore, if $\epsilon_1 > \epsilon_2$, expanding up to second order in ϵ, we obtain

$$\begin{cases} \epsilon_+ \simeq \epsilon_1 + \frac{\epsilon^2}{\epsilon_1 - \epsilon_2} \\ \epsilon_- \simeq \epsilon_2 - \frac{\epsilon^2}{\epsilon_1 - \epsilon_2} \end{cases}$$

This approximation is valid as long as

$$\epsilon^2 \ll (\epsilon_1 - \epsilon_2)^2$$

In the degenerate case $\epsilon_1 = \epsilon_2$, we obtain the exact result

$$\epsilon_\pm = \epsilon_1 \pm |\epsilon|$$

Fixing $\epsilon_1 + \epsilon_2$ and varying $\epsilon_1 - \epsilon_2$, we plot $\epsilon_\pm$.

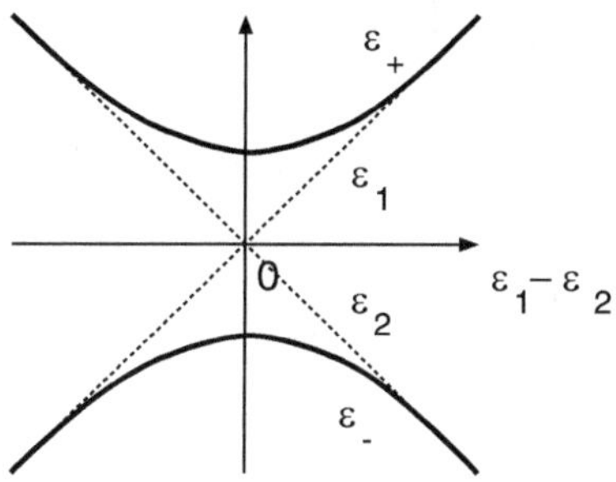

2. The n-th energy eigenvalue is given by

$$E_n = \hbar\omega\left(n + \frac{1}{2}\right) = \hbar\sqrt{\omega_0^2 + (\omega^2 - \omega_0^2)}\left(n + \frac{1}{2}\right)$$
$$\simeq \hbar\left(\omega_0 + \frac{1}{2}\frac{\omega^2 - \omega_0^2}{\omega_0} - \frac{1}{8}\frac{(\omega^2 - \omega_0^2)^2}{\omega_0^3}\right)\left(n + \frac{1}{2}\right)$$

We want to obtain this expansion using perturbation theory. Defining annihilation and creation operators by

$$\begin{cases} \hat{a} \equiv \sqrt{\frac{m\omega_0}{2\hbar}}\,\hat{x} + \frac{i}{\sqrt{2\hbar m\omega_0}}\,\hat{p} \\ \hat{a}^\dagger \equiv \sqrt{\frac{m\omega_0}{2\hbar}}\,\hat{x} - \frac{i}{\sqrt{2\hbar m\omega_0}}\,\hat{p} \end{cases}$$

we obtain

$$\epsilon\hat{V} = \frac{m(\omega^2 - \omega_0^2)}{2}\frac{\hbar}{2m\omega_0}\left(\hat{a} + \hat{a}^\dagger\right)^2 = \frac{\hbar(\omega^2 - \omega_0^2)}{4\omega_0}\left(\hat{a}^2 + 2\hat{a}^\dagger\hat{a} + (\hat{a}^\dagger)^2 + 1\right)$$

Hence, we obtain

$$\langle n|\epsilon\hat{V}|n\rangle = \frac{\hbar(\omega^2 - \omega_0^2)}{2\omega_0}\left(n + \frac{1}{2}\right)$$

and

$$\sum_{l \neq n} \frac{\langle n|\epsilon \hat{V}|l\rangle \langle l|\epsilon \hat{V}|n\rangle}{(n-l)\hbar\omega_0}$$

$$= \left(\frac{\hbar(\omega^2-\omega_0^2)}{4\omega_0}\right)^2 \frac{1}{2\hbar\omega_0} \left(\underbrace{\langle n-2|\hat{a}^2|n\rangle^2}_{=(n-1)n} - \underbrace{\langle n+2|(\hat{a}^\dagger)^2|n\rangle^2}_{=(n+2)(n+1)} \right)$$

$$= -\hbar \frac{(\omega^2-\omega_0^2)^2}{8\omega_0^3} \left(n+\frac{1}{2}\right)$$

Using these, the n-th eigenvalue is approximately

$$E_n \simeq \hbar \left(\omega_0 + \frac{\omega^2-\omega_0^2}{2\omega_0} - \frac{(\omega^2-\omega_0^2)^2}{8\omega_0^3} \right) \left(n+\frac{1}{2}\right)$$

as expected.

3. Since $x \sim \sqrt{\frac{\hbar}{m\omega}}$, in order for perturbation theory to be applicable, we must find

$$ux^3 \ll \hbar\omega \Longleftrightarrow u \ll \hbar\omega \left(\frac{m\omega}{\hbar}\right)^{\frac{3}{2}} = \frac{m^{\frac{3}{2}}\omega^{\frac{5}{2}}}{\sqrt{\hbar}}$$

Now,

$$\epsilon\hat{V} = u \left(\frac{\hbar}{2m\omega}\right)^{\frac{3}{2}} \left(a+\hat{a}^\dagger\right)^3$$

implies $\langle n|\epsilon\hat{V}|n\rangle = 0$, and there is no first-order perturbation. To compute the second-order perturbation, we use

$$\left(a+\hat{a}^\dagger\right)^3 = a^3 + 3(a^\dagger a^2 + a) + 3((a^\dagger)^2 a + a^\dagger) + (a^\dagger)^3$$

The terms needed for the second-order perturbation are

$$\begin{cases} \langle n|(a^\dagger)^3|n-3\rangle\langle n-3|a^3|n\rangle &= n(n-1)(n-2) \\ \langle n|(a^\dagger)^2 a + a^\dagger|n-1\rangle\langle n-1|a^\dagger a^2 + a|n\rangle &= n^3 \\ \langle n|a^\dagger a^2 + a|n+1\rangle\langle n+1|(a^\dagger)^2 a + a^\dagger|n\rangle &= (n+1)^3 \\ \langle n|a^3|n+3\rangle\langle n+3|(a^\dagger)^3|n\rangle &= (n+3)(n+2)(n+1) \end{cases}$$

Thus, the change to the n-th energy eigenvalue is approximately

$$\Delta E_n \simeq u^2\left(\frac{\hbar}{2m\omega}\right)^3\frac{1}{\hbar\omega}\left(\frac{n(n-1)(n-2)}{3}+9n^3\right.$$
$$\left.-9(n+1)^3-\frac{(n+3)(n+2)(n+1)}{3}\right)$$
$$=-u^2\left(\frac{\hbar}{2m\omega}\right)^3\frac{1}{\hbar\omega}\left(30n^2+30n+11\right)$$
$$=-\hbar\omega\frac{u^2\hbar}{m^3\omega^5}\frac{1}{8}\left(30n^2+30n+11\right)$$

4. (a) The energy due to the magnetic field is of the order $\mu_B B$. Since $\mu_B \simeq 6\times 10^{-9}\,\mathrm{eV/Gauss}$ (Exercise 3-1), for $\mu B \ll 1\,\mathrm{eV}$, we need $B \ll 10^8$ Gauss.
   (b) The first-order contribution from the vector potential is

$$\frac{e}{mc}\hat{\vec{p}}\cdot\vec{A}(\hat{\vec{r}})=\frac{e}{2mc}\hat{\vec{p}}\cdot\left(\vec{B}\times\hat{\vec{r}}\right)=\frac{\mu_B}{\hbar}\vec{B}\cdot\left(\hat{\vec{r}}\times\hat{\vec{p}}\right)=\mu_B\frac{\hat{\vec{L}}}{\hbar}\cdot\vec{B}$$

thus

$$\hat{H}=\frac{1}{2m}\hat{\vec{p}}^{\,2}-\frac{e^2}{\hat{r}}+\mu_B B\frac{1}{\hbar}\left(\hat{L}_z+2\hat{s}_z\right)$$

   (c) If we denote the z component of the orbital angular momentum as $\hbar l_z$ ($-l\le l_z\le l$), the energy change for the spin $\pm\frac{1}{2}$ state is given by

$$\mu B(l_z\pm 1)$$

5. The expectation value of V (to be interpreted as ϵV) is

$$\langle k|\hat{V}|k\rangle \equiv \int_{-\infty}^{\infty}dx\,|\psi_k(x)|^2\,V(x)=\int_0^a\frac{dx}{a}V(x)$$

which is a constant independent of the wave number. Since ψ_k and ψ_{-k} are degenerate, we need to calculate the off-diagonal element:

$$\langle -k|\hat{V}|k\rangle \equiv \int_{-\infty}^{\infty}dx\,\psi_{-k}^*(x)\,V(x)\psi_k(x)=\frac{1}{L}\int_{-\frac{L}{2}}^{\frac{L}{2}}dx\,V(x)\mathrm{e}^{2ikx}$$
$$=\frac{1}{L}\int_0^a dx\,V(x)\mathrm{e}^{2ikx}\sum_{n=-\frac{L}{2a}}^{\frac{L}{2a}}\mathrm{e}^{2ikna}$$

$$= \begin{cases} \int_0^a \frac{dx}{a} V(x) e^{2ikx} & ka = \pi, 2\pi, \cdots \\ 0 & \text{otherwise} \end{cases}$$

Let us define the Fourier coefficient of $V(x)$ as

$$\tilde{V}_k \equiv \int_0^a \frac{dx}{a} V(x) e^{-ikx} \quad (ka = 0, \pm 2\pi, \pm 4\pi, \cdots)$$

Unless ka is an integer multiple of π, the perturbation energy is a constant $\tilde{V}_0$. For $ka = \pm n\pi$, the perturbation energy becomes $\tilde{V}_0 \pm \left|\tilde{V}_{2n\pi}\right|$.

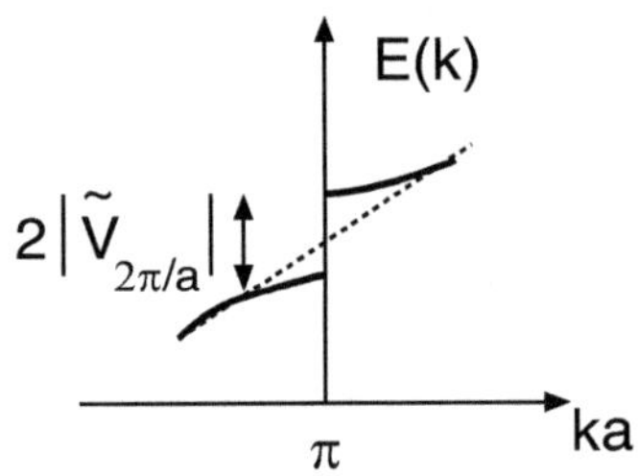

6. Near the origin $k = 0$, the second-order perturbation energy is given by

$$E^{(2)}(k) = \sum_{k' \neq k} \frac{\left|\frac{1}{L}\int_{-\frac{L}{2}}^{\frac{L}{2}} dx\, V(x) e^{i(k-k')x}\right|^2}{\frac{\hbar^2}{2m}\left(k^2 - (k')^2\right)}$$

where the integral is

$$\frac{1}{L}\int_{-\frac{L}{2}}^{\frac{L}{2}} dx\, V(x) e^{i(k-k')x} = \int_0^a \frac{dx}{a} V(x) e^{i(k-k')x} \frac{1}{\frac{L}{a}} \sum_{n=-\frac{L}{2a}}^{\frac{L}{2a}} e^{in(k-k')a}$$

$$= \begin{cases} \tilde{V}_{k'-k} & |k-k'|a = 2\pi, 4\pi, \cdots \\ 0 & \text{otherwise} \end{cases}$$

We obtain

$$E^{(2)}(k) = \frac{2m}{\hbar^2} \sum_{n \neq 0}^{\infty} \frac{\left|\tilde{V}_{\frac{2\pi n}{a}}\right|^2}{k^2 - \left(k + 2\pi n \frac{1}{a}\right)^2} = -\frac{4m}{\hbar^2} \sum_{n=1}^{\infty} \frac{\left|\tilde{V}_{\frac{2\pi n}{a}}\right|^2}{\left(\frac{2\pi n}{a}\right)^2 - 4k^2}$$

$$\simeq -\frac{4m}{\hbar^2}\sum_{n=1}^{\infty}\left|\tilde{V}_{\frac{2\pi n}{a}}\right|^2\left(\frac{1}{\left(\frac{2\pi n}{a}\right)^2}+\frac{4k^2}{\left(\frac{2\pi n}{a}\right)^4}\right)$$

for small k. Thus, the effective mass is

$$\frac{1}{m_{\text{eff}}} \simeq \frac{1}{m} - \frac{32m}{\hbar^4}\sum_{n=1}^{\infty}\left|\tilde{V}_{\frac{2\pi n}{a}}\right|^2\frac{1}{\left(\frac{2\pi n}{a}\right)^4}$$

$$= \frac{1}{m}\left(1 - 32\left(\frac{m}{\hbar^2\left(\frac{2\pi}{a}\right)^2}\right)^2\sum_{n=1}^{\infty}\frac{1}{n^4}\left|\tilde{V}_{\frac{2\pi n}{a}}\right|^2\right)$$

Clearly $m_{\text{eff}} > m$.

7. The matrix elements of the potential are obtained as

$$\langle k|\hat{V}|k'\rangle = \tilde{V}_{\frac{2n\pi}{a}}$$

$$\langle k'|\hat{V}|k\rangle = \tilde{V}_{-\frac{2n\pi}{a}}$$

$$\langle k|\hat{V}|k\rangle = \langle k'|\hat{V}|k'\rangle = \tilde{V}_0$$

where k and k' are related by

$$k' = k - \frac{2n\pi}{a}$$

Hence, the effective Hamiltonian is

$$\hat{H}_{\text{eff}} = \left(\frac{\hbar^2 k^2}{2m} + \tilde{V}_0\right)|k\rangle\langle k| + \left(\frac{\hbar^2 k'^2}{2m} + \tilde{V}_0\right)|k'\rangle\langle k'|$$

$$+ \tilde{V}_{\frac{2n\pi}{a}}|k\rangle\langle k'| + \tilde{V}_{-\frac{2n\pi}{a}}|k'\rangle\langle k|$$

Its eigenvalues are

$$\begin{cases} E_-(k) \equiv \frac{\hbar^2(k^2+k'^2)}{4m} + \tilde{V}_0 - \sqrt{\left(\frac{\hbar^2(k^2-k'^2)}{2m}\right)^2 + |\tilde{V}_{\frac{2n\pi}{a}}|^2} \\ E_+(k) \equiv \frac{\hbar^2(k^2+k'^2)}{4m} + \tilde{V}_0 + \sqrt{\left(\frac{\hbar^2(k^2-k'^2)}{2m}\right)^2 + |\tilde{V}_{\frac{2n\pi}{a}}|^2} \end{cases}$$

As shown in Example 11-5, if ka is away from an integral multiple of π, we obtain

$$\begin{cases} E_-(k) & \simeq \frac{\hbar^2 k^2}{2m} + \hat{V}_0 \\ E_+(k') & \simeq \frac{\hbar^2 (k')^2}{2m} + \hat{V}_0 \end{cases}$$

At $k = \frac{n\pi}{a}$, we find

$$\begin{cases} E_-\left(\frac{n\pi}{a}\right) & = \frac{\hbar^2}{2m}\left(\frac{n\pi}{a}\right)^2 + \hat{V}_0 - |\tilde{V}_{\frac{2n\pi}{a}}| \\ E_+\left(-\frac{n\pi}{a}\right) & = \frac{\hbar^2}{2m}\left(\frac{n\pi}{a}\right)^2 + \hat{V}_0 + |\tilde{V}_{\frac{2n\pi}{a}}| \end{cases}$$

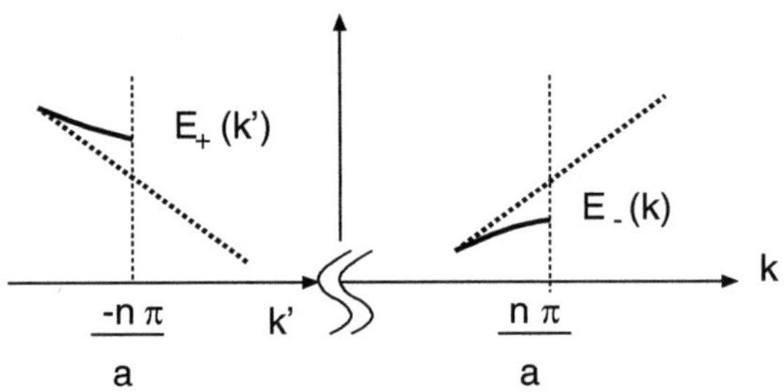

Exercises 11

1. **Model of van der Waals forces**—Consider two harmonic oscillators with the same angular frequency ω interacting with each other. The total Hamiltonian is given by

$$\hat{H} = \hat{H}_0 + \epsilon \hat{V}$$

where

$$\begin{cases} \hat{H}_0 & = \frac{\hat{p}_1^2 + \hat{p}_2^2}{2m} + \frac{m\omega^2}{2}\left(\hat{x}_1^2 + \hat{x}_2^2\right) \\ \epsilon \hat{V} & = g\hat{x}_1\hat{x}_2 \end{cases}$$

To find the ground state and its energy, we first use perturbation theory and then diagonalize the Hamiltonian exactly to make a comparison.

(a) Calculate the ground state energy to second order in g.
(b) Find the ground state $|\Psi\rangle$ to first order in g.
(c) Using the above result, compute the wave function $\langle x_1, x_2 \mid \Psi \rangle$, and compare the two cases where x_1 and x_2 have the same sign or opposite signs. Explain the result physically.

Hint: Is it more favorable for the two oscillators to be on the same side ($x_1x_2 > 0$) or the opposite side ($x_1x_2 < 0$)?

(d) Next, using the coordinates

$$\hat{Q} \equiv \frac{1}{2}\left(\hat{x}_1 + \hat{x}_2\right), \quad \hat{q} \equiv \frac{1}{2}\left(\hat{x}_1 - \hat{x}_2\right)$$

and their conjugate momenta, rewrite $\hat{H}$ and find the exact ground state and energy.

Comment: This model illustrates the essence of van der Waals forces acting between two electrically neutral atoms. For example, for two hydrogen atoms separated by a distance d, we find

$$\hbar\omega \sim Ry, \quad g \sim \frac{e^2}{d^3}$$

The interaction energy between the two electric dipoles is inversely proportional to d^3. From the two-oscillator model, two things should be clear:

i. Regardless of the sign of g, there is an attractive force.
ii. The interaction energy is proportional to $g^2 \propto \frac{1}{d^6}$.

2. **Helium atom**—Consider a system consisting of two electrons and a nucleus with charge Ze where $Z \gg 1$. We first apply perturbation theory, and then we take "boldly" the limit $Z \to 2$ for helium.
We first consider the Hamiltonian for a single electron under the Coulomb field of charge Ze:

$$\hat{H}_1 = \frac{\hat{\vec{p}}^{\,2}}{2m} - \frac{Ze^2}{\hat{r}}$$

(a) For the hydrogen atom with $Z = 1$, the normalized wave function of the ground state ($1s$) is given by

$$\frac{1}{\sqrt{\pi a^3}} \mathrm{e}^{-\frac{r}{a}},$$

where $a \equiv \frac{\hbar^2}{me^2}$ is the Bohr radius. Determine the normalized wave function of the $1s$ state of $\hat{H}_1$.

(b) What are the energy eigenvalues?

We next consider two electrons. We define $\hat{H}_2$ as the sum of two one-electron Hamiltonians as

$$\hat{H}_2 = \frac{\hat{\vec{p}}_1^{\,2}}{2m} - \frac{Ze^2}{\hat{r}_1} + \frac{\hat{\vec{p}}_2^{\,2}}{2m} - \frac{Ze^2}{\hat{r}_2}$$

Then the total Hamiltonian is given by

$$\hat{H} = \hat{H}_2 + \frac{e^2}{|\hat{\vec{r}}_1 - \hat{\vec{r}}_2|}$$

If $Z \gg 1$, the repulsion between the two electrons can be treated as a small perturbation.

(c) If both electrons are in the $1s$ orbital, what total spins are allowed?
(d) Calculate the shift of the ground state energy up to first order in perturbation. **Hint**: To compute an integral of the form $I \equiv \int dxdy\, f(x)K(x-y)g(y)$, it is useful to use Fourier transforms as

$$I = \int \frac{dk}{2\pi}\, \tilde{f}(-k)\tilde{K}(k)\tilde{g}(k)$$

where the tilde denotes the Fourier transform. For example,

$$\tilde{g}(k) \equiv \int dx\, \mathrm{e}^{-ikx} g(x)$$

(e) Calculate the total binding energy for $Z = 2$.
Comment: The experimental value for the binding energy is 78.99 eV. The first-order perturbation theory is a very crude approximation, and the good agreement we get with the experimental value is purely accidental.
(f) Determine the first ionization energy of He (the energy required to remove one electron).

The most precise treatment of the helium atom is based on the variational method. Perturbation theory is not a very good approximation.

3. **Fine structure**—We consider the relativistic corrections to the electron in a hydrogen atom.

(a) Using the Bohr model, compare the speed of the electron with the light velocity.

Using the relativistic Dirac equation for the electron, we find small relativistic effects given by the following Hamiltonian:

$$\hat{H} = \hat{H}_0 + \epsilon \hat{V}$$

where

$$\hat{H}_0 = \frac{\hat{\vec{p}}^{\,2}}{2m} - \frac{e^2}{\hat{r}}$$

$$\epsilon \hat{V} = \frac{e^2}{2m^2c^2} \frac{1}{\hat{r}^3} \hat{\vec{S}} \cdot \hat{\vec{L}} - \frac{1}{8m^3c^2} \left(\hat{\vec{p}}^{\,2}\right)^2$$

The physical meaning of each term in $\epsilon\hat{V}$ can be understood easily. The first term is the Zeeman effect: from the viewpoint of the electron, a magnetic field is created by the motion of the proton. The second term is a relativistic kinematic effect. From Einstein's formula for the kinetic energy, we obtain

$$\sqrt{p^2c^2+m^2c^4}=mc^2+\frac{p^2}{2m}-\frac{p^4}{8m^3c^2}+\cdots$$

(b) Estimate the magnitude of each term in $\epsilon\hat{V}$ and express it in units of eV.

(c) Denote the total angular momentum of the electron as $\hat{\vec{J}}\equiv\hat{\vec{L}}+\hat{\vec{s}}$. Enumerate the possible total angular momentum j and orbital angular momentum l for states with principal quantum number $n=1,2,\cdots$. Especially, for $n=2,3$, enumerate all possibilities using the notation nl_j. (e.g., $n=1$, $j=\frac{1}{2}$, $l=0$ is denoted as $1s_{\frac{1}{2}}$.)

(d) Calculate the first-order energy shift for the states with $l\neq 0$.
Comment: In fact, $\epsilon\hat{V}$ given above is correct only for $l\neq 0$. There is an additional correction to the $l=0$ states.
Hint: Use the following averages:

$$\left\langle n,l\left|\frac{a^2}{r^2}\right|n,l\right\rangle=\frac{1}{n^3\left(l+\frac{1}{2}\right)}$$
$$\left\langle n,l\left|\frac{a^3}{r^3}\right|n,l\right\rangle=\frac{1}{n^3l(l+1)\left(l+\frac{1}{2}\right)}$$

These will be calculated in problems Exercises 11-8 and 11-9. The average of $(\vec{p}^{\,2})^2$ can be obtained by calculating the square of $\frac{\vec{p}^{\,2}}{2m}=\hat{H}_0+\frac{e^2}{r}$.

(e) Show that the shift calculated above depends only on n and j, but not on l.

4. **Hyperfine structure**—The energy levels of the electron in a hydrogen atom are shifted further by the magnetic interaction with the nucleus. The proton is a particle with spin $\frac{1}{2}$, and its magnetic moment is expressed by the nuclear magneton $\mu_N\equiv\frac{e\hbar}{2m_pc}$ as $\mu_p\simeq 2.8\mu_N$ (where m_p is the proton mass).

   (a) Estimate the magnitude of the magnetic field in Gauss created by the proton at the distance of the Bohr radius, and estimate the magnitude of the magnetic interaction energy.
   Hint: The magnetic field created by a magnetic dipole varies inversely with the cube of the distance.

Thus, the hyperfine structure of the energy in a hydrogen atom is represented by the perturbation Hamiltonian

$$\hat{H}_{\rm hf}=\frac{A}{\hbar^2}\hat{\vec{s}}_e\cdot\hat{\vec{s}}_p$$

where $\hat{\vec{s}}$ represents the spin of the proton or electron and A is a positive energy estimated above.

(b) Find the eigenvalues of $\hat{H}_{\rm hf}$.
Hint: Consider the magnitude of the total spin angular momentum $(\hat{\vec{s}}_e + \hat{\vec{s}}_p)^2$.
(c) Determine the value of A in eV such that the difference between the two eigenvalues of $\hat{H}_{\rm hf}$ is the energy of a photon with the wavelength 21 cm.
Comment: The presence of hydrogen atoms in interstellar space was inferred from the observation of this light.

5. **Paschen-Back effect**—Consider the Hamiltonian $\hat{H} = \hat{H}_0 + \epsilon \hat{V}$ discussed in Example 11-4. Define the Hamiltonian for the spin-orbit interaction considered in Exercise 11-3 as

$$\hat{H}_{LS} \equiv \frac{e^2}{2m^2c^2} \frac{\hat{\vec{L}} \cdot \hat{\vec{s}}}{\hat{r}^3}$$

(a) How large should the magnetic field be so that we can regard $\hat{H}_{LS}$ as a small perturbation compared with $\epsilon \hat{V}$?
Comment: Of course, $\epsilon \hat{V}$ must be smaller than $\hat{H}_0$.
(b) Calculate the energy levels of the $2p$ state up to first order in spin-orbit interaction.
Hint: There are six states altogether.

6. **Band gap and effective mass**—Consider an electron with mass m satisfying the one-dimensional Schrödinger equation under the periodic potential $V(x) = V_0 \sum_{n=-\infty}^{\infty} \delta(x - na)$, where $V_0 > 0$ (Exercise 8-9). Bloch's theorem (Exercise 8-8) gives the wave function of the energy eigenstate with wave number k in the form

$$\psi(x) = e^{ikx} \phi_k(x)$$

where $\phi_k(x)$ is a periodic function satisfying $\phi_k(x + a) = \phi_k(x)$. We denote the energy eigenvalue as $E(k)$.

(a) Using the result of Example 11-5, compute $E\left(\frac{n\pi}{a} + 0\right)$ and $E\left(\frac{n\pi}{a} - 0\right)$ $(n = 1, 2, \cdots)$ to first order in V_0.
(b) Using the result of Example 11-6, compute $E(k = 0)$ and the effective mass $m_{\rm eff} = \dfrac{\hbar^2}{2\frac{d}{dk^2} E(k)\big|_{k=0}}$ to order V_0^2.
Hint: Use

$$\sum_{n=1}^{\infty} \frac{1}{n^2} = \frac{\pi^2}{6}, \quad \sum_{n=1}^{\infty} \frac{1}{n^4} = \frac{\pi^4}{90}$$

In Exercise 8-9 we have obtained

$$E(k) = \frac{\hbar^2}{2m}\kappa(k)^2$$

where $\kappa(k)$ is determined implicitly by

$$\cos ka = \cos\kappa a + \underbrace{\frac{mV_0 a}{\hbar^2}}_{\equiv\eta}\frac{\sin\kappa a}{\kappa a} \tag{F}$$

We reproduce the results of (a) and (b) using (F).

(c) Use (F) to compute $E\left(\frac{\pi}{a}+0\right)$ and $E\left(\frac{\pi}{a}-0\right)$ to order η. **Hint**: Let $\kappa a = \pi + \epsilon$, where ϵ is of order η. Substitute this κa into (F), and expand it to order ϵ^2 to determine κa.

(d) Likewise calculate $E(0)$ to order η^2. **Hint**: Solve (F) for $(\kappa a)^2$ by expanding the right-hand side to order $(\kappa a)^4$. Solve the resulting quadratic equation for $(\kappa a)^2$.

(e) Calculate the effective mass to order η^2. **Hint**: You need to calculate $\frac{m}{m_{\text{eff}}} = \frac{d(\kappa a)^2}{d(ka)^2}\Big|_{ka=0}$. Expand the right-hand side of (F) to order $(\kappa a)^6$, $\eta(\kappa a)^4$.

7. **Hydrogen molecular ion**—We wish to calculate the binding energy of a hydrogen molecular ion H_2^+. Assume we have two protons fixed on the z-axis at $z = \pm\frac{R}{2}$. The Hamiltonian of the electron is given by

$$\hat{H} = \frac{\hat{\vec{p}}^{\,2}}{2m} - \frac{e^2}{r_A} - \frac{e^2}{r_B} + \frac{e^2}{R}$$

where $r_A \equiv \sqrt{x^2+y^2+(z-R/2)^2}$ and $r_B \equiv \sqrt{x^2+y^2+(z+R/2)^2}$. The last term gives the electrostatic energy between the two protons.

We suppose the electron's wave function is given by the linear combination

$$\psi_\pm(\vec{r}) \equiv \psi_A(\vec{r}) \pm \psi_B(\vec{r})$$

where

$$\psi_A(\vec{r}) \equiv e^{-\frac{r_A}{a}}, \quad \psi_B(\vec{r}) \equiv e^{-\frac{r_B}{a}}$$

with the Bohr radius a.

(a) Plot the expectation value of $\hat{H}$ given by

$$E_\pm(R) \equiv \frac{\int d^3r\,\psi_\pm \hat{H}\psi_\pm}{\int d^3r\,\psi_\pm^2}$$

as a function of $\frac{R}{a}$. **Hint**: The integrals can be calculated exactly by using the elliptic coordinates

$$\xi \equiv \frac{r_A + r_B}{a}, \quad \eta \equiv \frac{r_A - r_B}{a}$$

Use $d^3r = \frac{\pi}{4}\left(\xi^2 - \eta^2\right) d\xi d\eta$.

(b) Minimize $E_\pm(R)$ with respect to R. Compare the result with the measured R which is close to $2a$. **Hint**: You will find $E_+(R) < E_-(R)$.

(c) Obtain the binding energy $E_+(R) + Ry$. **Hint**: $E_\pm(R) \xrightarrow{R\to\infty} -Ry$.

The following two questions calculate some matrix elements necessary for solving Exercise 11-3.[19] Here, we use atomic units, and the Bohr radius a is set to 1.

8. **Average of $\frac{1}{r^2}$**— To solve Exercise 11-3, it was necessary to find the average of $\frac{1}{r^2}$ for the ground state of a hydrogen atom. We will calculate it. The radial wave function $\chi_{n,l}(r)$ of the state with principal quantum number n and orbital angular momentum l satisfies the Schrödinger equation

$$\left(-\frac{d^2}{dr^2} - \frac{2}{r} + \frac{l(l+1)}{r^2}\right) \chi_{n,l}(r) = -\frac{1}{n^2}\, \chi_{n,l}(r) \tag{1}$$

Recall that $n' \equiv n - (l+1)$ gives the number of zeros of $\chi_{n,l}(r)$ at $r \neq 0$. Therefore, n' must be an integer, but l does not have to be. (Accordingly, n does not have to be an integer.)

(a) Differentiate equation (1) with respect to l while keeping n' constant to find the differential equation satisfied by $\frac{d}{dl}\chi_{n,l}(r)$.
Hint: Note that n varies with l.

(b) Integrate equation (1) to obtain

$$\int_0^\infty dr\, \chi_{n,l}(r) \left(-\frac{d^2}{dr^2} - \frac{2}{r} + \frac{l(l+1)}{r^2} + \frac{1}{n^2}\right) \frac{d\chi_{n,l}(r)}{dl} = 0$$

and derive the following result from it:

$$\left\langle n, l \left| \frac{1}{r^2} \right| n, l \right\rangle = \frac{1}{n^2\left(l + \frac{1}{2}\right)}$$

Note: The expectation value of $\frac{1}{r}$ depends only on n and not on l, but the expectation value of $\frac{1}{r^2}$ depends on both n and l.

[19] E. Abers, *Quantum Mechanics* (Prentice Hall) chapter 3.

9. **Average of other powers of** r—Using the expectation value of $\frac{1}{r^2}$ obtained in the previous problem, we will calculate the expectation values of other powers of r.

   (a) Multiply equation (1) by $r^k \frac{d}{dr}\chi_{n,l}(r)$, and integrate by parts with respect to r to express

   $$\int_0^\infty dr\, r^{k-1}\left(\frac{d}{dr}\chi_{n,l}(r)\right)^2 \tag{2}$$

   as a sum of the expectation values of powers of r.
   Hint: Use $\frac{d}{dr}\chi'^2 = 2\chi'\chi''$.

   (b) Multiply equation (1) by $r^{k-1}\chi_{n,l}(r)$, and integrate by parts with respect to r to obtain another expression for (2) as a sum of the expectation values of powers of r.
   Hint: This time, use $\frac{d}{dr}\chi^2 = 2\chi'\chi$.

   (c) Using the expectation value of $\frac{1}{r^2}$ obtained in Exercise 11-8 and the results from (9a) and (9b), calculate the following expectation values:

   $$\langle n,l|\,\frac{1}{r^3}\,|n,l\rangle\,, \quad \langle n,l|\,r\,|n,l\rangle$$

4.5 Time-Dependent Perturbation Theory

Two Points to Remember

1. Under the time-dependent interaction Hamiltonian $\hat{H}_{\text{int}}(t) = \hat{V}e^{-i\omega t}$, the transition probability per unit time from the initial state $|i\rangle$ to the final state $|f\rangle$ is given by Fermi's golden rule

$$w = \frac{2\pi}{\hbar}\left|\langle f|\,\hat{V}\,|i\rangle\right|^2 \rho(E_i + \hbar\omega)$$

   where ρ is the density of final states per unit energy.
2. Rabi's formula explains the principle of magnetic resonance.

4.5.1 Fermi's Golden Rule Gives the Transition Probability per Unit Time

Fermi's golden rule, which gives the transition probability per unit time, is the most important result in time-dependent perturbation theory. Under the time-dependent interaction Hamiltonian $\hat{H}_{\text{int}}(t) = \hat{V}e^{-i\omega t}$, the transition probability per unit time from the initial state $|i\rangle$ to the final state $|f\rangle$ is given by

$$\boxed{w = \frac{2\pi}{\hbar}\left|\langle f|\,\hat{V}\,|i\rangle\right|^2 \rho(E_i + \hbar\omega)}$$

where ρ is the density of final states per unit energy.

We consider the total Hamiltonian

$$\hat{H}(t) = \hat{H}_0 + \hat{H}_{\text{int}}(t) \tag{4.267}$$

which describes a system under time-dependent perturbation. The perturbative Hamiltonian is usually called the **interaction Hamiltonian**. Here are two examples.

Example 1: Magnetic Susceptibility in a Time-Dependent Magnetic Field

We consider the Hamiltonian of a magnetic moment under a time dependent magnetic field:

$$\hat{H}(t) = -\hat{\vec{\mu}} \cdot \vec{B}(t) = -g\mu_B \frac{\hat{\vec{s}}}{\hbar} \cdot \vec{B}(t) \tag{4.268}$$

where $\hat{\vec{s}}$ is the spin and $\hat{\vec{\mu}} = g\mu_B\hat{\vec{s}}/\hbar$ is the magnetic moment. If the magnetic field is given by

$$\vec{B}(t) = B_0\vec{e}_z + B_x(t)\vec{e}_x + B_y(t)\vec{e}_y \tag{4.269}$$

with its x and y components depending on time, the Hamiltonian can be written as

$$\hat{H}(t) = \underbrace{-\mu_B g\frac{1}{\hbar}\hat{s}_z B_0}_{\hat{H}_0} \underbrace{-\mu_B g\frac{1}{\hbar}\left(\hat{s}_x B_x(t) + \hat{s}_y B_y(t)\right)}_{\hat{H}_{\text{int}}(t)} \tag{4.270}$$

Example 2: Atom in an Electromagnetic Field

The total Hamiltonian of the electron in a hydrogen atom under an external electromagnetic field is given by

$$\begin{aligned}\hat{H}(t) &= \frac{1}{2m}\left(\hat{\vec{p}} + \frac{e}{c}\vec{A}(\hat{\vec{x}}, t)\right)^2 - \frac{e^2}{\hat{r}} \\ &= \underbrace{\frac{1}{2m}\hat{\vec{p}}^{\,2} - \frac{e^2}{\hat{r}}}_{\hat{H}_0} + \underbrace{\frac{e}{2mc}\left(\hat{\vec{p}}\cdot\vec{A} + \vec{A}\cdot\hat{\vec{p}}\right) + \frac{e^2}{2mc^2}\vec{A}^{\,2}}_{\hat{H}_{\text{int}}(t)}\end{aligned} \tag{4.271}$$

Let us begin the general theory. Given the total Hamiltonian $\hat{H}(t)$, the Schrödinger equation is given by

$$i\hbar\partial_t |\psi(t)\rangle = \hat{H}(t)|\psi(t)\rangle \tag{4.272}$$

We want to solve this using perturbation theory. We assume that the state of the system at $t = 0$ is an eigenstate of $\hat{H}_0$ with energy E_i:

$$|\psi(t=0)\rangle = |i\rangle \tag{4.273}$$

where $\hat{H}_0|i\rangle = E_i|i\rangle$. Let $\{|n\rangle\}$ be the orthonormal system consisting of eigenstates of $\hat{H}_0$ such that

$$\begin{cases} \hat{H}_0|n\rangle = E_n|n\rangle \\ \langle m \mid n\rangle = \delta_{mn} \end{cases} \tag{4.274}$$

Using this, we expand the solution $|\psi(t)\rangle$ of the Schrödinger equation as

$$|\psi(t)\rangle = \sum_n c_n(t)\,\mathrm{e}^{-i\frac{E_n}{\hbar}t}|n\rangle \tag{4.275}$$

Substituting this into Eq. (4.272), we obtain

$$\sum_n (i\hbar\dot{c}_n(t) + E_n c_n(t))\,\mathrm{e}^{-i\frac{E_n}{\hbar}t}|n\rangle = \sum_n c_n(t)\mathrm{e}^{-i\frac{E_n}{\hbar}t}\left(E_n|n\rangle + \hat{H}_{\text{int}}|n\rangle\right) \tag{4.276}$$

Taking the inner product of this equation with any eigenstate $|m\rangle$ of $\hat{H}_0$, we get

$$i\hbar\dot{c}_m(t)\mathrm{e}^{-i\frac{E_m}{\hbar}t} = \sum_n c_n(t)\mathrm{e}^{-i\frac{E_n}{\hbar}t}\langle m|\hat{H}_{\text{int}}(t)|n\rangle \tag{4.277}$$

which can be rewritten as

$$i\hbar\dot{c}_m(t) = \sum_n e^{i\frac{1}{\hbar}(E_m-E_n)t} \langle m| \hat{H}_{\text{int}}(t) |n\rangle c_n(t) \tag{4.278}$$

Note that this equation has the structure

$$\dot{c}_m(t) = \sum_n A_{mn}(t)c_n(t) \tag{4.279}$$

This type of equation can be solved formally using the notation of vectors and matrices as follows:

$$\vec{c}(t) = \mathbf{T}\exp\left[\int_0^t dt'\, A(t')\right]\vec{c}(0) \tag{4.280}$$

where the symbol **T** denotes **time ordering**, and the time-ordered exponential is defined by

$$\begin{aligned} &\mathbf{T}\exp\left[\int_0^t dt'\, A(t')\right] \\ &\equiv \mathbf{1} + \int_0^t dt_1\, A(t_1) + \int_0^t dt_1 \int_0^{t_1} dt_2\, A(t_1)A(t_2) \\ &\quad + \int_0^t dt_1 \int_0^{t_1} dt_2 \int_0^{t_2} dt_3\, A(t_1)A(t_2)A(t_3) + \cdots \end{aligned} \tag{4.281}$$

Let us check if Eq. (4.280) solves Eq. (4.279). First we check

$$\begin{aligned} &\frac{d}{dt}\mathbf{T}\exp\left[\int_0^t dt'\, A(t')\right] \\ &= A(t) + A(t)\int_0^t dt_1\, A(t_1) + A(t)\int_0^t dt_1 \int_0^{t_1} dt_2\, A(t_1)A(t_2) + \cdots \\ &= A(t)\,\mathbf{T}\exp\left[\int_0^t dt'\, A(t')\right] \end{aligned} \tag{4.282}$$

At $t = 0$, we find $\mathbf{T}\exp[\int_0^t dt'\, A(t')] = \mathbf{1}$. Hence, Eq. (4.280) is indeed the solution to Eq. (4.279) satisfying the desired initial condition $\vec{c}(0)$.

Thus, working out the m-th component of (4.280), we obtain the perturbative solution

$$\begin{aligned} c_m(t) &= c_m(0) \\ &\quad - \frac{i}{\hbar}\int_0^t dt_1 \sum_n e^{\frac{i}{\hbar}(E_m-E_n)t_1} \langle m| \hat{H}_{\text{int}}(t_1) |n\rangle c_n(0) \end{aligned}$$

$$-\frac{1}{\hbar^2}\int_0^t dt_1 \int_0^{t_1} dt_2 \sum_{n_1}\sum_{n_2} \mathrm{e}^{\frac{i}{\hbar}(E_m-E_{n_1})t_1} \langle m|\, \hat{H}_{\mathrm{int}}(t_1)\,|n_1\rangle$$

$$\times\, \mathrm{e}^{\frac{i}{\hbar}(E_m-E_{n_2})t_2} \langle n_1|\, \hat{H}_{\mathrm{int}}(t_2)\,|n_2\rangle\, c_{n_2}(0) + \cdots \tag{4.283}$$

In particular, the first-order approximation is obtained as

$$c_m(t) \simeq -\frac{i}{\hbar}\int_0^t dt_1 \sum_n \mathrm{e}^{\frac{i}{\hbar}(E_m-E_n)t_1} \langle m|\, \hat{H}_{\mathrm{int}}(t_1)\,|n\rangle\, c_n(0) \tag{4.284}$$

for $c_m(0) = 0$. Therefore, if the initial state is $n = i$ and the final state is $m = f \neq i$, the probability amplitude is given by

$$\boxed{\mathrm{Amp}_{f\leftarrow i} \simeq \frac{-i}{\hbar}\int_0^t dt_1\, \mathrm{e}^{\frac{i}{\hbar}(E_f-E_i)t_1} \langle f|\, \hat{H}_{\mathrm{int}}(t_1)\,|i\rangle} \tag{4.285}$$

Now, let us consider the interaction Hamiltonian[20]

$$\hat{H}_{\mathrm{int}}(t) = \hat{V}\mathrm{e}^{-i\omega t} \tag{4.286}$$

Using the above formula (4.285) to calculate the transition amplitude, we get

$$\begin{aligned}
\mathrm{Amp}_{f\leftarrow i}(t) &= \frac{-i}{\hbar}\int_0^t dt_1\, \mathrm{e}^{\frac{i}{\hbar}(E_f-E_i-\hbar\omega)t_1} \langle f|\, \hat{V}\,|i\rangle \\
&= \frac{-i}{\hbar}\frac{\mathrm{e}^{\frac{i}{\hbar}(E_f-E_i-\hbar\omega)t} - 1}{\frac{i}{\hbar}(E_f - E_i - \hbar\omega)} \langle f|\, \hat{V}\,|i\rangle \\
&= \frac{1}{i}\mathrm{e}^{\frac{i}{2\hbar}\Delta E\cdot t}\,\frac{\sin\frac{\Delta E\cdot t}{2\hbar}}{\frac{\Delta E}{2}} \cdot \langle f|\, \hat{V}\,|i\rangle
\end{aligned} \tag{4.287}$$

where

$$\Delta E \equiv E_f - E_i - \hbar\omega \tag{4.288}$$

Therefore, the transition probability is given by

$$\mathrm{Prob}_{f\leftarrow i} = \left|\mathrm{Amp}_{f\leftarrow i}(t)\right|^2 = \frac{\sin^2\frac{\Delta E\cdot t}{2\hbar}}{\left(\frac{\Delta E}{2}\right)^2}\left|\langle f|\, \hat{V}\,|i\rangle\right|^2 \tag{4.289}$$

[20] Of course, the hermitian conjugate must also be present since $\hat{H}_{\mathrm{int}}(t)$ is hermitian, but it contributes to transitions satisfying $E_f = E_i - \hbar\omega$.

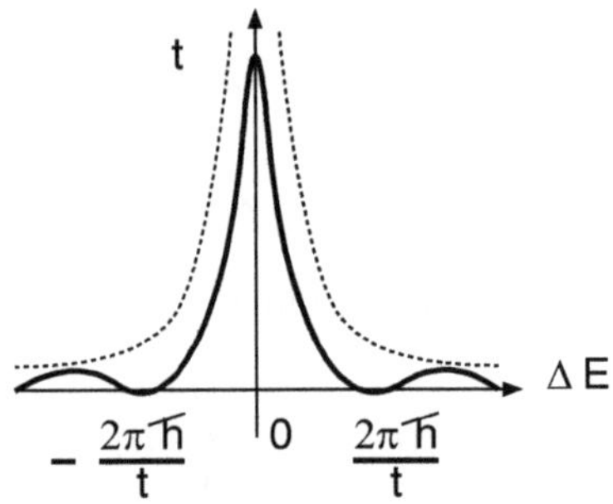

Fig. 4.12 $\frac{\sin^2 \frac{\Delta E \cdot t}{2\hbar}}{\left(\frac{\Delta E}{2\hbar}\right)^2 t}$ plotted as a function of ΔE. The width of the peak at $\Delta E = 0$ is $\frac{2\pi\hbar}{t}$

Note the periodic nature of the time dependence. As will be explained in the next section, this periodicity can be observed in Rabi oscillations. Typically, the contributions with different ΔE interfere, leading to a linear increase in the transition probability in time.

Now, let us plot

$$f_t(\Delta E) \equiv \frac{\sin^2 \frac{\Delta E \cdot t}{2\hbar}}{\left(\frac{\Delta E}{2\hbar}\right)^2 t}$$

as a function of ΔE. (See Fig. 4.12.) The width of the peak, $\frac{2\pi\hbar}{t}$, is a manifestation of the uncertainty relation for time and energy. As t increases, the peak becomes sharper.

The formula for the transition probability (4.289) is particularly important in the following two cases:

1. where the energy of the final state is continuously distributed—Let us assume that the squared matrix element

$$\left| \langle f | \hat{V} | i \rangle \right|^2 \tag{4.290}$$

is nearly constant over an energy range ϵ. If t is large enough so that

$$\frac{\hbar}{t} \ll \epsilon \tag{4.291}$$

the width of the peak in Fig. 4.12 can be taken to zero.

2. where the angular frequency ω of the perturbation is continuously distributed—Let us assume that the strength of the perturbation is nearly constant over the range $\frac{\epsilon}{\hbar}$ of ω. If t is large enough so that

$$\frac{\hbar}{t} \ll \epsilon \tag{4.292}$$

the width of the peak in Fig. 4.12 can be taken to zero.

Therefore, in both cases, $f_t(\Delta E)$ can be replaced by $f_\infty(\Delta E)$ where[21]

$$f_\infty(\Delta E) = \pi\delta\left(\frac{\Delta E}{2\hbar}\right) = 2\pi\hbar\delta(\Delta E) \tag{4.293}$$

Hence, (4.289) becomes

$$\text{Prob}_{f\leftarrow i}(t) = t\cdot\frac{2\pi}{\hbar}\left|\langle f|\,\hat{V}\,|i\rangle\right|^2\delta(E_f - E_i - \hbar\omega) \tag{4.294}$$

Note that the transition probability after time t is proportional to time t. Therefore, we can define the **transition probability per unit time**:

$$\boxed{w_{f\leftarrow i} \equiv \frac{\text{Prob}_{f\leftarrow i}(t)}{t} = \frac{2\pi}{\hbar}\left|\langle f|\,\hat{V}\,|i\rangle\right|^2\delta(E_f - E_i - \hbar\omega)} \tag{4.295}$$

The delta function implies **conservation of energy**. The external field oscillating with angular frequency ω provides energy $\hbar\omega$ to the initial state, resulting in the final state with energy $E_i + \hbar\omega$.

Of course, the delta function does not directly represent probability. As considered above, either the energy of the final state or the angular frequency of the perturbation is continuously distributed. Therefore, either E_f or ω must be integrated. By integration, the delta function is replaced by the density of the final states or the spectrum of the external perturbation. Here, let us consider the former case and introduce the **density of states** for the final state. The density of states represents the number of final states per unit energy and is defined by

$$\rho(E) \equiv \sum_f \delta(E_f - E)$$

If we denote the total number of states below energy E by $N(E)$, then the density of states is given by

$$\rho(E) = \frac{dN(E)}{dE} \tag{4.296}$$

This allows us to integrate the transition probability per unit time (4.295) over all possible final states:

$$w = \sum_f w_{f\leftarrow i} = \int dE\,\rho(E)\,w_{f\leftarrow i}$$

$$= \boxed{\frac{2\pi}{\hbar}\overline{\left|\langle f|\,\hat{V}\,|i\rangle\right|^2}\,\rho(E_i + \hbar\omega)} \tag{4.297}$$

21 $\int_{-\infty}^{\infty} dx\,\frac{\sin^2 Nx}{Nx^2} = \int_{-\infty}^{\infty} dx\,\frac{\sin^2 x}{x^2} = \pi$, so $\lim_{N\to\infty}\frac{\sin^2 Nx}{Nx^2} = \pi\delta(x)$.

Fig. 4.13 α Decay

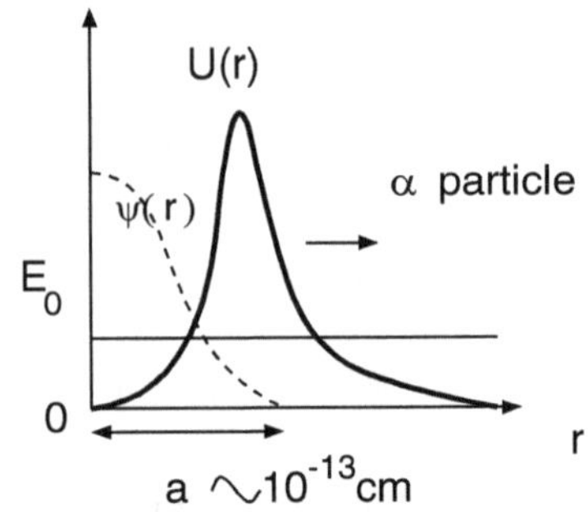

Here, the horizontal bar above the squared absolute value of the matrix element denotes the average over states with energy $E_i + \hbar\omega$. (We may need to average over the additional quantum numbers besides energy.) Equation (4.297) is a convenient formula known as **Fermi's golden rule**, named by E. Teller. The formula is worth memorizing.

Example 3: α Decay of Atomic Nuclei

There are three types of decay in atomic nuclei, α, β, and γ, but the α decay is the easiest to understand. In unstable atomic nuclei, α particles are trapped and feel a potential like the one in Fig. 4.13. The state $|\psi\rangle$ with energy E_0 is a metastable state and not a true bound state. The matrix element of the interaction Hamiltonian with momentum $\vec{p}$ can be approximated as a constant, ignoring the size of the atomic nucleus[22]

$$\left\langle \vec{p} \mid \hat{H}_{\text{int}} \mid \psi \right\rangle = g \tag{4.298}$$

On the other hand, the final state is the state of a free particle with momentum $\vec{p}$, so the density of states is given by

$$\frac{d^3p}{(2\pi)^3} \tag{4.299}$$

If expressed in terms of energy $E = \frac{p^2}{2m}$ and solid angle,

$$\frac{d^3p}{(2\pi)^3} = \frac{p^2 dp d\Omega}{(2\pi)^3} = \frac{mpdEd\Omega}{(2\pi)^3} \xrightarrow{\int d\Omega} \frac{mpdE}{2\pi^2} \tag{4.300}$$

we obtain the density of states per unit energy as

[22] That is, as long as $pa \ll \hbar$. The momentum dependence is of order $\frac{pa}{\hbar}$.

$$\rho(E) = \frac{mp}{2\pi^2} \tag{4.301}$$

Therefore, according to Fermi's golden rule, the probability of α decay per unit time is

$$\Gamma \equiv \frac{2\pi}{\hbar} g^2 \frac{mp_0}{2\pi^2} = \frac{g^2 mp_0}{\pi\hbar} \tag{4.302}$$

where p_0 is determined by $E_0 = \frac{p_0^2}{2m}$. See Sect. 4.6.7 for more details.

Example 4: Magnetic Resonance

In the presence of both a uniform magnetic field and a perpendicular oscillating magnetic field, the Hamiltonian of a magnetic moment μ of a spin $\frac{1}{2}$ particle is given by

$$\hat{H}(t) = \underbrace{\frac{-2\mu B_0}{\hbar}\hat{s}_z}_{\hat{H}_0} \underbrace{- \frac{2\mu}{\hbar} B(t)\hat{s}_x}_{\hat{H}_{\text{int}}(t)} \tag{4.303}$$

The eigenstates of $\hat{H}_0$ are $|\uparrow\rangle$ and $|\downarrow\rangle$, with energies

$$\hat{H}_0 |\uparrow\rangle = -\mu B_0 |\uparrow\rangle \tag{4.304}$$

$$\hat{H}_0 |\downarrow\rangle = \mu B_0 |\downarrow\rangle \tag{4.305}$$

Thus, the energy difference between them is

$$\Delta E = 2\mu B_0 \tag{4.306}$$

For the transition from state $|\uparrow\rangle$ to $|\downarrow\rangle$, the time t of perturbation must be long enough to distinguish the energy difference ΔE:

$$t \gg \frac{\hbar}{\Delta E} = \frac{\hbar}{2\mu B_0} \tag{4.307}$$

For the electron, $\mu = \mu_B \equiv \frac{e\hbar}{2mc}$, so this condition becomes

$$t \gg \frac{mc}{eB_0} = \frac{1}{\omega_c}, \tag{4.308}$$

i.e., t should be sufficiently longer than the period of the cyclotron motion (which is the same as the period of classical precession).

First, let us consider the case of a sinusoidally oscillating external magnetic field:

$$B(t) = 2B\cos\omega t = B\mathrm{e}^{-i\omega t} + B\mathrm{e}^{i\omega t} \tag{4.309}$$

If $\omega > 0$, only the term $\mathrm{e}^{-i\omega t}$ contributes to the transition from $|\uparrow\rangle$ to $|\downarrow\rangle$. The matrix element is calculated as

$$\langle\downarrow|\,\hat{H}_{\mathrm{int}}(t)\,|\uparrow\rangle = -\frac{2\mu_B B}{\hbar}\mathrm{e}^{-i\omega t}\underbrace{\langle\downarrow|\,\hat{s}_x\,|\uparrow\rangle}_{=\frac{\hbar}{2}} = -\mu_B B\mathrm{e}^{-i\omega t} \tag{4.310}$$

Therefore, the transition probability per unit time is

$$w = \frac{2\pi}{\hbar}(\mu_B B)^2\delta(\Delta E - \hbar\omega) \tag{4.311}$$

We now recall that the energy density of the oscillating magnetic field is

$$\frac{1}{8\pi}B(t)^2 = \frac{1}{8\pi}(2B\cos\omega t)^2 \overset{\text{time average}}{\Longrightarrow} \frac{1}{4\pi}B^2 \tag{4.312}$$

where the time average of $\cos^2\omega t$ is $\frac{1}{2}$. Suppose that the magnetic field has a continuous spectrum so that the energy density of the magnetic field within $[\omega, \omega + \Delta\omega]$ is given by

$$I(\omega)\Delta\omega \tag{4.313}$$

Especially, for the magnetic field oscillating at a constant frequency ω, we find

$$I(\omega') = \frac{B^2}{4\pi}\,\delta(\omega' - \omega)$$

Now, Eq. (4.311) can be rewritten as

$$w = \frac{8\pi^2}{\hbar^2}\mu_B^2\frac{B^2}{4\pi}\delta\left(\frac{\Delta E}{\hbar} - \omega\right)$$

By replacing the monochromatic spectral function $\frac{B^2}{4\pi}\delta\left(\frac{\Delta E}{\hbar} - \omega\right)$ by the continuous spectral function $I(\omega)$, we obtain the transition probability per unit time

$$w = \frac{8\pi^2}{\hbar^2}\mu_B^2 \cdot I\left(\frac{\Delta E}{\hbar}\right) \tag{4.314}$$

Example 5: 1s $\rightarrow$ 2p Transition

Let us consider a hydrogen atom under the influence of an electromagnetic field. The Hamiltonian is approximated as follows:

$$\hat{H} = \underbrace{\frac{\hat{\vec{p}}^{\,2}}{2m} - \frac{e^2}{\hat{r}}}_{=\hat{H}_0} + \frac{e}{mc}\hat{\vec{p}} \cdot \vec{A}(\hat{\vec{x}}, t) \tag{4.315}$$

where we have ignored the A^2 term. We assume that the external field is monochromatic:

$$\vec{A}(\vec{x}, t) = 2A\vec{\epsilon}\cos(\vec{k} \cdot \vec{x} - \omega t) \tag{4.316}$$

which satisfies the Coulomb gauge conditions:

$$kc = \omega, \quad \vec{\epsilon} \cdot \vec{k} = 0, \quad \vec{\epsilon}^{\,2} = 1 \tag{4.317}$$

We are interested in the probability of the transition:

$$1s \longrightarrow 2p$$

Since the energy difference between these two states is

$$\Delta E \equiv E_{2p} - E_{1s} = \frac{3}{4}Ry \tag{4.318}$$

the external field with $\hbar\omega = \Delta E$ is required for the transition.

According to Fermi's golden rule, the transition probability per unit time is given by

$$w_{2p \leftarrow 1s} = \frac{2\pi}{\hbar}\left|\frac{eA}{mc}\langle 2p|\, \mathrm{e}^{i\vec{k}\cdot\vec{x}}\vec{\epsilon} \cdot \hat{\vec{p}}\,|1s\rangle\right|^2 \cdot \delta(\Delta E - \hbar\omega) \tag{4.319}$$

Since $\hbar\omega = \Delta E \sim Ry \sim \frac{e^2}{a}$, we find

$$k = \frac{\omega}{c} = \frac{\hbar\omega}{\hbar c} \sim \frac{e^2}{\hbar c}\frac{1}{a} = \frac{\alpha}{a} \tag{4.320}$$

Thus,

$$ka \sim \alpha \ll 1\,, \tag{4.321}$$

i.e., the wavelength of the absorbed light is 137 times longer than the Bohr radius. Hence, in calculating the matrix element above, we can approximate

$$e^{i\vec{k}\cdot\vec{x}} \simeq 1 \tag{4.322}$$

This is called the **electric dipole approximation**.

Therefore, we obtain

$$\begin{aligned} w_{2p\leftarrow 1s} &\simeq \frac{2\pi}{\hbar}\frac{e^2A^2}{m^2c^2}\left|\langle 2p|\vec{\epsilon}\cdot\hat{\vec{p}}\,|1s\rangle\right|^2\cdot\delta(\Delta E-\hbar\omega) \\ &= \frac{2\pi}{\hbar}\frac{e^2A^2}{m^2c^2}\epsilon_i\epsilon_j\left\langle 2p\mid\hat{p}_i\mid 1s\right\rangle\langle 1s|\,\hat{p}_j\,|2p\rangle\cdot\delta(\Delta E-\hbar\omega) \end{aligned} \tag{4.323}$$

Summing over the three 2p states, we get

$$\sum_{m=\pm 1,0}\left\langle 2p,m\mid\hat{p}_i\mid 1s\right\rangle\left\langle 1s\mid\hat{p}_j\mid 2p,m\right\rangle = \frac{1}{3}\delta_{ij}\sum_m\sum_{k=1}^{3}\left|\left\langle 2p,m\mid\hat{p}_k\mid 1s\right\rangle\right|^2 \tag{4.324}$$

Since

$$\begin{aligned} \langle 2p,m|\,\hat{p}_k\,|1s\rangle &= \frac{im}{\hbar}\langle 2p,m|\left[\hat{H}_0,\hat{x}_k\right]|1s\rangle \\ &= \frac{im}{\hbar}\Delta E\,\langle 2p,m|\,\hat{x}_k\,|1s\rangle \\ &= im\omega\,\langle 2p,m|\,\hat{x}_k\,|1s\rangle \end{aligned} \tag{4.325}$$

we obtain

$$\begin{aligned} \frac{1}{3}\sum_m\sum_k\left|\left\langle 2p,m\mid\hat{p}_k\mid 1s\right\rangle\right|^2 &= (m\omega)^2\frac{1}{3}\sum_m\sum_k\left|\left\langle 2p,m\mid\hat{x}_k\mid 1s\right\rangle\right|^2 \\ &= \frac{m\hbar\omega}{2}\cdot\underbrace{\frac{2m\omega}{\hbar}\sum_{m=\pm 1,0}\frac{1}{3}\sum_{k=1}^{3}\left|\left\langle 2p,m\mid\hat{x}_k\mid 1s\right\rangle\right|^2}_{\equiv f\leq 1} \end{aligned} \tag{4.326}$$

We call f the **oscillator strength**. Thus, we obtain

$$w = \frac{2\pi}{\hbar}\frac{e^2A^2}{m^2c^2}\frac{m\hbar\omega}{2}f\cdot\delta(\Delta E-\hbar\omega) \tag{4.327}$$

As has been considered in Example 4, the energy density of the electromagnetic field is given by

$$\frac{1}{8\pi}(E^2+B^2)=\frac{1}{4\pi}E^2=\frac{1}{4\pi}\frac{\omega^2}{c^2}(2A)^2\underbrace{\frac{1}{2}}_{=\langle\cos^2\omega t\rangle}=\frac{A^2\omega^2}{2\pi c^2} \tag{4.328}$$

Rewriting Eq. (4.327) as

$$w=f\cdot\frac{2\pi^2e^2}{m\hbar\omega}\cdot\frac{A^2\omega^2}{2\pi c^2}\,\delta\left(\frac{\Delta E}{\hbar}-\omega\right), \tag{4.329}$$

and replacing the last factor with the energy density $I(\omega)$ per unit angular frequency, we obtain the result

$$w=f\cdot\frac{2\pi^2e^2}{m\hbar\omega}\cdot I(\omega)\Big|_{\omega=\frac{\Delta E}{\hbar}}=3\Gamma_{2p\to 1s}\frac{\pi^2c^3}{\omega^2}\frac{I(\omega)}{\hbar\omega} \tag{4.330}$$

for continuous energy spectrum. The initial factor of 3 represents the degeneracy of the 2p state, and

$$\Gamma_{2p\to 1s}=f\cdot\frac{2e^2\omega^2}{3mc^3}$$

is the decay width of the 2p state to be obtained in (5.149). In Eq. (5.156) of Sect. 5.2.3 we obtain $f\simeq 0.42$. The relation between w and $\Gamma_{2p\to 1s}$ given by Eq. (4.330) is an example of a general relation first obtained by Einstein. See Exercise 14-5.

Conditions for the Validity of Perturbation Theory

Before concluding this section, let us derive some conditions for the applicability of perturbation theory. For time-independent perturbation theory, it was necessary for the energy change to be small compared with the original energy, but for time-dependent perturbation theory, even if the energy change is not small compared with the original energy, it can still be treated as a perturbation.

Consider a transition from state i to another state f. If the transition probability per unit time is w, the uncertainty in energy of the state i is approximately

$$\delta E\sim\hbar w$$

due to the uncertainty relation between time and energy. This δE gives the energy resolution. If δE is larger than the energy difference ΔE between states i and f, then the states i and f cannot be distinguished, and discussing the transition between them loses its meaning. Therefore, the condition for treating the transition as small is

$$\delta E \ll \Delta E \Longrightarrow w \frac{\hbar}{\Delta E} \ll 1 \tag{4.331}$$

In the case of Example 4, $\Delta E = \mu B_0$, and $w \sim \frac{\mu^2}{\hbar^2} I$, so the condition for the validity of perturbation is

$$\frac{\mu I}{\hbar B_0} \ll 1$$

In the case of Example 5, $\Delta E \sim Ry$, $w \sim \frac{e^2 I}{m Ry}$. Thus,

$$\frac{e^2 \hbar I}{m Ry^2} \ll 1$$

ensures the applicability of perturbation theory. See Fig. 4.14.

4.5.2 *Rabi's Formula Explains the Principle of Magnetic Resonance*

Rabi's formula not only illustrates the essence of resonance phenomena, but also has wide-ranging applications, including precise measurements of magnetic moments and atomic clocks. Familiarity with the formula is an important asset for all physicists. Unusual for quantum mechanics, a classical analogy works well for the resonance.

In Example 4 of the previous section, we have considered the transition of magnetic moments under a non-oscillating external magnetic field. In this section,

Fig. 4.14 Conditions for the applicability of perturbation theory: $\Delta E \gg \delta E$

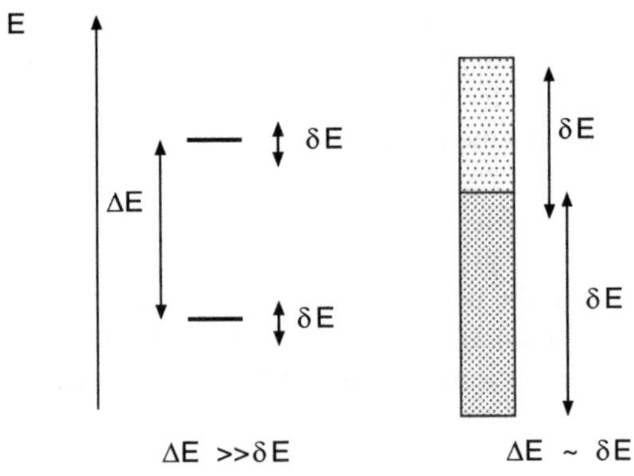

we consider in details a periodically oscillating external magnetic field. Let us denote the magnetic moment of a spin $\frac{1}{2}$ particle as $\frac{g}{2}\mu$. When a uniform magnetic field B_0 is applied along the z-direction, the Hamiltonian is given by

$$\hat{H}_0 = -\frac{g\mu}{\hbar}\hat{s}_z B_0 \tag{4.332}$$

Hence, the energy eigenstates are

$$\hat{H}_0 |\uparrow\rangle = -\frac{g}{2}\mu B_0 |\uparrow\rangle, \quad \hat{H}_0 |\downarrow\rangle = \frac{g}{2}\mu B_0 |\downarrow\rangle \tag{4.333}$$

The energy difference between these two states is

$$\Delta E \equiv \hbar\omega_0 \equiv g\mu B_0 \tag{4.334}$$

Under a magnetic field rotating perpendicularly to the z-axis

$$\vec{B}(t) = B(\cos\omega t, -\sin\omega t, 0) \tag{4.335}$$

the interaction Hamiltonian is given by

$$\begin{aligned}\hat{H}_{\text{int}}(t) &= \frac{g\mu B}{\hbar}\left(\hat{s}_x\cos\omega t - \hat{s}_y\sin\omega t\right)\\ &= \frac{g}{2}\frac{\mu B}{\hbar}\left((\hat{s}_+ + \hat{s}_-)\cos\omega t - \frac{1}{i}(\hat{s}_+ - \hat{s}_-)\sin\omega t\right)\\ &= \frac{g}{2}\frac{\mu B}{\hbar}\left(\mathrm{e}^{i\omega t}\hat{s}_+ + \mathrm{e}^{-i\omega t}\hat{s}_-\right) \quad (\hat{s}_\pm \equiv \hat{s}_x \pm i\hat{s}_y)\end{aligned} \tag{4.336}$$

Expressing the ket state of spin as

$$|\psi(t)\rangle = c_\uparrow(t)\,\mathrm{e}^{i\frac{\omega_0}{2}t}|\uparrow\rangle + c_\downarrow(t)\,\mathrm{e}^{-i\frac{\omega_0}{2}t}|\downarrow\rangle \tag{4.337}$$

we obtain the Schrödinger equation as

$$\begin{aligned}&i\hbar\dot{c}_\uparrow(t)\mathrm{e}^{i\frac{\omega_0}{2}t}|\uparrow\rangle + i\hbar\dot{c}_\downarrow(t)\mathrm{e}^{-i\frac{\omega_0}{2}t}|\downarrow\rangle\\ &= \frac{g}{2}\mu B\left(\mathrm{e}^{i\omega t - i\frac{\omega_0}{2}t}c_\downarrow(t)|\uparrow\rangle + \mathrm{e}^{-i\omega t + i\frac{\omega_0}{2}t}c_\uparrow(t)|\downarrow\rangle\right)\end{aligned} \tag{4.338}$$

Hence, comparing the coefficients, we get

$$i\hbar\dot{c}_\uparrow(t) = \frac{g}{2}\mu B\mathrm{e}^{i(\omega-\omega_0)t}c_\downarrow(t) \tag{4.339a}$$

$$i\hbar\dot{c}_\downarrow(t) = \frac{g}{2}\mu B\mathrm{e}^{-i(\omega-\omega_0)t}c_\uparrow(t) \tag{4.339b}$$

We can rewrite these as

$$e^{it\Delta\omega}\dot{c}_\uparrow(t) = -i\frac{\gamma}{2}c_\downarrow(t) \tag{4.340a}$$

$$\dot{c}_\downarrow(t) = -i\frac{\gamma}{2}e^{it\Delta\omega}c_\uparrow(t) \tag{4.340b}$$

where

$$\gamma \equiv \frac{g\mu B}{\hbar}, \qquad \Delta\omega \equiv \omega_0 - \omega \tag{4.341}$$

Therefore, differentiating (4.340a) with respect to time once more and eliminating $c_\downarrow$ using (4.340b), we obtain a second-order differential equation for $c_\uparrow$:

$$\ddot{c}_\uparrow + i\Delta\omega\dot{c}_\uparrow + \left(\frac{\gamma}{2}\right)^2 c_\uparrow = 0 \tag{4.342}$$

This is a linear ordinary differential equation with constant coefficients, so the solution can be sought in the form of $c_\uparrow(t) = e^{i\alpha t}$. Substituting this into Eq. (4.342), we obtain a quadratic equation for α:

$$-\alpha^2 - \alpha\Delta\omega + \frac{\gamma^2}{4} = 0 \tag{4.343}$$

Solving this, we obtain

$$\alpha = -\frac{\Delta\omega}{2} \pm \frac{1}{2}\sqrt{(\Delta\omega)^2 + \gamma^2} \tag{4.344}$$

Introducing

$$\Omega \equiv \sqrt{(\Delta\omega)^2 + \gamma^2} \tag{4.345}$$

we obtain

$$c_\uparrow(t) = e^{-\frac{it}{2}\Delta\omega}\left(c_+ e^{i\frac{\Omega}{2}t} + c_- e^{-i\frac{\Omega}{2}t}\right) \tag{4.346a}$$

$$c_\downarrow(t) = \frac{e^{\frac{it}{2}\Delta\omega}}{\gamma}\left[(\Delta\omega - \Omega)c_+ e^{i\frac{\Omega}{2}t} + (\Delta\omega + \Omega)c_- e^{-i\frac{\Omega}{2}t}\right] \tag{4.346b}$$

We take the initial condition as

$$c_\uparrow(0) = 1, \qquad c_\downarrow(0) = 0 \tag{4.347}$$

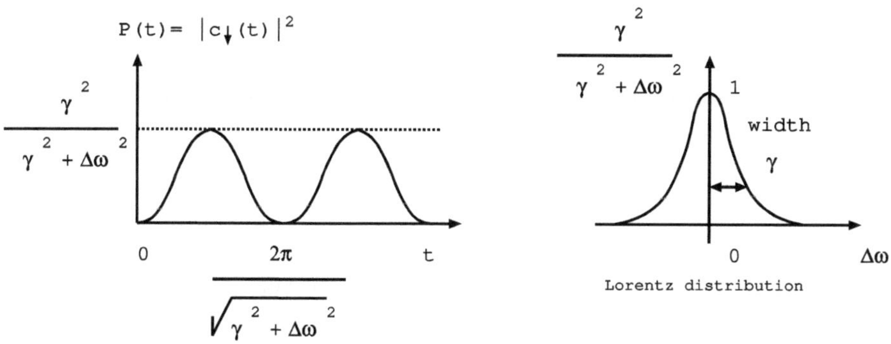

Fig. 4.15 Rabi resonance

so that we obtain

$$c_+ = \frac{\Delta\omega + \Omega}{2\Omega}, \quad c_- = -\frac{\Delta\omega - \Omega}{2\Omega} \tag{4.348}$$

We thus obtain

$$c_\uparrow(t) = \mathrm{e}^{-\frac{it}{2}\Delta\omega}\left(\cos\frac{\Omega}{2}t + i\frac{\Delta\omega}{\Omega}\sin\frac{\Omega}{2}t\right) \tag{4.349a}$$

$$c_\downarrow(t) = \mathrm{e}^{\frac{it}{2}\Delta\omega}\frac{\gamma}{\Omega}(-i)\sin\frac{\Omega}{2}t \tag{4.349b}$$

Recalling Eq. (4.345), we obtain

$$\boxed{|c_\downarrow(t)|^2 = \frac{\gamma^2}{\gamma^2 + (\Delta\omega)^2}\sin^2\frac{\sqrt{\gamma^2 + (\Delta\omega)^2}}{2}t} \tag{4.350}$$

This is called **Rabi's formula.** $|c_\uparrow(t)|^2$ is of course given by $1 - |c_\downarrow(t)|^2$. When the angular frequency satisfies the resonance condition $\omega = \omega_0$, the probability $|c_\downarrow(t)|^2$ can reach 1. Off resonance, it never reaches 1 (Fig. 4.15).

By measuring the amplitude of the probability $|c_\downarrow(t)|^2$, the resonance angular frequency can be determined with precision γ. Since $\omega_0 = \frac{g\mu B_0}{\hbar}$, the precise measurement of the particle's magnetic moment $\frac{g\mu}{2}$ can be obtained from the known magnetic field B_0.[23]

[23] Consult Feynman vol. III for more explanations.

Examples 12

1. **Density of states**—The number density of states $\rho(E)$ for a quantum system is defined so that the number of states in the small energy interval $[E, E + \Delta E]$ is equal to

$$\rho(E)\Delta E$$

To refine this definition, let us normalize the eigenstates $|\phi_n\rangle$ of a Hamiltonian $\hat{H}_0$ with energies E_n as

$$\langle\, \phi_m \mid \phi_n \,\rangle = \delta_{mn}$$

Then, the density of states is defined by

$$\rho(E) \equiv \sum_n \langle\phi_n|\, \delta(E - \hat{H}_0)\, |\phi_n\rangle = \sum_n \delta(E - E_n) \tag{1}$$

(a) For any function $f(x)$, show

$$\sum_n \langle\phi_n|\, f\left(\hat{H}_0\right) |\phi_n\rangle = \int_{-\infty}^{\infty} dE\, \rho(E) f(E)$$

(b) Consider a particle of mass m in a one-dimensional region $x \in [-L/2, L/2]$ with the periodic boundary condition. Calculate the density of states.
Hint: Use formula (1). When L is large, you can replace the sum over wave numbers by an integral. You will see that $\rho(E)$ is proportional to L.

(c) Calculate the density of states for a particle of mass m satisfying the periodic boundary condition in a three-dimensional cube of volume L^3.

(d) Calculate the density of states for a system consisting of two particles, one with mass m_1 and the other with mass m_2, in a three-dimensional periodic cube (i.e., torus) of volume L^3. **Hint**: Set the energy of the particle with mass m_1 as x^2 and the energy of the particle with mass m_2 as y^2, and then calculate the integral using cylindrical coordinates for x and y ($x = r\cos\theta$, $y = r\sin\theta$).

2. **Model of a particle decay**—Consider a system consisting of a harmonic oscillator with angular frequency ω and a string of length L. The Hamiltonian is given in the form

$$\hat{H} = \hat{H}_0 + \hat{V}$$

where $\hat{H}_0$ and $\hat{V}$ are given by creation and annihilation operators as follows:

$$\begin{cases} \hat{H}_0 &\equiv \hbar\omega\hat{a}^\dagger\hat{a} + \sum_{n\neq 0}\hbar\omega_n\hat{a}_n^\dagger\hat{a}_n \\ \hat{V} &\equiv \frac{g\sqrt{v}}{\sqrt{L}}\sum_{n\neq 0}\frac{1}{\sqrt{\omega_n}}\left(\hat{a}\hat{a}_n^\dagger + \hat{a}^\dagger\hat{a}_n\right) \end{cases}$$

where $\omega_n = \omega_{-n}$ is determined by the speed $v > 0$ of waves propagating along the string as

$$\omega_n \equiv \frac{2\pi|n|}{L}v \quad (n = \pm 1, \pm 2, \cdots)$$

(a) Determine the dimension of the constant g. **Hint**: $\hat{V}$ has the dimension of an energy. $\hat{a}$, $\hat{a}_n$ are dimensionless.
(b) Assuming that the harmonic oscillator is in the first excited state and the string is in the ground state, find the probability per unit time for the harmonic oscillator to transition to the ground state in the limit $L \to \infty$. **Hint**: Let $|\Omega\rangle$ be the ground state satisfying $\hat{a}|\Omega\rangle = \hat{a}_n|\Omega\rangle = 0$. The initial state is $\hat{a}^\dagger|\Omega\rangle$.
(c) Determine the conditions under which perturbation approximation can be used.

3. **Decay of the π^0 meson**—The π^0 meson decays into two photons:

$$\pi^0 \longrightarrow \gamma + \gamma$$

Denoting the unit polarization vectors of the photons as $\vec{\epsilon}_1$ and $\vec{\epsilon}_2$, the matrix element of the Hamiltonian describing the decay is given by

$$\begin{aligned} &\langle\gamma(\vec{p}_1, \vec{\epsilon}_1),\ \gamma(\vec{p}_2, \vec{\epsilon}_2)|\,\hat{H}_{\text{int}}\left|\pi^0(\vec{0})\right\rangle \\ &= \frac{C}{\sqrt{V}}\frac{\hbar^{\frac{3}{2}}}{(m_\pi c)^{\frac{5}{2}}}\delta_{\vec{p}_1+\vec{p}_2,\vec{0}}\,\vec{p}_1\cdot(\vec{\epsilon}_1\times\vec{\epsilon}_2) \end{aligned}$$

Note that the momentum $\vec{p}$ and polarization $\vec{\epsilon}$ of the photons must be orthogonal: $\vec{p}\cdot\vec{\epsilon} = 0$.

(a) Determine the dimension of the constant C.
(b) Determine the squared constant $|C|^2$, given that the lifetime of the π^0 meson is 8×10^{-17} s.
Hint: Sum over all possible polarization vectors.

4. **Precession of a classical spin**—In this problem, we will use classical mechanics to understand the precession of a magnetic moment in a magnetic field. Let $\vec{J}$ be the angular momentum. The magnetic moment is given by a constant multiple

$$\vec{\mu} = G\vec{J} \tag{2}$$

The magnetic field $\vec{B}$ exerts a torque $\vec{\mu} \times \vec{B}$ on the magnetic moment, and the equation of motion is given by

$$\frac{d}{dt}\vec{J} = \vec{\mu} \times \vec{B}$$

Using equation (2), we obtain the equation of motion of the magnetic moment as

$$\frac{d}{dt}\vec{\mu} = G\vec{\mu} \times \vec{B} \tag{3}$$

Before starting calculations, let us derive the equation of motion (3) using quantum mechanics.

(a) Consider the Hamiltonian

$$\hat{H} = -\hat{\vec{\mu}} \cdot \vec{B}$$

Assume any two states satisfying the Schrödinger equation as $|\phi(t)\rangle$ and $|\psi(t)\rangle$. Derive the equation of motion satisfied by the matrix elements

$$\vec{\mu}(t) \equiv \langle\phi(t)|\, \hat{\vec{\mu}}\, |\psi(t)\rangle$$

and compare it with equation (3).

We return to classical mechanics.

(b) When the magnetic field is given by a constant B_0 such that $\vec{B} = (0, 0, B_0)$, solve the equation of motion for $\vec{\mu}$.
Hint: Solve the equations of motion for $\mu_\pm \equiv \mu_x \pm i\mu_y$.
(c) For a constant B and $\vec{B}(t) = (B\cos\omega t, -B\sin\omega t, B_0)$, solve the equation of motion (3) for the case $\omega = GB_0$.
Hint: Note that when the resonance condition $\omega = GB_0$ holds, the magnetic field vector is rotating around the z-axis at the same angular velocity as the spin obtained in (b). Rewrite the equations of motion for $e^{\pm i\omega t}\mu_\pm$ and μ_z.
(d) Illustrate and explain the two eigenmodes obtained in (c).

Solutions to Examples 12

1. (a)

$$\sum_n \langle\phi_n|\, f(\hat{H}_0)\, |\phi_n\rangle = \sum_n f(E_n) = \sum_n \int dE\, f(E)\delta(E - E_n)$$
$$= \int dE\, f(E)\rho(E)$$

(b) Replace $\sum_n$ by the integral over the momentum (or wave number to be precise) space $L\int\frac{dk}{2\pi}$.

$$\rho(E) \equiv \sum_n \delta(E-E_n) = L\int \frac{dk}{2\pi}\delta\left(E-\frac{\hbar^2k^2}{2m}\right)$$

$$= 2L\frac{1}{2\pi}\frac{m}{\hbar^2 k} = \frac{L}{\pi\hbar}\sqrt{\frac{m}{2E}}$$

(c) Replace $\sum_n$ by the integral over the momentum space $L^3\int\frac{d^3k}{(2\pi)^3}$.

$$\rho(E) \equiv \sum_n \delta(E-E_n) = L^3\int \frac{d^3k}{(2\pi)^3}\delta\left(E-\frac{\hbar^2k^2}{2m}\right)$$

$$= L^3\frac{4\pi}{(2\pi)^3}k^2\frac{m}{\hbar^2 k} = L^3\frac{1}{2\pi^2\hbar^3}m^{\frac{3}{2}}\sqrt{2E}$$

(d)

$$\rho(E) \equiv \sum_n \delta(E-E_n) = L^6\int \frac{d^3k_1}{(2\pi)^3}\frac{d^3k_2}{(2\pi)^3}\delta\left(E-\frac{\hbar^2k_1^2}{2m_1}-\frac{\hbar^2k_2^2}{2m_2}\right)$$

Let $x_i \equiv \frac{\hbar^2 k_i^2}{2m_i}$ so that

$$\rho(E) = L^6\frac{1}{(2\pi^2)^2}\left(\frac{2\sqrt{m_1m_2}}{\hbar^2}\right)^3\int x_1^2dx_1x_2^2dx_2\,(E-x_1^2-x_2^2)$$

$$= L^6\frac{1}{(2\pi^2)^2}\left(\frac{2\sqrt{m_1m_2}}{\hbar^2}\right)^3\int_0^\infty rdr\int_0^{2\pi}d\theta\, r^4\sin^2\theta\cos^2\theta\,\delta(E-r^2)$$

$$= L^6\frac{1}{(2\pi^2)^2}\left(\frac{2\sqrt{m_1m_2}}{\hbar^2}\right)^3 4\pi\frac{E^2}{2} = L^6\frac{1}{\pi^3}\frac{(m_1m_2)^{\frac{3}{2}}}{\hbar^6}E^2$$

2. (a) The operator $\hat{V}$ has the dimension of energy. So does g.
   (b) The transition probability to the first excited state $\hat{a}_n^\dagger|\Omega\rangle$ of the mode n of a harmonic oscillator per unit time is given by

$$w_n = \frac{2\pi}{\hbar}\left|\langle\Omega|\,\hat{a}_n\hat{V}\hat{a}^\dagger\,|\Omega\rangle\right|^2\delta\left(\hbar(\omega-\omega_n)\right)$$

We find

$$\langle\Omega|\,\hat{a}_n\hat{V}\hat{a}^\dagger\,|\Omega\rangle = \frac{g\sqrt{v}}{\sqrt{L\omega_n}}\underbrace{\langle\Omega|\,\hat{a}_n\hat{a}\hat{a}_n^\dagger\hat{a}^\dagger\,|\Omega\rangle}_{=1} = g\sqrt{\frac{v}{L\omega_n}}$$

Summing w_n over all n, we obtain

$$w \equiv \sum_{n\neq 0} w_n = \frac{2\pi}{\hbar^2} g^2 \frac{1}{L} \sum_{n\neq 0} \frac{v}{\omega_n} \delta(\omega - \omega_n) = \frac{2\pi}{\hbar^2} g^2 \frac{v}{\omega} \frac{1}{L} \sum_{n\neq 0} \delta(\omega_n - \omega)$$

Using

$$\lim_{L\to\infty} \frac{1}{L} \sum_{n\neq 0} = \int_{-\infty}^{\infty} \frac{d\omega'}{2\pi v}$$

in the limit $L \to \infty$, we get

$$w = \frac{2\pi}{\hbar^2} g^2 \frac{v}{\omega} \int_{-\infty}^{\infty} \frac{d\omega'}{2\pi v} \delta(|\omega'| - \omega) = \frac{1}{\hbar^2} g^2 \frac{1}{\omega} \int_{-\infty}^{\infty} d\omega' \, \delta(|\omega'| - \omega)$$

$$= \boxed{\frac{2g^2}{\hbar^2 \omega}}$$

(c) The energy of the first excited state of a harmonic oscillator has an uncertainty of approximately $\hbar w$. This must be much smaller than the excitation energy $\hbar\omega$. Therefore, we find

$$w \ll \omega \Longleftrightarrow \boxed{|g| \ll \hbar\omega}$$

3. (a) Energy
   (b) The decay probability per unit time is given by

$$w = \frac{C^2 \hbar^3}{V (m_\pi c)^5} \sum_{\vec{\epsilon}_1, \vec{\epsilon}_2} V \int \frac{d^3 p_1}{(2\pi\hbar)^3} \left(\vec{p}_1 \cdot (\vec{\epsilon}_1 \times \vec{\epsilon}_2)\right)^2 \delta(p_1 c - m_\pi c^2/2)$$

For instance, if $\vec{p}_1$ points in the z-direction, we can consider $\vec{\epsilon}_1, \vec{\epsilon}_2$ to be along the x, y-axes or the y, x-axes, respectively. Hence, a coefficient of 2 arises from the sum over polarization vectors. Therefore, we obtain

$$w = \frac{C^2}{(m_\pi c)^5} \cdot 2 \cdot \frac{4\pi}{(2\pi)^3} \int_0^{\infty} dp_1 \, p_1^4 \, \delta(p_1 c - m_\pi c^2/2)$$

$$= \frac{C^2}{(m_\pi c)^5} \frac{1}{\pi^2} \frac{1}{c} \left(\frac{m_\pi c}{2}\right)^4 = \frac{C^2}{16\pi^2 \hbar m_\pi c^2}$$

Given $w \simeq \frac{1}{8\times 10^{-17}} \mathrm{s}^{-1}$, we obtain

$$C^2 \simeq 0.18 \, \mathrm{MeV}^2$$

4. (a) The time derivative of the matrix element can be obtained using the Schrödinger equation satisfied by the ket states:

$$\begin{aligned}\frac{d}{dt}\vec{\mu}(t) &= \left(\frac{d}{dt}\langle\phi(t)|\right)\hat{\vec{\mu}}\,|\psi(t)\rangle + \langle\phi(t)|\,\hat{\vec{\mu}}\frac{d}{dt}\,|\psi(t)\rangle \\ &= \frac{i}{\hbar}\,\langle\phi(t)|\,\hat{H}\hat{\vec{\mu}}\,|\psi(t)\rangle + \langle\phi(t)|\,\hat{\vec{\mu}}\frac{-i}{\hbar}\hat{H}\,|\psi(t)\rangle \\ &= \frac{i}{\hbar}\,\langle\phi(t)|\,[\hat{H},\hat{\vec{\mu}}]\,|\psi(t)\rangle\end{aligned}$$

where the commutator is

$$\begin{aligned}[\hat{H},\hat{\mu_i}] &= -\sum_j G^2 B_j \underbrace{[\hat{J}_j,\hat{J}_i]}_{i\hbar\sum_k \epsilon_{jik}\hat{J}_k} = -i\hbar G\sum_j \epsilon_{jik}B_j\hat{\mu}_k \\ &= -i\hbar G\left(\hat{\vec{\mu}}(t)\times\vec{B}\right)_i\end{aligned}$$

We thus obtain equation (3).

(b) The equations of motion are

$$\begin{cases}\frac{d}{dt}\mu_x &= GB_0\mu_y \\ \frac{d}{dt}\mu_y &= -GB_0\mu_x \\ \frac{d}{dt}\mu_z &= 0\end{cases} \quad\Longrightarrow\quad \frac{d}{dt}\mu_\pm = \mp iGB_0\mu_\pm$$

Therefore, μ_z remains unchanged. Using $\omega \equiv GB_0$, we have $\mu_\pm(t) = \mathrm{e}^{\mp i\omega t}\mu_\pm(0)$. Thus, when viewed from the above in the $-z$ direction, it undergoes a clockwise precession.

(c) The equations of motion are

$$\begin{cases}\frac{d}{dt}\mu_x &= G\left(B_0\mu_y + B\sin\omega t\cdot\mu_z\right) \\ \frac{d}{dt}\mu_y &= G\left(-B_0\mu_x + B\cos\omega t\cdot\mu_z\right) \\ \frac{d}{dt}\mu_z &= GB\left(-\mu_x\sin\omega t - \mu_y\cos\omega t\right)\end{cases}$$

These can be rewritten as

$$\begin{cases}\frac{d}{dt}\left(\mathrm{e}^{\pm i\omega t}\mu_\pm\right) &= \pm iGB\mu_z \\ \frac{d}{dt}\mu_z &= GB\frac{i}{2}\left(\mathrm{e}^{i\omega t}\mu_+ - \mathrm{e}^{-i\omega t}\mu_-\right)\end{cases}$$

The solution, with A and ϕ as arbitrary constants, is

$$\begin{cases}\mu_z &= A\cos\left(GBt+\phi\right) \\ \mathrm{e}^{\pm i\omega t}\mu_\pm &= c + iA\sin\left(GBt+\phi\right)\end{cases}$$

Hence,

$$\begin{cases} \mu_x = c\cos\omega t + A\sin(GBt+\phi)\sin\omega t \\ \mu_y = -c\sin\omega t + A\sin(GBt+\phi)\cos\omega t \end{cases}$$

The term proportional to c rotates in the xy-plane along with the rotating magnetic field $\vec{B}(t)$, while the remaining terms rotate perpendicular to it. The amplitude of the latter oscillates with the resonant angular frequency ω.

(d) (i) Mode 1 parallel to $\vec{B}(t)$

$$\vec{\mu}(t) = \frac{c}{B}\vec{B}(t)$$

(ii) Mode 2 perpendicular to $\vec{B}(t)$

$$\begin{cases} \mu_x(t) = A\sin(GBt+\phi)\sin\omega t \\ \mu_y(t) = A\sin(GBt+\phi)\cos\omega t \\ \mu_z(t) = A\cos(GBt+\phi) \end{cases}$$

Note that $\vec{\mu}$ passes through $\mu_z = \pm A$ at the poles of the spherical surface in the figure below.

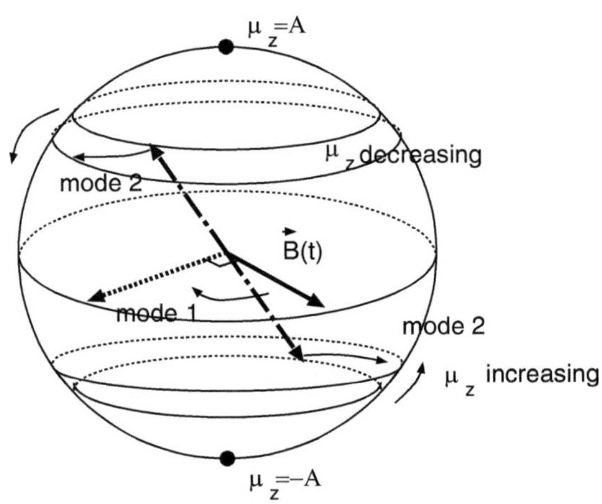

Mode 1 moves with the rotating $\vec{B}(t)$

Exercises 12

1. **Charged oscillator**—Consider a harmonic oscillator with an electric charge oscillating along the z-axis. The Hamiltonian is given by

$$\hat{H} = \frac{1}{2m}\left(\hat{p} + \frac{e}{c}A_z(t)\right)^2 + \frac{m\Omega^2}{2}\hat{z}^2$$

where

$$A_z(t) = 2A\cos\omega t$$

is the vector potential.

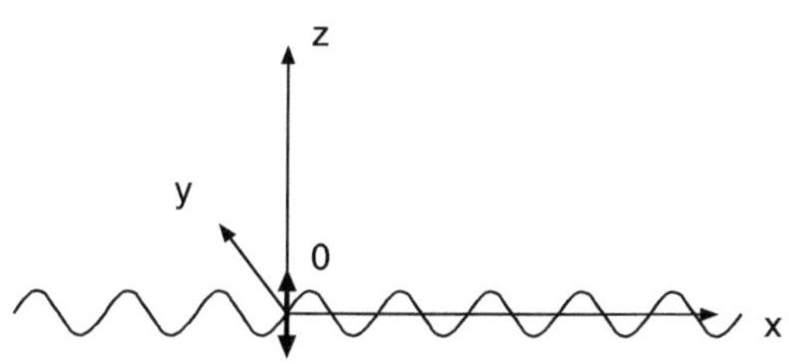

(a) Taking A as a real constant, the electromagnetic field is given by

$$E_z(x,t) = -B_y(x,t) = -2kA\sin(kx - \omega t)$$

Find the time average of the energy density.

(b) Show that in the first-order perturbation theory, the ground state of the harmonic oscillator cannot transition to any state other than the first excited state.

(c) Using the first-order perturbation theory, find the probability per unit time of the transition from the ground state to the first excited state:

$$w = \frac{A^2\omega^2}{2\pi c^2}\frac{2\pi^2 e^2}{m\hbar\omega}\delta(\omega - \Omega)$$

(d) Using the above result, find the transition probability per unit time under the presence of an electromagnetic field with the energy spectrum $I(\omega)$. **Hint**: The energy density of electromagnetic waves with angular frequency in the range $[\omega, \omega + \Delta\omega]$ is given by $I(\omega)\Delta\omega$.

(e) Explain why the factor f introduced by Eq. (4.326) is called the oscillator strength.

2. **Oscillator strength sum rule**—For the electron in a hydrogen atom, the **oscillator strength** of the bound state with principal quantum number n is defined as

$$f_n \equiv \frac{2mRy\left(1-\frac{1}{n^2}\right)}{\hbar^2}\frac{1}{3}\sum_{m=-1}^{1}\sum_{k=1}^{3}\left|\left\langle np, m\left|\hat{x}_k\right|1s\right\rangle\right|^2$$

For the state with positive energy E and angular momentum $l = 1$, it is defined as

$$f(E) \equiv \frac{2m(Ry+E)}{\hbar^2}\frac{1}{3}\sum_{k=1}^{3}\left|\langle E, m|\,\hat{x}_k\,|1s\rangle\right|^2$$

where the state $|E, m\rangle$ is normalized as

$$\left\langle E, m \mid E', m' \right\rangle = \delta(E-E')\delta_{m,m'}$$

(a) Using the commutation relations

$$[\hat{x}_i, \hat{p}_j] = i\hbar\delta_{ij}, \quad \hat{p}_k = \frac{im}{\hbar}[\hat{H}, \hat{x}_k]$$

derive the sum rule

$$\sum_{n=2}^{\infty} f_n + \int_0^{\infty} dE\, f(E) = 1$$

Hint: Within the state space with angular momentum $l = 1$, the completeness relation

$$\sum_{m=-1}^{1}\left(\sum_{n=2}^{\infty}|np, m\rangle\langle np, m| + \int_0^{\infty} dE\,|E, m\rangle\langle E, m|\right) = 1$$

holds.

Comment: From this sum rule and $f_n \geq 0$ & $f(E) > 0$, it follows that $f_n \leq 1$.

(b) Using the results of Exercise 12-1, interpret the sum rule physically.
Hint: What plays the role of a single charged oscillator in this case?

3. **Photoelectric effect**—We wish to calculate the probability of a $1s$ electron of an atom with valence Z transitioning to a free electron by absorbing light.

This phenomenon is known as the photoelectric effect and was first understood quantum mechanically by Einstein.[24]

Consider the electromagnetic field corresponding to the vector potential

$$A_z(\vec{x}, t) = 2A\cos(kx - \omega t), \quad A_x = A_y = 0,$$

where $\omega = kc$.

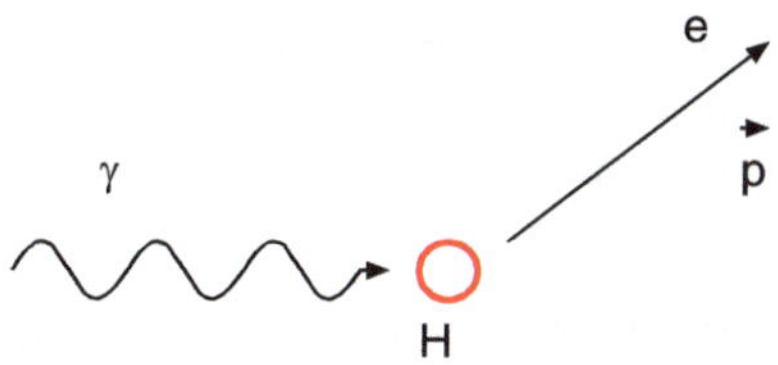

(a) We assume $\hbar\omega \sim Z\alpha mc^2$ where $\alpha \equiv \frac{e^2}{\hbar c} \simeq \frac{1}{137}$ is the fine structure constant. Compare the wavelengths of the emitted free electron, absorbed photon, and the radius of the $1s$ atomic orbit.
Hint: This shows that the electric dipole approximation cannot be used.

(b) Using Fermi's golden rule, find the transition probability per unit time, assuming

$$Z^2 Ry \ll \hbar\omega \ll mc^2$$

As the wave function of the free electron, use

$$\frac{1}{\sqrt{V}} e^{\frac{i}{\hbar}\vec{p}\cdot\vec{r}}$$

Show how the total volume, V, of the system gets canceled.
Hint: You need to find the Fourier transform of the wave function of the $1s$ state (an exponential function). One factor of V is obtained from the matrix element $\langle\vec{p}|\,\hat{H}_{\text{int}}\,|1s\rangle$, and the other comes from the density of the final states of the free electron.

(c) Find the cross section $\sigma(\omega)$. Express it in the form

$$\sigma(\omega) = r_{cl}^2 f\left(\frac{\hbar\omega}{mc^2}\right)$$

Hint: Remember that a "cross section" was introduced in classical electromagnetism (Sect. 2.2.2). Let w be the probability per unit time of the photoelectric effect. The energy absorbed by a $1s$ electron per unit time

[24] Absorption of X-rays by materials is mainly due to the photoelectric effect.

is given by $w\hbar\omega$. On the other hand, the energy of the electromagnetic field flowing per unit time per unit area is the energy density of the electromagnetic field multiplied by the speed of light c. In this case, the cross-section $\sigma(\omega)$ is given by

$$\sigma(\omega)[\mathrm{cm}^2] \equiv \frac{w \cdot \hbar\omega[\mathrm{erg}/s]}{\text{energy flow of the electromagnetic field}[\mathrm{erg}/(s \cdot \mathrm{cm}^2)]}$$

(d) Plot the cross section $\sigma(\omega)$ as a function of $\frac{\hbar\omega}{mc^2}$.
(e) Using the Z dependence of $\sigma(\omega)$, explain the principle of X-ray imaging. **Hint**: The atomic numbers of Ca and O are 20 and 8, respectively. The human body is mostly composed of water and bones.

4. **Precession of a classical spin** (Part 2)—Solve the equations of motion obtained in Example 12-4 for arbitrary ω that does not satisfy the resonance condition, and compare the result with Rabi's formula.
5. **Jaynes-Cummings Model**—Consider an electron inside a cavity with resonance angular frequency ω. The Hamiltonian of the cavity is given in the same form as that of a harmonic oscillator with eigenfrequency ω:

$$\hat{H}_{\text{cavity}} = \hbar\omega\hat{a}^\dagger\hat{a}\,,$$

where $\hat{a}, \hat{a}^\dagger$ are the annihilation and creation operators of the harmonic oscillator. The zero-point energy is ignored. Assume that the energy difference $\Delta E = \hbar\omega_0$ between the electron ground state $|g\rangle$ and its first excited state $|e\rangle$ is approximately equal to $\hbar\omega$, and the excitation of the higher energy states can be ignored. Thus, the total Hamiltonian of the system is given by

$$\hat{H} = \frac{1}{2}\hbar\omega_0\sigma_z + \hbar\omega\hat{a}^\dagger\hat{a} + g\left(\hat{a}\sigma_+ + \hat{a}^\dagger\sigma_-\right)\,,$$

where σ_z and $\sigma_\pm \equiv \sigma_x \pm i\sigma_y$ are the Pauli matrices.

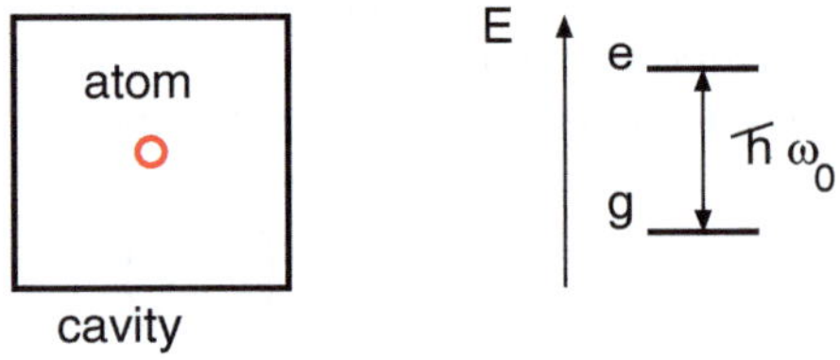

The electron in the atom has two low energy states $|g\rangle$, $|e\rangle$

(a) Physically interpret each term of the above Hamiltonian.
(b) Express the n-th excited state of the cavity as $|n\rangle$ $(n = 0, 1, 2, \cdots)$ and show that the Hamiltonian mixes $|n+1\rangle\,|g\rangle$ and $|n\rangle\,|e\rangle$.

(c) Find the energy eigenstates.
(d) Given that the state is $|n\rangle\,|e\rangle$ at time $t = 0$, find the probability that the state is $|n+1\rangle\,|g\rangle$ at time $t > 0$, and compare the result with Rabi's formula.

4.6 Supplementary Notes

4.6.1 *Mathematical Formalism of Quantum Mechanics*

The mathematical formalism of quantum mechanics relies heavily on vector spaces and linear operators. In this section, we provide basic definitions and a concise summary for finite dimensional discrete vector spaces. We will use the Dirac notation, which often simplifies the types of calculations encountered in quantum mechanics. Generalization to infinite dimensional spaces is straightforward.

Hilbert Spaces and Inner Products

1. Let $\mathcal{H}$ be an N-dimensional vector space over the complex numbers. We denote vectors in this space as *ket* vectors

$$|\psi\rangle\,,\,|\phi\rangle\,,\cdots\in\mathcal{H} \tag{4.351}$$

and the complex numbers as $\lambda,\,\mu,\,\cdots\in\mathbb{C}$.

2. $\mathcal{H}$ is a **linear vector space** if it is closed under scalar multiplication and vector addition:

$$|\lambda\psi\rangle = \lambda\,|\psi\rangle\in\mathcal{H}\quad\text{and}\quad|\psi+\lambda\phi\rangle = |\psi\rangle+\lambda|\phi\rangle\in\mathcal{H} \tag{4.352}$$

3. A vector space $\mathcal{H}$ is called a **Hilbert space** if it is endowed with an **inner product** $\langle\cdot|\cdot\rangle$ satisfying the following properties:

$$\langle\psi|\psi\rangle\geq 0\,,\qquad \langle\psi|\phi\rangle^* = \langle\phi|\psi\rangle \tag{4.353a}$$

$$\langle\psi|\phi_1+\lambda\phi_2\rangle = \langle\psi|\phi_1\rangle+\lambda\,\langle\psi|\phi_2\rangle\,,\qquad \langle\psi_1+\lambda\psi_2|\phi\rangle = \langle\psi_1|\phi\rangle+\lambda^*\,\langle\psi_2|\phi\rangle \tag{4.353b}$$

Here, a *bra* vector is obtained from the corresponding *ket* vector via the adjoint operation, which involves complex conjugation and transposition,

$$(\lambda\,|\psi\rangle)^\dagger = |\lambda\psi\rangle^\dagger = \langle\lambda\psi| = \lambda^*\,\langle\psi| \tag{4.354}$$

4. The inner product satisfies the **Schwarz inequality**:

$$|\langle\psi|\phi\rangle|^2 \le \langle\psi|\psi\rangle\langle\phi|\phi\rangle = ||\psi||^2||\phi||^2 \tag{4.355}$$

and the **triangle inequality**:

$$||\psi+\phi|| \le ||\psi|| + ||\phi|| \quad , \quad |(||\psi|| - ||\phi||)| \le ||\psi+\phi|| \tag{4.356}$$

where $||\psi|| \equiv \sqrt{\langle\psi|\psi\rangle}$ is the **norm** of a vector $|\psi\rangle \in \mathcal{H}$.

Linear Operators

1. Let $A : |\psi\rangle \to |\phi\rangle$ be an **operator**. A is a **linear operator** if

$$A\,|(\psi+\lambda\phi)\rangle = A\,|\psi\rangle + \lambda A\,|\phi\rangle \tag{4.357}$$

Note that, $(AB)\,|\psi\rangle = A\,(B|\psi\rangle)$ and $(\langle\phi|\,A)\,|\psi\rangle = \langle\phi|\,(A|\psi\rangle) = \langle\phi|\,A|\psi\rangle$.

2. The **hermitian conjugate (adjoint)** $A^\dagger$ of an operator A is defined as

$$\langle\phi|\,A^\dagger|\psi\rangle = \langle\psi|\,A|\phi\rangle^* \tag{4.358}$$

Let $\langle\phi|\,A = \langle\phi'|$ and $B|\psi\rangle = |\psi'\rangle$ so that $A^\dagger|\phi\rangle = |\phi'\rangle$ and $\langle\psi|\,B^\dagger = \langle\psi'|$. It is straightforward to show that

$$\langle\psi|\,(AB)^\dagger|\phi\rangle = \langle\phi|\,AB|\psi\rangle * = \langle\psi|\,B^\dagger A^\dagger|\phi\rangle, \tag{4.359}$$

which shows that $(AB)^\dagger = B^\dagger A^\dagger$.

3. An operator A is **hermitian** if $A = A^\dagger$ and **anti-hermitian** if $A = -A^\dagger$. Any operator A can be decomposed into its hermitian and anti-hermitian parts:

$$A = \frac{A + A^\dagger}{2} + \frac{A - A^\dagger}{2} \tag{4.360}$$

Note that even if both A and B are hermitian, their product AB is not necessarily hermitian.[25] Both $(A + B)$ and $i[A, B]$, however, are hermitian where the **commutator** of two operators is defined as

$$[A, B] = AB - BA \tag{4.361}$$

4. The **inverse** (A^{-1}) of an operator A is defined as

$$A^{-1}A = AA^{-1} = I \tag{4.362}$$

[25] This is because $(AB)^\dagger = B^\dagger A^\dagger = BA \neq AB$ in general.

where I is the **identity operator** satisfying

$$I|\psi\rangle = |\psi\rangle \quad , \quad \langle\psi| I = \langle\psi| \tag{4.363}$$

It is straightforward to show that

$$(A^{-1})^\dagger = (A^\dagger)^{-1} \tag{4.364}$$

5. An operator U is an **unitary operator** if it satisfies

$$UU^\dagger = U^\dagger U = I \Longleftrightarrow U^\dagger = U^{-1} \tag{4.365}$$

   Note that $(\forall\psi \in \mathcal{H})$ $||U\psi|| = ||\psi||$, or equivalently $\langle U\psi|U\psi\rangle = \langle\psi|\psi\rangle$. In other words, unitary operators preserve the inner product between the vectors.
6. We can also define the outer product of vectors as $|\psi\rangle\langle\phi|$. For a normalized vector $|\psi\rangle$, i.e., $\langle\psi|\psi\rangle = 1$, we define the corresponding **projector** as

$$\mathcal{P}_\psi = |\psi\rangle\langle\psi| , \qquad \mathcal{P}_\psi^2 = \mathcal{P}_\psi \tag{4.366}$$

   This helps one decompose any vector $|\phi\rangle$ into a vector parallel to $|\psi\rangle$ and that perpendicular to $|\psi\rangle$:

$$\begin{aligned} |\phi\rangle &= |\psi\rangle\langle\,\psi\mid\phi\,\rangle + (|\phi\rangle - |\psi\rangle\langle\,\psi\mid\phi\,\rangle) = |\psi\rangle\langle\,\psi\mid\phi\,\rangle + |\phi_\perp\rangle \\ &= \mathcal{P}_\psi\,|\phi\rangle + |\phi_\perp\rangle \end{aligned} \tag{4.367}$$

Basis Vectors and Matrix Representations

1. Let $|n\rangle = |1\rangle, |2\rangle, \ldots, |N\rangle$ be a finite basis of $\mathcal{H}$. $\{|n\rangle\}$ is **orthonormal** and **complete** if

$$\langle n|m\rangle = \delta_{nm} \quad , \quad \sum_{n=1}^{N} |n\rangle\langle n| = I \tag{4.368}$$

   respectively.
2. Note that, for any $|\psi\rangle \in \mathcal{H}$, we can write

$$|\psi\rangle = \sum_{n=1}^{N} c_n\,|n\rangle \quad , \quad \langle\psi| = \sum_{n=1}^{N} c_n^*\,\langle n| \tag{4.369}$$

3. The inner product can then be written as

$$\langle\psi|\psi\rangle = \sum_{n,m=1}^{N} c_n^* c_m \langle n|m\rangle = \sum_{n=1}^{N} |c_n|^2 \geq 0 \tag{4.370}$$

4. Let A be a linear operator, and $A|\psi\rangle = |\phi\rangle$ such that $|\psi\rangle = \sum_{n=1}^{N} c_n |n\rangle$ and $|\phi\rangle = \sum_{n=1}^{N} d_n |n\rangle$. We get

$$d_m = \langle m|\phi\rangle = \langle m| A |\psi\rangle = \sum_{n=1}^{N} c_n \langle m| A |n\rangle = \sum_{n=1}^{N} A_{mn} c_n \tag{4.371}$$

where

$$A_{mn} = \langle m| A |n\rangle \quad (m, n = 1, \cdots, N) \tag{4.372}$$

are the **elements of a matrix** representing the operator A in a given basis. Note also that, since $\langle m| A^\dagger |n\rangle = \langle n| A |m\rangle^*$, we have $(A^\dagger)_{mn} = A_{nm}^*$.

5. The matrix multiplication can then be written as

$$(AB)_{mn} = \langle m| AB |n\rangle = \sum_{l=1}^{N} \langle m| A |l\rangle \langle l| B |n\rangle = \sum_{l=1}^{N} A_{ml} B_{ln} \tag{4.373}$$

where we have used the completeness of the basis vectors.

6. The **trace** of an operator is the sum of its diagonal elements:

$$\text{Tr}A = \sum_{m=1}^{N} \langle m| A |m\rangle = \sum_{m=1}^{N} A_{mm} \tag{4.374}$$

Note that, $\text{Tr}(AB) = \text{Tr}(BA)$, $\text{Tr}(ABC) = \text{Tr}(CAB) = \text{Tr}(BCA)$

Eigenvalues and Eigenvectors of Operators

1. Let A be a linear operator. If $A|\psi\rangle = a|\psi\rangle$, then $|\psi\rangle$ is an **eigenvector**, and a is an **eigenvalue** of the operator A.
2. The eigenvalue problem

$$A|\psi\rangle = a|\psi\rangle \tag{4.375}$$

can be written as

$$(A - aI)|\psi\rangle = 0 \tag{4.376}$$

To get a nontrivial solution, i.e., $|\psi\rangle \neq 0$, we must find

$$\det(A - aI) = 0 \tag{4.377}$$

This equation is called the **characteristic equation** of the operator A, and the eigenvalues are found by solving it.

3. The eigenvalues of a hermitian operator A are real. Moreover, it is always possible to find a (non-unique) unitary matrix U (built out of the eigenvectors of A) such that $U^\dagger A U$ is diagonal.
4. If both A and B are hermitian operators and they commute ($[A, B] = 0$), there exist (at least) one basis constructed from the eigenvectors common to A and B.
5. Assume that A is a hermitian operator with eigenvalues a_n and corresponding eigenvectors $|a_n\rangle$. Then, A can be expressed via its **spectral decomposition** as

$$A = \sum_n a_n |a_n\rangle \langle a_n| \,, \tag{4.378}$$

where the eigenvectors $\{|a_n\rangle\}$ form a complete orthonormal basis:

$$\langle a_n|a_m\rangle = \delta_{nm}, \quad \sum_n |a_n\rangle \langle a_n| = I \tag{4.379}$$

6. The eigenvalues of a unitary operator U are complex numbers of unit modulus. That is, if $U|u\rangle = \lambda |u\rangle$, then $|\lambda| = 1$. This follows from norm preservation:

$$\| U |u\rangle \|^2 = \left\langle u|U^\dagger U|u\right\rangle = \langle u|u\rangle \Rightarrow |\lambda|^2 = 1 \tag{4.380}$$

7. Moreover, if a unitary operator U is non-degenerate (i.e., the eigenvalues are all distinct), its eigenvectors are mutually orthogonal. For two eigenvectors $U|u_n\rangle = \lambda_n |u_n\rangle$ and $U|u_m\rangle = \lambda_m |u_m\rangle$, we have

$$\langle u_n| \left(I - U^\dagger U\right) |u_m\rangle = \left(1 - \lambda_n^* \lambda_m\right) \langle u_n|u_m\rangle = 0 \tag{4.381}$$

Thus, if $\lambda_n \neq \lambda_m$, then $\langle u_n|u_m\rangle = 0$.

8. Even in the degenerate case, the eigenvectors can still be orthonormalized within each eigenspace.

Expectation Value, Probability, Uncertainty

1. Let A be a hermitian operator. For a normalized vector $|\psi\rangle$, the **expectation value** of A is defined as

$$\langle A \rangle_\psi = \langle \psi| A |\psi\rangle \tag{4.382}$$

2. Let $\{|a_n\rangle\}_n$ be an orthonormal, complete set of eigenvectors satisfying $A\,|a_n\rangle = a_n\,|a_n\rangle$. Using $|\psi\rangle = \sum_n |a_n\rangle\,\langle a_n|\psi\rangle$ and the spectral decomposition (4.378) we obtain

$$\langle\, A\,\rangle_\psi = \sum_n a_n\,|\langle a_n|\psi\rangle|^2 \tag{4.383}$$

Here, $|\langle a_n|\psi\rangle|^2$ is the **probability** of obtaining a_n when A is measured in $|\psi\rangle$. Note that, since we have normalized $|\psi\rangle$, the probabilities sum up to 1:

$$\sum_n |\langle a_n|\psi\rangle|^2 = \sum_n \langle\psi|a_n\rangle\,\langle a_n|\psi\rangle = \langle\psi|\psi\rangle = 1 \tag{4.384}$$

as expected.
3. The **uncertainty** ΔA_ψ of A in $|\psi\rangle$ is given by the standard deviation

$$\Delta A_\psi^2 = \left\langle\, (A - \langle\, A\,\rangle_\psi)^2\,\right\rangle_\psi = \left\langle\, A^2\,\right\rangle_\psi - \langle\, A\,\rangle_\psi^2 \tag{4.385}$$

4. Let A and B be hermitian operators that do not commute $[A, B] = iC$, where C is necessarily hermitian. Define

$$|\phi\rangle = \left(A - \langle A\rangle_\psi\right)|\psi\rangle, \quad |\chi\rangle = \left(B - \langle B\rangle_\psi\right)\psi$$

By using Schwarz inequality (4.355), it is straightforward to derive the **uncertainty relation**

$$\Delta A_\psi \cdot \Delta B_\psi \geq \frac{1}{2}\left|\langle\, C\,\rangle_\psi\right| \tag{4.386}$$

Generalization to Infinite Dimensions

1. Let us consider a continuous parameter x ranging over the interval $[a, b]$. To each point x we assign a basis vector $|x\rangle$. Clearly, the set $\{|x\rangle\,;\, a \leq x \leq b\}$ spans an infinite dimensional vector space. We can generalize the orthogonality and completeness relation (4.368) as

$$\langle x|x'\rangle = \delta(x - x') \quad , \quad \int_a^b |x\rangle\,\langle x|\,dx = 1 \tag{4.387}$$

where $\delta(x - x')$ is the delta function.
2. We can then expand any vector $|\psi\rangle$ in terms of $|x\rangle$:

$$|\psi\rangle = \int_a^b \psi(x)|x\rangle dx \tag{4.388}$$

The expansion coefficient $\psi(x)$ can be obtained as

$$\langle x|\psi\rangle = \langle x| \int_a^b \psi(x')\,|x'\rangle\,dx' = \int_a^b \psi(x')\,\underbrace{\langle x|x'\rangle}_{\delta(x-x')}\,dx' = \psi(x) \tag{4.389}$$

3. The inner product of two vectors can then be written as

$$\begin{aligned}\langle\phi|\psi\rangle &= \int_a^b \langle x'|\,\phi^*(x')dx' \int_a^b \psi(x)|x\rangle dx \\ &= \int_a^b \int_a^b \phi^*(x')\psi(x)\,\underbrace{\langle x'|x\rangle}_{\delta(x'-x)}\,dx'dx = \int_a^b \phi^*(x)\psi(x)dx\end{aligned} \tag{4.390}$$

4. The squared norm of a vector can then be written as

$$\langle\psi|\psi\rangle = \int_a^b \psi^*(x)\psi(x)dx \geq 0 \tag{4.391}$$

   The function $\psi(x)$ must be square integrable, i.e., $\psi \in L^2[a,b]$. Therefore, the Hilbert space is identified with the space of square integrable functions over $[a,b]$. Note also that, if the parameter x is a coordinate, then $\psi(x)$ is called a **wave function**.
5. Linear operators on this space $A : \psi \to \phi$ satisfy the same properties as in the discrete case (4.357). The matrix elements of an operator can also be obtained in a similar way as

$$A_{xx'} = \langle x|\,A\,|x'\rangle \tag{4.392}$$

6. As an example, let us consider the derivative operator

$$D|\psi\rangle = |\frac{d\psi}{dx}\rangle \tag{4.393}$$

   Since the ket vectors correspond to functions, we can write

$$\langle x|\,D|\psi\rangle = \left\langle x\middle|\frac{d\psi}{dx}\right\rangle = \frac{d\psi(x)}{dx} \tag{4.394}$$

   To obtain the matrix elements of the operator D, we insert the identity as

$$\int_a^b \langle x|\,D\,|x'\rangle\langle x'|\psi\rangle\,dx' = \int_a^b D_{xx'}\psi(x')dx' = \frac{d\psi(x)}{dx} \tag{4.395}$$

Hence, we obtain

$$D_{xx'} = \frac{d}{dx}\delta(x - x') \equiv \delta'(x - x') \tag{4.396}$$

Since the derivative of the delta function is an odd function, D is not hermitian:

$$D^*_{x'x} = \delta'(x' - x)^* = \delta'(x' - x) = -\delta'(x - x') = -D_{xx'} \tag{4.397}$$

By simply multiplying D by a pure imaginary number, we obtain

$$K = iD \tag{4.398}$$

that satisfies

$$K^*_{x'x} = \left(i\delta'(x' - x)\right)^* = -i\delta'(x' - x) = i\delta'(x - x') = K_{xx'} \tag{4.399}$$

Unlike for the discrete space, however, this does not guarantee that K is hermitian. Note that, for K to be hermitian, we must find

$$\langle\psi|\, K\,|\phi\rangle = \langle\phi|\, K\,|\psi\rangle^* \tag{4.400}$$

The left-hand side gives

$$\int_a^b \int_a^b \langle\psi|x\rangle\, \langle x|\, K\, |x'\rangle \langle x'|\phi\rangle = \int_a^b \int_a^b \psi^*(x) K_{xx'}\phi(x') = i \int_a^b \psi^*(x)\frac{d\phi(x)}{dx} \tag{4.401}$$

Compare this with the right-hand side, which gives

$$\begin{aligned}\left(i\int_a^b \phi^*(x)\frac{d\psi(x)}{dx}\right)^* &= -i\int_a^b \frac{d\psi^*(x)}{dx}\phi(x)\\ &= -i\; \psi^*(x)\phi(x)\big|_a^b + i\int_a^b \psi^*(x)\frac{d\phi(x)}{dx}\end{aligned} \tag{4.402}$$

Therefore, K is a hermitian operator only if the surface term[26]

$$-\, i\; \psi^*(x)\phi(x)\big|_a^b \tag{4.403}$$

vanishes.

[26] The surface term arises from integration by parts and must vanish for K to be hermitian. This typically requires that the wave functions vanish at the boundaries or satisfy periodicity.

7. Let us consider the spectral decomposition of a hermitian operator A with a continuous spectrum acting on an infinite dimensional Hilbert space,

$$A = \int a \, |a\rangle \langle a| \, da \,, \tag{4.404}$$

where $|a\rangle$ are eigenvectors satisfying

$$A \, |a\rangle = a \, |a\rangle \ , \tag{4.405}$$

and obeying the orthonormality and completeness relations:

$$\langle a|a'\rangle = \delta(a - a'), \quad \int |a\rangle \langle a| \, da = I. \tag{4.406}$$

This is the continuous analogue of the discrete spectral decomposition, where the sum over discrete eigenstates is replaced by an integral over a continuum of eigenstates.

8. Finally, suppose a hermitian operator A has an orthonormal, complete set of eigenvectors $|a\rangle$ with $A\,|a\rangle = a\,|a\rangle$. Similar to the discrete case, we can write the expectation value in a normalized $|\psi\rangle$ as

$$\langle \, A \, \rangle_\psi = \langle\psi| \, A \, |\psi\rangle = \int a \, |\langle a|\psi\rangle|^2 \, da \tag{4.407}$$

Here, $|\langle a|\psi\rangle|^2$ is the **probability density** of obtaining a, when A is measured in $|\psi\rangle$, and satisfies

$$\int |\langle a|\psi\rangle|^2 \, da = 1 \tag{4.408}$$

9. The extension from finite dimensional discrete spaces to infinite dimensional Hilbert spaces may appear natural and straightforward, but one must proceed with care to ensure that physical observables are well defined and hermitian.

4.6.2 On the Classical Virial Theorem

The derivation of the virial theorem learned in Example 10-1 has a nearly one-to-one correspondence with that of the well-known classical virial theorem in classical mechanics. In this note, we prove the classical virial theorem.

We consider a particle of mass m *moving in a finite region* under the potential $V(\vec{r})$. The equation of motion is

$$m \frac{d^2}{dt^2} \vec{r} = -\nabla V(\vec{r}) \tag{4.409}$$

We assume that the potential is given by a power of $\vec{r}$:

$$V(k\vec{r}) = k^a V(\vec{r}) \tag{4.410}$$

In the case of a Coulomb potential, $a = -1$, and in the case of a harmonic oscillator, $a = 2$.

We now consider the time derivative

$$\frac{d}{dt}(\vec{p}\cdot\vec{r}) \tag{4.411}$$

In quantum mechanics, this corresponds to the Heisenberg equation

$$\frac{d}{dt}\hat{\vec{p}}\cdot\hat{\vec{r}} = \frac{i}{\hbar}\left[\hat{H}, \hat{\vec{p}}\cdot\hat{\vec{r}}\right], \tag{4.412}$$

and this makes the starting point for Example 10-1. From Eq. (4.409), we obtain

$$\frac{d}{dt}(\vec{p}\cdot\vec{r}) = -\nabla V\cdot\vec{r} + \frac{1}{m}\vec{p}^{\,2} \tag{4.413}$$

Since Eq. (4.410) implies $\vec{r}\cdot\nabla V = aV$, we obtain

$$\frac{d}{dt}(\vec{p}\cdot\vec{r}) = -aV(\vec{r}) + \frac{1}{m}\vec{p}^{\,2} \tag{4.414}$$

Now, assuming that the particle moves in a finite region (corresponding to bound states in quantum mechanics), the time average of $\frac{d}{dt}(\vec{p}\cdot\vec{r})$ becomes zero:

$$\lim_{T\to\infty}\frac{1}{T}\int_0^T dt\,\frac{d}{dt}(\vec{p}\cdot\vec{r}) = \lim_{T\to\infty}\frac{1}{T}(\vec{p}\cdot\vec{r}(t=T) - \vec{p}\cdot\vec{r}(t=0)) = 0 \tag{4.415}$$

Therefore, the time average of the right-hand side of Eq. (4.414) also becomes zero:

$$-a\langle V\rangle + 2\frac{\langle\vec{p}^{\,2}\rangle}{2m} = 0 \tag{4.416}$$

From this, we obtain the relation between the average kinetic energy $K.E.$ and the average potential energy $P.E.$ as

$$\boxed{K.E. = \frac{a}{2}P.E.} \tag{4.417}$$

As has been found in Example 10-1, we obtain

$$K.E. = -\frac{1}{2}P.E. \tag{4.418}$$

in the case of a Coulomb potential, and

$$K.E. = P.E. \tag{4.419}$$

in the case of a harmonic oscillator.

There is a well-known application of the virial theorem in astronomy. The mass distribution within a globular cluster has been inferred from the relation between the velocity of a star and its distance from the center of the cluster. The presence of dark matter, invisible to optical observations, has been confirmed. We still do not know what constitutes the dark matter.

4.6.3 Wigner-Eckart Theorem

An important property of angular momentum is its rule of composition. When combining systems with angular momenta j_1 and j_2, the resulting system can have a total angular momentum j_3, which is restricted to $j_1 + j_2, j_1 + j_2 - 1, \cdots, |j_1 - j_2|$. Correspondingly, acting on a quantum state with angular momentum j_1 by an operator with angular momentum j_2, the angular momentum j_3 of the resulting quantum state is restricted to $j_1 + j_2, j_1 + j_2 - 1, \cdots, |j_1 - j_2|$. Furthermore, the matrix elements of the operator are determined by Clebsch-Gordan coefficients, independent of the choice of operator. This is the content of Wigner-Eckart theorem.

We introduce an operator $\hat{T}_m (m = \pm j_2, \pm(j_2 - 1), \cdots)$, which has angular momentum j_2. This is defined by the following commutation relations with the angular momentum operators $\hat{J}_i$:

$$\left[\hat{J}_+, \hat{T}_m\right] = \hbar\sqrt{(j_2 - m)(j_2 + m + 1)}\, \hat{T}_{m+1} \tag{4.420a}$$

$$\left[\hat{J}_-, \hat{T}_m\right] = \hbar\sqrt{(j_2 + m)(j_2 - m + 1)}\, \hat{T}_{m-1} \tag{4.420b}$$

$$\left[\hat{J}_z, \hat{T}_m\right] = \hbar m\, \hat{T}_m \tag{4.420c}$$

Examples $(j_2 = 1)$

- Position—$\hat{T}_{\pm 1} = \mp(\hat{x} \pm i\hat{y})$, $\hat{T}_0 = \hat{z}$
- Angular momentum—$\hat{T}_{\pm 1} = \mp(\hat{J}_x \pm i\hat{J}_y)$, $\hat{T}_0 = \hat{J}_z$

Examples $(j_2 = l = 0, 1, 2, \cdots)$

- l-th power of $\hat{\vec{x}}$—$\hat{T}_m \equiv \hat{r}^l\, Y_{l,m}(\hat{\theta}, \hat{\phi})$ $(m = \pm l, \pm(l-1), \cdots, \pm 1, 0)$

Denoting the eigenstates of angular momentum j_1 by $|j_1, m\rangle$ ($m = \pm j_1, \pm(j_1 - 1), \cdots$), we obtain

$$\hat{J}_z \, |j_1, m\rangle = \hbar m \, |j_1, m\rangle \tag{4.421}$$

The theorem of Wigner and Eckart gives

$$\boxed{\langle j_3, m_3| \, \hat{T}_m \, |j_1, m_1\rangle = \langle \, j_3 \, |T| \, j_1 \, \rangle \, \langle \, j_2, j_1; m_2, m_1 \mid j_2, j_1; \, j_3, m_3 \, \rangle} \tag{4.422}$$

where $\langle \, j_3 \, |T| \, j_1 \, \rangle$ is a constant independent of m and m_1, determined only by j_1 and j_3. Of course, the value of this constant depends on the operator $\hat{T}$; it can be different for two operators with the same j_2. The last factor $\langle \, j_2, j_1; m_2, m_1 \mid j_2, j_1; \, j_3, m_3 \, \rangle$ is the Clebsch-Gordan coefficient, a universal coefficient independent of the choice of the operator $\hat{T}$.

As a preparation for the proof of the theorem, let us review the Clebsch-Gordan coefficients first. Combining the states $|j_1, m_1\rangle$ and $|j_2, m_2\rangle$ with angular momenta j_1 and j_2, we obtain the state $|j_2, j_1; m_2, m_1\rangle \equiv |j_2, m_2\rangle \, |j_1, m_1\rangle$. The Clebsch-Gordan coefficient is defined by

$$|j_3, m_3\rangle = \sum_{m_1+m_2=m_3} |j_2, j_1; m_2, m_1\rangle \times \langle \, j_2, j_1; m_2, m_1 \mid j_2, j_1; \, j_3, m_3 \, \rangle \tag{4.423}$$

where the summation runs over all possible m_1 and m_2 satisfying $m_1 + m_2 = m_3$. The Clebsch-Gordan coefficient $\langle \, j_2, j_1; m_2, j_3 - m_2 \mid j_2, j_1; \, j_3, j_3 \, \rangle$ for the maximum $m_3 = j_3$ is uniquely determined except for the overall normalization, as $|j_3, j_3\rangle$ is orthogonal to $|j_1 + j_2, j_3\rangle$, $|j_1 + j_2 - 1, j_3\rangle$, $\cdots$, $|j_3 + 1, j_3\rangle$. The coefficients for the smaller m_3 ($< j_3$) can be determined from the recursion relation:

$$\begin{aligned}
&\hbar\sqrt{(j_3 + m_3)(j_3 - m_3 + 1)} \, \langle \, j_1, j_2; m_1, m_2 \mid j_1, j_2; \, j_3, m_3 - 1 \, \rangle \\
&= \langle j_1, j_2; m_1, m_2| \, \hat{J}_{3,-} \, |j_3, m_3\rangle \\
&= \langle j_1, j_2; m_1, m_2| \left(\hat{J}_{1,-} + \hat{J}_{2,-} \right) |j_3, m_3\rangle \\
&= \hbar\sqrt{(j_1 - m_1)(j_1 + m_1 + 1)} \, \langle \, j_1, j_2; m_1 + 1, m_2 \mid j_3, m_3 \, \rangle \\
&\quad + \hbar\sqrt{(j_2 - m_2)(j_2 + m_2 + 1)} \, \langle \, j_1, j_2; m_1, m_2 + 1 \mid j_3, m_3 \, \rangle
\end{aligned} \tag{4.424}$$

With this preparation, the proof of Eq. (4.422) is straightforward. Note that the matrix elements of the operator $\hat{T}$ satisfy the following recursion relation:

$$\begin{aligned}
&\hbar\sqrt{(j_3 + m_3)(j_3 - m_3 + 1)} \, \langle j_3, m_3 - 1| \, \hat{T}_{m_2} \, |j_1, m_1\rangle \\
&= \langle j_3, m_3| \, \hat{J}_+ \hat{T}_{m_2} \, |j_1, m_1\rangle
\end{aligned}$$

$$
\begin{aligned}
&= \langle j_3, m_3| \left([\hat{J}_+, \hat{T}_{m_2}] + \hat{T}_{m_2}\hat{J}_+ \right) |j_1, m_1\rangle \\
&= \hbar\sqrt{(j_2 - m_2)(j_2 + m_2 + 1)}\, \langle j_3, m_3| \hat{T}_{m_2+1} |j_1, m_1\rangle \\
&\quad + \hbar\sqrt{(j_1 - m_1)(j_1 + m_1 + 1)}\, \langle j_3, m_3| \hat{T}_{m_2} |j_1, m_1 + 1\rangle
\end{aligned} \tag{4.425}
$$

Evidently this recursion relation is identical to (4.424) satisfied by the Clebsch-Gordan coefficients. Since (4.424) determines the Clebsch-Gordan coefficients up to normalization, we conclude that

$$
\langle j_3, m_3| \hat{T}_{m_2} |j_1, m_1\rangle = \text{const} \times \langle\, j_2, j_1; m_2, m_1 \mid j_2, j_1; j_3, m_3 \,\rangle\ , \tag{4.426}
$$

where $m_3 = m_1 + m_2$. (End of Proof)

One of the most important applications of the Wigner-Eckart theorem is the selection rules for transitions among the energy levels of electrons in atoms. In the electric dipole approximation (valid if the wavelength is larger than the atomic size), the transition probability amplitude is proportional to

$$
\langle f| \vec{\epsilon} \cdot \sum_i \hat{\vec{x}}_i |i\rangle \tag{4.427}
$$

where the sum extends over all electrons in the atom and $\vec{\epsilon}$ is the polarization vector of the absorbed or emitted photon. Since the operator $\sum_i \hat{\vec{x}}_i$ corresponds to $j_2 = 1$, the final angular momentum j_f must satisfy

$$
j_f = j_i \pm 1,\ j_i \tag{4.428}
$$

and the matrix element is determined by the Clebsch-Gordan coefficients, apart from an overall proportionality constant.

Finally, let us consider giving the product $Y_{l,m}Y_{l',m'}$ of two spherical harmonics as a sum of spherical harmonics. For this purpose, let

$$
\hat{T}_m \equiv Y_{l,m}(\theta, \phi) \tag{4.429}
$$

and consider the matrix element

$$
\left\langle l'', m'' \right| \hat{T}_m \left| l', m' \right\rangle \tag{4.430}
$$

This matrix element can be expressed as

$$
\begin{aligned}
\left\langle l'', m'' \right| \hat{T}_m \left| l', m' \right\rangle &= \int d\cos\theta d\phi \left\langle\, l'', m'' \mid \theta, \phi \,\right\rangle \langle \theta, \phi| \hat{T}_m \left| l', m' \right\rangle \\
&= \int d\cos\theta d\phi\, Y^*_{l'',m''}(\theta, \phi) Y_{l,m}(\theta, \phi) Y_{l',m'}(\theta, \phi)
\end{aligned} \tag{4.431}
$$

Therefore, we obtain

$$Y_{l,m}(\theta,\phi)Y_{l',m'}(\theta,\phi) = \sum_{l''} Y_{l'',m+m'}(\theta,\phi)\left\langle l'', m''\right| \hat{T}_m \left|l', m'\right\rangle \tag{4.432}$$

Hence, we obtain

$$\begin{aligned} &Y_{l,m}(\theta,\phi)Y_{l',m'}(\theta,\phi) \\ &= \sum_{l''=|l-l'|,\cdots,l+l'} Y_{l'',m+m'}(\theta,\phi)\left\langle\, l'' \,|T|\, l' \,\right\rangle\left\langle\, l, l'; m, m' \mid l, l'; l'', m+m' \,\right\rangle \end{aligned} \tag{4.433}$$

The coefficient $\left\langle\, l'' \,|T|\, l' \,\right\rangle$ depends only on l, l', l'', but we cannot determine it using the Wigner-Eckart theorem alone.

4.6.4 Euler Angles and d-Functions

We denote the angular momentum operator as $\hat{\vec{J}}$ and the normalized eigenstates of $\hat{\vec{J}}^2$ and $\hat{J}_z$ as $|J, M\rangle$. We are interested in the matrix elements of an arbitrary rotation operator $\hat{R}$ given by

$$\left\langle J, M'\right| \hat{R} \left|J, M\right\rangle . \tag{4.434}$$

First, let us consider how $\hat{R}$ is expressed in terms of $\hat{\vec{J}}$. We recall that any rotation can be specified by three Euler angles. A general rotation can be obtained as follows:[27]

1. First, rotate the coordinate axes x, y, z around the z-axis by an angle α. The new coordinate system after the rotation is denoted as x', y', $z' = z$.
2. Next, rotate the coordinate axes x', y', z' around the y'-axis by an angle β. The new coordinate system after the rotation is denoted as x'', $y'' = y'$, z''.
3. Finally, rotate the x'', y'', z''-axes around the z''-axis by an angle γ.

Therefore, any rotation of a physical system (but not the coordinate axes) can be expressed as

$$\hat{R}(\alpha,\beta,\gamma) = \mathrm{e}^{-i\frac{\gamma}{\hbar}\hat{J}_{z''}}\mathrm{e}^{-i\frac{\beta}{\hbar}\hat{J}_{y'}}\mathrm{e}^{-i\frac{\alpha}{\hbar}\hat{J}_z} . \tag{4.435}$$

Let us rewrite this in terms of $\hat{J}_{x,y,z}$.

We first note

$$\hat{J}_{y'} = \mathrm{e}^{-i\frac{\alpha}{\hbar}\hat{J}_z}\hat{J}_y\mathrm{e}^{i\frac{\alpha}{\hbar}\hat{J}_z} . \tag{4.436}$$

---

[27] In some cases, the x-axis is used instead of the y-axis.

Similarly, we obtain

$$\hat{J}_{z''} = \mathrm{e}^{-i\frac{\beta}{\hbar}\hat{J}_{y'}}\hat{J}_{z'}\mathrm{e}^{i\frac{\beta}{\hbar}\hat{J}_{y'}} = \mathrm{e}^{-i\frac{\alpha}{\hbar}\hat{J}_z}\mathrm{e}^{-i\frac{\beta}{\hbar}\hat{J}_y}\hat{J}_z\mathrm{e}^{i\frac{\beta}{\hbar}\hat{J}_y}\mathrm{e}^{i\frac{\alpha}{\hbar}\hat{J}_z}\,. \tag{4.437}$$

Therefore,

$$\begin{aligned}\hat{R}(\alpha,\beta,\gamma) &= \mathrm{e}^{-i\frac{\alpha}{\hbar}\hat{J}_z}\mathrm{e}^{-i\frac{\beta}{\hbar}\hat{J}_y}\mathrm{e}^{-i\frac{\gamma}{\hbar}\hat{J}_z}\mathrm{e}^{i\frac{\beta}{\hbar}\hat{J}_y}\mathrm{e}^{i\frac{\alpha}{\hbar}\hat{J}_z}\,.\,\mathrm{e}^{-i\frac{\alpha}{\hbar}\hat{J}_z}\mathrm{e}^{-i\frac{\beta}{\hbar}\hat{J}_y}\mathrm{e}^{i\frac{\alpha}{\hbar}\hat{J}_z}\,.\,\mathrm{e}^{-i\frac{\alpha}{\hbar}\hat{J}_z}\\ &= \mathrm{e}^{-i\frac{\alpha}{\hbar}\hat{J}_z}\mathrm{e}^{-i\frac{\beta}{\hbar}\hat{J}_y}\mathrm{e}^{-i\frac{\gamma}{\hbar}\hat{J}_z}\,.\end{aligned} \tag{4.438}$$

Thus, when the physical system is rotated under fixed coordinate axes, the order of Euler angles is reversed. Therefore, the matrix elements of $\hat{R}$ are given by

$$\begin{aligned}\left\langle J, M'\right|\hat{R}(\alpha,\beta,\gamma)\left|J, M\right\rangle &= \left\langle J, M'\right|\mathrm{e}^{-i\frac{\alpha}{\hbar}\hat{J}_z}\mathrm{e}^{-i\frac{\beta}{\hbar}\hat{J}_y}\mathrm{e}^{-i\frac{\gamma}{\hbar}\hat{J}_z}\left|J, M\right\rangle\\ &= \mathrm{e}^{-iM'\alpha-iM\gamma}\left\langle J, M'\right|\mathrm{e}^{-i\frac{\beta}{\hbar}\hat{J}_y}\left|J, M\right\rangle\,.\end{aligned} \tag{4.439}$$

We define the d-function by

$$\boxed{d^J_{M,M'}(\theta) \equiv \left\langle J, M'\right|\mathrm{e}^{i\frac{\theta}{\hbar}\hat{J}_y}\left|J, M\right\rangle} \tag{4.440}$$

so that

$$\left\langle J, M'\right|\hat{R}(\alpha,\beta,\gamma)\left|J, M\right\rangle = \mathrm{e}^{-iM'\alpha-iM\gamma}d^J_{M,M'}(-\beta)\,. \tag{4.441}$$

For $M' = 0$ the d-functions can be written in terms of spherical harmonics $Y_{l,m}$, which are defined by

$$Y_{l,m}(\theta,\phi) = \mathrm{e}^{im\phi}\left\langle +z\right|\hat{R}_y(-\theta)\left|l, m\right\rangle\,. \tag{4.442}$$

Since

$$\left\langle +z\right| = \sum_{l=0}^{\infty}\underbrace{\langle +z\mid l,0\rangle}_{=Y_{l,0}(0,0)=\sqrt{\frac{2l+1}{4\pi}}}\cdot\left\langle l,0\right|\,, \tag{4.443}$$

we obtain

$$Y_{l,m}(\theta,\phi) = \mathrm{e}^{im\phi}\sqrt{\frac{2l+1}{4\pi}}\underbrace{\left\langle l,0\right|\hat{R}_y(-\theta)\left|l,m\right\rangle}_{=d^l_{m,0}(\theta)} \tag{4.444}$$

Hence, we get

$$\boxed{d^l_{m,0}(\theta) = \sqrt{\frac{4\pi}{2l+1}}\, Y_{l,m}(\theta,\phi)\mathrm{e}^{-im\phi}} \tag{4.445}$$

4.6.5 Quantization of Magnetic Monopoles

The magnetic field of a magnetic monopole with magnetic charge g is spherically symmetric and given by

$$\vec{B}(\vec{r}) = \frac{g}{r^2}\hat{r} = \frac{g}{r^3}\vec{r} \tag{4.446}$$

This magnetic field satisfies

$$\nabla \cdot \vec{B}(\vec{r}) = 4\pi g\, \delta^{(3)}(\vec{r}), \tag{4.447}$$

which contradicts Maxwell's equations. In quantum mechanics, however, a magnetic monopole can exist, though the magnetic charge cannot be arbitrary but must be quantized. We explain this result, first obtained by P. A. M. Dirac, in the following.[28]

To describe a magnetic monopole in terms of a vector potential, we follow Wu and Yang and divide the entire space into two regions. We denote the part of space excluding the negative half of the z-axis as R_+ and the part excluding the positive half of the z-axis as R_-. The vector potential $\vec{A}_+$ defined in R_+ gives the magnetic field produced by a solenoid placed along the negative z-axis, and the vector potential $\vec{A}_-$ defined in R_- gives the magnetic field produced by a solenoid placed along the positive z-axis. The solenoids are invisible in $R_\pm$, and if we choose the flux of the solenoids to be $4\pi g$ each, it appears as though a magnetic monopole with magnetic charge g exists (Fig. 4.16).

Now, consider any spherical surface of radius r centered on the magnetic monopole. We denote the hemisphere satisfying $z \geq 0$ as S_+ and the hemisphere satisfying $z \leq 0$ as S_-. Additionally, we denote the intersection of $S_\pm$ as the circle C. (See Fig. 4.17.) The flux passing through S_+ upward is given by

$$\int_{S_+} d\vec{S} \cdot \vec{B} = \oint_C \vec{A}_+ \cdot d\vec{x} = 2\pi g\,, \tag{4.448}$$

---

[28] P. A. M. Dirac, Proc. Roy. Soc. (London) **A 133**(1931)60. We follow a more modern method introduced in T.-T. Wu, C.-N. Yang, Nucl. Phys. **B107**(1976)365–380.

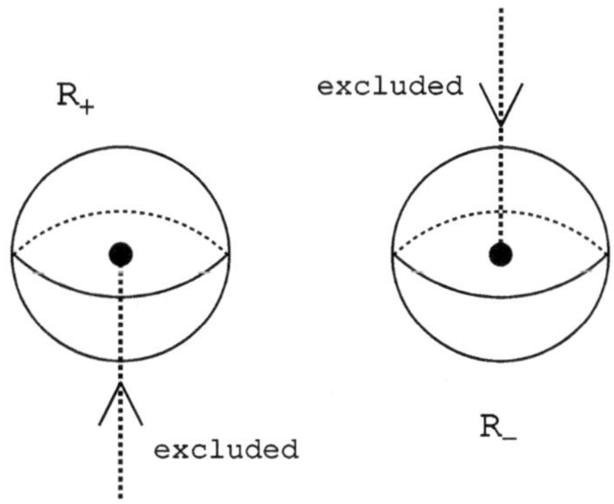

Fig. 4.16 Creating the entire space by combining $R_+ = \{(x, y, z) \mid (x, y, z) \neq (0, 0, z < 0)\}$ and $R_- = \{(x, y, z) \mid (x, y, z) \neq (0, 0, z > 0)\}$

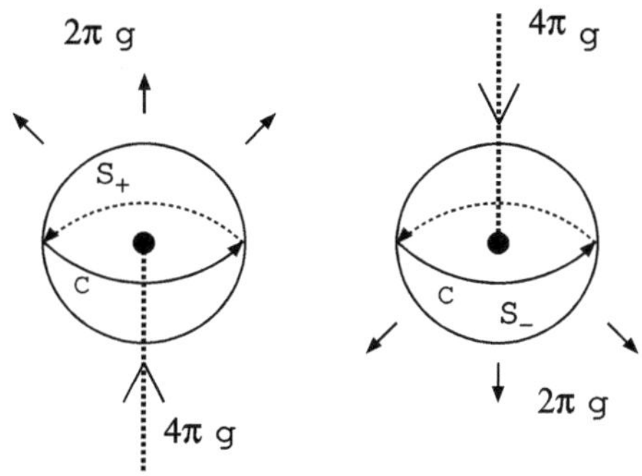

Fig. 4.17 The flux passing through $S_\pm$ is $2\pi g$

and the flux passing through S_- downward is given by

$$\int_{S_-} d\vec{S} \cdot \vec{B} = -\oint_C \vec{A}_- \cdot d\vec{x} = 2\pi g \tag{4.449}$$

Therefore, we find

$$\oint_C \left(\vec{A}_+ - \vec{A}_-\right) \cdot d\vec{x} = 4\pi g \tag{4.450}$$

The vector potentials $\vec{A}_+$ and $\vec{A}_-$ provide the same magnetic field near the vicinity of the circle C, so they must be connected by a gauge transformation:

$$\vec{A}_+ - \vec{A}_- = \nabla\phi\,, \tag{4.451}$$

where ϕ is a function defined along the circle C. From (4.450) we obtain

$$\oint_C d\vec{x} \cdot \nabla\phi = \Delta\phi = 4\pi g\,, \tag{4.452}$$

where $\Delta\phi$ is the change (or discontinuity) in ϕ around the circle C.

Now, let us consider a particle with charge e moving in the magnetic field of the monopole. The Hamiltonian for the particle is given by

$$H = \frac{1}{2m}\left(\vec{p} - \frac{e}{c}\vec{A}\right)^2 = -\frac{\hbar^2}{2m}\left(\nabla - i\frac{e}{\hbar c}\vec{A}\right)^2 \tag{4.453}$$

Let $\psi_+(\vec{x})$ be the wave function in R_+, and let $\psi_-(\vec{x})$ be that in R_-. The two must be related by a gauge transformation at the intersection of R_+ and R_- (i.e., the entire space excluding the z-axis) as

$$\psi_+(\vec{x}) = \exp\left(i\frac{e\phi(\vec{x})}{\hbar c}\right)\psi_-(\vec{x}) \tag{4.454}$$

The wave functions $\psi_\pm$ must remain single-valued along the circle C. This requires

$$\exp\left(i\frac{e\Delta\phi}{\hbar c}\right) = 1, \tag{4.455}$$

Using Eq. (4.452), we obtain

$$eg = \frac{n}{2}\hbar c \quad (n = 0, \pm 1, \pm 2, \cdots) \tag{4.456}$$

Given e, this implies the quantization of the magnetic charge g. Conversely, given g, this implies the quantization of the particle's electric charge:

$$e = \frac{n}{2}\frac{\hbar c}{g}. \tag{4.457}$$

4.6.6 Method of Propagators

In addition to the time independent perturbation theory summarized in Sect. 4.4, there is another version of perturbation theory due to Brillouin and Wigner. The purpose of this note is to introduce the Brillouin-Wigner perturbation theory using the method of propagators (also known as resolvents). Propagators become a very useful tool in quantum field theory.[29] The application of propagators to time-dependent perturbation theory is explained in Sect. 5.3.2.

As usual, we divide a Hamiltonian into two parts:

$$\hat{H} = \hat{H}_0 + \hat{V}, \tag{4.458}$$

[29] For applications of propagators in solid-state physics, R. D. Mattuck, *A guide to Feynman diagrams in the many-body problem* (Dover) is an enlightening and idiosyncratic reference.

where $\hat{H}_0$ is the Hamiltonian without interaction, and it is diagonalized as

$$\hat{H}_0 |n\rangle = \epsilon_n |n\rangle , \quad \langle\, m \mid n \,\rangle = \delta_{mn} \tag{4.459}$$

$\hat{V}$ is the time independent interaction Hamiltonian, and its matrix elements are written as

$$V_{nm} \equiv \langle n| \hat{V} |m\rangle \tag{4.460}$$

The propagator for energy E is defined as

$$\frac{1}{E - \hat{H}} \tag{4.461}$$

This can be formally expanded as a geometric series

$$\begin{aligned}\frac{1}{E-\hat{H}} &= \frac{1}{E-\hat{H}_0} + \frac{1}{E-\hat{H}_0}\hat{V}\frac{1}{E-\hat{H}_0} \\ &\quad + \frac{1}{E-\hat{H}_0}\hat{V}\frac{1}{E-\hat{H}_0}\hat{V}\frac{1}{E-\hat{H}_0} + \cdots\end{aligned} \tag{4.462}$$

The diagonal elements are given by

$$\begin{aligned}\langle n| \frac{1}{E-\hat{H}} |n\rangle &= \frac{1}{E-\epsilon_n} + \frac{1}{E-\epsilon_n} V_{nn} \frac{1}{E-\epsilon_n} \\ &\quad + \frac{1}{E-\epsilon_n} \sum_m V_{nm} \frac{1}{E-\epsilon_m} V_{mn} \frac{1}{E-\epsilon_n} + \cdots\end{aligned} \tag{4.463}$$

We now define the **self-energy** matrix $\Sigma^n(E)$ of the n-th state recursively as

$$\Sigma^n_{km}(E) \equiv V_{km} + \sum_{l \neq n} V_{kl} \frac{1}{E-\epsilon_l} \Sigma^n_{lm}(E) \tag{4.464}$$

where the summation excludes the n-th state. Expanding this in terms of the interaction matrix elements, we obtain

$$\Sigma^n_{km}(E) = V_{km} + \sum_{l \neq n} V_{kl} \frac{1}{E-\epsilon_l} V_{lm} + \sum_{l_1 \neq n} \sum_{l_2 \neq n} V_{kl_1} \frac{1}{E-\epsilon_{l_1}} V_{l_1 l_2} \frac{1}{E-\epsilon_{l_2}} V_{l_2 m} + \cdots \tag{4.465}$$

We emphasize that the sum over the intermediate states does not include n. Using the self-energy, the diagonal element of the propagator becomes a geometric series:

$$\langle n| \frac{1}{E-\hat{H}} |n\rangle = \frac{1}{E-\epsilon_n} + \frac{1}{E-\epsilon_n}\Sigma^n_{nn}(E)\frac{1}{E-\epsilon_n} + \frac{1}{E-\epsilon_n}\Sigma^n_{nn}(E)\frac{1}{E-\epsilon_n}\Sigma^n_{nn}(E)\frac{1}{E-\epsilon_n} + \cdots \tag{4.466}$$

Hence, we obtain

$$\langle n| \frac{1}{E-\hat{H}} |n\rangle = \frac{1}{E-\epsilon_n-\Sigma^n_{nn}(E)} \tag{4.467}$$

Now, let E_n be the n-th eigenvalue of the full Hamiltonian $\hat{H}$, corresponding to ϵ_n of $\hat{H}_0$. We denote the normalized eigenstate by $|E_n\rangle$. In the limit $E \to E_n$, the definition of the propagator gives

$$\langle n| \frac{1}{E-\hat{H}} |n\rangle \overset{E\to E_n}{\longrightarrow} \frac{1}{E-E_n} |\langle E_n \mid n \rangle|^2 \tag{4.468}$$

On the other hand, using Eq. (4.467) we obtain

$$\frac{1}{E-\epsilon_n-\Sigma^n_{nn}(E)} \overset{E\to E_n}{\longrightarrow} \frac{1}{E_n-\epsilon_n-\Sigma^n_{nn}(E_n)+\left(1-\Sigma^{n\prime}_{nn}(E_n)\right)(E-E_n)} \tag{4.469}$$

where the prime denotes a derivative with respect to E. Since these two expressions must match, we obtain the following two equalities:

$$E_n = \epsilon_n + \Sigma^n_{nn}(E_n) \tag{4.470}$$

and

$$|\langle E_n \mid n \rangle|^2 = \frac{1}{1-\Sigma^{n\prime}_{nn}(E_n)} \leq 1 \tag{4.471}$$

Substituting the perturbative expansion of $\Sigma^n_{nn}(E_n)$ into the first equality, we obtain a result of the Brillouin-Wigner perturbation theory:

$$\begin{aligned} E_n &= \epsilon_n + V_{nn} \\ &+ \sum_{m\neq n} V_{nm}\frac{1}{E_n-\epsilon_m}V_{mn} \\ &+ \sum_{m_1\neq n}\sum_{m_2\neq n} V_{nm_1}\frac{1}{E_n-\epsilon_{m_1}}V_{m_1m_2}\frac{1}{E_n-\epsilon_{m_2}}V_{m_2n} \\ &+ \cdots \end{aligned} \tag{4.472}$$

Solving this iteratively in powers of V's, we reproduce the usual perturbative expansion for E_n.[30] The Brillouin-Wigner expansion may result in a better approximation. For example, we truncate the right-hand side above at second order to obtain the following closed equation for E_n:

$$E_n = \epsilon_n + V_{nn} + \sum_{m \neq n} V_{nm} \frac{1}{E_n - \epsilon_m} V_{mn} \tag{4.473}$$

Solving this equation for E_n, we obtain a value that partially accounts for the higher-order contributions compared with the usual second-order perturbative approximation; hence it can and may lead to a better approximation.

To find the inner product between the eigenstate $|E_n\rangle$ and $|m\rangle$ ($m \neq n$), we consider the off-diagonal elements of the propagator:

$$\begin{aligned}
\langle m| \frac{1}{E - \hat{H}} |n\rangle &= \frac{1}{E - \epsilon_m} V_{mn} \frac{1}{E - \epsilon_n} + \frac{1}{E - \epsilon_m} \sum_l V_{ml} \frac{1}{E - \epsilon_l} V_{ln} \\
&\quad + \frac{1}{E - \epsilon_m} \sum_{l_1} V_{ml_1} \frac{1}{E - \epsilon_{l_1}} \sum_{l_2} V_{l_1 l_2} \frac{1}{E - \epsilon_{l_2}} V_{l_2 n} + \cdots \\
&= \frac{1}{E - \epsilon_m} \Sigma^n_{mn}(E) \frac{1}{E - \epsilon_n} \\
&\quad + \frac{1}{E - \epsilon_m} \Sigma^n_{mn}(E) \frac{1}{E - \epsilon_n} \Sigma^n_{nn}(E) \frac{1}{E - \epsilon_n} \\
&\quad + \frac{1}{E - \epsilon_m} \Sigma^n_{mn}(E) \frac{1}{E - \epsilon_n} \Sigma^n_{nn}(E) \frac{1}{E - \epsilon_n} \Sigma^n_{nn}(E) \frac{1}{E - \epsilon_n} + \cdots \\
&= \frac{1}{E - \epsilon_m} \Sigma^n_{mn}(E) \frac{1}{E - \epsilon_n - \Sigma^n_{nn}(E)}
\end{aligned} \tag{4.474}$$

Taking the limit $E \to E_n$, we find on one hand

$$\langle m| \frac{1}{E - \hat{H}} |n\rangle \overset{E \to E_n}{\longrightarrow} \langle\, m \mid E_n \,\rangle \langle\, E_n \mid n \,\rangle \frac{1}{E - E_n} \tag{4.475}$$

but on the other hand

$$\langle m| \frac{1}{E - \hat{H}} |n\rangle \overset{E \to E_n}{\longrightarrow} \frac{1}{E_n - \epsilon_m} \Sigma^n_{mn}(E_n) | \langle\, E_n \mid n \,\rangle |^2 \frac{1}{E - \epsilon_n} \tag{4.476}$$

Comparing these two, we obtain

$$\langle\, m \mid E_n \,\rangle \langle\, E_n \mid n \,\rangle = \frac{1}{E_n - \epsilon_m} \Sigma^n_{mn}(E_n) \,|\langle\, E_n \mid n \,\rangle|^2 \tag{4.477}$$

30 The usual perturbative expansion is due to Rayleigh and Schrödinger.

Thus, the Brillouin-Wigner perturbation theory gives

$$\begin{aligned}\frac{\langle\, m \mid E_n \,\rangle}{\langle\, n \mid E_n \,\rangle} &= \frac{1}{E_n - \epsilon_m}\Sigma^n_{mn}(E_n) \\ &= \frac{1}{E_n - \epsilon_m}\Big[V_{mn} + \sum_{k\neq n} V_{mk}\frac{1}{E_n - \epsilon_k}V_{kn} \\ &\quad + \sum_{k_1\neq n}\sum_{k_2\neq n} V_{mk_1}\frac{1}{E_n - \epsilon_{k_1}}V_{k_1k_2}\frac{1}{E_n - \epsilon_{k_2}}V_{k_2n} + \cdots\Big] \end{aligned} \tag{4.478}$$

Substituting the usual perturbative expansion into E_n, we reproduce the usual perturbative expansion for $\langle\, m \mid E_n \,\rangle$.

4.6.7 Resonances

We have given a very crude discussion of quantum tunneling at the end of Sect. 4.1.3. In the following we would like to give a more general derivation of the tunneling probability. To be concrete, we consider the Hamiltonian in the one-dimensional half space:

$$H = -\frac{\hbar^2}{2m}\frac{d^2}{dx^2} + V(x) \quad (x \geq 0) \tag{4.479}$$

where the potential is given by

$$V(x) = \begin{cases} 0 & (0 < x < a) \\ V_0 = \frac{\hbar^2\kappa^2}{2m} & (a < x < b) \\ 0 & (b < x) \end{cases} \tag{4.480}$$

We impose the boundary condition that the wave function vanish at $x = 0$. This Hamiltonian can be considered as that of a particle in a spherically symmetric potential in three dimensions, where the radius r replaces x.

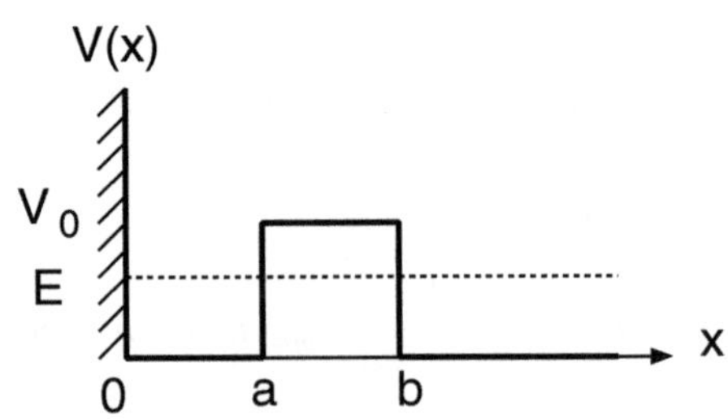

Potential barrier; the same as Fig. 4.5

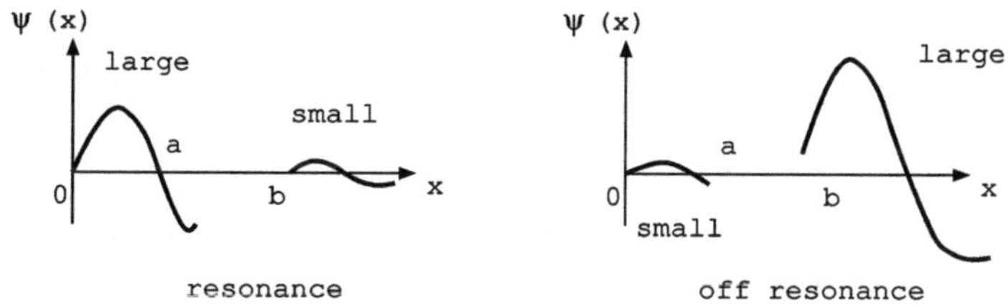

Fig. 4.18 $|A_k|$ is small for a resonance

For $V_0 \to +\infty$, we find bound states in $x \in [0, a]$ with energy eigenvalues $\frac{\hbar^2 k_n^2}{2m}$ and wave functions $\sin k_n x$, where $k_n = \frac{n\pi}{a}$ $(n = 1, 2, \cdots)$. For finite V_0, all these states become unstable, and we are left with no bound state. For V_0 large enough, however, the low-lying states satisfying $\frac{\hbar^2 k_n^2}{2m} \ll V_0$ remain almost stable with a long lifetime. These are called **resonance states** or **virtual levels**. In Sect. 4.1.3 we have estimated the lifetime of a generic bound state using the WKB approximation. We improve on the approximation.

For energy E less than V_0, the wave function has two modes in $x \in [a, b]$: one decreasing exponentially and the other increasing exponentially. Resonance states have only the exponentially decaying mode so that the amplitude of the wave function is much larger in $x \in [0, a]$ than in $x > b$ (Fig. 4.18).

Let $\psi_k(x)$ be the wave function of the energy eigenstate with energy $E_k = \frac{\hbar^2 k^2}{2m} < V_0$, i.e., $k < \kappa$. The wave function is given by

$$\psi_k(x) = \begin{cases} \sin kx & 0 < x < a \\ C_+ e^{\sqrt{\kappa^2 - k^2}\,(x-a)} + C_- e^{-\sqrt{\kappa^2 - k^2}\,(x-a)} & a < x < b \\ A_k e^{ik(x-b)} + A_k^* e^{-ik(x-b)} & x > b \end{cases} \tag{4.481}$$

By continuity of $\psi(x)$ and its derivative $\psi'(x)$, we can calculate the real coefficients $C_\pm$, and the complex coefficient A_k:

$$A_k = \frac{1}{2ik}\left\{\left(ik + \sqrt{\kappa^2 - k^2}\right) C_+ e^{\sqrt{\kappa^2 - k^2}(b-a)} + \left(ik - \sqrt{\kappa^2 - k^2}\right) C_- e^{-\sqrt{\kappa^2 - k^2}(b-a)}\right\} \tag{4.482}$$

where

$$C_+ = \frac{\sqrt{\kappa^2 - k^2}\sin ka + k\cos ka}{2\sqrt{\kappa^2 - k^2}} \tag{4.483a}$$

$$C_- = \frac{\sqrt{\kappa^2 - k^2}\sin ka - k\cos ka}{2\sqrt{\kappa^2 - k^2}} \tag{4.483b}$$

A_k depends on the exponentially enhanced C_+ and the exponentially suppressed C_-. Near the resonance states, we should find $|A_k| \ll 1$, which requires

$$|C_+| \ll |C_-| \tag{4.484}$$

Namely, the mode growing exponentially in $a < x < b$ is suppressed.

Assuming a very high potential wall satisfying

$$\kappa a \gg 1 \tag{4.485}$$

let us find the low-lying resonance states. $C_+ = 0$ gives

$$\tan ka = -\frac{k}{\sqrt{\kappa^2 - k^2}} \tag{4.486}$$

This is solved by

$$ka \simeq n\pi\left(1 - \frac{1}{\kappa a}\right) \quad (n = 1, 2, \cdots) \tag{4.487}$$

to the first order in $\frac{1}{\kappa a} \ll 1$. Here, the integer n must be of order 1 so that

$$\frac{n\pi}{\kappa a} \ll 1 \tag{4.488}$$

In Fig. 4.19 we plot $\frac{1}{|A_k|^2}$ for $0 < ka < 5\pi$ at $\kappa a = 5\pi$. We have chosen $\kappa(b-a) = \pi$ so that $e^{-\kappa(b-a)} \simeq 0.043$ is not too small. We observe that the width of the peak becomes wider as n increases.

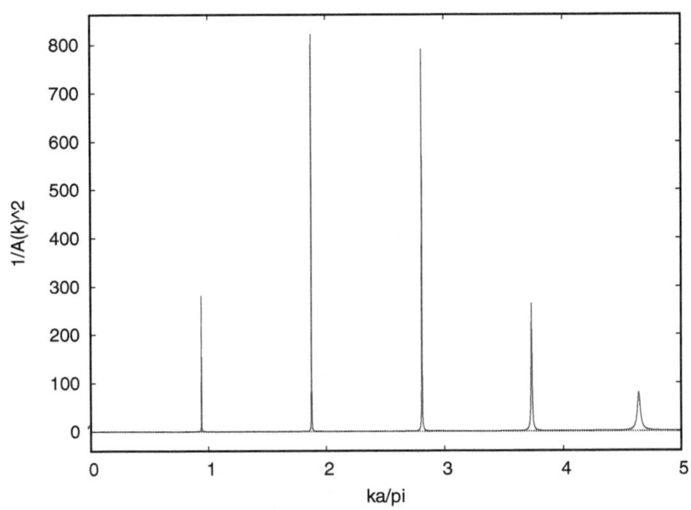

Fig. 4.19 Resonances

Now, let $k_n \simeq \frac{\pi n}{a}$ be a minimum of $|A_k|$. Near $k = k_n$, we can approximate A_k by

$$
\begin{aligned}
A_k &\simeq \frac{1}{2ik_n} e^{\sqrt{\kappa^2-k_n^2}(b-a)} \left(ik_n + \sqrt{\kappa^2 - k_n^2} \right) \\
&\quad \times \left\{ C_+(k) + \frac{ik_n - \sqrt{\kappa^2 - k_n^2}}{ik_n + \sqrt{\kappa^2 - k_n^2}} C_- e^{-2\sqrt{\kappa^2-k_n^2}(b-a)} \right\} \\
&= \frac{1}{2ik_n} e^{\sqrt{\kappa^2-k_n^2}(b-a)} \left(ik_n + \sqrt{\kappa^2 - k_n^2} \right) \\
&\quad \times \left\{ C_+(k) + \frac{2k_n^2 - \kappa^2 + 2ik_n\sqrt{\kappa^2 - k_n^2}}{\kappa^2} C_- e^{-2\sqrt{\kappa^2-k_n^2}(b-a)} \right\} \qquad (4.489)
\end{aligned}
$$

where

$$
C_+(k) \simeq \frac{1}{2} \sin ka + \frac{k_n}{2\sqrt{\kappa^2 - k_n^2}} \cos ka \qquad (4.490a)
$$

$$
C_- \simeq \frac{1}{2} \sin k_n a - \frac{k_n}{2\sqrt{\kappa^2 - k_n^2}} \cos k_n a \qquad (4.490b)
$$

We have fixed C_- at the value for $k = k_n$ so that only C_+ moves with k. We then obtain

$$
\begin{aligned}
|A_k|^2 &\simeq \frac{1}{4k_n^2} e^{2\sqrt{\kappa^2-k_n^2}(b-a)} \kappa^2 \left\{ \frac{4k_n^2(\kappa^2 - k_n^2)}{\kappa^4} C_-^2 e^{-4\sqrt{\kappa^2-k_n^2}(b-a)} \right. \\
&\quad \left. + \left(C_+(k) + \frac{2k_n^2 - \kappa^2}{\kappa^2} C_- e^{-2\sqrt{\kappa^2-k_n^2}(b-a)} \right)^2 \right\} \qquad (4.491)
\end{aligned}
$$

At $k = k_n$, we must find

$$
C_+(k_n) + \frac{2k_n^2 - \kappa^2}{\kappa^2} C_- e^{-2\sqrt{\kappa^2-k_n^2}(b-a)} = 0 \qquad (4.492)
$$

This determines k_n more precisely than (4.487). We find

$$
\frac{1}{a} C'_+(k_n) = \frac{1}{2} \cos k_n a - \frac{k_n}{2\sqrt{\kappa^2 - k_n^2}} \sin k_n a \qquad (4.493)
$$

$$
\overset{k_n a \simeq n\pi}{\simeq} \frac{1}{2}(-)^n \qquad (4.494)
$$

Hence, for $k \simeq k_n$, we obtain

$$|A_k|^2 \simeq \frac{1}{4k_n^2} e^{2\sqrt{\kappa^2-k_n^2}(b-a)} \kappa^2 \left\{ \frac{4k_n^2(\kappa^2 - k_n^2)}{\kappa^4} C_-^2 e^{-4\sqrt{\kappa^2-k_n^2}(b-a)} + \frac{1}{4}(k-k_n)^2 a^2 \right\} = \frac{(k-k_n)^2 + \Delta_n^2}{N_n} \tag{4.495}$$

where we have defined

$$\Delta_n \equiv \frac{4k_n\sqrt{\kappa^2 - k_n^2}}{\kappa^2 a} |C_-| e^{-2\sqrt{\kappa^2-k_n^2}(b-a)} \tag{4.496a}$$

$$N_n \equiv \frac{16k_n^2}{\kappa^2 a^2} e^{-2\sqrt{\kappa^2-k_n^2}(b-a)} \tag{4.496b}$$

Thus, near $k = k_n$, $\frac{1}{|A_k|^2}$ has the Lorentz distribution:

$$\frac{1}{|A_k|^2} \simeq \frac{N_n}{(k-k_n)^2 + \Delta_n^2} \tag{4.497}$$

Using $k_n a \simeq n\pi \left(1 - \frac{1}{\kappa a}\right)$, we obtain

$$C_- \simeq (-)^{n+1} \frac{n\pi}{2} \left(\frac{1}{\sqrt{\kappa^2 a^2 - (n\pi)^2}} + \frac{1}{\kappa a} \right) \tag{4.498}$$

Hence, we find

$$\Delta_n \simeq \frac{2}{a} \left(\frac{n\pi}{\kappa a}\right)^2 \left(1 + \frac{\sqrt{(\kappa a)^2 - (n\pi)^2}}{\kappa a} \right) e^{-2\sqrt{\kappa^2-(n\pi)^2/a^2}(b-a)} \tag{4.499}$$

The exponential factor makes the Lorentz distribution very narrow as expected. We also observe that the width Δ_n of the low-lying states are proportional to n^2; the width becomes wider for larger n.

We now approximate $\frac{1}{|A_k|^2}$ as the sum over the resonance states:

$$\frac{1}{|A_k|^2} \simeq \sum_n \frac{N_n}{(k-k_n)^2 + \Delta_n^2} \tag{4.500}$$

$\frac{1}{|A_k|^2}$ is negligible for k away from any of the resonances (Fig. 4.20).

Finally we must show how the width Δ_n is related to the tunneling probabilities. Let $\varphi(x)$ be an arbitrary initial state confined in the region $x < a$. We ask the

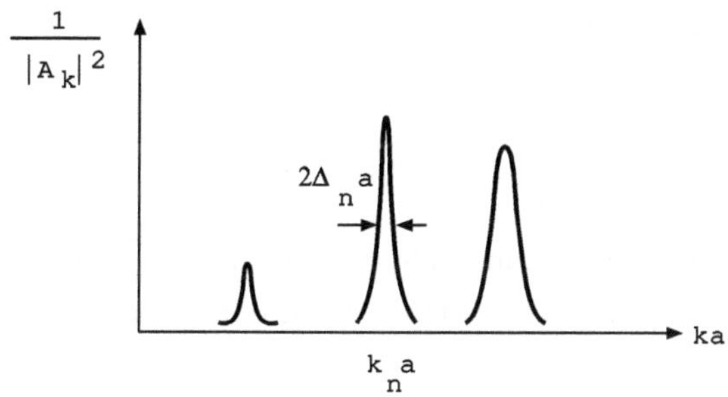

Fig. 4.20 The sum over the Lorentzian distributions approximates $\frac{1}{|A_k|^2}$ given in Fig. 4.19

probability amplitude that the state remains the same after time t:

$$\text{Amp}(t) \equiv \int_0^a dx\, \varphi(x) e^{-it\frac{1}{\hbar}H} \varphi(x) \tag{4.501}$$

We can expand an arbitrary wave function in terms of the normalized $\frac{1}{|A_k|}\psi_k$ (normalized at $x > b$):

$$\varphi(x) = \int_0^\infty \frac{dk}{\pi} \frac{1}{|A_k|^2} \psi_k(x) \int_0^\infty dy\, \psi_k^*(y) \varphi(y) \tag{4.502}$$

This gives

$$e^{-it\frac{1}{\hbar}H} \varphi(x) = \int_0^\infty \frac{dk}{\pi} \frac{1}{|A_k|^2} e^{-\frac{i}{\hbar}\frac{\hbar^2 k^2}{2m}t} \psi_k(x) \int_0^a dy\, \sin ky\, \varphi(y) \tag{4.503}$$

We thus obtain

$$\begin{aligned} \text{Amp}(t) &= \int_0^\infty \frac{dk}{\pi} \frac{1}{|A_k|^2} e^{-\frac{i}{\hbar}\frac{\hbar^2 k^2}{2m}t} \left| \int_0^a dy\, \sin ky\, \varphi(y) \right|^2 \\ &\simeq \sum_n \int_0^\infty \frac{dk}{\pi} \frac{N_n}{(k-k_n)^2 + \Delta_n^2} e^{-\frac{i}{\hbar}\frac{\hbar^2 k^2}{2m}t} \times \left| \int_0^a dy\, \sin ky\, \varphi(y) \right|^2 \end{aligned}$$

Since Δ_n is very narrow, we can replace $\sin ky$ by $\sin k_n y$ in the y integral and approximate k^2 by $k_n^2 + 2k_n(k - k_n)$. Since

$$\begin{aligned} &\int_0^\infty \frac{dk}{\pi} \frac{1}{(k-k_n)^2 + \Delta_n^2} e^{-i\frac{\hbar k_n}{m}(k-k_n)t} \\ &\simeq \int_{-\infty}^\infty \frac{dk}{\pi} \frac{1}{k^2 + \Delta_n^2} e^{-i\frac{\hbar k_n}{m}kt} = \frac{1}{\Delta_n} e^{-\frac{\hbar k_n}{m}\Delta_n t} \end{aligned} \tag{4.504}$$

we obtain

$$\text{Amp}(t) \simeq \sum_n \frac{N_n}{\Delta_n} e^{-\left(i\frac{\hbar k_n^2}{2m}+\frac{\hbar k_n \Delta_n}{m}\right)t} \times \left| \int_0^a dy \, \sin k_n y \, \varphi(y) \right|^2 \tag{4.505}$$

This implies that the n-th resonance state has the amplitude that decays as $\exp\left(-t\frac{\hbar k_n}{m}\Delta_n\right)$. Hence, the lifetime of each resonance state is

$$\tau_n \equiv \frac{m}{2\hbar k_n \Delta_n} \tag{4.506}$$

If the initial state φ has a non-vanishing overlap with the lowest resonance state, we can approximate

$$\text{Amp}(t) \approx \frac{N_1}{\Delta_1} e^{-\left(i\frac{\hbar k_n^2}{2m}+\frac{\hbar k_1 \Delta_1}{m}\right)t} \times \left| \int_0^a dy \, \sin k_1 y \, \varphi(y) \right|^2 \tag{4.507}$$

This implies that the lifetime does not depend much on the initial state as long as its overlap with $\sin k_1 x$ is non-vanishing, and it is given by the lifetime τ_1:

$$\begin{aligned} \frac{1}{\tau_1} &= \frac{2\hbar k_1 \Delta_1}{m} \\ &\simeq \frac{\hbar k_1}{ma} \cdot 4 \left(\frac{k_1}{\kappa}\right)^2 \left(1 + \frac{\sqrt{\kappa^2 - k_1^2}}{\kappa}\right) e^{-2\sqrt{\kappa^2 - k_1^2}\,(b-a)} \end{aligned} \tag{4.508}$$

This differs from the WKB estimate (4.96) by the factor

$$4\left(\frac{k_1}{\kappa}\right)^2 \left(1 + \frac{\sqrt{\kappa^2 - k_1^2}}{\kappa}\right)$$

Chapter 5
Quantum Mechanics II

In this final chapter, we introduce a little more advanced topics in quantum mechanics. Specifically, we overview low energy potential scattering of particles and the quantization of electromagnetic fields.

5.1 Potential Scattering

Three Points to Remember

1. Under the influence of a short-range potential, scattering states are expressed as the sum of an incident wave and a scattered wave. In the region free of the potential, the wave function is expressed as

$$\Psi_{\vec{p}}(\vec{r}) \to \mathrm{e}^{i\frac{\vec{p}\cdot\vec{r}}{\hbar}} + \frac{\mathrm{e}^{i\frac{pr}{\hbar}}}{r} f(\theta, \phi)$$

where $f(\theta, \phi)$ is the scattering amplitude.
2. In the Born approximation, the scattering amplitude is given by the Fourier transform $\tilde{V}$ of the potential V:

$$f(\theta, \phi) \simeq -\frac{m}{2\pi\hbar^2} \tilde{V}_{\frac{1}{\hbar}\Delta\vec{p}}\,, \quad \text{where } \Delta\vec{p} \equiv p\frac{\vec{r}}{r} - \vec{p}$$

3. At low energies, S-wave scattering is dominant. The scattering cross section σ_0 is given in terms of the phase shift $\delta_0(k)$ as

(continued)

H. Sonoda, K. Ülker, *Basic Physics for Graduate Students*,
https://doi.org/10.1007/978-981-95-6607-5_5

$$\sigma_0(k) = \frac{4\pi}{k^2} \sin^2 \delta_0(k),$$

where $k \equiv \frac{p}{\hbar}$ is the wave number.

5.1.1 Under the Influence of a Short-Range Potential, Scattering States Are Expressed as the Sum of Incident and Scattered Waves

Scattering cross section is a very important concept that everyone wants to be familiar with.

Consider a single particle moving under the influence of a potential $V(\vec{r})$. The Hamiltonian is given by

$$\hat{H} = \underbrace{\hat{H}_0}_{\equiv \frac{\hat{\vec{p}}^2}{2m}} + \hat{V}(\vec{r}) \tag{5.1}$$

The potential is assumed to have a finite range (Fig. 5.1). For example, a Yukawa-type potential

$$V(r) = g \frac{e^{-\kappa r}}{r} \tag{5.2}$$

has the range of $\frac{1}{\kappa}$. Consider particles with momentum $\vec{p}$ coming from far away where the potential does not reach. It is expected that the particles will be scattered by the potential, producing scattered waves. These scattered waves are outward waves, and inward waves are never generated. In fact, such scattering can be derived from the Schrödinger equation. Let us denote the wave function of an eigenstate with energy $E = \frac{\vec{p}^{\,2}}{2m}$ as $\Psi_{\vec{p}}(\vec{r})$. The Schrödinger equation is given by

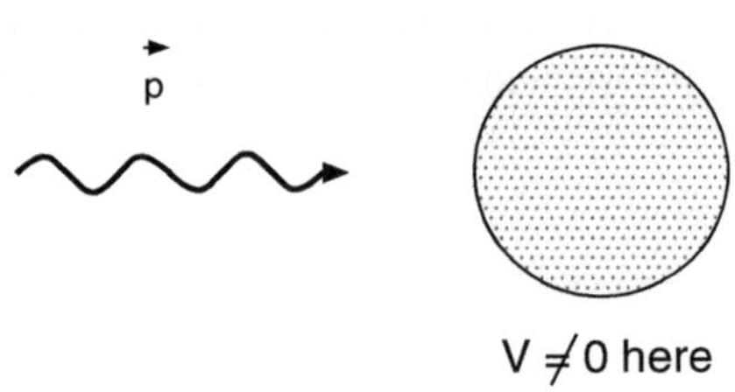

Fig. 5.1 Scattering of particles by a potential with a finite range

$$-\frac{\hbar^2}{2m}\nabla^2\Psi_{\vec{p}}(\vec{r}) + V(\vec{r})\Psi_{\vec{p}}(\vec{r}) = E\Psi_{\vec{p}}(\vec{r}) \tag{5.3}$$

In regions where the potential does not reach ($V = 0$), the free particle's Schrödinger equation is valid:

$$-\frac{\hbar^2}{2m}\nabla^2\Psi_{\vec{p}}(\vec{r}) = E\Psi_{\vec{p}}(\vec{r}) \tag{5.4}$$

Let us express the wave function as the sum of incident and scattered waves:

$$\Psi_{\vec{p}}(\vec{r}) = \mathrm{e}^{i\frac{\vec{p}\cdot\vec{r}}{\hbar}} + \psi_{\mathrm{scatt}}(\vec{r}) \tag{5.5}$$

The equation satisfied by the scattered wave $\psi_{\mathrm{scatt}}(\vec{r})$ is

$$\frac{\hbar^2}{2m}\nabla^2\psi_{\mathrm{scatt}}(\vec{r}) + E\psi_{\mathrm{scatt}}(\vec{r}) = V(\vec{r})\Psi_{\vec{p}}(\vec{r}) \tag{5.6}$$

Expressing the scattered wave in polar coordinates, we obtain

$$\psi_{\mathrm{scatt}}(\vec{r}) = \frac{\chi(r,\theta,\phi)}{r} \tag{5.7}$$

In the region where $V = 0$, the equation satisfied by $\chi(r, \theta, \phi)$ is

$$-\frac{\hbar^2}{2m}\frac{\partial^2}{\partial r^2}\chi + \underbrace{\frac{1}{2mr^2}\hat{L}^2}_{\xrightarrow{\text{large } r} 0}\chi = E\chi \tag{5.8}$$

As $r \to \infty$, the potential of centrifugal force can be ignored, and we eventually obtain

$$\chi(r,\theta,\phi) \approx \mathrm{e}^{i\frac{pr}{\hbar}} f_+(\theta,\phi) + \mathrm{e}^{-i\frac{pr}{\hbar}} f_-(\theta,\phi) \tag{5.9}$$

However, only outward waves should be produced by scattering; we must choose the plus sign (Fig. 5.2).

$$\chi(r,\theta,\phi) \approx \mathrm{e}^{i\frac{pr}{\hbar}} f(\theta,\phi) \tag{5.10}$$

Therefore, the asymptotic form of the wave function of the energy eigenstate is given by

$$\Psi_{\vec{p}}(\vec{r}) \xrightarrow{r\to\infty} \mathrm{e}^{i\frac{\vec{p}\cdot\vec{r}}{\hbar}} + \underbrace{\frac{\mathrm{e}^{i\frac{pr}{\hbar}}}{r} f(\theta,\phi)}_{\text{scattered wave}} \tag{5.11}$$

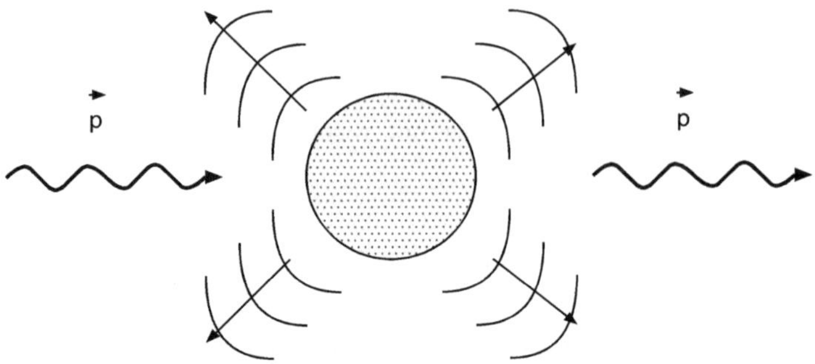

Fig. 5.2 Only outward waves are generated

Differential Scattering Cross Section

Let us consider the normalization of the incident wave $e^{i\frac{\vec{p}\cdot\vec{r}}{\hbar}}$. The wave function representing a single particle is given by

$$\frac{1}{\sqrt{V}}e^{i\frac{\vec{p}\cdot\vec{r}}{\hbar}}$$

where V is the volume of the entire space. With a fixed momentum, the position of the particle is completely uncertain. Usually, instead of adopting this normalization, we find it more convenient to consider $e^{i\frac{\vec{p}\cdot\vec{r}}{\hbar}}$ without $\frac{1}{\sqrt{V}}$. We are not eager to worry about the infinite volume V. It is important to note that even if the normalization is changed, we are still dealing with the scattering of a single particle, and we are never dealing with multiple particles at once. When we talk about the number of particles or the flow of particles, we are merely talking statistics; we always treat the scattering of an individual particle.

Now, the velocity of a particle with momentum $\vec{p}$ is $\frac{\vec{p}}{m}$, so the number of particles passing through per unit time and per unit area perpendicular to $\vec{p}$ is $\frac{\vec{p}}{m}$. On the other hand, the scattered wave $\frac{e^{i\frac{1}{\hbar}pr}}{r}f(\theta,\phi)$ represents a flow of particles per unit solid angle $\Delta\Omega$ given by

$$\underbrace{\frac{|f|^2}{r^2}}_{\text{particle density}}\underbrace{v}_{\text{speed}}\underbrace{r^2\Delta\Omega}_{\text{area}} = |f|^2 v\cdot\Delta\Omega \tag{5.12}$$

This represents the flow of $|f|^2 v$ particles per unit solid angle. We define the **differential cross section** as

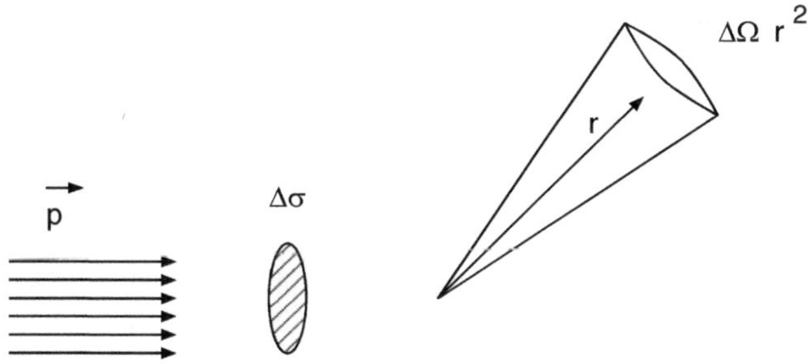

Fig. 5.3 Imagine that the particles passing through the cross-section $\Delta\sigma$ are scattered in the direction of $\Delta\Omega$

$$\boxed{\frac{d\sigma}{d\Omega}} \equiv \frac{\text{number of particles scattered per unit time and unit solid angle}}{\text{number of particles incident per unit time and unit area}}$$
$$= \frac{|f|^2 v}{v} = \boxed{|f|^2} \tag{5.13}$$

See Fig. 5.3. The **total cross section** is given by the integral of the differential cross section as

$$\sigma_{\text{total}} \equiv \int d\Omega \, \frac{d\sigma}{d\Omega} \tag{5.14}$$

which can be interpreted as the effective size of the potential as far as scattering is concerned.

5.1.2 In the Born Approximation, the Scattering Amplitude Is Given by the Fourier Transform of the Potential

The equation satisfied by the scattered wave $\psi_{\text{scatt}}(\vec{r})$ is obtained from (5.6) as

$$\frac{\hbar^2}{2m}\nabla^2\psi_{\text{scatt}} + \frac{p^2}{2m}\psi_{\text{scatt}} = V(\vec{r})\Psi_{\vec{p}}(\vec{r}) \tag{5.15}$$

The purpose of this section is to solve this equation using the method of a Green function. This method naturally provides a successive approximation method and is frequently used not only in quantum mechanics but also in electromagnetism. (Recall that we have introduced the retarded Green function in Sect. 2.2.1 to solve Maxwell's equations.)

First, let us define a Green function by

$$\left(\frac{\hbar^2}{2m}\nabla^2 + \frac{p^2}{2m}\right) G(\vec{r}) = \delta^{(3)}(\vec{r}) \tag{5.16}$$

(As in the case of electromagnetism, boundary conditions are required to determine the Green function uniquely. We will explain this shortly.) Using this, Eq. (5.15) can be formally solved as

$$\psi_{\text{scatt}}(\vec{r}) = \int d^3r'\, G(\vec{r}-\vec{r}\,')V(\vec{r}\,')\Psi_{\vec{p}}(\vec{r}\,') \tag{5.17}$$

It is convenient to consider the Fourier transform of the Green function. Let us denote the Fourier transform of $G(\vec{r})$ as

$$G(\vec{r}) = \int \frac{d^3q}{(2\pi)^3}\, e^{\frac{i}{\hbar}\vec{q}\cdot\vec{r}}\tilde{G}_{\vec{q}} \tag{5.18}$$

Then, Eq. (5.16) can be rewritten as

$$\frac{1}{2m}(p^2 - q^2)\tilde{G}_{\vec{q}} = 1 \tag{5.19}$$

Thus,

$$\tilde{G}_{\vec{q}} = \frac{1}{\frac{1}{2m}(p^2 - q^2)} \tag{5.20}$$

and we obtain

$$\begin{aligned} G(\vec{r}) &= \int \frac{d^3q}{(2\pi\hbar)^3}\frac{e^{\frac{i}{\hbar}\vec{q}\cdot\vec{r}}}{\frac{1}{2m}(p^2-q^2)} \\ &= \frac{2m}{\hbar^3}\int \frac{d^3q}{(2\pi)^3}\frac{e^{\frac{i}{\hbar}\vec{q}\cdot\vec{r}}}{p^2-q^2} \\ &= \frac{2m}{\hbar^3(2\pi)^2}\int_0^\infty q^2dq \int_{-1}^{1} d\cos\theta\, \frac{e^{i\frac{qr}{\hbar}\cos\theta}}{p^2-q^2} \\ &= \frac{2m}{\hbar^3(2\pi)^2}\int_{-\infty}^{\infty} dq\, \frac{q}{p^2-q^2}e^{i\frac{qr}{\hbar}}\frac{\hbar}{ir} \\ &= \frac{2m}{\hbar^2(2\pi)^2}\frac{1}{ir}\int_{-\infty}^{\infty} dq\, \underbrace{\frac{q}{p^2-q^2}}_{\text{poles at } q=\pm p}\, e^{i\frac{qr}{\hbar}} \end{aligned} \tag{5.21}$$

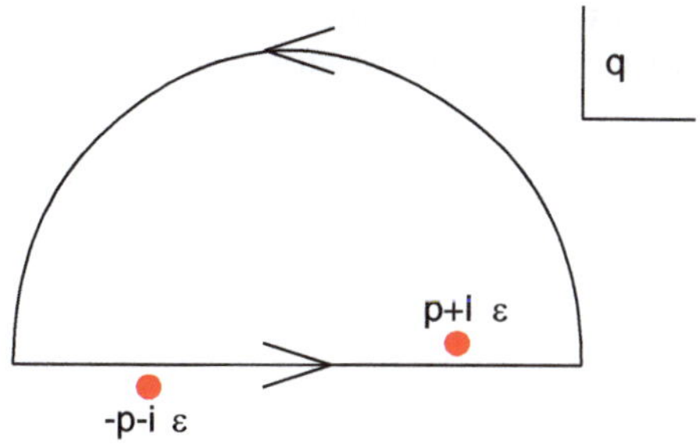

Fig. 5.4 Integration path in the complex plane

The integrand with respect to q has poles at $q = \pm p$, and as we have learned for the retarded Green function in Sect. 2.2.1, the boundary condition on the Green function dictates how these poles should be handled.

In this case, the boundary condition is that **the Green function has only outward waves**. This corresponds to shifting the pole at $q = p$ to the upper complex half-plane and the pole at $q = -p$ to the lower complex half-plane (Fig. 5.4). Therefore, the Green function is given by

$$G(\vec{r}) = \frac{2m}{\hbar^2 (2\pi)^2} \frac{1}{r} (-) \frac{1}{2} e^{\frac{i}{\hbar} pr}$$
$$= \boxed{(-) \frac{m}{2\pi \hbar^2} \frac{e^{\frac{i}{\hbar} pr}}{r}} \tag{5.22}$$

Thus, a formal solution to the scattered wave is

$$\boxed{\psi_{\text{scatt}}(\vec{r}) = -\frac{m}{2\pi \hbar^2} \int d^3 r' \frac{e^{\frac{i}{\hbar} p|\vec{r} - \vec{r}\,'|}}{|\vec{r} - \vec{r}\,'|} V(\vec{r}\,') \Psi_{\vec{p}}(\vec{r}\,')} \tag{5.23}$$

Let us check that the scattered wave obtained is an outward wave. For this purpose, we need to take $r \equiv |\vec{r}|$ to be much larger than the range of the potential (Fig. 5.5). Assuming $\vec{r}\,'$ is within the range of the potential, and $r \gg r'$, then we find

$$|\vec{r} - \vec{r}\,'| \simeq r - \frac{\vec{r}}{r} \cdot \vec{r}\,' \tag{5.24}$$

Therefore, the asymptotic form of the scattered wave is given by

$$\psi_{\text{scatt}}(\vec{r}) \stackrel{r \to \infty}{\longrightarrow} -\frac{m}{2\pi \hbar^2} \frac{e^{\frac{i}{\hbar} pr}}{r} \int d^3 r' \, e^{-\frac{i}{\hbar} p \hat{r} \cdot \vec{r}\,'} V(\vec{r}\,') \Psi_{\vec{p}}(\vec{r}\,') \tag{5.25}$$

where $\hat{r} \equiv \frac{\vec{r}}{r}$ is the unit vector in the direction of $\vec{r}$.

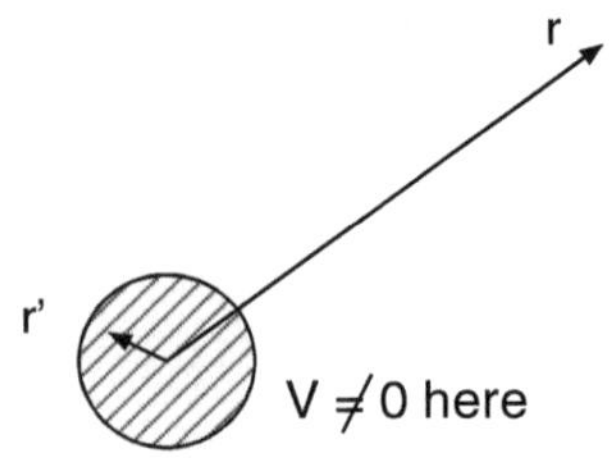

Fig. 5.5 r is much larger than the range of the potential

Thus, we have confirmed that the scattered wave is an outward wave. Furthermore, it is understood that the scattering amplitude is given by

$$f(\theta, \phi) = -\frac{m}{2\pi\hbar^2} \int d^3r'\, \mathrm{e}^{-\frac{i}{\hbar}p\hat{r}\cdot\vec{r}\,'}\, V(\vec{r}\,')\Psi_{\vec{p}}(\vec{r}\,') \tag{5.26}$$

This equation, however, includes the unknown wave function $\Psi_{\vec{p}}$ of the eigenstate, so it is not immediately useful as it is.

The simplest approximation for the scattering amplitude f is obtained by substituting the incident wave $\mathrm{e}^{\frac{i}{\hbar}\vec{p}\cdot\vec{x}}$ into $\Psi_{\vec{p}}$. This is called the **Born approximation**. Therefore, using the Born approximation, the scattering amplitude is given by

$$\boxed{f_{\text{Born}}(\theta, \phi) = -\frac{m}{2\pi\hbar^2} \tilde{V}_{\frac{1}{\hbar}(p\hat{r}-\vec{p})}} \tag{5.27}$$

where $\tilde{V}_{\vec{k}}$ is the Fourier transform of the potential $V(\vec{r})$:

$$\tilde{V}_{\vec{k}} \equiv \int d^3r\, \mathrm{e}^{-i\vec{k}\cdot\vec{r}}\, V(\vec{r}) \tag{5.28}$$

Thus, the differential scattering cross section is given by

$$\boxed{\frac{d\sigma}{d\Omega} \simeq \left(\frac{m}{2\pi\hbar^2}\right)^2 \left|\tilde{V}_{\frac{1}{\hbar}(p\hat{r}-\vec{p})}\right|^2} \tag{5.29}$$

Example: Yukawa Potential

$$V(\vec{r}) = g\frac{\mathrm{e}^{-\kappa r}}{r} \Longrightarrow \tilde{V}_{\vec{k}} = \frac{4\pi g}{k^2 + \kappa^2} \tag{5.30}$$

Therefore, the Born amplitude is

$$f_{\text{Born}}(\theta, \phi) = \frac{-m}{2\pi\hbar^2} \frac{4\pi g}{\frac{1}{\hbar^2}(p\hat{r} - \vec{p})^2 + \kappa^2} \tag{5.31}$$

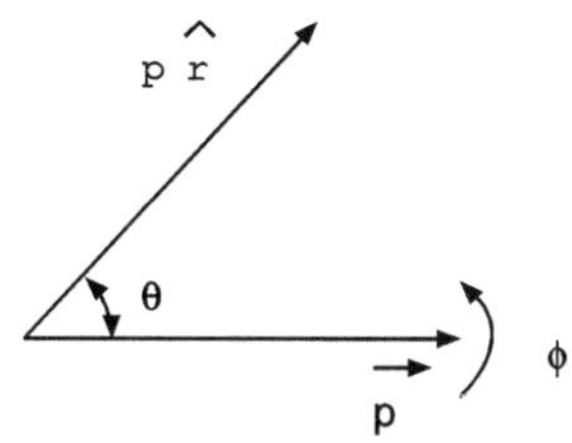

Fig. 5.6 Polar coordinates θ, ϕ

In terms of the angle θ between $\hat{r}$ and $\vec{p}$, the amplitude is given by

$$f_{\mathrm{Born}}(\theta, \phi) = \frac{-2mg}{2p^2(1 - \cos\theta) + \hbar^2\kappa^2} \tag{5.32}$$

which is independent of the angle ϕ around the $\vec{p}$ direction (Fig. 5.6). Thus, the differential scattering cross section is given by

$$\frac{d\sigma}{d\Omega} \simeq \frac{4m^2g^2}{\left\{2p^2(1 - \cos\theta) + \hbar^2\kappa^2\right\}^2} \tag{5.33}$$

From this, the total cross section is obtained as

$$\begin{aligned} \sigma_{\mathrm{total}} &= \int d\Omega \frac{d\sigma}{d\Omega} = \frac{4\pi m^2 g^2}{p^4} \frac{1}{\frac{\hbar^2\kappa^2}{2p^2}\left(2 + \frac{\hbar^2\kappa^2}{2p^2}\right)} \\ &= \frac{16\pi m^2 g^2}{\hbar^4\kappa^4} \frac{1}{1 + \frac{4p^2}{\hbar^2\kappa^2}} \longrightarrow \begin{cases} \frac{16\pi m^2 g^2}{\hbar^4\kappa^4} & \text{for} \quad p \ll \hbar\kappa \\ \frac{4\pi m^2 g^2}{\hbar^2\kappa^2 p^2} & \text{for} \quad p \gg \hbar\kappa \end{cases} \end{aligned} \tag{5.34}$$

See Fig. 5.7.

In the limit $\kappa \to 0$, the Yukawa potential becomes the Coulomb potential

$$V(\vec{r}) = \frac{g}{r}, \tag{5.35}$$

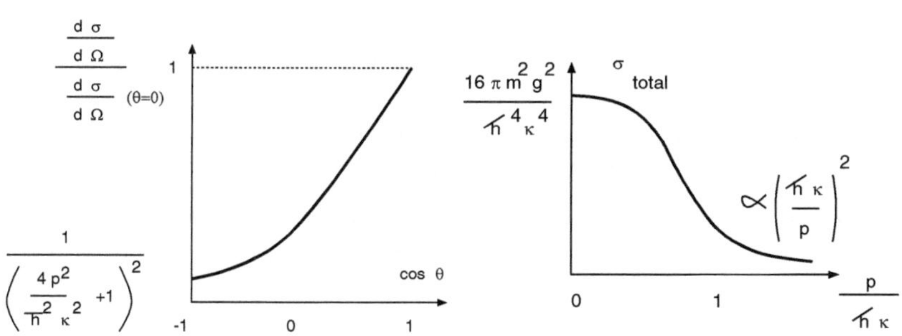

Fig. 5.7 Scattering differential cross section and total cross section for the Yukawa potential

and the differential cross section coincides with Rutherford's formula obtained in Example 13-1:

$$\frac{d\sigma}{d\Omega} = \frac{m^2 g^2}{p^4(1-\cos\theta)^2} = \frac{m^2 g^2}{4p^4 \sin^4 \frac{\theta}{2}} \tag{5.36}$$

This differential cross section becomes very large near the vanishing scattering angle:

$$\frac{d\sigma}{d\Omega} \xrightarrow{\theta \to 0} \frac{4m^2 g^2}{p^4 \theta^4} \tag{5.37}$$

Hence, the total cross section diverges. This divergence reflects the infinite range of the Coulomb potential.

When Is the Born Approximation Valid?

Clearly, a sufficient condition for the validity of the Born approximation is that $\Psi_{\vec{p}} \simeq e^{\frac{i}{\hbar}\vec{p}\cdot\vec{r}}$ within the range of the potential. Let us rewrite this condition in a clearer way. It is convenient to consider two cases: either the wavelength of the particle is longer or shorter than the range a of the potential. Denoting the particle's momentum as p, consider first the case $pa \ll \hbar$, i.e., when the wavelength is much longer than a. For the scattered wave given by Eq. (5.23) to be significantly smaller than 1 in the potential region, we must have

$$\frac{ma^2}{\hbar^2}|V| \ll 1 \tag{5.38}$$

where V represents the typical magnitude of the potential and $\frac{\hbar^2}{ma^2}$ represents the energy associated with the wavelength a. For the Yukawa-type potential mentioned earlier, with $a \sim \frac{1}{\kappa}$ and $V \sim g\kappa$, Eq. (5.38) yields

$$\frac{mg}{\hbar^2 \kappa} \ll 1 \tag{5.39}$$

This condition does not depend on the momentum. On the other hand, for $pa \gg \hbar$, i.e., when the wavelength is much shorter than a, the range a of the potential becomes less relevant. Since the wavelength is short, one can consider the particle classically, and the time the particle spends within the range a of the potential is roughly $\Delta t \equiv \frac{am}{p}$, during which the magnitude of the potential is roughly $\Delta E \equiv V$. Therefore, the phase shift of the wave function is approximately $\frac{1}{\hbar}\Delta E \cdot \Delta t$. If this shift is much smaller than 1, the Born approximation is expected to be good. Thus, the condition for the Born approximation is

$$\frac{\Delta E \cdot \Delta t}{\hbar} = \frac{V}{\hbar}\frac{am}{p} = \frac{maV}{\hbar p} = \frac{ma^2 V}{\hbar^2}\frac{\hbar}{pa} \ll 1 \tag{5.40}$$

Applying this to the Yukawa-type potential, this condition becomes

$$\frac{mg}{\hbar p} = \frac{mg}{\hbar^2 \kappa}\frac{\hbar \kappa}{p} \ll 1 \tag{5.41}$$

Hence, even if the Born approximation is poor at low energies, i.e., (5.39) is not satisfied, the approximation can be improved by increasing the momentum sufficiently so that condition (5.41) is met. In general the Born approximation is valid at high momentum.

5.1.3 At Low Energies, S-Wave Scattering Is Dominant

We consider only spherically symmetric potentials in this section. The orbital angular momentum of a scattered particle is conserved, allowing us to consider each component of the orbital angular momentum. Given a as the radius of the spherical domain of the potential, a classical consideration gives the highest orbital angular momentum involved in the scattering to be of the order of $l = pa$. Therefore, for small momentum

$$p \lesssim \frac{\hbar}{a} \tag{5.42}$$

the S-wave scattering with $l = 0$ becomes dominant (Fig. 5.8).

For the S-wave, replacing the wave function $\Psi_{\vec{p}}(\vec{r})$ by $\frac{\chi_0(r)}{r}$, we obtain the Schrödinger equation

$$-\frac{\hbar^2}{2m}\frac{d^2}{dr^2}\chi_0(r) + V(r)\chi_0(r) = \frac{\hbar^2 k^2}{2m}\chi_0(r) \tag{5.43}$$

The boundary condition for $\chi_0(r)$ is

$$\chi_0(r)\Big|_{r=0} = 0 \tag{5.44}$$

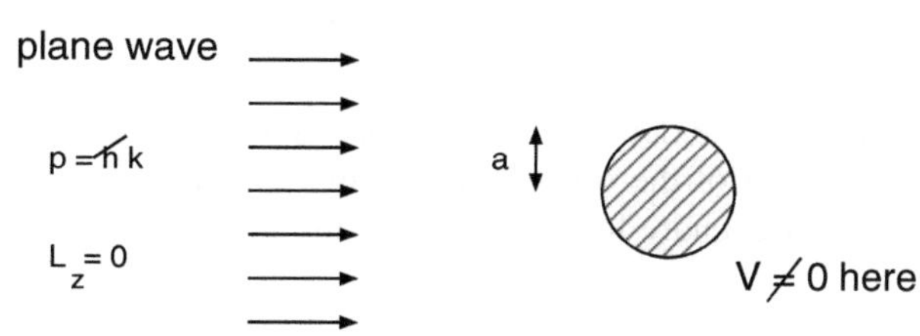

Fig. 5.8 S-wave scattering

In the region untouched by the potential, $\chi_0(r)$ satisfies the equation of a free particle

$$-\frac{d^2}{dr^2}\chi_0(r) = k^2\chi_0(r) \tag{5.45}$$

so $\chi_0(r)$ is given by a linear combination of $\mathrm{e}^{\pm ikr}$.

If $\chi_0(r)$ is assumed to be a solution to Eq. (5.43) satisfying the boundary condition $\chi_0(0) = 0$, then its complex conjugate $\chi_0^*(r)$ is also a solution to (5.43), satisfying the same boundary condition. Therefore, for any phase ϕ, $\chi_0(r) + \mathrm{e}^{i\phi}\chi_0^*(r) = \mathrm{e}^{i\frac{\phi}{2}}(\mathrm{e}^{-i\frac{\phi}{2}}\chi_0(r) + c.c.)$ also satisfies the same boundary condition. Consequently, $\chi_0(r)$ can generally be written as

$$\chi_0(r) = (\text{complex coefficient}) \times \underbrace{\chi_0^r(r)}_{\text{real function}} \tag{5.46}$$

Thus, outside the range of the potential, we find

$$\chi_0(r) = c_0\left(-\mathrm{e}^{-ikr} + \mathrm{e}^{ikr+2i\delta_0}\right) = c_0\mathrm{e}^{i\delta_0}\cdot 2i\sin(kr+\delta_0) \tag{5.47}$$

where c_0 is complex. The phase $\delta_0(k)$ is a function of the wave number k, called a **phase shift**.

Now, assuming that the incident wave travels along the z-axis, and considering only the S-wave for the scattered wave, the asymptotic form of the eigenfunction $\Psi_{\vec{p}}$ is given by

$$\boxed{\Psi_{\vec{p}}(\vec{r}) \longrightarrow \mathrm{e}^{ikz} + \frac{f_0(k)}{r}\mathrm{e}^{ikr} \quad \text{as} \quad r\to\infty} \tag{5.48}$$

The scattered wave is spherically symmetric, and the differential scattering cross section is independent of angle, given by

$$\frac{d\sigma}{d\Omega} = |f_0(k)|^2 \tag{5.49}$$

The incident wave, which is a plane wave, can be decomposed into components of each orbital angular momentum (referred to as partial waves) as follows:

$$\mathrm{e}^{ikz} = \sum_{l=0}\frac{\varphi_l(r)}{r}P_l(\cos\theta) \tag{5.50}$$

Since the incident wave is invariant under rotations around the z-axis, the z-component of the orbital angular momentum is zero. To determine the coefficient of each Legendre function, we can use the orthogonality of the Legendre functions, which yields

$$\frac{\varphi_l(r)}{r} = \frac{2l+1}{2}\int_{-1}^{1} d\cos\theta\; P_l(\cos\theta)\mathrm{e}^{ikr\cos\theta} \tag{5.51}$$

Therefore, for the S-wave $\varphi_0(r)$ with $P_0 = 1$, we obtain

$$\begin{aligned}\varphi_0(r) &= r\frac{1}{2}\frac{\mathrm{e}^{ikr} - \mathrm{e}^{-ikr}}{ikr} \\ &= \frac{1}{2ik}\left(\mathrm{e}^{ikr} - \mathrm{e}^{-ikr}\right)\end{aligned} \tag{5.52}$$

Hence, considering only the S-wave component of the incident wave in the asymptotic form (5.48), we obtain

$$\begin{aligned}\chi_0(r) &\longrightarrow \frac{1}{2ik}\left(\mathrm{e}^{ikr} - \mathrm{e}^{-ikr}\right) + f_0\mathrm{e}^{ikr} \\ &= \frac{1}{2ik}\left(-\mathrm{e}^{-ikr} + \mathrm{e}^{ikr}(1 + 2ikf_0)\right)\end{aligned} \tag{5.53}$$

Comparing this with Eq. (5.47), we find that the phase shift $\delta_0(k)$ is given by

$$\mathrm{e}^{2i\delta_0(k)} = 1 + 2ikf_0(k) \tag{5.54}$$

Consequently,

$$f_0(k) = \frac{\mathrm{e}^{2i\delta_0(k)} - 1}{2ik} = \frac{\mathrm{e}^{i\delta_0(k)}\sin\delta_0(k)}{k} \tag{5.55}$$

From this, we obtain the total scattering cross section for the S-wave as

$$\boxed{\sigma_0(k) = 4\pi|f_0(k)|^2 = \frac{4\pi}{k^2}\sin^2\delta_0(k)} \tag{5.56}$$

Example: Hard Sphere

$$V(r) = \begin{cases}\infty & \text{if} \quad 0 < r < a \\ 0 & \text{if} \quad r > a\end{cases} \tag{5.57}$$

In the region $r > a$, one needs to solve the equation for a free particle

$$-\frac{d^2}{dr^2}\chi_0(r) = k^2\chi_0(r) \tag{5.58}$$

with the boundary condition

$$\chi_0(a) = 0 \tag{5.59}$$

This gives

$$\chi_0(r) = \text{const} \times \sin k(r - a) \tag{5.60}$$

and thus, the phase shift is

$$\delta_0(k) = -ka \tag{5.61}$$

Therefore, the S-wave scattering cross section is

$$\sigma_0(k) = \frac{4\pi}{k^2} \sin^2 ka \overset{k\to 0}{\longrightarrow} 4\pi a^2 > \pi a^2 \tag{5.62}$$

The cross section obtained in the limit of zero wave number is not the geometric cross section πa^2 but rather the entire surface area of the sphere with radius a. When ka is approximately 1, the contribution of the P-wave becomes important.

Appendix: Derivation of the Born Approximation Formula Using Fermi's Golden Rule

The Born approximation formula (5.29) for scattering can also be derived using the time-dependent perturbation theory. Considering the initial state as an eigenstate with momentum $\vec{p}_i$ and the final state as an eigenstate with momentum $\vec{p}_f$, the probability of transition per unit time by the potential V is given by Fermi's golden rule as

$$dw = \frac{2\pi}{\hbar} \left| \langle \vec{p}_f | \, \hat{V} \, | \vec{p}_i \rangle \right|^2 d\Phi(E_i) \tag{5.63}$$

where

$$\begin{aligned} \langle \vec{p}_f | \, \hat{V} \, | \vec{p}_i \rangle &\equiv \int d^3 r \, \mathrm{e}^{-\frac{i}{\hbar}\vec{p}_f\cdot\vec{r}} V(\vec{r}) \mathrm{e}^{\frac{i}{\hbar}\vec{p}_i\cdot\vec{r}} \\ &= \int d^3 r \, \mathrm{e}^{-\frac{i}{\hbar}(\vec{p}_f - \vec{p}_i)\cdot\vec{r}} V(\vec{r}) = \tilde{V}_{\frac{1}{\hbar}(\vec{p}_f - \vec{p}_i)} \end{aligned} \tag{5.64}$$

is the Fourier transform of the potential and $d\Phi(E_i)$ is the density of final states at energy $E_i = \frac{p_i^2}{2m}$, given by

$$d\Phi(E_i) = \frac{d^3 p_f}{(2\pi\hbar)^3} \delta\left(\frac{p_f^2}{2m} - E_i\right) = \frac{m p_i}{(2\pi\hbar)^3} d\Omega_f \tag{5.65}$$

We thus obtain

$$dw = \frac{mp_i}{(2\pi\hbar^2)^2} \left| \tilde{V}_{\frac{1}{\hbar}(\vec{p}_f - \vec{p}_i)} \right|^2 d\Omega_f \tag{5.66}$$

The differential scattering cross section is obtained by dividing dw by the flux v as

$$d\sigma = \frac{1}{v} dw = \left(\frac{m}{2\pi\hbar^2} \right)^2 \left| \tilde{V}_{\frac{1}{\hbar}(\vec{p}_f - \vec{p}_i)} \right|^2 d\Omega_f \tag{5.67}$$

This is the Born approximation formula (5.29).

Examples 13

1. **Rutherford formula**—Consider a particle of mass m moving under the Coulomb potential $V(r) = \frac{e^2}{r}$. The purpose of this problem is to calculate the scattering cross section using classical physics.

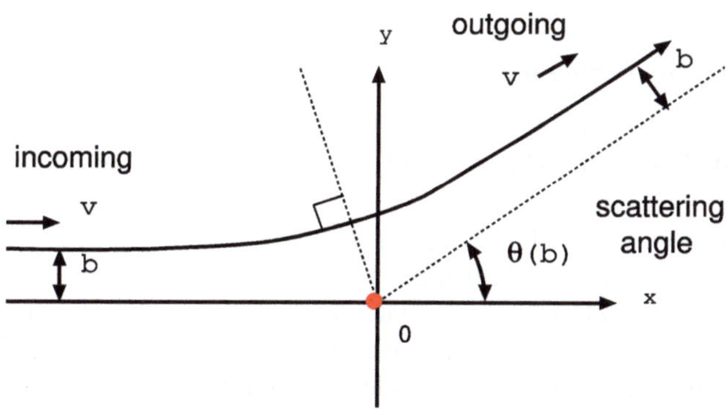

(a) Express the angular momentum M in the z-direction in terms of the impact parameter b and the initial velocity v.

(b) Solve the classical equations of motion to express the scattering angle $\theta(b)$ as a function of b.
Hint: The trajectory is a hyperbola determined by the total energy $E \equiv \frac{mv^2}{2}$ and the angular momentum M.

(c) Explain geometrically why the differential scattering cross section $\frac{d\sigma}{d\cos\theta}$ is given by

$$\frac{d\sigma}{d\cos\theta} = 2\pi b \frac{db}{d\cos\theta}$$

(d) Calculate the differential scattering cross section $\frac{d\sigma}{d\cos\theta}$, and plot its dependence on $\cos\theta$.[1]

2. **Spherical well potential (Part 1)**—Consider a particle of mass m under a spherically symmetric potential

$$V(r) = \begin{cases} V_0 & (r < R) \\ 0 & (r > R) \end{cases}$$

(a) Using the Born approximation, calculate the differential cross section.
(b) Using the above result, calculate the total cross section.
(c) For low energy where $\frac{pR}{\hbar} \ll 1$, evaluate the magnitude of the scattering wave in the region $r < R$, and derive the condition for the validity of the Born approximation, i.e., the scattering wave is significantly smaller compared with 1.

3. **Spherical well potential (Part 2)**—We continue on the previous problem. Let $k \equiv \frac{p}{\hbar}$ be the wave number of the particle. For small k, the phase shift $\delta_0(k)$ of the S-wave is given by

$$\delta_0(k) \simeq -ka$$

We call a a **scattering length.**

(a) For both $V_0 > 0$ and $V_0 < 0$, calculate the scattering length.
Hint: To obtain $\delta_0(k)$, assume $k^2 \ll \frac{m|V_0|}{\hbar^2}$ and also that $\delta_0(k)$ is proportional to k.
(b) Explain qualitatively the sign of $\delta_0(k)$.
Hint: Attraction or repulsion?
(c) Using the results of (a), calculate the total cross section to the order V_0^2.
(d) Using the results of (a), find the limit of the total cross section as $V_0 \to +\infty$.
Hint: Compare with the example of the hard sphere discussed in the text.

Solutions to Examples 13

1. (a) $M = -mvb$
(b) Using polar coordinates r, θ, we obtain

[1] There is a simple reason why the classical Rutherford formula is also correct quantum mechanically. Generally, $\frac{d\sigma}{d\cos\theta}$ is expected to depend only on the particle's mass m, energy E, scattering angle θ, $\hbar$, and e^2. Now the scattering cross section is proportional to e^4. It cannot depend on $\hbar$ since no dimensionless quantity can be formed from $\hbar$ and m; thus the scattering cross section cannot depend on $\hbar$.

$$\begin{cases} M &= mr^2\dot{\theta} = -mvb \\ E &= \frac{m}{2}\dot{r}^2 + \frac{M^2}{2mr^2} + \frac{e^2}{r} = \frac{m}{2}v^2 \end{cases}$$

Along the orbit, the point closest to the origin has $r = r_{\min}$, where $\dot{r} = 0$:

$$\frac{M^2}{2mr_{\min}^2} + \frac{e^2}{r_{\min}} = \frac{m}{2}v^2 \Longrightarrow r_{\min} = \frac{e^2 + \sqrt{v^2M^2 + e^4}}{mv^2}$$

Therefore,

$$\begin{aligned} \pi - \theta(b) &= 2(\theta(t = -\infty) - \theta(r = r_{\min})) \\ &= 2\int_{r_{\min}}^{\infty} dr\,\frac{d\theta}{dr} = 2\int_{r_{\min}}^{\infty} dr\,\frac{\frac{d\theta}{dt}}{\frac{dr}{dt}} \\ &= 2\int_{r_{\min}}^{\infty} dr\,\frac{vb}{r^2\sqrt{v^2 - \frac{M^2}{m^2r^2} - \frac{2e^2}{mr}}} \\ &= \pi - 2\arcsin\frac{e^2}{\sqrt{e^4 + v^2M^2}} \end{aligned}$$

Thus, $\theta(b) = 2\arcsin\frac{e^2}{\sqrt{e^4+m^2v^4b^2}}$. From this,

$$\sin\frac{\theta(b)}{2} = \frac{e^2}{\sqrt{e^4 + m^2v^4b^2}}, \qquad \cos\frac{\theta(b)}{2} = \frac{mv^2b}{\sqrt{e^4 + m^2v^4b^2}}$$

$$\Longrightarrow \cos\theta(b) = \frac{m^2v^4b^2 - e^4}{m^2v^4b^2 + e^4}$$

(c) The area of the annulus between b and $b + db$ is $2\pi b db$, and thus $d\sigma = 2\pi b db$.

(d) Using the result from (b), we obtain

$$\begin{aligned} \frac{d\sigma}{d\cos\theta} &= 2\pi b\frac{db}{d\cos\theta} = 2\pi b\frac{(e^4 + m^2v^4b^2)^2}{4e^4\cdot m^2v^4b} \\ &= \boxed{\frac{\pi}{2}\frac{e^4}{m^2v^4\sin^4\frac{\theta}{2}} = \frac{2\pi e^4}{m^2v^4(1-\cos\theta)^2}} \end{aligned}$$

which is the famous Rutherford formula.

2. (a) The Fourier transform of the potential is

$$\tilde{V}_k = V_0 \int d^3x\, \mathrm{e}^{i\vec{k}\cdot\vec{r}}\theta(R-r) = 2\pi V_0 \int_0^R r^2 dr \int_{-1}^1 d\cos\theta\, \mathrm{e}^{ikr\cos\theta}$$

$$= 2\pi V_0 \int_0^R r^2 dr\, \frac{\mathrm{e}^{ikr} - \mathrm{e}^{-ikr}}{ikr} = 4\pi V_0 R^3 \left(\frac{\sin kR}{(kR)^3} - \frac{\cos kR}{(kR)^2}\right)$$

Thus, in the Born approximation the differential cross section is

$$\frac{d\sigma_B}{d\Omega} = \frac{m^2}{(2\pi\hbar^2)^2}(4\pi V_0 R^3)^2 \left(\frac{\sin kR}{(kR)^3} - \frac{\cos kR}{(kR)^2}\right)^2$$

$$= 4R^2 \left(\frac{V_0}{\frac{\hbar^2}{mR^2}}\right)^2 \left(\frac{\sin kR}{(kR)^3} - \frac{\cos kR}{(kR)^2}\right)^2$$

where $k \equiv 2\frac{p}{\hbar}\sin\frac{\theta}{2}$.

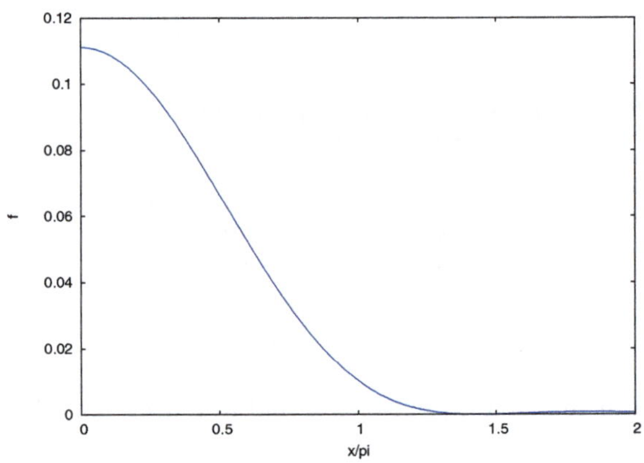

$f(x) \equiv \left(\frac{\sin x}{x^3} - \frac{\cos x}{x^2}\right)^2$, where the x-axis represents x/π

(b) Using $\cos\theta = 1 - 2\sin^2\frac{\theta}{2}$, we obtain $d\cos\theta = 4\sin\frac{\theta}{2} d\sin\frac{\theta}{2}$. Therefore,

$$\sigma_B = 2\pi \int_{-1}^1 d\cos\theta\, \frac{d\sigma_B}{d\Omega} = \frac{2\pi}{\left(\frac{p}{\hbar}\right)^2} \int_0^{2\frac{p}{\hbar}} kdk\, \frac{d\sigma_B}{d\Omega}$$

$$= \frac{2\pi}{\left(\frac{p}{\hbar}\right)^2} 4R^2 \left(\frac{V_0}{\frac{\hbar^2}{mR^2}}\right)^2 \int_0^{2\frac{p}{\hbar}R} dx\, x \left(\frac{\sin x}{x^3} - \frac{\cos x}{x^2}\right)^2$$

$$= 32\pi R^2 \left(\frac{V_0}{\frac{\hbar^2}{mR^2}}\right)^2 \frac{1}{x^2}\left(\frac{\sin 2x}{4x^3} + \frac{\cos 2x}{8x^4} - \frac{1}{4x^2} - \frac{1}{8x^4} + \frac{1}{4}\right)_{x=2\frac{pR}{\hbar}}$$

$$\xrightarrow{p\to 0} \frac{16\pi}{9} R^2 \left(\frac{V_0}{\frac{\hbar^2}{mR^2}}\right)^2$$

(c)

$$V_0 \ll \frac{\hbar^2}{mR^2}$$

3. (a) The Schrödinger equation for the S-wave is

$$-\frac{d^2}{dr^2}\chi(r) + \frac{2mV(r)}{\hbar^2}\chi(r) = k^2\chi(r)$$

so that

$$-\frac{d^2}{dr^2}\chi(r) = \begin{cases} \left(k^2 - \frac{2mV_0}{\hbar^2}\right)\chi(r) & (r < R) \\ k^2\chi(r) & (r > R) \end{cases}$$

For $r > R$, $\chi(r) = \sin(kr + \delta_0(k))$. As $k \to 0$, $\delta_0 \simeq -ka$, so $\chi(r) \simeq \sin k(r - a)$.

- For $V_0 > 0$—Let $\kappa \equiv \sqrt{\frac{2mV_0}{\hbar^2}}$, then

$$a = R\left(1 - \frac{\tanh \kappa R}{\kappa R}\right) > 0$$

- For $V_0 < 0$—Let $K \equiv \sqrt{\frac{-2mV_0}{\hbar^2}}$, then

$$a = R\left(1 - \frac{\tan KR}{KR}\right)$$

If $KR < \frac{\pi}{2}$, then $a < 0$.

(b) For $V_0 < 0$, the wave number inside $r < R$ becomes larger, causing an increase in phase change. Hence, $\delta_0 > 0$, i.e., $a < 0$. Conversely, for $V_0 > 0$, the phase change decreases, resulting in $\delta_0 < 0$, i.e., $a > 0$.

(c) As $x \to 0$, $\frac{\tan x}{x} \to 1 + \frac{1}{3}x^2$, thus

$$\lim_{k\to 0} \sigma_{\text{total}} = 4\pi a^2 \overset{\kappa R \ll 1}{\longrightarrow} 4\pi R^2 \frac{1}{9}\frac{4m^2 V_0^2}{\hbar^4} R^2 = \frac{16\pi m^2 V_0^2 R^6}{9\hbar^4}$$

agreeing with the result from the Born approximation.

(d) As $\kappa \to \infty$, $a \to R$, and thus $\sigma_{\text{total}} \to 4\pi R^2$.

Exercises 13

1. **Born approximation**—The proton is not a point particle; to a good approximation its electric charge e is distributed spherically within a finite region. Let $\rho(r)$ be the spherically symmetric charge density of a proton. The potential energy experienced by the point-like electron is given by

$$V(\vec{r}) = -e \int d^3 r' \, \frac{\rho(r')}{|\vec{r} - \vec{r}\,'|}$$

   (a) Using the Born approximation, calculate the differential cross section and the total cross section.
   (b) Explain how to determine the charge density $\rho(r)$ from the differential cross section.
   Hint: R. Hofstadter used this method to measure the charge densities of various nuclei in the 1950s.

2. **Spherical well potential (Part 3)**—Consider once again the spherical well potential given by

$$V(r) = \begin{cases} V_0 & (r < R) \\ 0 & (r > R) \end{cases}$$

   Assume that

$$V_0 < 0$$

   (a) Using the uncertainty principle, estimate the minimum value of $-V_0$ required for the existence of bound states.
   (b) For larger values of $-V_0$ where bound states exist, calculate the phase shift $\delta_0(k)$ for the S-wave, and plot the cross section $\sigma_0(k)$ as a function of k.
   Hint: It may be helpful to find $\tan \delta_0(k)$.

3. **Delta-shell potential**—Consider the spherically symmetric potential given by

$$V(r) = g\delta(r - R)$$

where $R>0$. In general the phase shift $\delta_0(k)$ is ambiguous by an integral multiple of π. Here, we demand

$$\delta_0(k) \xrightarrow{k\to 0} -ka$$

to remove this ambiguity.

(a) Calculate the scattering length a.
(b) Determine the condition on g for the existence of bound states.
(c) Consider the case with bound states. Calculate the phase shift $\delta_0(k)$ and the cross section $\sigma_0(k)$, and plot the k-dependence of $\sigma_0(k)$.
(d) Interpret physically the local maxima of $\sigma_0(k)$.
(e) Calculate $-\frac{1}{\pi}\delta_0(+\infty)$.
Comment: According to Levinson's theorem, this should match the number of bound states. Is the theorem correct?

4. **Electron-electron scattering**—Consider the scattering of two identical electrons in the center-of-mass frame. Let $\mu \equiv \frac{m}{2}$ be the reduced mass, and let $\vec{r}$ be the relative coordinate. The total Hamiltonian is given by

$$\hat{H} = \frac{1}{2\mu}\hat{\vec{p}}^{\,2} - \frac{e^2}{\hat{r}}$$

where $\hat{\vec{p}}$ is the momentum conjugate to $\hat{\vec{r}}$. Spin dependence of the interaction is neglected.

Let $f(\theta)$ be the scattering amplitude. Since the two electrons are indistinguishable, we cannot distinguish between scattering angles θ and $\pi - \theta$. Assuming the total spin is 1, the orbital part of the wave function must be antisymmetric, and thus the scattering amplitude is given by

$$f_{\text{anti}}(\theta) \equiv f(\theta) - f(\pi - \theta)$$

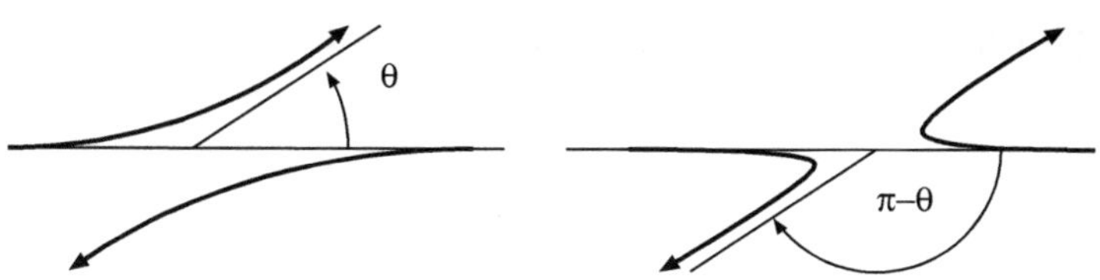

Indistinguishable

(a) Plot the differential scattering cross section $\frac{d\sigma_a}{d\cos\theta} = 2\pi\,|f_{\text{anti}}(\theta)|^2$.

On the other hand, assuming the total spin is 0, the orbital part of the wave function must be symmetric, and thus the scattering amplitude is given by

$$f_{\text{sym}}(\theta) \equiv f(\theta) + f(\pi - \theta)$$

(b) Plot the differential cross section $\frac{d\sigma_s}{d\cos\theta} = 2\pi |f_{sym}(\theta)|^2$.

(c) What happens to the scattering cross section if one electron has spin up and the other has spin down?
Hint: The Coulomb interaction causes no spin flip. Can't we distinguish the two electrons?

5.2 Quantization of the Electromagnetic Field

Three Points to Remember

1. A continuous medium satisfying a wave equation has an infinite number of eigenmodes, all of which are harmonic oscillators. Quantizing the medium amounts to quantizing each of these oscillators.
2. The electromagnetic field consists of an infinite number of harmonic oscillators, each characterized by a wave vector $\vec{k}$ and a polarization vector $\vec{\varepsilon}$. The n-th excited state of an oscillator is interpreted as a state with n photons of momentum $\hbar\vec{k}$ and energy $\hbar kc$.
3. The fundamental processes of photons are absorption and emission. The probability of absorption is proportional to the number of the existing photons; that of emission to the number of existing photons plus one.

5.2.1 *Quantizing a Continuous Medium Amounts to Quantizing Each Eigenmode as a Harmonic Oscillator*

As a preparation for quantizing the electromagnetic field in the next section, we explain how to quantize a string.

We consider transverse waves propagating along a one-dimensional string. The string lies along the x-axis and is fixed at $x = 0$ and $x = L$. Denoting the displacement in the y-direction as $y(x, t)$, we obtain the wave equation

$$\partial_t^2 y(x, t) - v^2 \partial_x^2 y(x, t) = 0 \tag{5.68}$$

Here, the velocity v of propagation is given by $v^2 = \frac{T}{\rho}$ with the tension T and mass density ρ of the string. (Refer to Example 2-6 for derivation.) Due to the boundary conditions $y(0,t) = y(L,t) = 0$, the displacement can be expanded into Fourier modes as follows:

$$y(x,t) = \sqrt{\frac{2}{L}} \sum_{n=1}^{\infty} y_n(t) \sin \frac{\pi n}{L} x \tag{5.69}$$

The wave equation gives

$$\ddot{y_n}(t) = -v^2 \left(\frac{\pi n}{L}\right)^2 y_n(t) \tag{5.70}$$

This is the equation of motion for a harmonic oscillator with the angular frequency

$$\omega_n \equiv \frac{\pi n}{L} \cdot v \tag{5.71}$$

Now, the total energy of the string is expressed as

$$\begin{aligned} E &= \frac{1}{2} \int_0^L dx \left[\rho \left(\frac{\partial y}{\partial t}\right)^2 + T \left(\frac{\partial y}{\partial x}\right)^2\right] \\ &= \frac{1}{2} \sum_{n=1}^{\infty} \left[\rho\, \dot{y_n}^2 + T \left(\frac{\pi n}{L}\right)^2 y_n^2\right] \\ &= \rho \frac{1}{2} \sum_{n=1}^{\infty} \left[\dot{y_n}^2 + \omega_n^2 y_n^2\right] \end{aligned} \tag{5.72}$$

Comparing this with the energy of a harmonic oscillator with mass m and angular frequency ω

$$E = \frac{m}{2}\left(\dot{x}^2 + \omega^2 x^2\right) \tag{5.73}$$

we obtain the following rule (or replacement) of quantization:

$$y_n \to \hat{y}_n = \sqrt{\frac{\hbar}{2\rho\omega_n}} \left(\hat{a}_n + \hat{a}_n^\dagger\right) \tag{5.74}$$

$$\rho \dot{y}_n \to \hat{p}_n = \sqrt{\frac{\hbar \rho \omega_n}{2}} \frac{1}{i} \left(\hat{a}_n - \hat{a}_n^\dagger\right) \tag{5.75}$$

where the annihilation and creation operators satisfy the commutation relation

$$[\hat{a}_n, \hat{a}_m^\dagger] = \delta_{nm} \tag{5.76}$$

The annihilation operator $\hat{a}_n$ decreases the excitation energy by $\hbar\omega_n$, and the creation operator $\hat{a}_n^\dagger$ increases it by the same amount. We interpret each excitation as the presence of a particle (we may call it a phonon) carrying energy $\hbar\omega_n$. Hence, $\hat{a}_n^\dagger$ acting twice creates two particles of the same energy $\hbar\omega_n$.

By substituting operators (5.74), (5.75) into energy (5.72), we obtain the Hamiltonian

$$\hat{H} = \sum_{n=1}^{\infty} \hbar\omega_n \left(\hat{a}_n^\dagger \hat{a}_n + \frac{1}{2}\right) \tag{5.77}$$

The ground state $|\Omega\rangle$, called the *vacuum state*, satisfies

$$\hat{a}_n |\Omega\rangle = 0$$

for all the annihilation operators $\hat{a}_n$. Therefore, the ground state has the zero-point energy

$$E_0 = \sum_{n=1}^{\infty} \frac{1}{2}\hbar\omega_n = \frac{\hbar\pi v}{2L} \sum_{n=1}^{\infty} n \tag{5.78}$$

which diverges. However, this divergence does not pose any physical problem. What is physically important is the energy differences, which are always finite. (Besides, the wave number $\frac{\pi n}{L}$ has a natural upper limit, since the string consists of atoms separated by a finite distance.) Excited states are given by

$$\left(\prod_{n=1}^{\infty} \left(\hat{a}_n^\dagger\right)^{N_n}\right) |\Omega\rangle \tag{5.79}$$

The state contains N_n particles with energy $\hbar\omega_n$. All integers $N_n = 0, 1, 2, 3, \cdots$ are allowed; the excitations are bosonic particles.

Using creation and annihilation operators, the displacement operator is expressed as

$$\hat{y}(x) = \sqrt{\frac{2}{L}} \sum_{n=1}^{\infty} \sqrt{\frac{\hbar}{2\rho\omega_n}} \left(\hat{a}_n + \hat{a}_n^\dagger\right) \sin\frac{\pi n}{L}x \tag{5.80}$$

Considering the state $|n\rangle \equiv \hat{a}_n^\dagger |\Omega\rangle$ with one particle in the n-th eigenmode, the matrix element of the displacement operator $\hat{y}(x)$ is

$$\langle n \mid \hat{y}(x) \mid \Omega\rangle = \langle \Omega \mid \hat{y}(x) \mid n\rangle = \sqrt{\frac{2}{L}} \sqrt{\frac{\hbar}{2\rho\omega_n}} \sin\frac{\pi n}{L}x \tag{5.81}$$

Interpreted physically, this gives the wave function of the particle.

$L \to \infty$ Limit

Taking the thermodynamic limit $L \to \infty$, we can handle an infinitely long string lying on the semi-axis $x > 0$. Fixing T and ρ, and taking the limit $L \to \infty$, the wave number $k_n \equiv \frac{\pi n}{L}$ and the eigen angular frequency $\omega_n \equiv k_n v$ become continuous variables $k, \omega \equiv kv$. We define

$$\hat{a}_k \equiv \lim_{L\to\infty} \sqrt{L}\,\hat{a}_n \tag{5.82}$$

which satisfy the commutation relations

$$[\hat{a}_k, \hat{a}_{k'}^\dagger] = \pi\delta(k-k') = 2\lim_{L\to\infty}\int_0^L dx\ \sin kx \sin k'x \tag{5.83}$$

In the continuum limit, the displacement of the string is given by

$$\begin{aligned}
\hat{y}(x) &= \lim_{L\to\infty}\sqrt{\frac{2}{L}}\frac{1}{\sqrt{\rho}}\sum_{n=1}^{\infty}\sqrt{\frac{\hbar}{2\omega_n}}\left(\hat{a}_n + \hat{a}_n^\dagger\right)\sin \pi n\frac{x}{L} \\
&= \lim_{L\to\infty}\sqrt{\frac{2}{L}}\frac{1}{\sqrt{\rho}}\int_0^\infty \frac{dk}{\pi}L\sqrt{\frac{\hbar}{2\omega}}\frac{1}{\sqrt{L}}\left(\hat{a}_k + \hat{a}_k^\dagger\right)\sin kx \\
&= \frac{1}{\sqrt{\rho}}\int_0^\infty \frac{dk}{\pi}\sqrt{\frac{\hbar}{2\omega}}\sin kx\left(\hat{a}_k + \hat{a}_k^\dagger\right)
\end{aligned} \tag{5.84}$$

Furthermore, the Hamiltonian is given by

$$\hat{H} = \int_0^\infty \frac{dk}{\pi}\hbar\omega\left(\hat{a}_k^\dagger\hat{a}_k + \frac{L}{2}\right) \tag{5.85}$$

The zero-point energy per unit length diverges as

$$\int_0^\infty \frac{dk}{\pi}\frac{\hbar\omega}{2} \tag{5.86}$$

due to the existence of an infinite number of eigenmodes per unit length.

Periodic Boundary Conditions

We now consider the periodic boundary conditions

$$(\forall x) \quad y(x+L,t) = y(x,t) \tag{5.87}$$

Then, the Fourier expansion of the displacement of the string is given by

$$y(x,t) = \frac{1}{\sqrt{L}} \sum_{n=-\infty}^{\infty} e^{i\frac{2\pi n x}{L}} y_n(t) \tag{5.88}$$

where $y(x,t)$ is real, so complex numbers $y_n(t)$ must satisfy

$$y_{-n}(t) = y_n(t)^* \tag{5.89}$$

Computing the energy, we find

$$\begin{aligned} E &= \frac{1}{2}\int_{-\infty}^{\infty} dx \left[\rho\,(\partial_t y(x,t))^2 + T\,(\partial_x y(x,t))^2\right] \\ &= \frac{1}{2}\sum_{n=-\infty}^{\infty}\left(\rho \dot{y}_{-n}\dot{y}_n + T\left(\frac{2\pi n}{L}\right)^2 y_{-n}y_n\right) \\ &= \frac{1}{2}\rho \dot{y}_0^2 + \rho\sum_{n=1}^{\infty}\left(|\dot{y}_n|^2 + \omega_n^2 |y_n|^2\right) \end{aligned} \tag{5.90}$$

where

$$\omega_n \equiv \frac{2\pi n}{L} v \tag{5.91}$$

The mode $n = 0$ represents the motion of the entire string. Defining

$$Y \equiv \frac{y_0}{\sqrt{L}}, \qquad M \equiv \rho L \tag{5.92}$$

we obtain

$$\frac{\rho}{2}\dot{y}_0^2 = \frac{M}{2}\dot{Y}^2 \tag{5.93}$$

which represents the kinetic energy of the transverse motion of the entire string with total mass M. For $n \geq 1$, if we separate y_n into real and imaginary parts,

$$y_n = \frac{1}{\sqrt{2}}(\alpha_n + i\beta_n) \tag{5.94}$$

the energy becomes

$$E = \frac{1}{2}M\dot{Y}^2 + \frac{\rho}{2}\sum_{n=1}^{\infty}\left(\dot{\alpha}_n^2 + \dot{\beta}_n^2 + \omega_n^2(\alpha_n^2 + \beta_n^2)\right) \tag{5.95}$$

Since we are not interested in the overall motion of the string, we ignore the first term.

Quantization proceeds similarly as before:

$$\alpha_n \to \sqrt{\frac{\hbar}{2\rho\omega_n}}\left(\hat{a}_n^{(1)} + \hat{a}_n^{(1)\dagger}\right) \tag{5.96}$$

$$\rho\dot{\alpha}_n \to \sqrt{\frac{\hbar\rho\omega_n}{2}}\frac{1}{i}\left(\hat{a}_n^{(1)} - \hat{a}_n^{(1)\dagger}\right) \tag{5.97}$$

$$\beta_n \to \sqrt{\frac{\hbar}{2\rho\omega_n}}\left(\hat{a}_n^{(2)} + \hat{a}_n^{(2)\dagger}\right) \tag{5.98}$$

$$\rho\dot{\beta}_n \to \sqrt{\frac{\hbar\rho\omega_n}{2}}\frac{1}{i}\left(\hat{a}_n^{(2)} - \hat{a}_n^{(2)\dagger}\right) \tag{5.99}$$

For the same n, $\hat{a}_n^{(1)}$ and $\hat{a}_n^{(2)}$ are two independent annihilation operators. The commutation relations are given by

$$\left[\hat{a}_n^{(1)}, \hat{a}_{n'}^{(1)\dagger}\right] = \left[\hat{a}_n^{(2)}, \hat{a}_{n'}^{(2)\dagger}\right] = \delta_{n,n'} \tag{5.100}$$

For $n \geq 1$, we find from Eq. (5.94)

$$y_n \to \hat{y}_n \equiv \frac{1}{\sqrt{2}}\sqrt{\frac{\hbar}{2\rho\omega_n}}\left(\hat{a}_n^{(1)} + i\hat{a}_n^{(2)} + \hat{a}_n^{(1)\dagger} + i\hat{a}_n^{(2)\dagger}\right) \tag{5.101}$$

$$\rho\dot{y}_n \to \hat{p}_n \equiv \frac{1}{\sqrt{2}}\sqrt{\frac{\hbar\rho\omega_n}{2}}\frac{1}{i}\left(\hat{a}_n^{(1)} + i\hat{a}_n^{(2)} - \hat{a}_n^{(1)\dagger} - i\hat{a}_n^{(2)\dagger}\right) \tag{5.102}$$

We rearrange the annihilation and creation operators as

$$\hat{a}_n \equiv \frac{1}{\sqrt{2}}\left(\hat{a}_n^{(1)} + i\hat{a}_n^{(2)}\right) \tag{5.103}$$

$$\hat{a}_{-n} \equiv \frac{1}{\sqrt{2}}\left(\hat{a}_n^{(1)} - i\hat{a}_n^{(2)}\right) \tag{5.104}$$

such that the commutation relations become

$$\left[\hat{a}_n, \hat{a}_{n'}^\dagger\right] = \delta_{n,n'}, \quad \left[\hat{a}_n, \hat{a}_{n'}\right] = \left[\hat{a}_n^\dagger, \hat{a}_{n'}^\dagger\right] = 0 \tag{5.105}$$

For any $n \neq 1$, we obtain

$$\hat{y}_n = \sqrt{\frac{\hbar}{2\rho\omega_n}} \left(\hat{a}_n + \hat{a}^\dagger_{-n}\right) \tag{5.106}$$

$$\hat{p}_n = \sqrt{\frac{\hbar\rho\omega_n}{2}} \frac{1}{i} \left(\hat{a}_n - \hat{a}^\dagger_{-n}\right) \tag{5.107}$$

which automatically satisfies

$$\hat{y}_{-n} = \hat{y}_n^\dagger \tag{5.108}$$

Using the new annihilation and creation operators, the Hamiltonian becomes

$$\hat{H} = \sum_{n\neq 0} \hbar\omega_n \left(\hat{a}_n^\dagger \hat{a}_n + \frac{1}{2}\right) \tag{5.109}$$

Similar to the case of a string with fixed ends, $\hat{a}_n^\dagger$ creates a particle with energy $\hbar\omega_n$ and wave number $k_n = \frac{2\pi n}{L}$.

In the thermodynamic limit $L \to \infty$, we obtain an infinitely long string spanning the entire x-axis. Replacing the sum over n by a continuous integral over wave numbers, we obtain

$$\hat{y}(x) = \frac{1}{\sqrt{\rho}} \int_{-\infty}^{\infty} \frac{dk}{2\pi} \sqrt{\frac{\hbar}{2\omega}} e^{ikx} \left(\hat{a}_k + \hat{a}^\dagger_{-k}\right) \tag{5.110}$$

$$\hat{H} = \int_{-\infty}^{\infty} \frac{dk}{2\pi} \hbar\omega \left(\hat{a}_k^\dagger \hat{a}_k + \frac{L}{2}\right) \tag{5.111}$$

where $\hat{a}_k$ is the annihilation operator for the excitation with energy $\hbar\omega = \hbar k v$ and $\hat{a}_k^\dagger$ is its creation operator. The commutation relation is given by

$$[\hat{a}_k, \hat{a}_{k'}^\dagger] = 2\pi\,\delta(k - k') \tag{5.112}$$

In contrast to the string with fixed ends, the infinite (or periodic) string is invariant under translations along the x-axis, i.e., $y(x, t) \to y(x + \Delta x, t)$, which implies the conservation of momentum. Both $\hat{a}_k$ and $\hat{a}^\dagger_{-k}$ operators decrease the total momentum of the string by $\hbar k$. The conservation laws for energy and momentum of the string are summarized in Sect. 5.3.1.

5.2.2 Electromagnetic Field (or the Radiation Field) Consists of an Infinite Number of Harmonic Oscillators, and the n-th Excited State of an Oscillator Corresponds to the Existence of n Photons

Let us recall the energy of the electromagnetic field. In the Coulomb gauge $\nabla \cdot \vec{A}(\vec{x}, t) = 0$, the vector potential is expanded as

$$\vec{A}(\vec{x}, t) = \int \frac{d^3k}{(2\pi)^3} \mathrm{e}^{i\vec{k}\cdot\vec{x}} \sum_{i=1,2} \vec{\varepsilon}_{\vec{k}}{}^{(i)} A_{\vec{k}}^{(i)}(t) \tag{5.113}$$

where the polarization vectors $\vec{\varepsilon}_{\vec{k}}{}^{(i)}$ are unit vectors orthogonal to $\vec{k}$ and to each other:

$$\vec{\varepsilon}_{\vec{k}}{}^{(1)} \cdot \vec{\varepsilon}_{\vec{k}}{}^{(2)} = 0 \tag{5.114}$$

We choose the polarization vectors to be the same for wave vectors $\vec{k}$ and $-\vec{k}$:

$$\vec{\varepsilon}_{-\vec{k}}{}^{(i)} = \vec{\varepsilon}_{\vec{k}}{}^{(i)} \tag{5.115}$$

so that the vector potential is real:

$$A_{-\vec{k}}^{(i)}(t) = \left(A_{\vec{k}}^{(i)}(t)\right)^* \tag{5.116}$$

In the absence of charged matter, the Maxwell equation reduces to the free wave equation

$$\left(\frac{1}{c^2}\partial_t^2 - \nabla^2\right)\vec{A} = 0 \tag{5.117}$$

The total energy of the free electromagnetic field is given by

$$\begin{aligned} \mathcal{E} &= \frac{1}{8\pi} \int d^3x \left(\vec{E}^2 + \vec{B}^2\right) \\ &= \frac{1}{8\pi} \int d^3x \left[\left(-\frac{1}{c}\partial_t \vec{A}\right)^2 + (\nabla \times \vec{A})^2\right] \end{aligned} \tag{5.118}$$

Substituting Eq. (5.113) into the wave Eq. (5.117), we get

$$\partial_t^2 A_{\vec{k}}^{(i)}(t) = -\omega^2 A_{\vec{k}}^{(i)}(t) \quad (\text{where } \omega \equiv kc) \tag{5.119}$$

which is the equation of a simple harmonic oscillator. Expressing the total energy of the electromagnetic field using the Fourier modes, we obtain

$$\mathcal{E} = \frac{1}{4\pi c^2}\frac{1}{2}\int \frac{d^3k}{(2\pi)^3} \sum_{i=1,2} \left(\dot{A}^{(i)}_{\vec{k}}\dot{A}^{(i)}_{-\vec{k}} + \omega^2 A^{(i)}_{\vec{k}} A^{(i)}_{-\vec{k}}\right) \tag{5.120}$$

The quantization of the electromagnetic field follows the quantization of a periodic string discussed toward the end of the previous section. The eigenmodes of the electromagnetic field are quantized by the following rule of substitution:

$$A^{(i)}_{\vec{k}}(t) \longrightarrow \hat{A}^{(i)}_{\vec{k}} \equiv \sqrt{4\pi c^2}\sqrt{\frac{\hbar}{2\omega}}\left(\hat{a}^{(i)}_{\vec{k}} + \hat{a}^{(i)\dagger}_{-\vec{k}}\right) \tag{5.121}$$

$$\frac{1}{4\pi c^2}\dot{A}^{(i)}_{\vec{k}}(t) \longrightarrow \hat{P}^{(i)}_{\vec{k}} \equiv \frac{1}{\sqrt{4\pi c^2}}\sqrt{\frac{\hbar\omega}{2}}\frac{1}{i}\left(\hat{a}^{(i)}_{\vec{k}} - \hat{a}^{(i)\dagger}_{-\vec{k}}\right) \tag{5.122}$$

where the creation and annihilation operators satisfy the commutation relation

$$\left[\hat{a}^{(i)}_{\vec{k}}, \hat{a}^{(j)\dagger}_{\vec{l}}\right] = \delta_{ij}(2\pi)^3\delta^{(3)}\left(\vec{k} - \vec{l}\right) \tag{5.123}$$

By now the physical interpretation of the quantization is obvious: $\hat{a}^{(i)}_{\vec{k}}$ is the annihilation operator of a photon (light quantum) of energy $\hbar kc$, momentum $\hbar\vec{k}$, and polarization vector $\vec{\varepsilon}_{\vec{k}}^{\;(i)}$.

Therefore, the quantized electromagnetic field is given by

$$\vec{A}(\vec{x}, t)$$

$$\longrightarrow \hat{\vec{A}}(\vec{x}) = \sqrt{4\pi c^2}\int \frac{d^3k}{(2\pi)^3} e^{i\vec{k}\cdot\vec{x}}\sqrt{\frac{\hbar}{2\omega}} \sum_{i=1,2} \vec{\varepsilon}_{\vec{k}}^{\;(i)}\left(\hat{a}^{(i)}_{\vec{k}} + \hat{a}^{(i)\dagger}_{-\vec{k}}\right) \tag{5.124}$$

and the total energy of the electromagnetic field by

$$\mathcal{E} \longrightarrow \hat{H}_{\text{rad}} = \int \frac{d^3k}{(2\pi)^3}\hbar\omega \sum_{i=1,2}\left(\hat{a}^{(i)\dagger}_{\vec{k}}\hat{a}^{(i)}_{\vec{k}} + \frac{1}{2}V\right) \tag{5.125}$$

where V is the volume of space. Similarly, the total momentum of the electromagnetic field is

$$\hat{\vec{P}} = \int \frac{d^3k}{(2\pi)^3}\hbar\vec{k} \sum_{i=1,2} \hat{a}^{(i)\dagger}_{\vec{k}}\hat{a}^{(i)}_{\vec{k}} \tag{5.126}$$

To summarize so far,

- Annihilation operator $\hat{a}_{\vec{k}}^{(i)}$ removes one photon with energy $\hbar\omega$, momentum $\hbar\vec{k}$, and polarization $\vec{\varepsilon}_{\vec{k}}^{\ (i)}$
- Creation operator $\hat{a}_{\vec{k}}^{(i)\dagger}$ adds one photon with the above propertics

The ground state (vacuum state) satisfies

$$\hat{a}_{\vec{k}}^{(i)} |\Omega\rangle = 0 \tag{5.127}$$

for any wave vector $\vec{k}$ and polarization i. Excited states are expressed in the form

$$\left(\prod_{\vec{k},i} \left(\hat{a}_{\vec{k}}^{(i)\dagger}\right)^{N_{\vec{k},i}}\right) |\Omega\rangle \tag{5.128}$$

where $N_{\vec{k},i}$ denotes the number of photons in mode $\vec{k}, i$. There is no restriction on $N_{\vec{k},i}$ since the photons are bosons.

The zero-point energy per unit volume, $\int \frac{d^3k}{(2\pi)^3}\hbar\omega$, is infinite. This energy is physical. By using conductors to change the boundary conditions of the electromagnetic field, we can alter the eigenmodes and their energies. Consequently, the zero-point energy also changes, and this change can be measured in laboratories. This phenomenon is known as the **Casimir effect**, named after its discoverer.

5.2.3 *The Fundamental Processes of Photons Are Absorption and Emission*

Consider the electron in a hydrogen atom. The total Hamiltonian, when the atomic nucleus is treated as a fixed external charge, is given by

$$\begin{aligned}\hat{H} &= \hat{H}_{\text{el}} + \hat{H}_{\text{Coulomb}} + \hat{H}_{\text{rad}} \\ &= \frac{1}{2m}\left(\hat{\vec{p}} + \frac{e}{c}\hat{\vec{A}}(\hat{\vec{x}})\right)^2 + \mu\frac{2\hat{\vec{s}}}{\hbar}\cdot\hat{\vec{B}}(\hat{\vec{x}}) - \frac{e^2}{\hat{r}} \\ &\quad + \int \frac{d^3k}{(2\pi)^3}\hbar\omega\left(\sum_{i=1,2}\hat{a}_{\vec{k}}^{(i)\dagger}\hat{a}_{\vec{k}}^{(i)} + V\right)\end{aligned} \tag{5.129}$$

To use perturbation theory, we can separate the total Hamiltonian into the unperturbed Hamiltonian

$$\hat{H}_0 \equiv \frac{1}{2m}\hat{\vec{p}}^{\,2} - \frac{e^2}{\hat{r}} + \int \frac{d^3k}{(2\pi)^3}\hbar\omega\left(\sum_i \hat{a}_{\vec{k}}^{(i)\dagger}\hat{a}_{\vec{k}}^{(i)} + V\right) \tag{5.130}$$

and the interaction Hamiltonian

$$\begin{aligned}\hat{H}_{\text{int}} &\equiv \frac{e}{mc}\hat{\vec{p}}\cdot\hat{\vec{A}}(\hat{\vec{x}}) + \underbrace{\frac{1}{2m}\frac{e^2}{c^2}\hat{\vec{A}}^{\,2} + \mu\frac{2\hat{\vec{s}}}{\hbar}\cdot\hat{\vec{B}}}_{\text{higher-order perturbation}} \\ &\simeq \frac{e}{mc}\hat{\vec{p}}\cdot\hat{\vec{A}}(\hat{\vec{x}}) \end{aligned} \tag{5.131}$$

Thus, the interaction Hamiltonian is approximated as

$$\boxed{\hat{H}_{\text{int}} \simeq \frac{e}{mc}\sqrt{4\pi c^2}\int\frac{d^3k}{(2\pi)^3}\sqrt{\frac{\hbar}{2\omega}}\,\mathrm{e}^{i\vec{k}\cdot\hat{\vec{x}}}\sum_{i=1,2}(\hat{\vec{p}}\cdot\vec{\varepsilon}_{\vec{k}}^{\,(i)})\left(\hat{a}_{\vec{k}}^{(i)} + \hat{a}_{-\vec{k}}^{(i)\dagger}\right)} \tag{5.132}$$

where $\vec{k}\cdot\hat{\vec{x}}$ and $\vec{\varepsilon}_{\vec{k}}^{\,(i)}\cdot\hat{\vec{p}}$ commute because the wave vector $\vec{k}$ and the polarization vector $\vec{\varepsilon}_{\vec{k}}^{\,(i)}$ are orthogonal:

$$\left[\vec{k}\cdot\hat{\vec{x}}\,,\;\vec{\varepsilon}_{\vec{k}}^{\,(i)}\cdot\hat{\vec{p}}\right] = 0 \tag{5.133}$$

The interaction Hamiltonian has two physical manifestations. To understand them, let us consider the transition from the initial state $|\Psi_i\rangle$ to the final state $|\Psi_f\rangle$. The transition probability per unit time at first order in perturbation is given by Fermi's golden rule as

$$w_{f\leftarrow i} = \frac{2\pi}{\hbar}\left|\langle\Psi_f|\,\hat{H}_{\text{int}}\,|\Psi_i\rangle\right|^2\delta(E_f - E_i) \tag{5.134}$$

We can express the state vectors as the direct product of the electron's state and that of the electromagnetic field:

$$|\Psi_i\rangle = |e_i\rangle\,|\Psi_{\text{rad},i}\rangle \tag{5.135}$$

$$|\Psi_f\rangle = |e_f\rangle\,|\Psi_{\text{rad},f}\rangle \tag{5.136}$$

Here, $|e_i\rangle$, $|e_f\rangle$ are the electron's states, and $|\Psi_{\text{rad},i}\rangle$, $|\Psi_{\text{rad},f}\rangle$ are those of the electromagnetic field. Therefore, from Eq. (5.132), we obtain

$$\langle\Psi_f|\,\hat{H}_{\text{int}}\,|\Psi_i\rangle = \frac{e}{mc}\sqrt{4\pi c^2}\int\frac{d^3k}{(2\pi)^3}\sqrt{\frac{\hbar}{2\omega}}$$
$$\times\sum_{i=1,2}\langle e_f|\,\mathrm{e}^{i\vec{k}\cdot\hat{\vec{x}}}\,\hat{\vec{p}}\cdot\vec{\varepsilon}_{\vec{k}}^{\,(i)}\,|e_i\rangle\cdot\langle\Psi_{\text{rad},f}|\,\hat{a}_{\vec{k}}^{(i)}+\hat{a}_{-\vec{k}}^{(i)\dagger}\,|\Psi_{\text{rad},i}\rangle \tag{5.137}$$

Defining

$$\vec{A}_{\vec{k}} \equiv \sum_i \vec{\varepsilon}_{\vec{k}}^{\,(i)}\langle\Psi_{\text{rad},f}|\,\hat{a}_{\vec{k}}^{(i)}+\hat{a}_{-\vec{k}}^{(i)\dagger}\,|\Psi_{\text{rad},i}\rangle \tag{5.138}$$

and the classical vector potential

$$\vec{A}(\vec{x}) = \int\frac{d^3k}{(2\pi)^3}\mathrm{e}^{i\vec{k}\cdot\vec{x}}\,\vec{A}_{\vec{k}}(\vec{x}) \tag{5.139}$$

the matrix element (5.137) matches the matrix element of the quantized electron moving in the classical electromagnetic field $\vec{A}(\vec{x})$, as we have seen in Sect. 4.5.1. Therefore, the effects of quantizing the electromagnetic field are all contained within the matrix element

$$\langle\Psi_{\text{rad},f}|\,\hat{a}_{\vec{k}}^{(i)}+\hat{a}_{-\vec{k}}^{(i)\dagger}\,|\Psi_{\text{rad},i}\rangle$$

Absorption Processes

First, consider

$$\langle\Psi_{\text{rad},f}|\,\hat{a}_{\vec{k}}^{(i)}\,|\Psi_{\text{rad},i}\rangle$$

The annihilation operator $\hat{a}_{\vec{k}}^{(i)}$ removes one photon with momentum $\hbar\vec{k}$, energy $\hbar\omega$, and polarization $\vec{\varepsilon}_{\vec{k}}^{\,(i)}$. Therefore, this matrix element vanishes unless $\Psi_{\text{rad},i}$ contains at least one such photon. Hence, in this transition, one photon is absorbed by the electron. If the initial state contains $N_{\vec{k}}^{(i)}$ photons with the same quantum numbers $\vec{k}$, $\vec{\varepsilon}_{\vec{k}}^{\,(i)}$, then

$$w_{f\leftarrow i} \propto \left|\langle\Psi_{\text{rad},f}|\,\hat{a}_{\vec{k}}^{(i)}\,|\Psi_{\text{rad},i}\rangle\right|^2 \propto N_{\vec{k}}^{(i)} \tag{5.140}$$

The final state remains the same as the initial state except for the absorbed photon.

Emission Processes

Next, consider

$$\left\langle \Psi_{\mathrm{rad},f} \right| \hat{a}^{(i)\dagger}_{-\vec{k}} \left| \Psi_{\mathrm{rad},i} \right\rangle$$

Since $\hat{a}^{(i)\dagger}_{-\vec{k}}$ is a creation operator, the matrix element vanishes unless the final state $\Psi_{\mathrm{rad},f}$ contains a photon with quantum numbers $-\vec{k}, \vec{\varepsilon}_{\vec{k}}^{\;(i)}$. Therefore, one photon is emitted in this transition. If the initial state contains $N^{(i)}_{-\vec{k}}$ photons with these quantum numbers, then

$$w_{f\leftarrow i} \propto \left| \left\langle \Psi_{\mathrm{rad},f} \right| \hat{a}^{(i)\dagger}_{-\vec{k}} \left| \Psi_{\mathrm{rad},i} \right\rangle \right|^2 \propto N^{(i)}_{-\vec{k}} + 1 \tag{5.141}$$

Even if $N^{(i)}_{-\vec{k}} = 0$, the matrix element does not vanish. The final state remains the same as the initial state except for the emitted photon.

Though the 1 and $N^{(i)}_{-\vec{k}}$ in Eq. (5.141) are inseparable, they have been interpreted separately:

- **Spontaneous emission**—this refers to 1. As an example, consider the transition from the $2p$ state to the $1s$ state of the electron in a hydrogen atom. Even if there are initially no photons, a photon is spontaneously emitted, and the electron can transition from the unstable state $2p$ to the stable state $1s$.
- **Stimulated emission**—this refers to $N^{(i)}_{-\vec{k}}$. If there are already photons with the same quantum numbers, their presence induces the emission of another photon. This is what is happening in lasers that continuously produce photons with the same quantum numbers.

The presence of both spontaneous and stimulated emissions in radiation was discovered by Einstein. (See Exercise 14-5.)

Example: $2p \to 1s$ Transition

- Initial state $|\Psi_i\rangle = |2p, l_z\rangle\, |\Omega\rangle$
- Final state $\left|\Psi_f\right\rangle = |1s\rangle \left|\vec{k}, \vec{\varepsilon}\right\rangle$

Here, $l_z = 0, \pm 1$ represents the magnetic quantum number, and $|\vec{k}, \vec{\varepsilon}\rangle$ represents the state of one photon in the electromagnetic field with momentum $\hbar\vec{k}$ and polarization $\vec{\varepsilon}$. The energy of $1s$ is $-Ry$, and the energy of $2p$ is $-\frac{1}{4}Ry$, so the energy difference is $\frac{3}{4}Ry$. According to Fermi's golden rule, the transition probability per unit time is given by (Fig. 5.9)

$$w_{\vec{k},\vec{\varepsilon}} = \frac{2\pi}{\hbar} \left| \left\langle \Psi_f \right| \hat{H}_{\mathrm{int}} \left| \Psi_i \right\rangle \right|^2 \delta\left(\hbar\omega - \frac{3}{4}Ry\right) \tag{5.142}$$

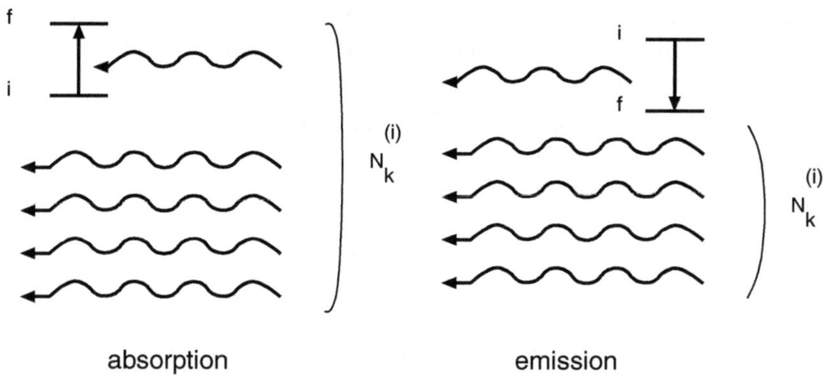

Fig. 5.9 The probability of absorption is proportional to $N_k^{(i)}$, and that of emission is proportional to $N_k^{(i)} + 1$

Therefore, integrating over $\vec{k}$ and summing over $\vec{\varepsilon}$, the decay width (inverse lifetime) is given by

$$\begin{aligned}\Gamma &\equiv \frac{2\pi}{\hbar}\int\frac{d^3k}{(2\pi)^3}\sum_{\vec{\varepsilon}}\left|\langle\Psi_f|\,\hat{H}_{\text{int}}\,|\Psi_i\rangle\right|^2\delta\left(\hbar\omega-\frac{3}{4}Ry\right)\\ &=\frac{2\pi}{\hbar}\int\frac{k^2dkd\Omega_\gamma}{(2\pi)^3}\sum_{\vec{\varepsilon}}\left|\langle\Psi_f|\,\hat{H}_{\text{int}}\,|\Psi_i\rangle\right|^2\delta\left(\hbar\omega-\frac{3}{4}Ry\right)\\ &=\frac{\omega^2}{\hbar^2c^3(2\pi)^2}\int d\Omega_\gamma\sum_{\vec{\varepsilon}}\left|\langle\Psi_f|\,\hat{H}_{\text{int}}\,|\Psi_i\rangle\right|^2\end{aligned}\tag{5.143}$$

where Ω_γ is the solid angle of the emitted photon.

Now, using

$$\left\langle\vec{k},\vec{\varepsilon}\,\middle|\,\hat{a}^\dagger_{\vec{k}}\,|\Omega\rangle\right. = (2\pi)^3\delta(\vec{k}'-\vec{k})\tag{5.144}$$

the matrix element is calculated as

$$\begin{aligned}&\langle\Psi_f|\,\hat{H}_{\text{int}}\,|\Psi_i\rangle\\ &=\frac{e}{mc}\sqrt{4\pi c^2}\int\frac{d^3k'}{(2\pi)^3}\sqrt{\frac{\hbar}{2\omega'}}\,\langle 1s|\,\mathrm{e}^{-i\vec{k}'\cdot\hat{\vec{x}}}(\hat{\vec{p}}\cdot\vec{\varepsilon})\,|2p,l_z\rangle\,(2\pi)^3\delta(\vec{k}'-\vec{k})\\ &=\frac{e}{mc}\sqrt{4\pi c^2}\sqrt{\frac{\hbar}{2\omega}}\,\langle 1s|\,\mathrm{e}^{-i\vec{k}\cdot\hat{\vec{x}}}(\hat{\vec{p}}\cdot\vec{\varepsilon})\,|2p,l_z\rangle\end{aligned}\tag{5.145}$$

Here, the required matrix elements for the electron have already been calculated for the transition from 1s to 2p in an external electromagnetic field in Sect. 4.5.1. Recall that two steps are involved in the calculation:

1. Electric dipole approximation—The product of the photon's wave number k and the Bohr radius a is given by

$$ka \sim \alpha \simeq \frac{1}{137}$$

so that $e^{i\vec{k}\cdot\hat{\vec{x}}} \simeq 1$ can be used.
2. Substitution of a momentum by a coordinate—Eq. (4.325) gives

$$\langle 1s|\,\hat{\vec{p}}\,|2p, l_z\rangle = -im\omega\,\langle 1s|\,\hat{\vec{x}}\,|2p, l_z\rangle$$

From this, taking the average over the three $2p$ states, we obtain

$$\frac{1}{3}\sum_{l_z=0,\pm1}\left|\langle 1s|\,e^{-i\vec{k}\cdot\hat{\vec{x}}}(\hat{\vec{p}}\cdot\vec{\varepsilon})\,|2p, l_z\rangle\right|^2 \simeq \frac{m^2\omega^2}{3}\sum_{l_z=0,\pm1}\left|\langle 1s|\,\hat{\vec{x}}\cdot\vec{\varepsilon}\,|2p, l_z\rangle\right|^2 \tag{5.146}$$

which, due to spherical symmetry, does not depend on the polarization. Therefore, taking the polarization vector in the z direction, we obtain

$$\sum_{l_z=0,\pm1}\left|\langle 1s|\,\hat{\vec{x}}\cdot\vec{\varepsilon}\,|2p, l_z\rangle\right|^2 = \left|\langle 1s|\hat{z}|2p, 0\rangle\right|^2 = \frac{\hbar}{2m\omega}f \tag{5.147}$$

Here, f is the oscillator strength defined in (4.326):

$$f \equiv \frac{2m\omega}{\hbar}\left|\langle 2p, 0|\hat{z}|1s\rangle\right|^2 \tag{5.148}$$

Putting together Eqs. (5.143), (5.145), and (5.147), we obtain

$$\begin{aligned}\Gamma_{2p\to 1s} &\simeq \frac{\omega^2}{\hbar^2 c^3 (2\pi)^2}\frac{e^2}{m^2c^2}4\pi c^2\frac{\hbar}{2\omega}\frac{m^2\omega^2}{3}\frac{\hbar}{2m\omega}f\underbrace{\int d\Omega_\gamma}_{=4\pi}\underbrace{\sum_{\vec{\varepsilon}}}_{=2}\\ &= \frac{2}{3}\omega\frac{e^2}{\hbar c}\frac{\hbar\omega}{mc^2}f = f\cdot\frac{2e^2\omega^2}{3mc^3}\end{aligned} \tag{5.149}$$

In Exercise 4-5 the friction coefficient of a classical charged harmonic oscillator has been obtained as $\gamma = \frac{2e^2\omega^3}{3mc^3}$. Hence, we can write the above result as

$$\Gamma_{2p\to 1s} = \gamma f \tag{5.150}$$

Using

$$\hbar\omega = \frac{3}{4}Ry = \frac{3}{8}\frac{\hbar\alpha c}{a} = \frac{3}{8}\alpha^2 mc^2 \tag{5.151}$$

we can rewrite

$$\Gamma_{2p\to 1s} = \frac{3f}{32}\alpha^3\frac{\alpha c}{a} \tag{5.152}$$

Finally, let us calculate the oscillator strength. Using the wave functions

$$\varphi_{2p,0}(\vec{x}) = \frac{1}{4\sqrt{2\pi a^3}}\frac{z}{a}\mathrm{e}^{-\frac{r}{2a}}, \quad \varphi_{1s}(\vec{x}) = \frac{1}{\sqrt{\pi a^3}}\mathrm{e}^{-\frac{r}{a}} \tag{5.153}$$

we obtain

$$\begin{aligned}\langle 2p,0|\hat{z}|1s\rangle &= \int d^3r\, z\varphi_{2p,0}\varphi_{1s}\\ &= \frac{1}{4\sqrt{2}\pi a^4}2\pi \underbrace{\int_{-1}^{1} d\cos\theta\, \cos^2\theta}_{=\frac{2}{3}} \underbrace{\int_0^\infty dr\, r^4\mathrm{e}^{-\frac{3r}{2a}}}_{=\left(\frac{2a}{3}\right)^5\cdot 4!}\\ &= \frac{a}{3\sqrt{2}}\left(\frac{2}{3}\right)^5 4!\end{aligned} \tag{5.154}$$

Hence, we find

$$f = \frac{2m\omega}{\hbar}\frac{a^2}{18}\left(\frac{2}{3}\right)^{10}(4!)^2 \tag{5.155}$$

Using Eq. (5.151), we obtain

$$f = \frac{1}{24}\left(\frac{2}{3}\right)^{10}(4!)^2 \simeq 0.42 \tag{5.156}$$

Therefore, using Eq. (5.152), we obtain

$$\Gamma_{2p\to 1s} = \alpha^3\frac{\alpha c}{a}\cdot\frac{3}{32}\frac{1}{24}\left(\frac{2}{3}\right)^{10}(4!)^2 = \left(\frac{2}{3}\right)^8\alpha^3\frac{\alpha c}{a} \simeq \underline{1.6\times 10^{-9}\mathrm{s}} \tag{5.157}$$

It is worth noting that

$$\Gamma_{2p\to 1s} = \text{const}\times\alpha^3\frac{\alpha c}{a} \tag{5.158}$$

is a result that can be derived from semiclassical arguments, as given in Exercises 4-4 and 4-5. The effort put into the above calculations was primarily focused on the dimensionless constant coefficient. Dependence on the physical constants is often understood through qualitative arguments, and much of the effort is directed toward computing constants.

Examples 14

1. $2p \to 1s$ **transition**—We have calculated the total probability of the $2p \to 1s$ transition for the electron in a hydrogen atom. Here, let us consider the electron in the initial state with angular momentum $\hat{L}_z = \hbar$. Find the differential transition probability per unit time for a photon emitted in the solid angle Ω. Furthermore, integrating over the directions, confirm that the total transition probability per unit time is the same as what has been obtained in the text.
   Note—Only calculate the parts that differ from the calculation in the text. There is no need to redo everything.
2. **Photon momentum**—Let us show that the momentum of a photon with energy $\hbar\omega$ is given by $\hbar\vec{k}$ (where the wave vector $\vec{k}$ satisfies $|\vec{k}|c = \omega$). In Exercise 3-5, we have derived that the energy flow of an electromagnetic field is given by the Poynting vector

$$\vec{S} = \frac{c}{4\pi}\vec{E} \times \vec{B}$$

   (a) For a relativistic particle with mass m and velocity $\vec{v}$, derive that the energy flow is the momentum multiplied by c^2. **Hint**: The energy of the particle is $\frac{mc^2}{\sqrt{1-\frac{v^2}{c^2}}}$, including the rest mass energy, and the momentum is $\frac{m\vec{v}}{\sqrt{1-\frac{v^2}{c^2}}}$.

   The result derived above holds true for any matter (whether continuous or discontinuous) in relativity. Therefore, the momentum density of an electromagnetic field is given by $\frac{1}{c^2}\vec{S}$.

   (b) Given the quantized vector potential

$$\hat{\vec{A}}(\vec{x}) = \sqrt{4\pi c^2}\int \frac{d^3k}{(2\pi)^3}\sqrt{\frac{\hbar}{2\omega}}\,\mathrm{e}^{i\vec{k}\cdot\vec{x}}\sum_{i=1,2}\vec{\varepsilon}_{\vec{k}}^{\,(i)}\left(\hat{a}_{\vec{k}}^{(i)} + \left(\hat{a}_{-\vec{k}}^{(i)}\right)^\dagger\right)$$

   and the Hamiltonian

$$\hat{H} = \int \frac{d^3k}{(2\pi)^3}\hbar\omega\sum_{i=1,2}\left(\hat{a}_{\vec{k}}^{(i)}\right)^\dagger \hat{a}_{\vec{k}}^{(i)}$$

compute the electromagnetic fields

$$\hat{\vec{E}}(\vec{x}) = -\partial_t \hat{\vec{A}} = -i\frac{1}{\hbar}\left[\hat{H},\, \hat{\vec{A}}\right]$$

$$\hat{\vec{B}}(\vec{x}) = \nabla \times \vec{A}$$

(c) Compute the total momentum of the electromagnetic field

$$\hat{\vec{P}} \equiv \frac{1}{4\pi c}\int d^3x\, \hat{\vec{E}} \times \hat{\vec{B}}$$

and interpret the result physically.

3. **Heisenberg equations**—The Hamiltonian of a system consisting of N electrons and an electromagnetic field is given as the sum of three parts:

$$\hat{H} = \hat{H}_{\text{el}} + \hat{H}_{\text{Coulomb}} + \hat{H}_{\text{rad}}$$

Each part is defined as follows:

$$\begin{cases} \hat{H}_{\text{el}} & \equiv \sum_{i=1}^{N} \frac{1}{2m}\left(\vec{p}_i + \frac{e}{c}\vec{A}(\vec{x}_i)\right)^2 \\ \hat{H}_{\text{Coulomb}} & \equiv e^2 \sum_{i<j} \frac{1}{|\vec{x}_i - \vec{x}_j|} \\ \hat{H}_{\text{rad}} & \equiv \int \frac{d^3k}{(2\pi)^3}\, \hbar\omega \sum_{i=1,2} a_{\vec{k}}^{(i)\dagger} a_{\vec{k}}^{(i)} \end{cases}$$

where the hat ˆ for operators is omitted. The Heisenberg equation for an operator $\mathcal{O}$ is given by

$$\frac{d}{dt}\mathcal{O} = \frac{i}{\hbar}[\hat{H}, \mathcal{O}]$$

(a) From the Heisenberg equation

$$\partial_t \vec{A}(\vec{x}) = \frac{i}{\hbar}[\hat{H}, \vec{A}(\vec{x})]$$

compute the electric field

$$\vec{E}(\vec{x}) \equiv -\frac{1}{c}\partial_t \vec{A}(\vec{x}) - e\sum_{i=1}^{N} \frac{\vec{x} - \vec{x}_i}{|\vec{x} - \vec{x}_i|^3}$$

(b) Derive the equation of motion for the electron:

$$\frac{d^2\vec{x}_i}{dt^2} = -e\left(\vec{E}(\vec{x}_i) + \frac{1}{c}\frac{d\vec{x}_i}{dt} \times \vec{B}(\vec{x}_i)\right)$$

(c) Find the time derivative of the electric field $\frac{1}{c}\partial_t \vec{E} = \frac{i}{\hbar c}[\hat{H}, \vec{E}(\vec{x})]$, and compare it with Maxwell's equations.

Solutions to Examples 14

1. First, under the electric dipole approximation, we obtain

$$\frac{d\Gamma}{d\Omega_\gamma} = \frac{e^2\omega^3}{2\pi\hbar c^3}\sum_{\vec{\varepsilon}} |\langle 1s|\,\vec{x}\cdot\vec{\varepsilon}\,|2\,\mathrm{p}, m = 1\rangle|^2$$

Taking the sum over polarization vectors, this depends only on the angle Θ between Ω_γ and the z-axis. Now, taking the direction of Ω_γ as $(0, \sin\Theta, \cos\Theta)$, two mutually orthogonal polarization vectors can be chosen as

$$\begin{cases} \vec{\varepsilon}^{\,(1)} = (1, 0, 0) \\ \vec{\varepsilon}^{\,(2)} = (0, -\cos\Theta, \sin\Theta) \end{cases} \Longrightarrow \begin{cases} \vec{x}\cdot\vec{\varepsilon}^{\,(1)} = x \\ \vec{x}\cdot\vec{\varepsilon}^{\,(2)} = -y\cos\Theta + z\sin\Theta \end{cases}$$

Using the wave functions

$$\varphi_{2p,1}(\vec{x}) = \frac{1}{8\sqrt{\pi a^3}}\frac{x+iy}{a}\mathrm{e}^{-\frac{r}{2a}}, \quad \varphi_{1s}(\vec{x}) = \frac{1}{\sqrt{\pi a^3}}\mathrm{e}^{-\frac{r}{a}}$$

we obtain

$$\langle 1s|\,\vec{x}\cdot\vec{\varepsilon}^{\,(1)}\,|2p, 1\rangle = a\frac{2^7}{3^5}, \quad \langle 1s|\,\vec{x}\cdot\vec{\varepsilon}^{\,(2)}\,|2p, 1\rangle = a\frac{2^7}{3^5}(-i)\cos\Theta$$

Thus,

$$\frac{d\Gamma}{d\Omega_\gamma} = \frac{e^2\omega^3}{2\pi\hbar c^3}a^2\frac{2^{14}}{3^{10}}\left(1+\cos^2\Theta\right) = \frac{1}{\pi}\frac{2^4}{3^7}\frac{\alpha^4 c}{a}\left(1+\cos^2\Theta\right)$$

showing preference for the photon emission along the z-axis. Finally, integrating over the solid angle, we obtain

$$\Gamma = \int d\Omega_\gamma \frac{d\Gamma}{d\Omega_\gamma} = \left(\frac{2}{3}\right)^8\frac{\alpha^4 c}{a}$$

agreeing with the result obtained in the text. The lifetime of the 2p state does not depend on its orientation.

2. (a) The relativistic energy is $\frac{mc^2}{\sqrt{1-\frac{v^2}{c^2}}}$, so the energy flow is $\vec{v}\frac{mc^2}{\sqrt{1-\frac{v^2}{c^2}}} = c^2\frac{m\vec{v}}{\sqrt{1-\frac{v^2}{c^2}}}$, which is the momentum multiplied by c^2.

   (b)

$$\frac{i}{\hbar}\left[\hat{H},\hat{a}\right] = -i\omega\hat{a}, \quad \frac{i}{\hbar}\left[\hat{H},\hat{a}^\dagger\right] = i\omega\hat{a}^\dagger$$

thus

$$\begin{cases} \vec{E}(\vec{x}) = \sqrt{4\pi c^2}\int \frac{d^3k}{(2\pi)^3}\, e^{i\vec{k}\cdot\vec{x}}\sqrt{\frac{\hbar}{2\omega}}\frac{i\omega}{c}\sum_{i=1,2}\vec{\varepsilon}_{\vec{k}}^{\,(i)}\left(a_{\vec{k}}^{(i)} - a_{-\vec{k}}^{(i)\dagger}\right) \\ \vec{B}(\vec{x}) = \sqrt{4\pi c^2}\int \frac{d^3k}{(2\pi)^3}\, e^{i\vec{k}\cdot\vec{x}}\sqrt{\frac{\hbar}{2\omega}}i\vec{k}\times\sum_{i=1,2}\vec{\varepsilon}_{\vec{k}}^{\,(i)}\left(a_{\vec{k}}^{(i)} + a_{-\vec{k}}^{(i)\dagger}\right) \end{cases}$$

where the hat is omitted.

   (c) Each photon carries momentum $\vec{k}$.

$$\vec{P} = \int \frac{d^3k}{(2\pi)^3}\hbar\vec{k}\sum_{i=1}^{2} a_{\vec{k}}^{(i)\dagger}a_{\vec{k}}^{(i)}$$

3. (a)

$$\vec{A}(\vec{x}) = \sqrt{4\pi c^2}\int \frac{d^3k}{(2\pi)^3}\, e^{i\vec{k}\cdot\vec{x}}\sqrt{\frac{\hbar}{2\omega}}\sum_{i=1,2}\vec{\varepsilon}_{\vec{k}}^{\,(i)}\left(a_{\vec{k}}^{(i)} + a_{-\vec{k}}^{(i)\dagger}\right)$$

then

$$\begin{aligned} \frac{\partial}{\partial t}\vec{A}(\vec{x}) &= \frac{i}{\hbar}[\hat{H},\vec{A}(\vec{x})] = \frac{i}{\hbar}[\hat{H}_{\text{rad}},\vec{A}(\vec{x})] \\ &= -i\sqrt{4\pi c^2}\int \frac{d^3k}{(2\pi)^3}\, e^{i\vec{k}\cdot\vec{x}}\sqrt{\frac{\hbar\omega}{2}}\sum_{i=1,2}\vec{\varepsilon}_{\vec{k}}^{\,(i)}\left(a_{\vec{k}}^{(i)} - a_{-\vec{k}}^{(i)\dagger}\right) \end{aligned}$$

Thus

$$\begin{aligned} \vec{E}(\vec{x}) = {} & i\sqrt{4\pi}\int \frac{d^3k}{(2\pi)^3}\, e^{i\vec{k}\cdot\vec{x}}\sqrt{\frac{\hbar\omega}{2}}\sum_{i=1,2}\vec{\varepsilon}_{\vec{k}}^{\,(i)}\left(a_{\vec{k}}^{(i)} - a_{-\vec{k}}^{(i)\dagger}\right) \\ & - e\sum_{i=1}^{N}\frac{\vec{x}-\vec{x}_i}{|\vec{x}-\vec{x}_i|^3} \end{aligned}$$

(b) First, we obtain

$$\frac{d\vec{x}_i}{dt} = \frac{i}{\hbar}[\hat{H}_{\rm el}, \vec{x}_i] = \frac{1}{m}\left(\vec{p}_i + \frac{e}{c}\vec{A}(\vec{x}_i)\right)$$

Differentiating this once more, we obtain

$$\begin{aligned} m\frac{d^2\vec{x}_i}{dt^2} &= \frac{i}{\hbar}[\hat{H}_{\rm el} + \hat{H}_{\rm Coulomb}, \vec{p}_i] + \frac{i}{\hbar}[\hat{H}_{\rm el} + \hat{H}_{\rm rad}, \frac{e}{c}\vec{A}(\vec{x}_i)] \\ &= \frac{i}{\hbar}\frac{1}{2m}\left[\left(\vec{p}_i + \frac{e}{c}\vec{A}(\vec{x}_i)\right)^2, \vec{p}_i\right] + \frac{i}{\hbar}e^2\sum_{j\neq i}\left[\frac{1}{|\vec{x}_j - \vec{x}_i|}, \vec{p}_i\right] \\ &+ \frac{i}{\hbar}\frac{1}{2}\frac{e}{mc}\left[\left(\vec{p}_i + \frac{e}{c}\vec{A}(\vec{x}_i)\right)^2, \vec{A}(\vec{x}_i)\right] + \frac{i}{\hbar}\left[\hat{H}_{\rm rad}, \frac{e}{c}\vec{A}(\vec{x}_i)\right] \\ &= -e\left(\vec{E}(\vec{x}_i) + \frac{1}{c}\frac{d\vec{x}_i}{dt} \times \vec{B}(\vec{x}_i)\right) \end{aligned}$$

which gives the Lorentz force. Here, the magnetic field is given by

$$\vec{B}(\vec{x}) = \sqrt{4\pi c^2}\int \frac{d^3k}{(2\pi)^3}\, {\rm e}^{i\vec{k}\cdot\vec{x}}\sqrt{\frac{\hbar}{2\omega}}\sum_{i=1,2} i\vec{k} \times \vec{\varepsilon}_{\vec{k}}^{\ (i)}\left(a_{\vec{k}}^{(i)} + a_{-\vec{k}}^{(i)\dagger}\right)$$

(c)

$$\begin{aligned} \frac{1}{c}\partial_t\vec{E}(\vec{x}) &= \frac{i}{\hbar c}[\hat{H}, \vec{E}(\vec{x})] \\ &= \sqrt{4\pi}\int \frac{d^3k}{(2\pi)^3}\, {\rm e}^{i\vec{k}\cdot\vec{x}}\sqrt{\frac{\hbar\omega}{2}}\frac{\omega}{c}\sum_{i=1,2}\vec{\varepsilon}_{\vec{k}}^{\ (i)}\left(a_{\vec{k}}^{(i)} + a_{-\vec{k}}^{(i)\dagger}\right) \\ &- \frac{e}{c}\sum_{i=1}^{N}\frac{i}{\hbar}\left[\frac{\vec{x} - \vec{x}_i}{|\vec{x} - \vec{x}_i|^3}, \hat{H}_{\rm el}\right] \end{aligned}$$

Using

$$\nabla^2\frac{1}{r} = -\nabla\cdot\frac{\vec{x}}{r^3} = -4\pi\delta^{(3)}(\vec{x})$$

we obtain

$$\frac{i}{\hbar}\left[\frac{\vec{x} - \vec{x}_i}{|\vec{x} - \vec{x}_i|^3}, \hat{H}_{\rm el}\right] = -4\pi\frac{d\vec{x}_i}{dt}\delta^{(3)}(\vec{x} - \vec{x}_i)$$

Since

$$\nabla \times \vec{B}(\vec{x}) = \sqrt{4\pi} \int \frac{d^3k}{(2\pi)^3} e^{i\vec{k}\cdot\vec{x}} \sqrt{\frac{\hbar\omega}{2}} \frac{\omega}{c} \sum_{i=1,2} \vec{\varepsilon}_{\vec{k}}{}^{(i)} \left(a_{\vec{k}}^{(i)} + a_{-\vec{k}}^{(i)\dagger} \right)$$

we obtain

$$\frac{1}{c} \partial_t \vec{E}(\vec{x}) = \nabla \times \vec{B}(\vec{x}) + 4\pi \frac{e}{c} \sum_{i=1}^{N} \delta^{(3)}(\vec{x} - \vec{x}_i) \frac{d\vec{x}_i}{dt}$$

This agrees with one of Maxwell's equations in the presence of a current density $\vec{j}(\vec{x}) \equiv -e \sum_i \delta^{(3)}(\vec{x} - \vec{x}_i) \frac{d\vec{x}_i}{dt}$.

Exercises 14

1. **Quantum field of massive particles in one dimension**—Consider a field $y(x, t)$ of a continuous medium in one-dimensional space. This satisfies the wave equation

$$\left(\partial_t^2 - v^2 \partial_x^2 + \omega_0^2 \right) y(x, t) = 0$$

The total energy of the medium is given by

$$E = \int_{-\infty}^{\infty} dx \, \frac{1}{2} \left((\partial_t y)^2 + v^2 (\partial_x y)^2 + \omega_0^2 \, y^2 \right)$$

   (a) Give an example of such a mechanical system from classical physics.
   Hint: How about the example of connected pendulums? (Example 2-5)
   (b) Decompose the field $y(x, t)$ into eigenmodes.
   (c) Quantize each eigenmode as a quantum harmonic oscillator.
   (d) Consider the physical meaning of quantization.
   Hint: The relation between the momentum and energy of a particle $p^2c^2 + m^2c^4 = E^2$ is called Einstein's relation. What is the analog here?

2. **Quantization of Fermi particles**—The aim of this problem is to introduce the quantization of Fermi particles.
We introduce a continuous field $\psi(\vec{x})$ and its hermitian conjugate

$$\hat{\psi}(\vec{x}) = \int \frac{d^3p}{(2\pi)^3} e^{\frac{i}{\hbar}\vec{p}\cdot\vec{x}} \, \hat{b}_{\vec{p}}, \quad \hat{\psi}^\dagger(\vec{x}) = \int \frac{d^3p}{(2\pi)^3} e^{-\frac{i}{\hbar}\vec{p}\cdot\vec{x}} \, \hat{b}_{\vec{p}}^\dagger$$

Here, $\hat{b}$ and $\hat{b}^\dagger$ are the annihilation and creation operators of Fermi particles, respectively, satisfying the anticommutation relations

$$\{\hat{b}_{\vec{p}},\, \hat{b}^\dagger_{\vec{q}}\} \equiv \hat{b}_{\vec{p}}\hat{b}^\dagger_{\vec{q}} + \hat{b}^\dagger_{\vec{q}}\hat{b}_{\vec{p}} = (2\pi\hbar)^3\delta^{(3)}(\vec{p}-\vec{q}), \quad \{\hat{b}_{\vec{p}},\, \hat{b}_{\vec{q}}\} = \{\hat{b}^\dagger_{\vec{p}},\, \hat{b}^\dagger_{\vec{q}}\} = 0 \tag{1}$$

As in the case of harmonic oscillators, the ground state $|\Omega\rangle$ is defined by

$$(\forall\vec{p}) \qquad \hat{b}_{\vec{p}}\,|\Omega\rangle = 0$$

Then,

$$|\vec{p}_1, \cdots, \vec{p}_n\rangle \equiv \hat{b}^\dagger_{\vec{p}_1}\cdots\hat{b}^\dagger_{\vec{p}_n}\,|\Omega\rangle$$

gives a state consisting of Fermi particles with momenta $\vec{p}_1, \cdots, \vec{p}_n$.

(a) Calculate the anticommutator $\{\hat{\psi}(\vec{x}),\, \hat{\psi}^\dagger(\vec{y})\}$.

(b) Explain how the anticommutation relations (1) give the Fermi statistics. In other words, explain if any two of $\vec{p}_1, \cdots, \vec{p}_n$ are the same, we find

$$|\vec{p}_1, \cdots, \vec{p}_n\rangle = 0$$

(c) Find the Hamiltonian $\hat{H}$ that satisfies the commutation relations

$$[\hat{H},\, \hat{b}_{\vec{p}}] = -\epsilon_p\hat{b}_{\vec{p}}, \quad [\hat{H},\, \hat{b}^\dagger_{\vec{p}}] = \epsilon_p\hat{b}^\dagger_{\vec{p}}$$

where $\epsilon_p \equiv \frac{\vec{p}^{\,2}}{2m}$. **Hint**: Use

$$\left[\hat{b}^\dagger_{\vec{p}}\hat{b}_{\vec{p}}\,,\, \hat{b}_{\vec{q}}\right] = \hat{b}^\dagger_{\vec{p}}\{\hat{b}_{\vec{p}}\,,\, \hat{b}_{\vec{q}}\} - \{\hat{b}^\dagger_{\vec{p}}\,,\, \hat{b}_{\vec{q}}\}\hat{b}_{\vec{p}} = -(2\pi\hbar)^3\delta^{(3)}(\vec{p}-\vec{q})\hat{b}_{\vec{p}}$$

(d) Express $\hat{H}$ as an integral over space using $\hat{\psi}(\vec{x})$, $\hat{\psi}^\dagger(\vec{x})$, and their spatial derivatives.

(e) Calculate the time derivative

$$\partial_t\hat{\psi}(\vec{x}) \equiv \frac{i}{\hbar}\left[\hat{H},\, \hat{\psi}(\vec{x})\right]$$

3. **Decay of Rydberg atoms**—We consider the transition from the state of a hydrogen atom with principal quantum number $n \gg 1$ and orbital angular momentum $l = n - 1$.

   (a) Show that in the electric dipole approximation, transition can occur only to the state with principal quantum number $n-1$ and orbital angular momentum $l = n - 2$.

Hint: $\vec{r}\, Y_{l,m}(\theta, \phi)$ in terms of $Y_{l\pm 1,*}$.

(b) Show that the state with magnetic quantum number $m = n - 1$ can transition only to the state with magnetic quantum number $n - 2$.

(c) Find the probability per unit time for the transition from principal quantum number n to $n - 1$.

Hints: Since the decay probability does not depend on the magnetic quantum number, it suffices to find the transition probability for the $m = n - 1$ state. From Example 10-2, the wave function for principal quantum number n and orbital quantum number $l = n - 1$ is given by

$$\psi(\vec{r}) = (\text{normalization constant}) \cdot r^{n-1} \mathrm{e}^{-r/a} Y_{n-1,m}(\theta, \phi)$$

where $m = -(n-1), \cdots, n-1$. Furthermore, from Exercise 9-3, we have $Y_{l,l} = c_l \left(\frac{x+iy}{r}\right)^l$. Hence,

$$\int d\cos\theta d\phi\, Y^*_{n-2,n-2} \frac{x - iy}{r} Y_{n-1,n-1}$$

$$= \frac{c_{n-2}}{c_{n-1}} \int d\cos\theta d\phi\, |Y_{n-1,n-1}|^2 = \frac{c_{n-2}}{c_{n-1}} = \sqrt{\frac{n-1}{n-\frac{1}{2}}}$$

4. **Contribution of the magnetic moment**—In considering the transition of the $2p$ electron in a hydrogen atom, we have ignored the interaction between the magnetic moment and the magnetic field

$$\hat{H}_{\text{mag}} = \mu \frac{\hat{\vec{s}}}{\frac{\hbar}{2}} \cdot \hat{\vec{B}}(\hat{\vec{x}})$$

where $\mu \simeq \frac{e\hbar}{2mc}$ (m is the electron's mass), $\hat{\vec{s}}$ is the electron's spin, and $\vec{B}$ is the magnetic field. Estimate the contribution of this term to the $2p \to 1s$ transition, and confirm that it is negligible. **Hint**: Compare $\hat{H}_{\text{mag}}$ with $\frac{e}{mc}\hat{\vec{p}} \cdot \hat{\vec{A}}(\vec{x})$, where we estimate $\frac{1}{\lambda}\vec{A} \sim \vec{B}$ (λ is the wavelength of the emitted photon).

5. **Einstein's A, B coefficients**—Consider $N \gg 1$ atoms with two energy levels E_1 and $E_2 \equiv E_1 + \Delta E$ ($\Delta E > 0$). Let N_i be the number of atoms with energy E_i. In thermal equilibrium at temperature T, N_1 and N_2 are related by

$$\frac{N_2}{N_1} = \mathrm{e}^{-\frac{\Delta E}{k_B T}}$$

The energy per unit volume of the electromagnetic field receives contributions from all angular frequencies of electromagnetic waves. Let $I(\omega)\Delta\omega$ be the contributions from the electromagnetic waves between frequencies ω and $\omega + \Delta\omega$.

(a) Recall Planck's formula for $I(\omega)$.
Comment: This is worth remembering.

Assume that one photon is emitted in the transition from the state E_2 to the state E_1, and conversely that one photon is absorbed in the transition from the state E_1 to the state E_2. The probability per unit time for a transition from state E_1 to state E_2 is proportional to $I(\Delta E/\hbar)$:

$$w_{2\leftarrow 1} = B_{2\leftarrow 1} I(\Delta E/\hbar)$$

where the coefficient $B_{2\leftarrow 1}$ is an intrinsic property of the atom and is considered independent of temperature. (The temperature dependence is in I.) On the other hand, the transition from state E_2 to state E_1 per unit time is given by

$$w_{1\leftarrow 2} = A + B_{1\leftarrow 2} I(\Delta E/\hbar)$$

where A is for spontaneous transition and is distinguished from the induced transitions proportional to I. Neither the coefficient A nor $B_{1\leftarrow 2}$ depend on the temperature. In thermal equilibrium, $N_{1,2}$ are independent of time, and we find

$$w_{2\leftarrow 1} N_1 = w_{1\leftarrow 2} N_2 \tag{2}$$

(b) Explain the physical meaning of Eq. (2).
(c) Using Planck's formula for $I(\omega)$ and Eq. (2), find the coefficients A, $B_{1\leftarrow 2}$, and $B_{2\leftarrow 1}$.
(d) Show that $w_{1\leftarrow 2}$ is proportional to the average number of photons with energy ΔE plus 1.
Hint: This is a result expected from the matrix elements of photon creation operators.

6. **Bloch-Nordsieck Model**—This problem is based on a very famous paper by Bloch and Nordsieck, aiming to understand the quantum state of electromagnetic fields under steady currents.[2]
Consider a current flowing along the z-axis. Its current density is given by

$$\vec{j}(\vec{x}) = (0, 0, I\delta(x)\delta(y))$$

Consequently, the interaction Hamiltonian is given by

$$\hat{H}_{\text{int}} = \int d^3x \, \vec{j}(\vec{x}) \cdot \hat{\vec{A}}(\vec{x})$$

[2] F. Bloch, A. Nordsieck, Phys. Rev. **52**(1937) 54.

where $\hat{\vec{A}}(\vec{x})$ is the vector potential satisfying the Coulomb gauge condition

$$\hat{\vec{A}}(\vec{x}) = \sqrt{4\pi c^2} \int \frac{d^3k}{(2\pi)^3} \sqrt{\frac{\hbar}{2\omega}}\, \mathrm{e}^{i\vec{k}\cdot\vec{x}} \sum_{i=1,2} \vec{\varepsilon}_{\vec{k}}^{\,(i)} \left(\hat{a}_{\vec{k}}^{(i)} + \hat{a}_{-\vec{k}}^{(i)\dagger}\right)$$

with $\omega \equiv kc$, $\vec{\varepsilon}_{\vec{k}}^{\,(i)} = \vec{\varepsilon}_{-\vec{k}}^{\,(i)} \perp \vec{k}$.

Let $\hat{H}_0$ be the energy of the free electromagnetic fields

$$\hat{H}_0 \equiv \int \frac{d^3k}{(2\pi)^3} \hbar\omega \sum_{i=1,2} \hat{a}_{\vec{k}}^{(i)\dagger} \hat{a}_{\vec{k}}^{(i)}$$

Then, the total Hamiltonian is given by

$$\hat{H} \equiv \hat{H}_0 + \hat{H}_{\text{int}}$$

(a) Show that the ground state is a coherent state, i.e., an eigenstate of the annihilation operators $\hat{a}_{\vec{k}}^{(i)}$.
Hint: Refer to Exercise 8-5 for coherent states. Use the relation

$$\mathrm{e}^{c\hat{a}+d\hat{a}^\dagger} = \mathrm{e}^{cd}\mathrm{e}^{d\hat{a}^\dagger}\mathrm{e}^{c\hat{a}}$$

for the annihilation and creation operators satisfying the commutation relation $[\hat{a}, \hat{a}^\dagger] = 1$.

(b) Calculate the average value of the vector potential in the ground state.

$$\vec{A}(x, y) \equiv \left\langle \Omega \middle| \hat{\vec{A}}(\vec{x}) \middle| \Omega \right\rangle$$

(independent of z-coordinate).

The ground state is not an eigenstate of the photon number. However, the average number of photons can be defined. Let n be the number of photons with the angular frequency between ω and $\omega + \Delta\omega$ per unit length along the z-axis. We consider the probability distribution of n.

(c) Find the average value $\bar{n}$ of n.
(d) Show that the probability distribution of n is given by the Poisson distribution

$$P_n(\omega, \omega + \Delta\omega) = \mathrm{e}^{-\bar{n}} \frac{\bar{n}^n}{n!}$$

Hint: If each of $n_1, \ldots, n_N$ follows a Poisson distribution with mean $\bar{n}_i$, then their sum $\sum_{i=1}^N n_i$ also follows a Poisson distribution with mean $\sum_{i=1}^N \bar{n}_i$.

Let $\omega_0 > 0$ be any angular frequency. We consider a finite region $0 < z < L$ along the z-axis.

(e) Show that the probability of the absence of photons with energy less than $\hbar\omega_0$ is zero.
Comment: That is, no matter how small ω_0 is, there are always photons with energy less than $\hbar\omega_0$.

(f) Show that the probability of the absence of photons with energy larger than $\hbar\omega_0$ is zero.
Comment: That is, no matter how large ω_0 is, there are always photons with energy greater than $\hbar\omega_0$.

Note that what we call photons here are not real photons. They merely represent the change to the ground state due to a current; these photons cannot be observed directly, i.e., they are not absorbed by other charged particles. It is often said that charged particles are surrounded by a "cloud of light," and the photons discussed here constitute that cloud.

7. **Casimir effect**—Consider two infinitely large parallel plates at distance a from each other. The plates are perfectly conducting so that the electric field is orthogonal to the plates. Following the original article by Casimir,[3] let us derive the force per unit area acting on the plates.

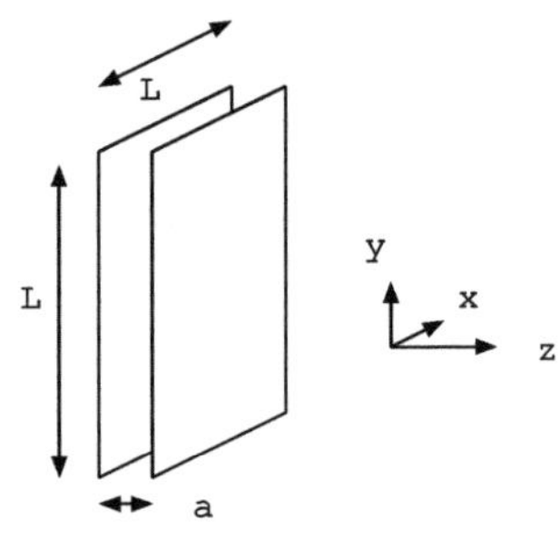

Casimir effect for two parallel plates

(a) Let $\vec{k} = (k_x, k_y, k_z)$ be the wave vector of standing waves between the two plates, where

$$k_z = n\frac{\pi}{a} \quad (n = 0, 1, 2, \cdots)$$

For $n \neq 0$, there are two modes of polarization, but for $n = 0$, there is only one. Explain.

[3] H. B. G. Casimir, "On the attraction between two perfectly conducting plates," Proc. Kon. Ned. Akad. Wet. **51**: 793, available from wiki on Casimir effect.

For the wave vector $\vec{k} = (k_x, k_y, n\pi/a)$, the corresponding zero point energy is

$$\frac{1}{2}\hbar c\sqrt{k_x^2 + k_y^2 + n^2\frac{\pi^2}{a^2}}$$

Hence, for very large L, the total zero point energy between the two plates is

$$E(a) = \hbar c \int_0^\infty \frac{L}{\pi}dk_x \int_0^\infty \frac{L}{\pi}dk_y \left(\frac{1}{2}\sqrt{k_x^2 + k_y^2} + \sum_{n=1}^{\infty}\sqrt{k_x^2 + k_y^2 + n^2\frac{\pi^2}{a^2}}\right)$$

Using $\kappa \equiv \sqrt{k_x^2 + k_y^2}$, we can write

$$E(a) = \hbar c\frac{L^2}{\pi^2}\frac{\pi}{2}\int_0^\infty \kappa d\kappa \left(\frac{\kappa}{2} + \sum_{n=1}^{\infty}\sqrt{\kappa^2 + n^2\frac{\pi^2}{a^2}}\right)$$

This is to be compared with the zero point energy in the absence of the plates for the volume aL^2, given by

$$\begin{aligned} E_0(a) &= \hbar c \int_0^\infty \frac{L}{\pi}dk_x \int_0^\infty dk_y\frac{L}{\pi}\int_0^\infty \frac{a}{\pi}dk_z\sqrt{\vec{k}^2} \\ &= \hbar c\frac{L^2}{\pi^2}\frac{\pi}{2}\int_0^\infty \kappa d\kappa \int_0^\infty \frac{a}{\pi}dk_z\sqrt{\kappa^2 + k_z^2} \end{aligned}$$

(b) Explain $E_0(a)$.

The difference is the change in the zero point energy due to the plates and is given by

$$\begin{aligned} \delta E &\equiv E(a) - E_0(a) \\ &= \hbar c\frac{L^2}{\pi^2}\frac{\pi}{2}\int_0^\infty \kappa d\kappa \left(\frac{\kappa}{2} + \sum_{n=1}^{\infty}\sqrt{\kappa^2 + n^2\frac{\pi^2}{a^2}} - \int_0^\infty dk_z\sqrt{\kappa^2 + k_z^2}\right) \\ &= L^2\hbar c\frac{\pi^2}{4a^3}\int_0^\infty du\left(\frac{1}{2}\sqrt{u} + \sum_{n=1}^{\infty}\sqrt{u + n^2} - \int_0^\infty dn\sqrt{u + n^2}\right) \end{aligned}$$

where $u \equiv \frac{\kappa^2 a^2}{\pi^2}$ and $n \equiv \frac{k_z a}{\pi}$ are dimensionless variables.

At first sight we are not sure if δE makes sense; we are not sure if the integrand is finite, and even if it is finite, we are not sure if it can be integrated over u. To make sense of the expression for δE, we introduce a converging function

$$f(\sqrt{u + n^2}/U)$$

satisfying

$$f(0) = 1, \quad f(s) \overset{s\to+\infty}{\longrightarrow} 0$$

We then define

$$F(n) \equiv \int_0^\infty du\, f(\sqrt{u+n^2}/U)\sqrt{u+n^2} \overset{w=u+n^2}{=} \int_{n^2}^\infty dw\, f(\sqrt{w}/U)\sqrt{w}$$

which is convergent due to the damping of $f(s)$ at large s. We modify δE to

$$\delta E = L^2 \hbar c \frac{\pi^2}{4a^3}\left[\frac{1}{2}F(0) + \sum_{n=1}^{\infty} F(n) - \int_0^\infty dn\, F(n)\right]$$

In the limit $U \to +\infty$, this reduces to the original δE. We take the limit $U \to +\infty$ only at the end of the calculations.

We now use the Euler-Maclaurin formula, which relates an integral and an infinite sum:

$$\frac{1}{2}F(0) + \sum_{n=1}^{\infty} F(n) - \int_0^\infty dn\, F(n) = -\frac{1}{12}F'(0) + \frac{1}{24\times 30}F'''(0) + \cdots$$

where the coefficients on the right-hand side are determined by the Bernoulli numbers ($B_2 = \frac{1}{6}$, $B_4 = -\frac{1}{30}$). In our case we find

$$F'(n) = -2n^2 f(n/U)$$

Hence,

$$F'(0) = 0, \quad F'''(0) = -4f(0) = -4$$

and the higher-order derivatives are inversely proportional to U and vanishing as $U \to \infty$.

(c) Show, in the limit $U \to +\infty$,

$$\frac{1}{L^2}\delta E = -\hbar c \frac{\pi^2}{24\times 30}\frac{1}{a^3} \propto \frac{1}{a^3}$$

(d) Derive the force per unit area. Is it attractive or repulsive?
(e) For $a = 10^{-3}$ mm, derive the force for the area a^2. Is it measurable?

5.3 Supplementary Notes

5.3.1 Summary of the Energy-Momentum Tensor for a String

The energy-momentum tensor is a very useful concept for deriving the conservation laws of energy and momentum. It consists of the density and flux of energy and momentum at each point in space.

Let us consider the one-dimensional string considered in Sect. 5.2.1. The Lagrangian density is given by

$$\mathcal{L} = \frac{\rho}{2}(\partial_t y)^2 - \frac{T}{2}(\partial_x y)^2 , \tag{5.161}$$

where ρ is the mass density and T is the tension. We have shown that the Euler–Lagrange equation is the wave equation

$$\partial_t^2 y - v^2 \partial_x^2 y = 0 , \quad \text{where} \quad v^2 \equiv \frac{T}{\rho} \tag{5.162}$$

As opposed to the fixed boundary condition $y(0, t) = y(L, t) = 0$, we impose the periodic boundary condition

$$y(x + L, t) = y(x, t) \tag{5.163}$$

Then, the Lagrangian $L[y(x, t)] \equiv \int_0^L dx\, \mathcal{L}(x, t)$ is invariant under translations both in space and in time.[4] Noether's theorem in Sect. 1.1.3 implies that energy and momentum are conserved.

We define the four components of the energy-momentum tensor as follows:

$$T^{00} \equiv \frac{\partial \mathcal{L}}{\partial(\partial_t y)} \partial_t y - \mathcal{L} \tag{5.164a}$$

$$T^{01} \equiv -\frac{\partial \mathcal{L}}{\partial(\partial_t y)} \partial_x y \tag{5.164b}$$

$$T^{10} \equiv \frac{\partial \mathcal{L}}{\partial(\partial_x y)} \partial_t y \tag{5.164c}$$

$$T^{11} \equiv -\frac{\partial \mathcal{L}}{\partial(\partial_x y)} \partial_x y + \mathcal{L} \tag{5.164d}$$

Energy and momentum conservation laws are then expressed as

$$\partial_t T^{0i} + \partial_x T^{1i} = 0 \quad (i = 0, 1) \tag{5.165}$$

[4] L is invariant under an infinitesimal transformation $y(x, t) \to y(x + \epsilon_1, t + \epsilon_0)$.

$\int_0^L dx\, T^{00}(x,t)$ is the total energy and $\int_0^L dx\, T^{01}$ is the total momentum. Let us look at each component of the energy-momentum tensor.

1. **Energy density**

$$T^{00} = \frac{1}{2}\rho(\partial_t y)^2 + \frac{1}{2}T(\partial_x y)^2 \tag{5.166}$$

2. **Energy flux** (energy flowing per unit time)

$$T^{10} = -T\partial_x y \partial_t y \tag{5.167}$$

3. **Energy conservation law**

$$\partial_t T^{00} + \partial_x T^{10} = \rho\partial_t y \underbrace{\partial_t^2 y}_{=\frac{T}{\rho}\partial_x^2 y} + T\partial_x y \partial_x \partial_t y - T\partial_t y \partial_x^2 y - T\partial_x \partial_t y \partial_x y = 0 \tag{5.168}$$

4. **Momentum density**

$$T^{01} = -\rho\partial_t y \partial_x y = \frac{1}{v^2}T^{10} \tag{5.169}$$

5. **Momentum flux** (momentum flowing per unit time)

$$T^{11} = \frac{1}{2}T(\partial_x y)^2 + \frac{1}{2}\rho(\partial_t y)^2 \tag{5.170}$$

6. **Momentum conservation law**

$$\partial_t T^{01} + \partial_x T^{11} = -\underbrace{\rho\partial_t^2 y}_{=T\partial_x^2 y}\partial_x y - \rho\partial_t y \partial_x \partial_t y + T\partial_x y \partial_x^2 y + \rho\partial_t y \partial_x \partial_t y = 0 \tag{5.171}$$

7. **Total energy** (quantized)

$$\int_0^L dx T^{00} \overset{L\to\infty}{\longrightarrow} \int \frac{dk}{2\pi}\hbar\omega\left(\hat{a}_k^\dagger \hat{a}_k + \frac{L}{2}\right) \tag{5.172}$$

8. **Total momentum** (quantized)

$$\int_0^L dx\, T^{01} \overset{L\to\infty}{\longrightarrow} -\rho\int dx \frac{1}{\sqrt{\rho}}\int \frac{dk}{2\pi}\sqrt{\frac{\hbar\omega}{2}}e^{ikx}\frac{1}{i}\left(\hat{a}_k - \hat{a}_{-k}^\dagger\right)$$
$$\times \frac{1}{\sqrt{\rho}}\int \frac{dk'}{2\pi}\sqrt{\frac{\hbar}{2\omega}}e^{ik'x}ik'\left(\hat{a}_{k'} + \hat{a}_{-k'}^\dagger\right)$$

$$= \int \frac{dk}{2\pi}\frac{\hbar k}{2} \underbrace{\left(\hat{a}_k - \hat{a}^\dagger_{-k}\right)\left(\hat{a}_{-k} + \hat{a}^\dagger_k\right)}_{\hat{a}_k\hat{a}^\dagger_k - \hat{a}^\dagger_{-k}\hat{a}_{-k} = \hat{a}^\dagger_k\hat{a}_k - \hat{a}^\dagger_{-k}\hat{a}_{-k}+1}$$

$$= \int \frac{dk}{2\pi}\hbar k\, \hat{a}^\dagger_k\hat{a}_k \tag{5.173}$$

5.3.2 Perturbation Theory with a Retarded Green Function

In Sect. 5.1, we have discussed potential scattering using the method of Green function. Here, we will introduce a method to discuss potential scattering using time-dependent perturbation theory. We use the results obtained here in the following Sect. 5.3.3.

Let $\hat{H} = \hat{H}_0 + \hat{V}$ be a time-independent total Hamiltonian, $\hat{H}_0$ be the Hamiltonian for the system without scattering, and $\hat{V}$ be the perturbation Hamiltonian causing scattering. The **retarded Green function** or **retarded propagator** is defined as[5]

$$K_R(t) \equiv \theta(t)\mathrm{e}^{-i\frac{t}{\hbar}\hat{H}} \tag{5.174}$$

This can be expressed using a Fourier integral as

$$K_R(t) = \int_{-\infty}^{\infty} \frac{dE}{2\pi}\mathrm{e}^{-i\frac{t}{\hbar}E}\frac{i}{E - \hat{H} + i\epsilon} \tag{5.175}$$

The propagator can be expanded formally as a geometric series

$$\begin{aligned} \frac{i}{E - \hat{H} + i\epsilon} &= \frac{i}{E - \hat{H}_0 + i\epsilon} + \frac{i}{E - \hat{H}_0 + i\epsilon}(-i)\hat{V}\frac{i}{E - \hat{H}_0 + i\epsilon} \\ &+ \frac{i}{E - \hat{H}_0 + i\epsilon}(-i)\hat{V}\frac{i}{E - \hat{H}_0 + i\epsilon}(-i)\hat{V}\frac{i}{E - \hat{H}_0 + i\epsilon} + \cdots \end{aligned} \tag{5.176}$$

The retarded propagator without interaction is written as

$$K_{R,0}(t) = \theta(t)\mathrm{e}^{-i\frac{t}{\hbar}\hat{H}_0} \tag{5.177}$$

[5] For applications of propagators to time-independent perturbation theory, see Sect. 4.6.6.

Substituting the above geometric series into (5.175), we obtain

$$
\begin{aligned}
K_R(t) &= K_{R,0}(t) + \int_{-\infty}^{\infty} dt_1\, K_{R,0}(t-t_1)\frac{-i}{\hbar}\hat{V} K_{R,0}(t_1) \\
&\quad + \int_{-\infty}^{\infty} dt_1 \int_{-\infty}^{\infty} dt_2\, K_{R,0}(t-t_1)\frac{-i}{\hbar}\hat{V} K_{R,0}(t_1-t_2)\frac{-i}{\hbar}\hat{V} K_{R,0}(t_2) + \cdots \\
&= K_{R,0}(t) + \int_0^t dt_1\, K_{R,0}(t-t_1)\frac{-i}{\hbar}\hat{V} K_{R,0}(t_1) \qquad (5.178) \\
&\quad + \int_0^t dt_1 \int_0^{t_1} dt_2\, K_{R,0}(t-t_1)\frac{-i}{\hbar}\hat{V} K_{R,0}(t_1-t_2)\frac{-i}{\hbar}\hat{V} K_{R,0}(t_2) + \cdots
\end{aligned}
$$

This result can be interpreted intuitively. For example, the last integral represents the system evolving freely from time $t = 0$ to t_2, then being affected by V at t_2, evolving freely again from t_2 to t_1, being affected by V again at t_1, and finally evolving freely from t_1 to t.

Let $|i\rangle$ be the initial state at time $-\frac{T}{2}$, and $|f\rangle$ be the final state at time $\frac{T}{2}$.

$$
\hat{H}_0 |i\rangle = E_i |i\rangle, \quad \hat{H}_0 |f\rangle = E_f |f\rangle \qquad (5.179)
$$

Then the transition probability amplitude is given by

$$
A_{fi} \equiv \mathrm{e}^{\frac{i}{\hbar}E_f\frac{T}{2}} \langle f| K_R(T) |i\rangle\, \mathrm{e}^{\frac{i}{\hbar}E_i\frac{T}{2}} \qquad (5.180)
$$

where the phase of the amplitude is chosen such that $A_{fi} = \delta_{fi}$ in the absence of interaction.

We then obtain

$$
\begin{aligned}
A_{fi} &= \delta_{fi} + \int_0^T dt_1\, \mathrm{e}^{-\frac{i}{\hbar}E_f(\frac{T}{2}-t_1)}\frac{-i}{\hbar}V_{fi}\mathrm{e}^{-\frac{i}{\hbar}E_i(t_1-\frac{T}{2})} \\
&+ \int_0^T dt_1 \int_0^{t_1} dt_2\, \mathrm{e}^{-\frac{i}{\hbar}E_f(\frac{T}{2}-t_1)}\sum_n \frac{-i}{\hbar}V_{fn}\mathrm{e}^{-\frac{i}{\hbar}E_n(t_1-t_2)}\frac{-i}{\hbar}V_{ni}\mathrm{e}^{-\frac{i}{\hbar}E_i(t_2-\frac{T}{2})} + \cdots \\
&= \delta_{fi} + \int_{-\frac{T}{2}}^{\frac{T}{2}} dt_1\, \mathrm{e}^{\frac{i}{\hbar}(E_f-E_i)t_1}\frac{-i}{\hbar}V_{fi} \qquad (5.181) \\
&\quad + \int_{-\frac{T}{2}}^{\frac{T}{2}} dt_1\, \mathrm{e}^{\frac{i}{\hbar}(E_f-E_i)t_1}\int_0^{\frac{T}{2}+t_1} d\tau \sum_n \frac{-i}{\hbar}V_{fn}\frac{-i}{\hbar}V_{ni}\mathrm{e}^{\frac{i}{\hbar}(E_i-E_n)\tau} + \cdots
\end{aligned}
$$

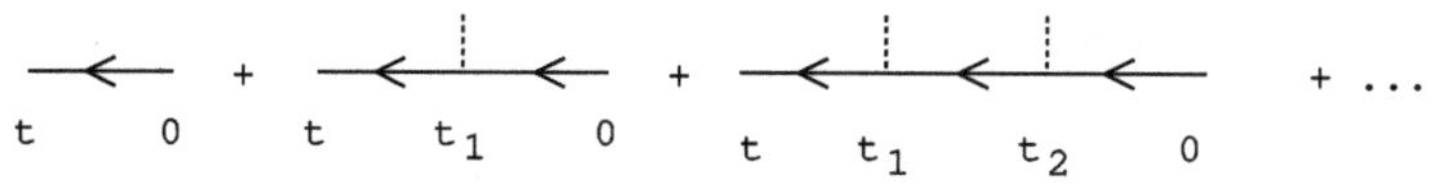

Fig. 5.10 Intuitive understanding of perturbative expansion of the retarded propagator. Dashed lines represent V

Taking the limit $T \to \infty$, we find

$$
\begin{aligned}
S_{fi} &\equiv \lim_{T\to\infty} A_{fi} \\
&= \delta_{fi} + 2\pi\delta(E_f - E_i)\left[(-i)V_{fi} + \sum_n (-i)V_{fn}\frac{i}{E_i - E_n + i\epsilon}(-i)V_{ni} + \cdots\right]
\end{aligned}
\tag{5.182}
$$

(See Fig. 5.10.) In general, the higher-order terms are given by

$$
2\pi\delta(E_f - E_i)\sum_{n_1,\cdots,n_k}(-i)V_{fn_1}\frac{i}{E_i - E_{n_1} + i\epsilon}(-i)V_{n_1n_2}\cdots \frac{i}{E_i - E_{n_k} + i\epsilon}(-i)V_{n_ki}
\tag{5.183}
$$

Therefore, the transition amplitude is written as

$$
S_{fi} = \delta_{fi} + 2\pi\delta(E_f - E_i)iT_{fi}
\tag{5.184}
$$

and the probability for $f \neq i$ is

$$
P_{fi} = |S_{fi}|^2 = 2\pi\delta(0)\cdot 2\pi\delta(E_f - E_i)|T_{fi}|^2
\tag{5.185}
$$

where

$$
2\pi\hbar\delta(0) = \int_{-\frac{T}{2}}^{\frac{T}{2}} dt\, e^{\frac{i}{\hbar}Et}\Big|_{E=0} = T
\tag{5.186}
$$

Hence, the transition probability per unit time is

$$
w_{fi} \equiv \frac{1}{T}P_{fi} = \frac{2\pi}{\hbar}\delta(E_f - E_i)|T_{fi}|^2
\tag{5.187}
$$

Therefore, up to the first order of perturbation, Fermi's golden rule is recovered as

$$
w_{fi} \simeq \frac{2\pi}{\hbar}\delta(E_f - E_i)\,|V_{fi}|^2
\tag{5.188}
$$

and up to the second order

$$w_{fi} \simeq \frac{2\pi}{\hbar}\delta(E_f - E_i)\left|V_{fi} + \sum_n \frac{V_{fn}V_{ni}}{E_i - E_n + i\epsilon}\right|^2 \tag{5.189}$$

We use this result in the following Sect. 5.3.3.

5.3.3 Kramers-Heisenberg Dispersion Relation

Consider the scattering of light by an atom, or more precisely, the scattering of light by an electron in the atom. The Hamiltonian, representing the potential energy of the electron as V, is given by

$$\hat{H} = \frac{1}{2m}\left(\vec{p} + \frac{e}{c}\vec{A}(\vec{x})\right)^2 + V(\vec{x}_i) + \hat{H}_{\text{rad}} \tag{5.190}$$

where $\hat{H}_{\text{rad}}$ is the Hamiltonian of the free radiation field. The Hamiltonian can be divided into two parts: $\hat{H}_0$ that does not contain the interaction between the electron and the radiation field and $\hat{H}_I$ that does:

$$\hat{H}_0 = \underbrace{\frac{1}{2m}\vec{p}^{\,2} + V(\vec{x})}_{=\hat{H}_{e,0}} + \hat{H}_{\text{rad}} \tag{5.191}$$

$$\hat{H}_I = \underbrace{\frac{e}{mc}\vec{p}\cdot\vec{A}(\vec{x})}_{=\hat{H}_1} + \underbrace{\frac{e^2}{mc^2}\frac{1}{2}\vec{A}(\vec{x})^2}_{=\hat{H}_2} \tag{5.192}$$

where $\vec{A}$ satisfies the Coulomb gauge condition

$$\nabla\cdot\vec{A} = 0 \tag{5.193}$$

We expand $\vec{A}(\vec{x})$ in terms of creation and annihilation operators of photons as

$$\vec{A}(\vec{x}) = \sqrt{4\pi c^2}\int \frac{d^3k}{(2\pi)^3}\, e^{i\vec{k}\cdot\vec{x}}\sqrt{\frac{\hbar}{2\omega}}\sum_{i=1,2}\vec{\varepsilon}^{\,(i)}_{\vec{k}}\left(\hat{a}^{(i)}_{\vec{k}} + \hat{a}^{(i)\dagger}_{-\vec{k}}\right) \tag{5.194}$$

where the unit polarization vector $\vec{\varepsilon}^{\,(i)}_{\vec{k}}$ is orthogonal to $\vec{k}$ due to the Coulomb gauge condition:

$$\vec{k}\cdot\vec{\varepsilon}^{\,(i)}_{\vec{k}} = 0 \tag{5.195}$$

Now, let $|n\rangle$ be the normalized eigenstates of the electron Hamiltonian $H_{e,0}$, ignoring the interaction with the radiation field:

$$H_{e,0}\,|n\rangle = E_n\,|n\rangle \tag{5.196}$$

Consider the process where a photon of energy $\hbar\omega$ is scattered by an electron in state n_0:

$$\underbrace{e^-}_{|n_0\rangle} + \underbrace{\gamma}_{\vec{k},\vec{\varepsilon}} \longrightarrow \underbrace{e^-}_{|n_0\rangle} + \underbrace{\gamma}_{\vec{k}',\vec{\varepsilon}'} \tag{5.197}$$

Here, $\vec{k}, \vec{\varepsilon}$ and $\vec{k}', \vec{\varepsilon}'$ represent the wave vectors and unit polarization vectors of the photons before and after the scattering, respectively. Assuming that the electron state remains unchanged, the energy conservation law implies $|\vec{k}| = |\vec{k}'| = \omega/c$. Therefore, the initial and final states of the scattering are given by

$$|\Psi_i\rangle = |n_0\rangle \left|\vec{k}, \vec{\varepsilon}\right\rangle, \qquad |\Psi_f\rangle = |n_0\rangle \left|\vec{k}', \vec{\varepsilon}'\right\rangle \tag{5.198}$$

Since the vector potential is a linear combination of creation and annihilation operators of photons, the scattering amplitude at the lowest order of perturbation is given as follows:

$$iT_{fi} = -i\left[\langle\Psi_f|\,\hat{H}_2\,|\Psi_i\rangle + \sum_n \frac{\langle\Psi_f|\,\hat{H}_1\,|\Psi_n\rangle\,\langle\Psi_n|\,\hat{H}_1\,|\Psi_i\rangle}{E_i - E_n + i\epsilon}\right] \tag{5.199}$$

Assuming that the wavelength of the photon is sufficiently longer than the size of the atom, the electric dipole approximation

$$e^{i\vec{k}\cdot\vec{x}} \simeq 1 \tag{5.200}$$

can be used. Then, the first term of the scattering amplitude is given by

$$\langle\Psi_f|\,\hat{H}_2\,|\Psi_i\rangle = \frac{2\pi e^2}{\omega}\frac{\hbar}{m}\vec{\varepsilon}\cdot\vec{\varepsilon}' \tag{5.201}$$

Two types of intermediate states contribute to the second term. The contribution from the intermediate states without photons $|\Psi_n\rangle = |n\rangle\,|\Omega\rangle$ (energy E_n) is

$$\frac{2\pi e^2}{\omega}\sum_n \frac{\langle n_0|\,\frac{1}{m}\vec{p}\cdot\vec{\varepsilon}'\,|n\rangle\,\langle n|\,\frac{1}{m}\vec{p}\cdot\vec{\varepsilon}\,|n_0\rangle}{\omega - \frac{1}{\hbar}(E_n - E_{n_0}) + i\epsilon} \tag{5.202}$$

and the contribution from the intermediate states containing two photons $|\Psi_n\rangle = |n\rangle\left|\vec{k}, \vec{\varepsilon}; \vec{k}', \vec{\varepsilon}'\right\rangle$ (energy $E_n + 2\hbar\omega$) is

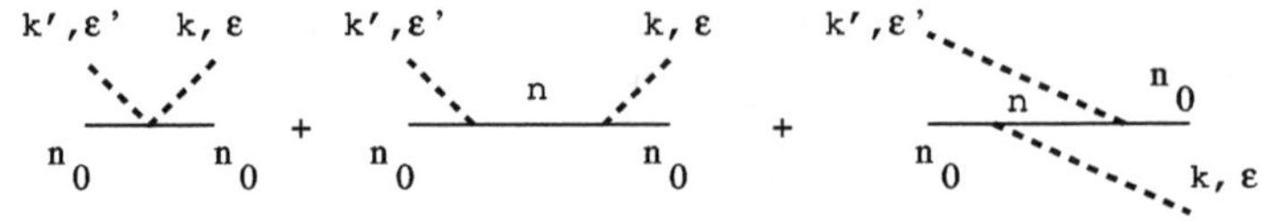

Fig. 5.11 Three contributions to iT_{fi}; read from right to left

$$-\frac{2\pi e^2}{\omega}\sum_n \frac{\langle n_0|\,\frac{1}{m}\vec{p}\cdot\vec{\varepsilon}\,|n\rangle\,\langle n|\,\frac{1}{m}\vec{p}\cdot\vec{\varepsilon}'\,|n_0\rangle}{\omega+\frac{1}{\hbar}(E_n-E_{n_0})-i\epsilon} \tag{5.203}$$

Here, the electric dipole approximation has been used. Therefore, the total amplitude is given by

$$iT_{fi} = -i\frac{2\pi e^2}{\omega}\left[\frac{\hbar}{m}\vec{\varepsilon}\cdot\vec{\varepsilon}'\right. \tag{5.204}$$
$$\left.+\sum_n\left\{\frac{\langle n_0|\,\frac{1}{m}\vec{p}\cdot\vec{\varepsilon}'\,|n\rangle\,\langle n|\,\frac{1}{m}\vec{p}\cdot\vec{\varepsilon}\,|n_0\rangle}{\omega-\frac{1}{\hbar}(E_n-E_{n_0})+i\epsilon}-\frac{\langle n_0|\,\frac{1}{m}\vec{p}\cdot\vec{\varepsilon}\,|n\rangle\,\langle n|\,\frac{1}{m}\vec{p}\cdot\vec{\varepsilon}'\,|n_0\rangle}{\omega+\frac{1}{\hbar}(E_n-E_{n_0})-i\epsilon}\right\}\right]$$

The diagrams in Fig. 5.11 provide an intuitive understanding of the three contributions to the amplitude.

As we have done in Eq. (4.325) and Exercise 12-2 of Sect. 4.5, the matrix elements of momentum $\vec{p}$ can be expressed by the matrix elements of position $\vec{x}$:

$$\frac{1}{m}\vec{p} = \frac{i}{\hbar}\left[H_{e,0},\vec{x}\right] \tag{5.205}$$

We thus obtain

$$\langle n_0|\,\frac{1}{m}\vec{p}\,|n\rangle = \frac{i}{\hbar}(E_{n_0}-E_n)\,\langle n_0|\,\vec{x}\,|n\rangle \tag{5.206a}$$

$$\langle n|\,\frac{1}{m}\vec{p}\,|n_0\rangle = \frac{i}{\hbar}(E_n-E_{n_0})\,\langle n|\,\vec{x}\,|n_0\rangle \tag{5.206b}$$

Using (5.206) for $\vec{p}\cdot\vec{\varepsilon}'$, we obtain

$$\sum_n\left\{\frac{\langle n_0|\,\frac{1}{m}\vec{p}\cdot\vec{\varepsilon}'\,|n\rangle\,\langle n|\,\frac{1}{m}\vec{p}\cdot\vec{\varepsilon}\,|n_0\rangle}{\omega-\frac{1}{\hbar}(E_n-E_{n_0})+i\epsilon}-\frac{\langle n_0|\,\frac{1}{m}\vec{p}\cdot\vec{\varepsilon}\,|n\rangle\,\langle n|\,\frac{1}{m}\vec{p}\cdot\vec{\varepsilon}'\,|n_0\rangle}{\omega+\frac{1}{\hbar}(E_n-E_{n_0})-i\epsilon}\right\}$$
$$= i\sum_n \langle n_0|\,\vec{x}\cdot\vec{\varepsilon}'\,|n\rangle\,\frac{\frac{1}{\hbar}(E_{n_0}-E_n)}{\omega-\frac{1}{\hbar}(E_n-E_{n_0})+i\epsilon}\,\langle n|\,\frac{1}{m}\vec{p}\cdot\vec{\varepsilon}\,|n_0\rangle$$
$$-\,i\sum_n \langle n_0|\,\frac{1}{m}\vec{p}\cdot\vec{\varepsilon}\,|n\rangle\,\frac{\frac{1}{\hbar}(E_n-E_{n_0})}{\omega+\frac{1}{\hbar}(E_n-E_{n_0})-i\epsilon}\,\langle n|\,\vec{x}\cdot\vec{\varepsilon}'\,|n_0\rangle$$

$$= \frac{i}{m} \langle n_0 | \underbrace{\left[\vec{x} \cdot \vec{\varepsilon}' \, , \, \vec{p} \cdot \vec{\varepsilon} \right]}_{i\hbar \vec{\varepsilon} \cdot \vec{\varepsilon}'} | n_0 \rangle$$

$$- i\omega \sum_n \langle n_0 | \vec{x} \cdot \vec{\varepsilon}' | n \rangle \frac{1}{\omega - \frac{1}{\hbar}(E_n - E_{n_0}) + i\epsilon} \langle n | \frac{1}{m} \vec{p} \cdot \vec{\varepsilon} | n_0 \rangle$$

$$+ i\omega \sum_n \langle n_0 | \frac{1}{m} \vec{p} \cdot \vec{\varepsilon} | n \rangle \frac{1}{\omega + \frac{1}{\hbar}(E_n - E_{n_0}) - i\epsilon} \langle n | \vec{x} \cdot \vec{\varepsilon}' | n_0 \rangle \qquad (5.207)$$

Using (5.206) for $\vec{p} \cdot \vec{\varepsilon}$, the above equation transforms into

$$\sum_n \left\{ \cdots \right\} = -\frac{\hbar}{m} \vec{\varepsilon} \cdot \vec{\varepsilon}'$$

$$+ \omega \sum_n \langle n_0 | \vec{x} \cdot \vec{\varepsilon}' | n \rangle \frac{\frac{1}{\hbar}(E_n - E_{n_0})}{\omega - \frac{1}{\hbar}(E_n - E_{n_0}) + i\epsilon} \langle n | \vec{x} \cdot \vec{\varepsilon} | n_0 \rangle$$

$$- \omega \sum_n \langle n_0 | \vec{x} \cdot \vec{\varepsilon} | n \rangle \frac{\frac{1}{\hbar}(E_{n_0} - E_n)}{\omega + \frac{1}{\hbar}(E_n - E_{n_0}) - i\epsilon} \langle n | \vec{x} \cdot \vec{\varepsilon}' | n_0 \rangle$$

$$= -\frac{\hbar}{m} \vec{\varepsilon} \cdot \vec{\varepsilon}' + \langle n_0 | \underbrace{\left[\vec{x} \cdot \vec{\varepsilon} \, , \, \vec{x} \cdot \vec{\varepsilon}' \right]}_{=0} | n_0 \rangle \qquad (5.208)$$

$$+ \omega^2 \sum_n \left\{ \frac{\langle n_0 | \vec{x} \cdot \vec{\varepsilon}' | n \rangle \langle n | \vec{x} \cdot \vec{\varepsilon} | n_0 \rangle}{\omega - \frac{1}{\hbar}(E_n - E_{n_0}) + i\epsilon} - \frac{\langle n_0 | \vec{x} \cdot \vec{\varepsilon} | n \rangle \langle n | \vec{x} \cdot \vec{\varepsilon}' | n_0 \rangle}{\omega + \frac{1}{\hbar}(E_n - E_{n_0}) - i\epsilon} \right\}$$

Thus, the total amplitude can be expressed as

$$iT_{fi} = -i2\pi e^2 \omega \sum_n \left\{ \frac{\langle n_0 | \vec{x} \cdot \vec{\varepsilon}' | n \rangle \langle n | \vec{x} \cdot \vec{\varepsilon} | n_0 \rangle}{\omega - \frac{1}{\hbar}(E_n - E_{n_0}) + i\epsilon} - \frac{\langle n_0 | \vec{x} \cdot \vec{\varepsilon} | n \rangle \langle n | \vec{x} \cdot \vec{\varepsilon}' | n_0 \rangle}{\omega + \frac{1}{\hbar}(E_n - E_{n_0}) - i\epsilon} \right\} \qquad (5.209)$$

In Dirac's textbook, this equation is called the **Kramers-Heisenberg dispersion relation**, but the name is usually saved for the rewritten formula to be given below.

Now, assuming that $|n_0\rangle$ is a state with zero orbital angular momentum, we can take the average over the direction of $\vec{x}$. The above result becomes

$$iT_{fi} = -i2\pi e^2 \omega \left(\vec{\varepsilon} \cdot \vec{\varepsilon}' \right) \frac{1}{3} \sum_{j=1}^{3} \sum_n | \langle n | x_j | n_0 \rangle |^2$$

$$\times \left\{ \frac{1}{\omega - \frac{1}{\hbar}(E_n - E_{n_0}) + i\epsilon} - \frac{1}{\omega + \frac{1}{\hbar}(E_n - E_{n_0}) - i\epsilon} \right\}$$

$$= -i2\pi e^2\omega\left(\vec{\varepsilon}\cdot\vec{\varepsilon}'\right)\sum_n \frac{1}{3}\sum_{j=1}^{3} |\langle n| x_j |n_0\rangle|^2$$

$$\times \frac{\frac{2}{\hbar}(E_n - E_{n_0})}{\omega^2 - \frac{1}{\hbar^2}(E_n - E_{n_0})^2 + i\epsilon(E_n - E_{n_0})} \tag{5.210}$$

We can rewrite this as

$$iT_{fi} = -i2\pi e^2\omega\left(\vec{\varepsilon}\cdot\vec{\varepsilon}'\right)\sum_n \frac{1}{3}\sum_{j=1}^{3} |\langle n| x_j |n_0\rangle|^2 \frac{\frac{2}{\hbar}(E_n - E_{n_0})}{\omega^2 - \frac{1}{\hbar^2}(E_n - E_{n_0})^2 + i\epsilon} \tag{5.211}$$

Now, let us recall the oscillator strength introduced in Sect. 4.5.1 (Eq. (4.326) and Exercise 12-2):

$$f_n \equiv \frac{2m}{\hbar^2}(E_n - E_{n_0})\frac{1}{3}\sum_j |\langle n| x_j |n_0\rangle|^2 \tag{5.212}$$

In Exercise 12-2, we have considered the case where n_0 is the ground state, but the sum rule

$$\sum_n f_n = 1 \tag{5.213}$$

can be derived for the excited states as well. In this case, if $E_n < E_{n_0}$, then $f_n < 0$. Using the oscillator strength, the scattering amplitude can be expressed as

$$\boxed{iT_{fi} = -i2\pi e^2\frac{\hbar}{m}\omega\left(\vec{\varepsilon}\cdot\vec{\varepsilon}'\right)\sum_n \frac{f_n}{\omega^2 - \frac{1}{\hbar^2}(E_n - E_{n_0})^2 + i\epsilon}} \tag{5.214}$$

This is the result commonly known as the **Kramers-Heisenberg dispersion relation**. Regarding the oscillator strength, we have shown, in Sect. 4.5.1, that the probability of photon absorption is proportional to the oscillator strength, using the 1s →2p transition as an example. Furthermore, in Sect. 5.2.3, we have shown that the probability of spontaneous emission of photons is also proportional to the oscillator strength, using the 2p → 1s transition as an example.

The dispersion relation derived above played an important role in early quantum theory. Kramers and Heisenberg derived it using Bohr's correspondence principle even before the mathematical formulation of quantum mechanics. As shown in Exercise 5-3, the dielectric constant of a dielectric consisting of uniformly distributed classical harmonic oscillators of charge e and angular frequency Ω is given by

$$\epsilon(\omega) - 1 = -\frac{4\pi ne^2}{m}\frac{1}{\omega^2 - \Omega^2} \tag{5.215}$$

where n is the number density of the oscillators. (We have set $\gamma = 0$.) Kramers and Heisenberg showed, using the correspondence principle, that this should be replaced by

$$\epsilon(\omega) - 1 = -\frac{4\pi n e^2}{m} \sum_n \frac{f_n}{\omega^2 - \Omega_n^2} \qquad \left(\Omega_n \equiv \frac{1}{\hbar}(E_n - E_{n_0})\right) \tag{5.216}$$

In Exercise 4-4 we have derived the scattering cross-section of light by a classical harmonic oscillator as

$$\sigma_{cl} = \frac{8\pi}{3} r_e^2 \left(\frac{\omega^2}{\omega^2 - \Omega^2}\right)^2 \tag{5.217}$$

where

$$r_e \equiv \frac{e^2}{mc^2} \tag{5.218}$$

is the classical electron radius. Hence, quantum mechanically, we obtain

$$\sigma = \frac{8\pi}{3} r_e^2 \omega^4 \left(\sum_n \frac{f_n}{\omega^2 - \Omega_n^2}\right)^2 \tag{5.219}$$

This final result can, of course, be derived from Eq. (5.214) without using the correspondence principle. We can simply use Fermi's golden rule. The transition probability per unit time is given by

$$dw_{fi} = \frac{2\pi}{\hbar} |T_{fi}|^2 d\Phi \tag{5.220}$$

where $d\Phi$ is the density of final states of photons with energy $\hbar\omega$:

$$d\Phi = \frac{d^3k}{(2\pi)^3} \delta(\hbar kc - \hbar\omega) = \frac{1}{\hbar c^3} \frac{\omega^2 d\Omega}{(2\pi)^3} \tag{5.221}$$

Therefore, we obtain

$$dw_{fi} = \frac{\omega^2}{\hbar^2 c^3 (2\pi)^2} |T_{fi}|^2 d\Omega \tag{5.222}$$

The number of photons arriving per unit area per unit time is c, so the scattering cross section is given by

$$\frac{d\sigma}{d\Omega} = \frac{1}{c} \frac{dw_{fi}}{d\Omega} = \frac{\omega^2}{\hbar^2 c^4 (2\pi)^2} |T_{fi}|^2 \tag{5.223}$$

Substituting Eq. (5.214) into the above, we obtain

$$\frac{d\sigma}{d\Omega} = r_e^2 \omega^4 \left(\vec{\varepsilon} \cdot \vec{\varepsilon}'\right)^2 \left| \sum_n \frac{f_n}{\omega^2 - \frac{1}{\hbar^2}(E_n - E_{n_0})^2 + i\epsilon} \right|^2 \tag{5.224}$$

Summing over the polarization vectors $\vec{\varepsilon}'$ and integrating over the solid angle Ω, we obtain

$$\sum_{\vec{\varepsilon}'} \int d\Omega \left(\vec{\varepsilon} \cdot \vec{\varepsilon}'\right)^2 = 2\pi \int_{-1}^{1} d\cos\theta \, \sin^2\theta = \frac{8\pi}{3} \tag{5.225}$$

Therefore, we obtain the total cross Sect. (5.219).

A famous application of Eq. (5.219) is the **Rayleigh scattering**. Considering the scattering of photons with energy (visible light) much lower than the binding energy of electrons (UV light), we obtain

$$\omega \ll |\Omega_n| \tag{5.226}$$

This leads to

$$\sigma \simeq \frac{8\pi}{3} r_e^2 \omega^4 \left(\sum_n \frac{f_n}{\Omega_n^2} \right)^2 \propto \omega^4 \tag{5.227}$$

which implies that the scattering cross section increases as the frequency increases (and thus as the wavelength decreases). This explains why the sky is blue (more scattering of blue light) during the day and red (less scattering of red light) at dawn and dusk.

Index

H. Sonoda, K. Ülker, *Basic Physics for Graduate Students*,
https://doi.org/10.1007/978-981-95-6607-5

FSC
www.fsc.org
MIX
Papier aus verantwortungsvollen Quellen
Paper from responsible sources
FSC® C105338

If you have any concerns about our products,
you can contact us on
ProductSafety@springernature.com

In case Publisher is established outside the EU,
the EU authorized representative is:
Springer Nature Customer Service Center GmbH
Europaplatz 3, 69115 Heidelberg, Germany

Printed by Libri Plureos GmbH
in Hamburg, Germany